D0295069

Please return on or before the last
date stamped below.
... are made for late

SIGNAL TRANSDUCTION BY REACTIVE OXYGEN AND NITROGEN SPECIES:
PATHWAYS AND CHEMICAL PRINCIPLES

Signal Transduction by Reactive Oxygen and Nitrogen Species: Pathways and Chemical Principles

Edited by

Henry Jay Forman
University of Alabama at Birmingham,
Birmingham, Alabama, U.S.A.

Jon Fukuto
University of California at Los Angeles,
Los Angeles, California, U.S.A.

and

Martine Torres
Childrens Hospital Los Angeles Research Institute,
Department of Pediatrics, Keck School of Medicine,
University of Southern California,
Los Angeles, California, U.S.A.

KLUWER ACADEMIC PUBLISHERS
DORDRECHT / BOSTON / LONDON

A C.I.P. Catalogue record for this book is available from the Library of Congress.

571
·74
S

CARDIFF UNIVERSITY

★ 1 6 JUN 2003 ★

PRIFYSGOL CAERDYDD

ISBN 1-4020-1117-2 (HB)
ISBN 1-4020-1121-0 (E-book)

Published by Kluwer Academic Publishers,
P.O. Box 17, 3300 AA Dordrecht, The Netherlands.

Sold and distributed in North, Central and South America
by Kluwer Academic Publishers,
101 Philip Drive, Norwell, MA 02061, U.S.A.

In all other countries, sold and distributed
by Kluwer Academic Publishers,
P.O. Box 322, 3300 AH Dordrecht, The Netherlands.

Cover: One of the earliest color illustrations of a South American mayfly:
Siphlonella guttta, as illustrated by Pictet (1843-45)

Printed on acid-free paper

All Rights Reserved
© 2003 Kluwer Academic Publishers
No part of this work may be reproduced, stored in a retrieval system, or transmitted
in any form or by any means, electronic, mechanical, photocopying, microfilming, recording
or otherwise, without written permission from the Publisher, with the exception
of any material supplied specifically for the purpose of being entered
and executed on a computer system, for exclusive use by the purchaser of the work.

Printed in the Netherlands

Contents

Contributors

Rebecca S. Arnold
Department of Pathology and Laboratory Medicine
Emory University School of Medicine
Atlanta, GA

Cecilia Barone
Department of Pathology
University of Alabama at Birmingham
Birmingham, AL

Garry Buettner
Free Radical Research Institute & ESR Facility
The University of Iowa
Iowa City, IA

Elena Chiarpotto
Department of Clinical and Biological Sciences
University of Turin at S. Luigi Gonzaga Hospital
Turin, Italy

Michael Collins
Department of Environmental Health Sciences
University of California
Los Angeles School of Public Health
Los Angeles, CA

Timothy P. Dalton
Department of Environmental Health
University Cincinnati Medical Center
Cincinnati, OH

Victor M. Darley-Usmar
Department of Pathology
University of Alabama at Birmingham
Birmingham, AL

Barry L. Fanburg
Department of Medicine
Pulmonary & Critical Care Division
Tufts - New England Medical Center
Boston, MA

Leopold Flohé
Department of Biochemistry
Technical University of Braunschweig
Braunschweig, Germany

Henry Jay Forman
Department of Environmental Health Sciences
Univerity of Alabama at Birmingham
Birmingham, AL

Ulrich Förstermann
Department of Pharmacology
Johannes Gutenberg University
Mainz, Germany

Jon Fukuto
Department of Pharmacology
University of California
Los Angeles, CA

Mark Garfinkel
Department of Environmental Health Sciences
Univerity of Alabama at Birmingham
Birmingham, AL

Cecilia Giulivi
Department of Chemistry
University of Minnesota
Duluth, MN

Tzipora Goldkorn
One Shields Avenue
University of California at Davis
Davis, CA

Arne Holmgren
Department of Medical Biochemistry
Medical Nobel Institute for Biochemistry
Stockholm, Sweden

Hidenori Ichijo
Laboratory of Cell Signaling
Graduate School of Pharmaceutical Sciences
The University of Tokyo
Tokyo, Japan

Sang Won Kang
Center for Cell Signaling Research and Division of Molecular Life Sciences
Ewha Women's University
Seoul, Korea

Hartmut Kleinert
Department of Pharmacology
Johannes Gutenberg University
Mainz, Germany

Irene E. Kochevar
Wellman Laboratories of Photomedicine
Harvard Medical School
Massachusetts General Hospital
Boston, MA

Jaeyul Kwon
Holland Laboratory
American Red Cross
Rockville, MD

David Lambeth
Department of Pathology
Emory University School of Medicine
Atlanta, GA

Jack R. Lancaster , Jr.
Center for Free Radical Biology
University of Alabama at Birmingham
Birmingham, AL

Lisa M. Landino
Department of Chemistry
The College of William and Mary
Williamsburg , VA

Seung-Rock Lee
Center for Cell Signaling Research and Division of Molecular Life Sciences
Ewha Women's University
Seoul, Korea

He Lei
Department of Environmental Health
University Cincinnati Medicine. Center
Cincinnati, OH

Gabriella Leonarduzzi
Department of Clinical and Biological Sciences
University of Turin at S. Luigi Gonzaga Hospital
Turin, Italy

Anna-Liisa Levonen
Department of Pathology
University of Alabama at Birmingham
Birmingham, AL

Huige Li
Department of Pharmacology,
University: Johannes Gutenberg University
Mainz, Germany

Joseph Loscalzo
Department of Medicine
Boston University School of Medicine
Boston, MA

Antonio F. Machado
Department of Environmental and Occupational Health
California State University at Northridge
Northridge, CA

Atsushi Matsuzawa
Laboratory of Cell Signaling
Graduate School of Pharmaceutical Sciences
The University of Tokyo
Tokyo, Japan

Edward A. Medina
One Shields Avenue
University of California at Davis
Davis, CA

Hideki Nishitoh
Laboratory of Cell Signaling
Graduate School of Pharmaceutical Sciences
Tokyo Medical and Dental University
Tokyo, Japan

Merry Jo Oursler
Department of Biology
University of Minnesota
Duluth, MN

Giuseppi Poli
Department of Clinical and Biological Sciences
University of Turin at S. Luigi Gonzaga Hospital
Turin, Italy

Leslie B. Poole
Department of Biochemistry
Wake Forest University School of Medicine
Winston-Salem, NC

Alvaro Puga
Department of Environmental Health
University Cincinnati Medicine. Center
Cincinnati, OH

Tommer Ravid
One Shields Avenue
University of California at Davis
Davis, CA

Sue Goo Rhee
Laboratory of Cell Signaling
NHLBI
NIH
Bethesda, MD

Douglas Ruden
Department of Environmental Health Sciences
Univerity of Alabama at Birmingham
Birmingham, AL

Freya Q. Schafer
Free Radical and Radiation Biology
The University of Iowa
Iowa City, IA

Petra M. Schwarz
Department of Pharmacology
Johannes Gutenberg University
Mainz, Germany

WilliamJ. Scott , Jr.
Division of Developmental Biology
Children's Hospital Research Foundation
Cincinnati,

Howard G. Shertzer
Department of Environmental Health
University Cincinnati Medicine. Center
Cincinnati, OH

Sruti Shiva
Department of Pathology
University of Alabama at Birmingham
Birmingham, AL

Kohsuke Takeda
Laboratory of Cell Signaling
Graduate School of Pharmaceutical Sciences
Tokyo Medical and Dental University
Tokyo, Japan

Victor J. Thannickal
Division of Pulmonary and Critical Care Medicine
University of Michigan Medical Center
Ann Arbor , MI

Martine Torres
Department of Pediatrics
Keck School of Medicine
University of Southern California
Childrens Hospital Los Angeles Research Institute
Los Angeles, CA

Kap-Seok Yang
Center for Cell Signaling Research and Division of Molecular Life Sciences
Ewha Women's University
Seoul, Korea

Xue Zhang
Department of Environmental Health Sciences
Univerity of Alabama at Birmingham
Birmingham, AL

Acknowledgements

The editors would like to thank the contributors for their cooperation and willingness to comply with the short time frame we gave them for completion of their chapters. We also thank our colleagues for helpful discussions and our families for their support.

Preface

Henry Jay Forman, Jon Fukuto and Martine Torres

"Research is to see what everybody else has seen and to think what nobody else has thought."

-- *Albert Szent-Gyorgyi*

Several years ago, one of us put together a book that dealt with various aspects of oxidative stress and introduced the concept of signal transduction by oxidants. Since then, the interest in the mechanisms by which reactive oxygen and nitrogen species (ROS/RNS) can modulate the cell's response has tremendously grown, paralleling the intense efforts towards identifying new signaling pathways in which phosphorylation/dephosphorylation events take center stage. Evidence is now mounting that production of these species by the cells is required for their function from growth to apoptosis and numerous signaling pathways have been identified where the participation of ROS and RNS is apparent (see Chapters 11-14, 16 and 18). Thus, the field is no more limited to the group of free radical aficionados who have pioneered this area of research but has now gone mainstream. While it is satisfactory for those of us who have been working on this topic for a long time, it has the risk of becoming the "fashionable" motto where those molecules, still mysterious to some, become responsible for everything and anything. In a way, it is reminiscent of the discovery of the phorbol ester receptor, that is to say protein kinase C (PKC) in 1977 [1], a major breakthrough in signal transduction that sparked a flurry of papers. Almost everything seemed to be PKC-dependent at that time. Little did we know that PKC come in various flavors, some of which have nothing to do with the initial definition of the enzyme as a lipid and calcium-dependent

H. J. Forman, J. Fukuto, and M. Torres (eds.), Signal Transduction by Reactive Oxygen and Nitrogen Species: Pathways and Chemical Principles, ©2003, Kluwer Academic Publishers. Printed in the Netherlands

kinase, that there are many other kinases, since then discovered and involved in complex signaling pathways, and that the "specific" inhibitors used in many studies were not that specific. Nevertheless, redox signaling has gained credence and is now on the map to stay. We profess the hope that this book will help researchers avoid some pitfalls by providing the current state of knowledge in the area of redox signaling, including controversies when they exist and future directions and by including information on physiologically relevant chemistry that can be applied across signaling systems. Although the name "reactive species" seems to imply high and non-discriminative reactivity, chemistry principles may help identify where specificity may occur, a particularly.critical concept in signaling, which needs to be better understood in redox signaling.

   To put things in perspective, we would like to quickly recount the principal findings that led to today's state of research. It is hard to believe that oxygen, this essential element of aerobic life, was not discovered before the late 18th century when Sheele, Priestly and Lavoisier independently isolated gaseous oxygen, initially branded as "fire air". The generally accepted theory of combustion or "Phlogiston theory" continued to exist for some time and was even defended by Priestly for the rest of his life. It took more than a century before the mechanism by which oxygen supports aerobic life was revealed and several competing theories arose early in the last century. Michaelis first proposed that all biological reductions including that of oxygen were univalent [2]. Thus, the production of superoxide anion ($O_2^{\cdot-}$) was indirectly predicted several decades before any demonstration of its existence in biological systems. Nonetheless, "oxygen activation" whereby oxygen is reduced to water in a concerted reaction in which no intermediates are formed, was for a long period of time considered as the major mechanism of oxygen consumption. The concept was in fact validated by the discovery of cytochrome c oxidase, an enzyme that transfers four electrons to oxygen to produce two molecules of water [3] without the release of intermediates. This led to the general assumption that reduced oxygen species were irrelevant in biology. However, the discovery of hydrogenases by Wieland in 1925 refuted Michaelis's theory as these enzymes catalyze the two-electron reduction of oxygen to form hydrogen peroxide (H_2O_2) [4]. This was the first direct evidence for the potential of H_2O_2 production in biology, although Thenard had discovered as early as 1818 that animal tissues could decompose H_2O_2. Perhaps the over century long delay in acceptance of the reality of two-electron reduction of oxygen occurred, in part, because catalase, the mysterious component in Thenard's preparation, was not recognized as a unique enzyme until 1901 [5].

   During the 1940's and 1950's, a large number of flavoproteins and metalloproteins were shown to reduce oxygen by two electrons to produce H_2O_2. The discovery that xanthine oxidase, a ubiquitous enzyme, could produce $O_2^{\cdot-}$ provided the first clue that this free radical might be of importance in biology [6]. Nevertheless, the existence of $O_2^{\cdot-}$ was still not fully accepted as many thought that this reaction was a laboratory curiosity due to protein modification during isolation. However, the "nail in the coffin" for $O_2^{\cdot-}$ came in 1969 when McCord and Fridovich discovered an enzyme whose sole purpose was to remove $O_2^{\cdot-}$, i.e. superoxide dismutase (SOD). First isolated from erythrocytes, it was soon after characterized as a ubiquitous enzyme in eukaryotic organisms and their tissues [7]. Subsequently, SOD isoforms were found in all aerobic organisms, giving credence to the idea that $O_2^{\cdot-}$ is generated in biological systems [8]. The search for $O_2^{\cdot-}$ generating systems showed that, in addition to xanthine oxidase, a number of flavoproteins and metalloproteins could catalyze univalent reduction of oxygen [9-11]. In 1973, Babior and colleagues reported that such a flavoprotein, present in neutrophils, was able to produce $O_2^{\cdot-}$ at the expense of NAPDH in cells "on command", i.e. upon phagocytosis of bacteria, and that this production explained the drastic cyanide-insensitive oxygen consumption associated with phagocytosis [12]. Furthermore, this NADPH oxidase was an essential part of the host defense against bacterial infections as cells from patients with chronic granulomatous disease were deficient in $O_2^{\cdot-}$ production and bacterial killing, the first proof by nature of the *beneficial* role of free radicals in biology. It took several more years to discover that the enzyme was in fact formed of several components, which were in separate compartments in resting cells. Stimulation results in translocation of the cytosolic proteins (p47^{phox}/p67^{phox}/p40^{phox}) to the plasma membrane where they bind to the flavocytochrome (p22^{phox}/gp91^{phox}) in a stable complex, competent for electron transfer (see Chapter 6). Unfortunately, soluble agents can also activate this enzyme and its products (i.e. $O_2^{\cdot-}$ and H_2O_2) are not released then within the confine of the phagolysosomes but in the surrounding tissues resulting in damage, hence the *detrimental* and double edge sword image long associated with $O_2^{\cdot-}$ and free radicals in biology.

In the late 1980s and early 1990s, several lines of evidence coming from three distinct fields merged and led to the demonstration of the endogenous generation of nitric oxide (NO) by endothelial cells and the finding that NO, a small diatomic free radical could activate guanylate cyclase, resulting in a dramatic rise in cGMP in the adjacent smooth muscle tissue and ensuing vasorelaxation [13-15]. This established NO as a critical regulator of vascular tone and as a signaling intermediate, the first demonstration of such role for a reactive species [16]. The importance of these findings was affirmed by the awarding of the Nobel Prize in Physiology or Medicine to Drs. Lou Ignarro, Ferid Murad and Robert Furchgott in 1998 for their discovery. The role of NO in physiology is not limited to smooth muscle relaxation as activated macrophages, epithelial cells, and other cell types can also produce NO. Biosynthesis of NO occurs via enzymatic oxidation of the amino acid L-arginine [17-19]. Several nitric oxide synthases (NOS) were identified that had some tissue specificity and particular properties such as the inducible character of the enzyme found in immune cells (iNOS) (see Chapter 7). In these cells, NO appears to participate in the elimination of various infectious agents, as demonstrated in the iNOS knockout model. However, the exact mechanism by which NO exerts its effect in the immune system and in functions other than vasodilation in other cells remains uncertain, as cGMP production does not account for all its effects. This remains one of the most intriguing and active topics in nitrogen oxide biology. NO is also produced by neuronal cells via an analogous biosynthetic pathway [20]. However, as in the immune response, the exact mechanism through which NO functions in the nervous system is not well defined. Understanding the complex chemistry of NO may help elucidate these mechanisms (see Chapter 4).

The finding that a free radical could participate in the production of a second messenger and in the activation of a signaling pathway was of paramount importance as it opened up minds to the idea that the role of reactive species may be more extensive in normal physiology. This idea had previously been put forth about a quarter century ago when exogenous H_2O_2 was shown to mimic the action of insulin growth factor [21] and, a few years later, when Mukherjeee and coworkers showed that insulin and nerve growth factor stimulated endogenous H_2O_2 production [22]. At that time, the research emphasis was on delineating the involvement of ROS in various pathologies and the identification in biological systems of novel redox agents, resulting from the interaction between ROS and RNS, dictated by their particular chemistry. Another drawback for the expansion of the redox field was the lack of clear understanding of how and where endogenous ROS were produced. Generation of H_2O_2 by mitochondria was discovered in the 1960's [23,24] and in the mid 1970's, several groups demonstrated that H_2O_2 generation by the mitochondrial electron transport chain occurred through the obligatory univalent reduction of oxygen to O_2^- [25-27]. Thus, a so-called leak from the mitochondrial chain is frequently cited as being responsible for increased cellular ROS. However, the production of O_2^- (the generally assumed reaction being via semiquinone autooxidation) has an equilibrium constant that does not favor O_2^- production, and is thereby thermodynamically unfavorable[13]. This means that the O_2^- production is not spontaneous, as often stated, but can only occur if the equilibrium is shifted by coupling to a second reaction such as the dismutation of O_2^- to H_2O_2. Thus, O_2^- is a very transient intermediate, and H_2O_2 production by mitochondria is catalyzed by mitochondrial SOD [16];

$$QH^{\bullet} + O_2 \rightleftharpoons Q + H^+ + O_2^{\bullet -}$$

$$2H^+ + 2O_2^{\bullet -} \xrightarrow{\quad SOD \quad} H_2O_2 + O_2$$

Interestingly, this is one of the few situations in biology in which SOD has been demonstrated to cause an increase in H_2O_2 generation [17]. The production of H_2O_2 by mitochondria has been shown to be dependent upon oxygen concentration; however, an increase in H2O2 can only be observed in a range of O_2 concentrations well above normal physiology [28,29]. In what would appear to be a contradiction of the known dependence on oxygen concentration, increased production of ROS by mitochondria was suggested to be involved in hypoxic signaling [30,31]. Thus, it was not entirely surprising to see that an alternative explanation appeared soon after that did not involve ROS. These recent studies have indicated that activation of the HIFα transcription factor, which modulates most hypoxic adaptation, is accomplished through enzymatic hydroxylation of proline and asparagine that signals for degradation and interaction with the transcriptional apparatus, respectively [32-35]. Thus, as society learned after the 1960's, radicals are not everywhere!

Nonetheless, there is more to mitochondrial H_2O_2 production than its dependence upon oxygen concentration and the role of NO as a regulator of mitochondrial activity and the discovery of a mitochondrial NOS (see Chapters 15 and 17) has provided further insights. Low levels of NO were shown to bind tightly to cytochrome c oxidase, which may increase oxygen at the mitochondrial inner membrane and possibly increase O_2^- production through reduction of Complex III [36] (see Chapter 15). Nevertheless, if NO is present in higher amounts, it may react with O_2^-, a reaction that is faster than enzyme-catalyzed O_2^- dismutation [16], and peroxynitrite will be produced (Chapter 4). It has also been suggested that NO and peroxynitrite inhibit Complex III in a similar manner to antimycin A and thereby promote O_2^- and H_2O_2 production [36,37] (see Chapter 17). Nevertheless, the question remains as to whether mitochondrial H_2O_2 production is regulated in a manner consistent with a role in signaling and further studies will be needed to understand the relationship between NO and regulated production of ROS by mitochondria. In the mean time, the discovery that homologs of $gp91^{phox}$ or NOX proteins are expressed in many cell types and that agents such as angiotensin can induce the regulated production of O_2^- in non-phagocytic cells has given further credence to the role of ROS in signaling (Chapter 6)..

The book previously edited by one of us was entitled "Oxidative Stress and Signal Transduction". The title for this new edition was changed to reflect our perception of redox signaling as events that occur when low levels of ROS are produced and when the targeting of signaling intermediates by reactive oxygen and nitrogen species is specific, transient and required for information to flow through a specific signaling pathway. The involvement of ROS in the EGF and PDGF signaling pathways seems to imply such definition. In contrast, we view signaling during oxidative stress as a response to cell injury, possibly with limited specificity as to the type of stress. Oxidation of a protein cysteine to a sulfinic acid (SO_2H) or a sulfonic acid (SO_3H), or oxidation of bases in DNA are modifications that involve oxidation but are either irreparable or require multiple enzymatic steps for repair that do not involve redox chemistry. Such damage to cellular constituents can stimulate signaling pathways leading to repair or even adaptation; however, these pathways may also be stimulated by damage that is independent of oxidation. Thus, the difference between oxidative stress signaling and redox signaling is not defined by whether cells die because physiological signaling may lead to cell death, albeit regulated as during development (described in Chapters 12, 19 and 20) but rather by the specificity of the response being due to redox chemistry rather than a recognition of damaged cellular constituents. Nonetheless, the boundary between redox chemistry and oxidative stress signaling is sometimes blurred as when a lipid peroxidation product, such as 4-hydroxy-2,3-nonenal, acts as a second messenger (see Chapter 10). The next big challenge for the field of redox signaling will be the identification of the chemical alterations imposed upon signaling proteins by ROS/RNS (or products derived from their action) and how such modification can affect the biological activity of the target, whether it is a kinase, a phosphatase or others. In addition, showing specificity will also be critical. One site of action of ROS/RNS that has long been recognized is the heme iron of enzymes, such as in the interaction of H_2O_2 with catalase and of NO with guanylate cyclase. The interaction of NO with cytochrome oxidase (see Chapter 15) or of H_2O_2 and RNS with cyclooxygenase (see Chapter 13) have also been described to affect signaling. Not surprisingly, thiol chemistry also plays a major role (See Chapters 1-3, 5 and 9). As H_2O_2 does not significantly react with protonated thiols, the oxidation of thiols by H_2O_2 most likely involves thiolates (-S$^-$) to produce a sulfenic acid (-SOH):

$$Protein - S^- + H_2O_2 \rightarrow Protein - SOH + OH^-$$

This requirement for a thiolate, only present in particular electrostatic fields, and the partial oxidation may provide specificity and reversibility that both characterize a signaling pathway. Glutathione can conjugate to protein thiols through two mechanisms resulting in formation of protein mixed disulfides:

$$GSSG + Protein - S^- \rightleftharpoons Protein - SSG + GS^-$$
$$GSH + Protein - SO^- \rightarrow Protein - SSG + OH^-$$

Thiols can also react with NO to form S-nitroso (S-NO) adducts. This reaction is often called "S-nitrosylation." Others refer to that process as "S-nitrosation," but this implies addition of a nitrosium ion NO^+ when the mechanism of formation of S-NO is still uncertain (Chapter 8). It could be argued then that S-nitrosylation implies addition of NO^- and that a more proper terminology would be "S-nitroso-ylation," which does not imply any particular mechanism but just the addition of an NO residue. As "S-nitrosylation" seems to have gained acceptance as the descriptive term for the formation of S-NO, this should be the common terminology. In fact, the posttranslational modifications regulating the activity of signaling proteins are usually described by the suffix, "ylation", as in phosphorylation, farnesylation, or ribosylation. Thus, we propose that the formation of a sulfenic acid be called "S-hydroxylation," and that "glutathionylation" be used to refer to the formation of mixed disulfides for consistency with other posttranslational modifications involved in signaling.

The last few years have seen exciting development in the area of signal transduction and redox signaling. We anticipate that the coming years will see the "consecration" of reactive species as signaling entities and that further studies will help better understand how dysregulation of ROS/RNS production may alter physiological pathways and lead to disease states. We are grateful to all the authors of this book for their generous contribution and salute their past and future efforts for advancing research in redox signaling.

References

1. Takai, Y., A. Kishimoto, M. Inoue and Y. Nishizuka. 1977. Studies on a cyclic nucleotide-independent protein kinase and its proenzyme in mammalian tissues. I. Purification and characterization of an active enzyme from bovine cerebellum. *J. Biol. Chem.* **252**:7603-7609.
2. Michaelis, L. 1946. Fundamentals of oxidation and reduction. In *Currents in Biochemical Research* (Green, D. E., ed.), pp. 207-227, Wiley, New York.
3. Warburg, O. and E. Negelein. 1929. Uber das Absorptionsspektrum des Atmungsferments. *Biochem. Z.* **214**:64-100.
4. Wieland, H. 1925. *Justus Liebigs Ann. Chem.* **445**:181.
5. Saunders, B. C., A. G. Holmes-Siedle and B. P. Stark. 1964. *Peroxidase*, Butterworth,Washington, D.C.
6. Fridovich, I. and P. Handler. 1962. Xanthine oxidase. V. Differential inhibition of the reduction of various electron acceptors. *J. Biol. Chem.* **237**:916-921.
7. McCord, J. M. and I. Fridovich. 1969. Superoxide dismutase: an enzymic function for erythrocuprein (hemocuprein). *J. Biol. Chem.* **244**:6049-6055.
8. Fridovich, I. 1989. Superoxide dismutases. An adaptation to a paramagnetic gas. *J. Biol. Chem.* **264**:7761-7764.
9. Massey, V., S. Strickland, S. G. Mayhew, L. G. Howell, P. C. Engel, R. G. Matthews, M. Schulman and P. A. Sullivan. 1969. The production of superoxide anion radicals in the

reaction of reduced flavins and flavoproteins with molecular oxygen. *Biochem. Biophys. Res. Commun.* **36**:891.

10. Nakamura, S. 1970. Initiation of sulfite oxidation by spinach ferredoxin-NADP reductase and ferredoxin system: a model experiment on the superoxide anion radical production by metalloflavoproteins. *Biochem. Biophys. Res. Commun.* **41**:177-183.

11. Misra, H. P. and I. Fridovich. 1971. The generation of superoxide radical during autoxidation of ferredoxins. *J. Biol. Chem.* **246**:6886.

12. Babior, B. M., R. S. Kipnes and J. T. Curnutte. 1973. The production by leukocytes of superoxide, a potential bactericidal agent. *J. Clin. Invest* **52**:741.

13. Furchgott, R. F. 1999. Endothelium-derived relaxing factor: Discovery, eraly studies and identification as nitric oxide (Nobel lecture). *Angew. Chem. Int. Ed.* **38**:1870-1880.

14. Murad, F. 1999. Discovery of some of the biological effects of nitric oxide and its role in cell signaling (Nobel Lecture). *Angew Chem. Int. Ed., 38, 1856-1868.* **38**:1856-1868.

15. Ignarro, L. J. 1999. Nitric oxide: A unique endogenous signaling molecule in vascular biology (Nobel Lecture). *Angew. Chem. Int. Ed.* **38**:1882-1892.

16. Kerwin, J. F., Jr., J. R. Lancaster, Jr. and P. L. Feldman. 1995. Nitric oxide: A new paradigm for second messengers,. *J. Med. Chem.* **38**:4343-4362.

17. Hicks, M. and J. M. Gebicki. 1979. A spectrophotometric method for the determination of lipid hydroperoxides. *Anal. Biochem.* **99**:249-253.

18. Hibbs, J. B., Jr., R. R. Taintor, Z. Vavrin and E. M. Rachlin. 1988. Nitric oxide: a cytotoxic activated macrophage effector molecule. *Biochem. Biophys. Res. Commun.* **157**:87-94.

19. Stuehr, D. J., S. S. Gross, I. Sakuma, R. Levi and C. F. Nathan. 1989. Activated murine macrophages secrete a metabolite of arginine with the bioactivity of endothelium-derived relaxing factor and the chemical reactivity of nitric oxide. *J. Exp. Med.* **169**:1011-1020.

20. Garthwaite, J. and C. L. Boulton. 1995. Nitric oxide signaling in the central nervous system. *Annu. Rev. Physiol.* **57**:683-706.

21. Czech, M. P. 1976. Differential effects of sulfhydryl reagents on activation and deactivation of the fat cell hexose transport system. *J. Biol. Chem.* **251**:1164-1170.

22. Mukherjee, S. P. and C. Mukherjee. 1982. Similar activities of nerve growth factor and its homologue proinsulin in intracellular hydrogen peroxide production and metabolism in adipocytes. Transmembrane signalling relative to insulin-mimicking cellular effects. *Biochem Pharmacol* **31**:3163-3172.

23. Jensen, P. K. 1966. Antimycin-insensitive oxidation of succinate and reduced nicotinamide- adenine dinucleotide in electron-transport particles. I. pH dependency and hydrogen peroxide formation. *Biochim. Biophys. Acta* **122**:157-166.

24. Hinkle, P. C., R. A. Butow, E. Racker and B. Chance. 1967. Partial resolution of the enzymes catalyzing oxidative phosphorylation. XV. Reverse electron transfer in the flavin-cytochrome beta region of the respiratory chain of beef heart submitochondrial particles. *J. Biol. Chem.* **242**:5169-5173.

25. Forman, H. J. and J. A. Kennedy. 1974. Role of superoxide radical in mitochondrial dehydrogenase reactions. *Biochem. Biophys. Res. Commun.* **60**:1044-1050.

26. Loschen, G., A. Azzi, C. Richter and L. Flohe. 1974. Superoxide radicals as precursors of mitochondrial hydrogen peroxide. *FEBS Letters* **42**:68.

27. Boveris, A. and E. Cadenas. 1975. Mitochondrial production of superoxide anions and its relationship to the antimycin insensitive respiration. *FEBS Letters* **54**:311.

28. Boveris, A. and B. Chance. 1973. The mitochondrial generation of hydrogen peroxide. General properties and effect of hyperbaric oxygen. *Biochem. J.* **134**:707-716.

29. Freeman, B. A. and J. D. Crapo. 1981. Hyperoxia increases oxygen radical production in rat lungs and lung mitochondria. *J. Biol. Chem.* **256**:10986-10992.

30. Huang, L. E., Z. Arany, D. M. Livingston and H. F. Bunn. 1996. Activation of hypoxia-inducible transcription factor depends primarily upon redox-sensitive stabilization of its alpha subunit. *J. Biol. Chem.* **271**:32253-32259.

31. Chandel, N. S., D. S. McClintock, C. E. Feliciano, T. M. Wood, J. A. Melendez, A. M. Rodriguez and P. T. Schumacker. 2000. Reactive oxygen species generated at mitochondrial complex III stabilize hypoxia-inducible factor-1α during hypoxia: a mechanism of O_2 sensing. *J. Biol. Chem.* **275**:25130-25138.
32. Jaakkola, P., D. R. Mole, Y. M. Tian, M. I. Wilson, J. Gielbert, S. J. Gaskell, A. Kriegsheim, H. F. Hebestreit, M. Mukherji, C. J. Schofield, P. H. Maxwell, C. W. Pugh and P. J. Ratcliffe. 2001. Targeting of HIF-α to the von Hippel-Lindau ubiquitylation complex by O_2-regulated prolyl hydroxylation. *Science* **292**:468-472.
33. Ivan, M., K. Kondo, H. Yang, W. Kim, J. Valiando, M. Ohh, A. Salic, J. M. Asara, W. S. Lane and W. G. Kaelin, Jr. 2001. HIFα targeted for VHL-mediated destruction by proline hydroxylation: implications for O2 sensing. *Science* **292**:464-468.
34. Lando, D., D. J. Peet, D. A. Whelan, J. J. Gorman and M. L. Whitelaw. 2002. Asparagine hydroxylation of the HIF transactivation domain a hypoxic switch. *Science* **295**:858-861.
35. Freedman, S. J., Z. Y. Sun, F. Poy, A. L. Kung, D. M. Livingston, G. Wagner and M. J. Eck. 2002. Structural basis for recruitment of CBP/p300 by hypoxia-inducible factor-1 α . *Proc. Natl. Acad. Sci. USA* **99**:5367-5372.
36. Poderoso, J. J., M. C. Carreras, C. Lisdero, N. Riobo, F. Schopfer and A. Boveris. 1996. Nitric oxide inhibits electron transfer and increases superoxide radical production in rat heart mitochondria and submitochondrial particles. *Arch Biochem Biophys* **328**:85-92.
37. Radi, R., M. Rodriguez, L. Castro and R. Telleri. 1994. Inhibition of mitochondrial electron transport by peroxynitrite. *Arch Biochem Biophys* **308**:89-95.

Chapter 1

REDOX STATE AND REDOX ENVIRONMENT IN BIOLOGY

Freya Q. Schafer and Garry R. Buettner

1.1 Introduction

1.1.1 Cells have a reducing environment

Cells and tissues must maintain a reducing environment to survive. This reducing environment provides the electrochemical gradient needed for electron flow. This movement of electrons provides the energy needed to build and maintain cellular structures and associated functions. An array of redox couples is responsible for the electron transfer. Some of these redox couples are linked to each other to form sets of related couples. Sets of couples can be independent from other sets if activation energies for reactions are high and there are no enzyme systems to link them kinetically. The redox environment of a cell is a reflection of the state of these couples. Bücher was the originator of studies that addressed cellular redox biochemistry.[1] His laboratory developed approaches to determine the states of various redox couples in cells and was the first to estimate the actual cellular reduction potentials for the $NAD^+/NADH$ and $NADP^+/NADPH$ couples[1].

---

[1] **Abbreviations used:** BSO, L-Buthionine-SR-sulfoximine; CoQ, Coenzyme Q, ubiquinone; E, E^o, $E^{o\prime}$; ... , Reduction potential at nonstandard conditions, standard condition (pH=0), standard conditions (pH=7); E_{hc} , Half-cell reduction potential; GPx, Glutathione peroxidase; GR, Glutathione (disulfide) reductase; $Grx(SH)_2$/GrxSS, Glutaredoxin, glutaredoxin disulfide; GSH, GSSG , Glutathione, glutathione disulfide; NAC, N-acetyl-L-cysteine; NADH, NAD^+, Nicotinamide adenine dinucleotide, oxidized form; NADPH, $NADP^+$, $NADP^\bullet$, Nicotinamide adenine dinucleotide phosphate, oxidized

H. J. Forman, J. Fukuto, and M. Torres (eds.), Signal Transduction by Reactive Oxygen and Nitrogen Species: Pathways and Chemical Principles, ©2003, Kluwer Academic Publishers. Printed in the Netherlands

It is now realized that the biological state of the cell changes with the reducing environment of the cell.[2] Many transcription factors are sensitive to redox changes in the cell. Thus, changes in the cellular redox environment can initiate signaling cascades.[3] The signaling by redox mechanisms relies on the chemistry that is driven by changes in the electrochemical potential of redox couples in the cell. Below we discuss how the electrochemical potentials of these redox couples are determined and indicate how these potentials influence cell signaling.

1.2 Redox State and Redox Environment: A Definition

Redox state is a term that has been used to describe the ratio of the interconvertible oxidized and reduced forms of a specific redox couple. For example, Sir Hans Krebs focused on the NAD^+/NADH couple and defined the redox state of this couple in a cell to be $[NAD^+]_{free}$/$[NADH]_{free}$.[4,5,6] In recent years, the term redox state has been used not only to describe the state of a particular redox pair, but also to more generally describe the redox environment of a cell. This more general use of the term redox state is not very well defined and differs considerably from historical uses. We suggest that the term redox environment be used when a general description of a linked set of redox couples is intended.

> *The redox state of a redox couple is defined by the half-cell reduction potential and the reducing capacity of that couple.* [2]

> *The redox environment of a linked set of redox couples as found in a biological fluid, organelle, cell, or tissue is the summation of the products of the reduction potential and reducing capacity of the linked redox couples present.*

> *Reducing capacity refers to the "size" of the pool (concentration) of reducing equivalents available, i.e. the strength of the redox buffer.* [2]

The above definitions attempt to clarify the difference between the redox state of a specific redox couple and the redox environment of a cell or tissue. When describing a specific redox couple the term redox state is appropriate. When describing changes in biological systems that involve various redox couples the term redox state should not be used, rather the term redox environment should be used.

1.3 The Nernst Equation

The redox couples in cells and tissues can be viewed as electrochemical cells. The Nernst equation allows one to determine the voltage of an electrochemical cell (ΔE) taking the Gibbs energy change (ΔG) and the mass action expression (Q) into account (Eqns. 1-3).

$$\Delta G^\circ = -n F \Delta E^\circ ,\qquad\qquad (1)$$

form, radical; ROS, Reactive Oxygen Species; SOD, Superoxide dismutase; Trx(SH)$_2$/ TrxSS, Thioredoxin, thioredoxin disulfide.

where n is the number of electrons exchanged in the chemical process, F is the Faraday constant, and $\Delta E°$ is the electromotive force under standard conditions, *i.e.*, the difference in the standard reduction potentials of the two half-cells involved in the process. The superscript ° implies the thermodynamic standard state[2],[7]

Using $\Delta G° = -nF\Delta E°$, $\Delta G = -nF\Delta E = \Delta G° + RT \ln Q$ and equation 6, the voltage of an electrochemical cell can be expressed as:

$$\Delta E = \Delta E° - \frac{RT}{nF} \ln Q \qquad \textbf{Nernst Equation} \qquad (2)$$

where, R is the gas constant ($R = 8.314 \; J \; K^{-1} \; mol^{-1}$), T the temperature (in Kelvin), and F the Faraday constant ($F = 9.6485 \times 10^4 \; C \; mol^{-1}$). This will yield results in volts.

The Nernst equation at $T = 25°C$ (298.15 K), using 2.303 as the conversion factor for ln into $\log_{10}$, can be written as:

$$\Delta E = \Delta E° - \frac{59.1 \; mV}{n} \log Q \qquad (3)$$

Thus, the Nernst equation can be used to determine the electrochemical potential between two redox couples. The electromotive force from these redox pairs (ΔE) is determined by subtracting the reduction potential of the species that is oxidized E_1 from the species that is reduced E_2:

$$\Delta E = E_2 - E_1 \qquad (4)$$

Redox-couple 1	$Red_1 \longrightarrow Ox_1 + e^-$	
Redox-couple 2	$Ox_2 + e^- \longrightarrow Red_2$	
Redox-reaction	$Red_1 + Ox_2 \rightleftharpoons Ox_1 + Red_2$	(5)

Standard Conditions: E°, G° ... ° imply 1 molal solution (unit activity); 1 atm pressure for gases; T = 298 K or 25°C and pH = 0. If a non-standard condition is to be used as a reference state, such as the pH being 7, then a prime mark (') is added to these notations, *i.e.* E°', G°', *etc*. All E° and E°' are measured against the normal hydrogen electrode. This electrode by convention is defined to have E° and E°' = 0 mV. A potential for the half reaction: $Ox + ne^- \rightarrow Red$ is a reduction potential; an oxidation potential corresponds to Red $\rightarrow$ Ox + ne$^-$ as a half reaction. Here all values of E are reduction potentials unless otherwise noted.

$$\Delta E = \left(E^{\circ}_2 - \frac{59.1 \text{ mV}}{n} \log \frac{[\text{Red}_2]}{[\text{Ox}_2]} \right) - \left(E^{\circ}_1 - \frac{59.1 \text{ mV}}{n} \log \frac{[\text{Red}_1]}{[\text{Ox}_1]} \right) \qquad (6)$$

If ΔE is zero, there is no net electron flow. When ΔE is not zero, the sign determines the direction of electron flow, *i.e.*, the direction of the redox reaction. The greater the magnitude of ΔE, the greater the thermodyndamic "pressure" pushing the reaction. Thus, as the redox state of couples such as GSSG/2GSH and NADP$^+$/NADPH change, they can force changes in other redox pairs, for example signaling proteins. These changes then can lead to biological consequences, such as proliferation, differentiation, apoptosis *etc.*

1.4   Example Redox Reactions

1.4.1 1e$^-$-processes

Paraquat (PQ^{2+}) is a widely used pesticide, a nonselective contact herbicide, that has high pulmonary toxicity with no known antidote. Cellular enzyme systems readily reduce PQ^{2+} by one-electron reactions to PQ$^{\bullet+}$. This radical is very reducing; for the redox pair

$$\text{PQ}^{2+}/\text{PQ}^{\bullet+} \quad E^{\circ\prime} = -448 \text{ mV} \quad \text{pH 7.0.} \qquad (7)$$

PQ$^{\bullet+}$ has a strong tendency to donate its electron to other species, such as oxygen to produce superoxide,

$$k = 8 \times 10^8 \text{ M}^{-1} \text{ s}^{-1} \quad [89]$$
$$\text{PQ}^{\bullet+} + \text{O}_2 \xrightarrow{\hspace{1.5cm}} \text{PQ}^{2+} + \text{O}_2^{\bullet-} \qquad (8)$$

There is a large thermodynamic driving force for this reaction as $E^{\circ\prime}_{\text{O}_2/\text{O}_2^{\bullet-}} = -160$ mV.[10,11,7] Thus, at standard conditions, $\Delta E^{\circ\prime} = (-160 \text{ mV}) - (-448 \text{ mV}) = +288$ mV. However, in the real world the concentrations of the species will not be 1 M; to make an estimate of the effective potential driving this reaction we use Eqn 6 with n = 1, *i.e.*;

$$\Delta E = (-160 - 59.1 \log [\text{O}_2^{\bullet-}]/[\text{O}_2]) - (-448 - 59.1 \log [\text{PQ}^{\bullet+}]/[\text{PQ}^{2+}]) \qquad \text{in mV (9)}$$

If the steady-state level of superoxide in a cell is 10^{-10} M and [O$_2$] is 10 μM (10^{-5} M), and the ratio of [PQ$^{\bullet+}$]/[PQ^{2+}] is 1/10 000, then $\Delta E = +345$ mV. This positive potential is consistent with the observed rapid formation of superoxide by PQ$^{\bullet+}$.

The high flux of superoxide due to PQ$^{\bullet+}$ will lead to a high flux of hydrogen peroxide and consequently a large demand for reducing equivalents from GSH and NADPH, species that in general are two-electron donors. This will undoubtedly result in a major change in the redox environment of cells and tissues exposed to paraquat, contributing to the toxicity of paraquat.

1.4.2 2e⁻ -processes

Most redox reactions in biology are two-electron processes. Two-electron processes avoid formation of reactive intermediates such as free radicals. As with the 1e⁻ processes, there is a thermodynamic hierarchy for the two-electron redox reactions, Table 1.[12,13,14]

Table 1: Two-Electron Reduction Potentials

Redox Couple (2-electron reductions)	$E^{\circ\prime}$/mV at 25°C	Reference
Xanthine/hypoxanthine, H^+	−371	13,15
Uric acid/xanthine, H^+	−360	13
DTTox (cyclic disulfide), $2H^+$/DTT (Dithiotreitol)	−317[a]	16
NAD^+, H^+/NADH	−316	17,13
$NADP^+$, H^+/NADPH	−315	16,11
Lipoic acid, $2H^+$/Dihydrolipoic acid	−290	11,18
GSSG, $2H^+$/2GSH (Glutathione)	−240	19,20
Cys-S-S-Cys, $2H^+$/2Cys-SH (Cystiene)	−230	11,21
TrxSS, $2H^+$/Trx(SH₂) (Thioredoxin)	-240[b]	18
GrxSS, $2H^+$/Grx(SH₂) (Glutaredoxin)	-218	18
FMN, $2H^+$/FMNH₂	−219	22
FAD, $2H^+$/FADH₂	−219	21
Riboflavin, $2H^+$/Leuco-riboflavin	−200	13,12
Acetaldehyde, $2H^+$/ethanol	−197	13
Pyruvate, $2H^+$/lactate	−183	11
Oxaloacetate, $2H^+$/malate	−166	23
Methylene blue, $2H^+$/Leuco-methylene blue	+11	11, 12
Dehydroascorbate, H^+/Ascorbate	+54	24
Ubiquinone (CoQ),$2H^+$/ubihydroquinone (CoQH₂)	+84	25
O_2, $2H^+$/ H_2O_2	+ 300	7
NO_3^-/NO_2^-	+421	26
H_2O_2, $2H^+$/2H₂O	+1320	7

[a] This value assumes the reduction potential of NAD^+ is −316 mV at pH 7.0.
[b] This value can range from −270 to −124 mV. The value −240 mV is taken as typical for mammalian thioredoxin.

Some of the most important redox couples in biology are the glutathione system (GSSG/2GSH), the nicotinamide adenine dinucleotide phosphate system (NADP$^+$/NADPH) and the thioredoxin system (TrxSS/Trx(SH)$_2$). Glutathione is considered to be the major thiol-disulfide redox buffer of the cell.[27] NADPH is a major source of electrons for reductive biosynthesis and the source of electrons for the glutathione and thioredoxin systems. Thioredoxin is another important thiol-system in the cell. Among its many functions, thioredoxin reduces cystine moieties in the DNA-binding sites of several transcription factors and is therefore important in gene expression.[28,29] All three redox couples transfer electrons through 2e$^-$-processes, but the Nernst equation is different in each case. Note, the reduction potential for the two 1e$^-$-steps cannot just be added because the value of E for a process depends not only on the number of electrons but also on the Gibbs energy changes for each step. For example:

a. NAD$^+$ + e$^-$ $\qquad$ → NAD$^\bullet$ E$^{o\prime}$ = −913 mV

b. NAD$^\bullet$ + e$^-$ + H$^+$ → NADH E$^{o\prime}$ = +282 mV

c. NAD$^+$ + 2e$^-$ + H$^+$→ NADH E$^{o\prime}$ = −316 mV at 25°C, pH 7.0 $\qquad$ (10)

For simplicity one can view the −316 mV as the average of the two one-electron reduction reactions. This value results because in reaction "c" two electrons are involved.

1.4.2.1 The NADP$^+$/NADPH system

The reduction half-reaction for the NADP$^+$, H$^+$/NADPH couple is:

$$NADP^+ + 2e^- + H^+ \longrightarrow NADPH$$

The Nernst equation for this process has the form:

$$E_{hc} = -315 - \frac{59.1}{2} \log \frac{[NADPH]}{[NADP^+]} \quad mV \text{ at } 25°C, pH\ 7.0 \ ^3 \qquad (11)$$

As an example, if we assume that in a cell NADPH is 80 μM and NADP$^+$ is 0.8 μM, then

$$E_{hc} = -315 - \frac{59.1}{2} \log \frac{[80 \times 10^{-6}\,M]}{[0.8 \times 10^{-6}\,M]} \quad mV \qquad (12)$$

Because the concentration units divide out, only the ratio of the concentrations of the two species in the redox pair are needed to determine E$_{hc}$ of this couple.

---

[3] In this chapter details on adjusting Nernst equation calculations for different pH values are not presented. A detailed discussion can be found in Ref[2].

Thus, the actual units need not be molarity as is typically entered into the Nernst equation, but any form of concentration can be used as long as the units divide out.

In this example the ratio of NADPH/NADP$^+$ is 100:1, which is on the order of that found in various animal tissues. [30]

$$E_{hc} = -315 - \frac{59.1}{2} \log \frac{100}{1} = -374 \text{ mV} \tag{13}$$

The very negative reduction potential of –374 mV supports the idea that the NADP$^+$/NADPH couple is a major driving force for maintaining the reducing environment in cells and tissues.

1.4.2.2   The GSSG/2GSH system

In case of the GSSG, 2H$^+$/2GSH couple knowing only the ratio [GSH]/[GSSG] does not allow the determination of the half-cell reduction potential for this couple. The reduction half-reaction for this couple is:

$$GSSG + 2H^+ + 2e^- \longrightarrow 2GSH \tag{14}$$

Note that **one** molecule of GSSG forms **two** molecules of GSH. Therefore, [GSH] will enter the Nernst equation as a squared term:

$$E_{hc} = -240 - \frac{59.1}{2} \log \frac{[GSH]^2}{[GSSG]} \text{ mV at 25°C, pH 7.0.} \tag{15}$$

As an example, if we assume that the concentration of GSH is 5 mM and that of GSSG is 50 μM, then the Nernst equation would be:

$$E_{hc} = -240 - \frac{59.1}{2} \log \frac{[5 \times 10^{-3} \text{ M}]\ [5 \times 10^{-3} \text{ M}]}{[5 \times 10^{-5} \text{ M}]} \tag{16}$$

Because [GSH] enters this equation as a squared term, the units will not divide out as in the case of NADPH. In this example if we enter the ratio of [GSH]/[GSSG] into the equation, the absolute molar concentration of GSH must still be known to completely specify the redox state of this couple:

$$E_{hc} = -240 - \frac{59.1}{2} \log \frac{(100\ [GSH])}{1} \text{ mV} \tag{17}$$

Thus, in cases where reduction of the oxidized half of the couple results in the formation of two reduced molecules, as with GSH, the molar concentration needs to be known.

The GSSG/2GSH couple is thought of as the major redox buffer in the cell, while the more reducing NADP$^+$/NADPH couple with its lower concentration but more negative reduction potential provides a strong reducing force. Because GSH is present at mM levels in the cell, it can serve as a buffer for oxidative or reducing events, thereby keeping the cellular redox environment stable. The redox state of the GSSG/2GSH couple is now being considered as a possible reflection of the biological state of the cell.[31,2] Signaling events probably contribute to the development of a particular redox state for this couple. Changes in the GSH pool might trigger signals for adaptive responses, such as proliferation, differentiation or death. Thus, the redox state of the GSSG/2GSH couple could both be a result of as well as a trigger for signaling cascades. Research in the next years will need to investigate the interdependence of the GSSG/2GSH couple and cell signaling events.

1.4.2.3   The Thioredoxin, TrxSS/Trx(SH)$_2$, system

The thioredoxins are a family of low molecular weight ($\approx$11-12 kDa) dicysteine proteins that in contrast to the GSH system use a macromolecular structure to transfer electrons. Their redox chemistry is characterized by two intra-molecular sulfhydryl groups that can be oxidized to form internal disulfide bonds. The reduction half-reaction for this redox pair is:

$$T{\overset{\displaystyle S}{\underset{\displaystyle S}{|}}} + 2e^- + 2H^+ \longrightarrow T{\overset{\displaystyle SH}{\underset{\displaystyle SH}{\diagdown}}} \tag{18}$$

One molecule TrxSS forms **one** molecule Trx(SH)$_2$ and the Nernst equation for the thioredoxin half-cell potential has the same form as that of the NADP$^+$/NADPH couple.

$$E_{hc} = E^{o\prime} - \frac{59.1}{2} \log \frac{[T(SH)_2]}{[TSS]} \text{ mV at } 25°C, \text{ pH } 7.0 \tag{19}$$

The thioredoxins are more specific than the glutathione system. They are known to react with approximately 20-30 biomolecules that are involved in regulatory and catalytic processes.[32] Thioredoxin modulates the activity of transcription factors such as Ref-1 (that regulates AP-1), NF-κB and the glucocorticoid receptor.[33] It also plays a role in the regulation of stress kinases, such as ASK1, thereby regulating stress signaling cascades.

1.4.2.4   When push comes to shove

The two electron processes discussed above are an example of linked cellular redox couples. Both the Trx and GSH-systems use NADPH as a source of reducing equivalents; thus, they are thermodynamically connected to each other. **Figure 1** clearly shows why NADPH is the thermodynamic driving force for both the thioredoxin and GSH systems. Because of the much higher concentration of GSH compared to NADPH and Trx(SH)$_2$, it has been the focus for research on how the redox environment of cells is connected to the biological state of the cell and cell signaling processes. As seen in the Nernst equation, simply reporting the total GSH or the ratio GSH/GSSG does not convey complete information on the redox state of this couple. One must specify both E$_{hc}$ for this couple and [GSH].

Cellular Redox Systems

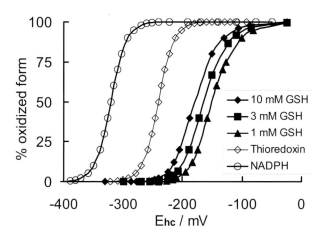

Figure 1. The $NADP^+/NADPH$ couple provides the reducing equivalents needed for the thioredoxin and GSH system.

The thioredoxin $E^{\circ\prime}$ is taken as -240 mV. For the $NADP^+/NADPH$ couple to be an efficient source of reducing equivalents for the GSH system these two systems should not be at thermodynamic equilibrium. This clearly is the case as the potential of the $NADP^+/NADPH$ couple is considered to be on the order of -375 mV in the cell while that of the GSSG/2GSH couple is on the order of -240 mV. [2] Thus, these two redox pairs, which are connected by glutathione disulfide reductase, appear to be out of equilibrium by a factor of 1000 or more. This is to be expected. If they were in equilibrium, then there would be no driving force to maintain an appropriate pool of GSH with a minimum amount of GSSG.

To convey all this information in a simple way we suggest use of the following notation for the status of a redox pair, such as GSSG/2GSH,

$$\{E_{hc}(GSH); [GSH]\}, \ e.g. \ \{-187 \text{ mV (GSH); 3.5 mM}\}.$$

Here -187 mV is the actual half-cell reduction potential of the GSSG/2GSH couple in the setting of interest and 3.5 mM is the concentration of GSH, the reduced species in the couple. With this information all elements of this redox pair are known.

1.6 Practical Approaches for Determining Redox Environment

As seen above, the half-cell potential of redox couples is pH-dependent, temperature-dependent, and sometimes requires molar concentrations rather than concentration ratios. Thus, to determine the E_{hc} of the GSSG/2GSH couple in cells the intra cellular pH, the temperature and the volume of the cell needs to be assessed. Intracellular pH can be measured using fluorescence probes with a flow cytometer.[34,35,36,37,38] The volume of the cell can be determined using a Coulter Counter. Once the volume is known, the concentration of the species in the redox couple can be determined.

If these techniques are not available, a simpler approach using the resting state of a cell as reference, can be used to estimate changes in redox potentials. An example for the GSSG/2GSH couple is given by Antunes *et al.*[39] In this approach the GSH and GSSG levels are measured in the "resting" state of the cell and then upon treatment. These values can then be entered into the Nernst equation as formulated in Eqn. 20.

$$\Delta E_{hc} = E_{hc(treatment)} - E_{hc(resting\ state)}$$

$$\Delta E_{hc} = \{E^{\circ\prime} - (RT/nF) \ln ([GSH]^2/[GSSG])_{(treatment)}\} - \{E^{\circ\prime} - (RT/nF) \ln ([GSH]^2/[GSSG])_{(resting\ state)}\}$$

$$\Delta E_{hc} = \frac{RT}{nF} \ln \{(GSSG]/[GSH]^2)_{(treatment)} \bullet ([GSH]^2/[GSSG])_{(resting\ state)}\} \quad (20)$$

Using this method, all the concentration units for the glutathione couple divide out. Thus, any means of expressing concentration can be used as long as those for GSSG in the two states are the same units and the units for the GSH levels in the two states are the same. This approach will not yield a half-cell potential for the redox couple, but rather it provides a measure of the change in the redox state of this couple, which should reflect the change in the overall cellular redox environment. Please note that in this approach two principal assumptions are made:

1.   The volume of the cells does not change with treatment, and
2.   The intracellular pH is the same in the resting and treatment states.

This approach can provide a great deal of quantitative information when studying the toxicological effects of substances on cells.

1.7 Redox Environment, Considerations and Measurements

1.7.1 Redox Environment and the Redox State of the GSH Couple

It is now being recognized that the redox state of the glutathione system may be involved in the determination of the biological state of the cell.[2,30,40] In general it appears that high levels of GSH and the usually associated low E_{hc} are affiliated with proliferation; increases in E_{hc} will slow proliferation and increase differentiation; additional increases in E_{hc} can suppress differentiation and bring about apoptosis or if severe, necrosis.

Several examples in the literature now exist that demonstrate quantitatively that an increase in E_{hc} can result in apoptosis. Antunes *et al.* found that initiation of apoptosis through H_2O_2 was accompanied by a shift in the redox environment towards more oxidizing E_{hc}.[38] Cai *et al.* have demonstrated that toxicants that increase E_{hc} of the GSSG/2GSH couple by $\approx$70 mV from control cells can initiate apoptosis.[41]

Findings in the GSH-redox field often seem confusing and contradicting. One treatment might change the biological status of one cell line or tissue while bringing about no response in another cell line. One of the reasons might be that the concentration of GSH varies from cell line to cell line. As seen in Figure 1, to make a substantial change in the redox environment of an organism that has 10 mM GSH *vs.* 1 mM GSH requires a much greater oxidative assault. Thus, when examining the influence of redox stress on biological events one should take the total concentration, or more importantly, the redox-buffer capacity into account.

1.7.2 Redox Environment and Signaling

A signal is like a speck of methylene blue entering water. If the small speck of dye is put into a small puddle of water, an intense color is observed, but if it is put into the ocean the initial small bit of evidence that methylene blue is there is soon gone as it is diluted; no color change will be visible. The GSH buffer can be viewed similarly. If a signal enters a cellular GSH buffer with low capacity it will be "visible" and result in activation of transcription factors or initiation of signaling cascades. If the GSH buffer capacity is very high, the signals might be dampened and never reach their endpoint. Thus, the GSH buffer could influence the strength and duration of a signal, but might not determine which individual signaling process occurs.[42] The concept is similar when investigating overall changes in biological status, such as a shift from proliferation to cell death. If the redox state of the GSSG/2GSH couple is close to the border where apoptosis can occur, then a weak signal (small change in GSH concentration) can initiate this shift. Thus, it is important to consider the capacity of the cellular redox-buffer.

Another concern is that all processes are considered and appropriately examined. An example is provided by Hansen *et al.* in their studies on the modulation of differentiation by changes in the GSH.[43] Treatment of rabbit limb bud cultures with BSO lead to apparent inhibition of differentiation, while treatment with N-acetylcysteine (NAC) appears to restore differentiation, an observation that seemed inconsistent with the generalization on GSH levels outlined in section 7.1 above. However, when these investigators considered each stage of this process, their observations were consistent with the generalizations. Their experience points to a need for careful examination of processes before conclusions can be appropriately made.

Hansen *et al.* also compared rat tissue with rabbit tissue and found contrasting results. Because rat tissue had 43% more GSH than the rabbit tissue, the differing results could be due to a difference in GSH buffer capacity as mentioned above.

Many researchers consider the initiation of signaling processes to be a local event that require local gradients in the electrochemical potential of the redox pairs involved in the signaling process. Absolute quantitation of local cellular gradients of the redox state of specific redox couples is not yet possible. Some beginning approaches to qualitative imaging are underway. A most interesting study is the use of electron paramagnetic imaging of redox gradients in tumors.[44] Using a mouse model and the rate of nitroxide destruction, this study was able to provide spatially resolved, quantitative data on the redox environment of normal and tumor tissue and the influence of GSH on tissue redox status. This study clearly demonstrates that tumors have significant heterogeneity in their redox environment compared with normal tissue. In addition the RIF-1 tumor model used had about 4-fold greater GSH than normal tissue and this was reflected in the image. These initial studies point to more general approaches that may developed to examine quantitatively redox status and redox gradients in cells and tissues.

Acknowledgements:

This work was supported by NIH grants CA 66081 and CA 81090.

References

1. Bücher, T. and M. Klingenberg. 1958. Wege des Wasserstoffs in der lebendigen Organisation. *Angew. Chem.* **70:**552-570.

2. Schafer, F. Q. and G. R. Buettner. 2000. Redox environment of the cell as viewed through the redox state of the glutathione disulfide/glutathione couple. *Free Radic. Biol. Med.* **30:**1191-1212.

3. Sun, Y. and L. W. Oberley. 1996. Redox regulation of transcriptional activators. *Free Radic Biol Med.* **21:**335-348.

4. Krebs, H. A. 1967. The redox state of nicotinamide adenine dinucleotide in the cytoplasm and mitochondria of rat liver. *Advances in Enzyme Regulation.* **5:**409-434.

5. Krebs, H. A. and T. Gascoyne. 1968. The redox state of the nicotinamide-adenine dinucleotides in rat liver homogenates. *Biochem. J.* **108:**513-520.

6. Krebs, H. A. and R. L. Veech. 1969. Equilibrium relations between pyridine nucleotides and adenine nucleotides and their roles in the regulation of metabolic processes. *Advances in Enzyme Regulation.* **7:**397-413.

7. Koppenol, W. H. and J. Butler. 1985. Energetics of interconversion reactions of oxyradicals. *Adv. Free Radic. Biol. Med.* **1:** 91-131.

8. Farrington, J. A., M. Ebert, E. J. Land and K. Fletcher. 1973. Bipyridylium quaternary salts and related compounds. V. Pulse radiolysis studies of the reaction of paraquat radical with oxygen. Implications for the mode of action of bipyridyl herbicides. *Biochim. Biophys. Acta.* **314:**372-381.

9. Buettner, G. R. 1993. The pecking order of free radicals and antioxidants: Lipid peroxidation, α-tocopherol, and ascorbate. *Arch. Biochem. Biophys.*, **300:**535-543.

10. Koppenol, W. H. 1997. The chemical reactivity of radicals. In: *Free Radical Toxicology* (Wallace, K.B., ed.), pp. 3-14, Taylor and Francis, London.

11. Clark, W. M. ed. 1960. *Oxidation-Reduction Potentials of Organic Systems.* Williams and Wilkens, Baltimore, MD.

12. Lardy, H. A. ed. 1949. *Respiratory Enzymes.* Burgess, Minneapolis, MN.

13. Burton, K. 1957. Free energy data of biological interest. *Ergeb. Physiol. Biol. Chem. Exptl. Pharmakol.* **49:**275.

14. Weber, G. 1961. Absorption bands and molar absorption coefficients of substances of biochemical interest. *Biochemist's Handbook.* (Long, C. ed.), pp. 81-82, Spon. London.

15. Cleland, W. W. 1964. Dithiothreitol, a new protective reagent for SH groups. *Biochemistry.* **3:**480-482.

16. Rodkey, F. L. and J. A. Donovan. 1959. Oxidation-reduction potentials of the diphosphopyridine nucleotide system. *J. Biol. Chem.* **234:**677-680.

17. Ke, B. 1957. Polarographic behavior of a-lipoic acid. *Biochim. Biophys. Acta.* **25:**650-651.

18. Aslund, F., K. D. Berndt and A. Holmgren. 1997. Redox potentials of glutaredoxins and other thiol-disulfide oxidoreductases of the thioredoxin superfamily determined by direct protein-protein redox equilibria. *J. Biol. Chem.* **272:**30780-30786.

19. Gilbert, H. F. 1995. Thiol/disulfide exchange equilibria and disulfide bind stability. *Methods in Enzymol.* (Packer L. ed.) **251:**8-28. Academic Press, San Diego, CA.

20. Fruton, J. S. and H. T. Clark. 1934. Chemical reactivity of cystine and its derivatives. *J. Biol. Chem.* **106:**667-691.

21. Lowe, H. J. and W. M. Clark. 1956. Studies on oxidation-reduction. XXIV. Oxidation - reduction potentials of flavin adenine dinucleotide. *J. Biol. Chem.* **221:**983-992.

22. Burton, K. and T. H. Wilson. 1953. The free-energy changes for the reduction of diphosphopyridine nucleotide and the dehydrogenation of L-malate and L-glycerol 1-phosphate. *Biochem. J.* **54:** 86.

23. Williams, M. H. and J. K. Yandell. 1982. Outer-sphere electron transfer reactions of ascorbate anions. *Aust. J. Chem.* **35:**1133-1144.

24. De Vries, S., J. A. Berden and E. C. Slater. 1980. Properties of a semiquinone anion located in the QH2:cytochrome c oxidoreductase segment of the mitochondrial respiratory chain. *FEBS Lett.* **122:**143-148.

25. Koppenol, W. H., J. J. Moreno, W. A. Pryor, H. Ischiropoulos and J. S. Beckman. 1992. Peroxynitrite, a cloaked oxidant formed by nitric oxide and superoxide. *Chem. Res. Toxicol.* **5:**834-842;

26. Gilbert, H. F. 1990. Molecular and cellular aspects of thiol-disulfide exchange. In *Advances in Enzymology*, (Meister, A., ed.), pp.69-173, Wiley Interscience, New York.

27. Matthews, J. R., N. Wakasugi, J. L. Virelizier, J. Yodoi and R. T. Hay. 1992. Thioredoxin regulates the DNA binding activity of NF-κB by reduction of a disulfide bond involving cysteine 62. *Nucl. Acid Res.* **20:** 821-3830.

28. Okamoto, T., H. Ogiwara, T. Hayashi, A. Mitsui, T. Kawabe and J. Yodoi. 1992. Human thioredoxin/adult T cell leukemia-derived factor activates the enhancer binding protein of human immunodeficiency virus type 1 by thiol redox control mechanism. *Int. Immunol.* **4:**811-819.

29. Veech, R. L., L. V. Eggleston and H. A. Krebs. 1969. The redox state of free nicotinamide-adenine dinuleotide phosphate in the cytoplasm of rat liver. *Biochem. J.* **155:**609-619.

30. Kirlin, W. G., J. Cai, S. A. Thompson, D. Diaz, T. J. Kavanagh and D. P. Jones. 1999. Glutathione redox potential in response to differentiation and enzyme inducers. *Free Radic. Biol. Med.* **27:**1208-1218.

31. Follmann H. and I. Haeberlein. (1995/1996). Thioredoxin: universal, yet specific thiol-disulfide redox cofactors. *BioFactors.* **5:**147-156.

32. Adler V., Y. Zhimin, K. D. Tew.and Z. Ronai Z. 1999. Role of redox potential and reactive oxygen species in stress signaling. *Oncogene.* **18:**6104-6111.

33. Darzynkiewicz, Z., H. A. Crissman and J. P. Robinson eds. 1994. *Methods in Cell Biology,* **41** (Part A, Chap.9), Flow Cytometry (2nd Edition). Academic Press, Inc, San Diego.

34. Musgrove, E. A. and D. W. Hedley. 1990. Measurement of intracellular pH. *Methods in Cell Biology.* **33:**59-69.

35. Kleyman, R. T. and E. J. Cragoe. 1988. Amiloride and its analogs as tools in the study of ion transport. *J. Membrane Biol.* **105:**1-21.

Chapter 2

SULFUR AND SELENIUM CATALYSIS AS PARADIGMS FOR REDOX REGULATIONS

Leopold Flohé

2.1 Introduction

Redox regulation of enzymatic activities is a most topical, but by no means a young field of research. From its beginnings in the late sixties of the last century, the concept emerged that exposed SH groups of enzymes might react with the most abundant cellular redox-mediator GSH to modulate activity. In 1967, the Horecker group reported fructose diphosphatase activity to be regulated by thiol/disulfide exchange,[1] and shortly thereafter glucose-6-phosphate dehydrogenase was recognized to be activated by oxidized glutathione (GSSG).[2,3] But despite the Kosowers' startling appeal[4] "Lest I forget thee, glutathione", the concept of enzyme regulation by thiol/disulfide exchange developed slowly. Reviews from the mid eighties listed less than a dozen metabolic pathways that might be regulated this way.[5,6] Instead, protein phosphorylation and dephosphorylation became the principle that apparently dominated metabolic regulation.[7,8] More recently, however, many of the phosphorylation cascades turned out to be modulated by redox phonemena[9] and people started wondering how signaling cascades that appeared to be clearly defined by protein phosphorylation might be affected by the redox status of the cell.[10]

Since, redox regulation has become fashionable again. Apart from a few cases, however, the molecular basis of the regulatory events still remains obscure. The frequently heard statement that a signaling cascade is activated by "reactive oxygen species" is too nebulous to be revealing. In fact, the most reactive oxygen species,

H. J. Forman, J. Fukuto, and M. Torres (eds.), Signal Transduction by Reactive Oxygen and Nitrogen Species: Pathways and Chemical Principles, ©2003, Kluwer Academic Publishers. Printed in the Netherlands

hydroxy or alkoxy radicals, would promiscuously react with every protein. They do not meet the criterium of specificity that is mandatory for the fine-tuning of metabolic processes.   Less reactive compounds like O_2^-, H_2O_2 or lipid hydroperoxides might be considered instead. Alternatively, enzymes are often claimed to be regulated by the "GSH/GSSG ratio". This seemingly more precise explanation of the phenomenon is not satisfactory either. Single molecules, not ratios react. The idea of ratios or potentials being relevant to regulation appears simplistic in mixing up thermodynamic feasibilities with real chemical events in a biological system, that never is in equilibrium but is determined by the kinetics of specific reactions. Also, a rapid equilibration between, e. g., the glutathione system and protein SH groups would at best yield a redox buffer. What would instead be required for regulation are independently catalysed on and off reactions, as is analogously achieved in protein phosphorylation cascades by specific kinases and phosphatases.

In short, we have to know the conditions that allow a particular redox-active compound to react with an enzyme, how the reaction is reversed and what the catalytic consequences are, before we can reasonably declare process to be redox-regulated. Needless to stress our inability to name a single example that might meet these criteria. We therefore will present the chemical processes that conceivably modify proteins by redox-active compounds. Some mechanisms of enzymes relying on sulfur and selenium catalysis will be discussed to demonstrate which of the chemical options are commonly used in biological systems. It is hoped that this detour will pave the way to explore the chemical principles that link unspecific oxidative processes to specific enzymatic reactions.

2.2 The Keys Players: Hydroperoxides, Glutathione, Thioredoxins and Thiol-Dependent Peroxidases

Non-destructive and reversible oxidative modifications may be achieved by H_2O_2 and other hydroperoxides with or without the aid of redox mediators such as GSH or thioredoxins. The most common oxidant *in vivo* is H_2O_2 that, in inflammatory conditions, is amply formed by dismutation of $\cdot O_2^-$ released from activated phagocytes[11] but physiologically also by intracellular NADPH oxidases[12] and other enzymatic systems. Although some H_2O_2 appears to be produced constitutively, receptor-mediated H_2O_2 formation appears to be more common. Typical examples are TNFα-induced mitochondrial $\cdot O_2^-$ formation[13] or cytoplasmic increase of H_2O_2 upon growth factor receptor stimulation.[9] Similarly, the formation of lipid-hydroperoxides usually depends on the activation of phospholipases followed by activation of lipoxygenases, both groups of enzymes being silenced in unstimulated cells.[14]

ROOH thus formed can oxidize thiols. The most abundant thiol in mammalian cells, GSH, is not readily oxidized by ROOH because of its high pK near 9.2. Its oxidation to GSSG has to be catalyzed by any of the selenium-containing glutathione peroxidase (GPx) isozymes[15] or a type VI peroxiredoxin according to eq. (1)

$$\overset{\textit{GPx-1, -2, -3, -4, Prx VI}}{2GSH + ROOH \longrightarrow GSSG + ROH + H_2O} \qquad (1)$$

An alternative reductant of ROOH is reduced thioredoxin that contains a redox active CGPC motif, wherein the first C is solvent-exposed and largely dissociated at physiological pH. It therefore reacts with H_2O_2 faster than GSH, but nevertheless the thioredoxin oxidation is catalyzed by at least three distinct peroxiredoxins[16] and by GPx-3. [17]

$$Trx\ (SH)_2 + ROOH \xrightarrow{\quad Prx\ I,\ II,\ III,\ GPx\text{-}3 \quad} TrxS_2 + ROH + H_2O \qquad (2)$$

The fate of a particular hydroperoxide will largely depend on the site of production and the subcellular concentration of the diverse peroxidases, and accordingly the products of the peroxidase reaction will be either GSSG or $TrxS_2$. Both disulfides are readily reduced by the related flavoproteins, glutathione reductase (GR) or thioredoxin reductases (TR) the latter being a selenoprotein in mammals.[18] In addition, the intracellular GSSG concentration is kept low by a specific export system (reviewed in [5]) and also $TrxS_2$ appears to be exported.[19]

The multiplicity of peroxidases using thiols as reductants hardly complies with the assumption that they simply back up each other in fulfilling a common biological task such as the detoxification of hydroperoxides. In this respect, also their molar efficiencies differ too extremely to allow a mutual competition. The most efficient ones are the selenium-containing GPx-type enzymes which react with hydroperoxides with rate constants beyond $10^7 M^{-1}s^{-1}$. The corresponding rate constants for the peroxiredoxin-type enzymes of mammals have not yet been worked out, but may be assumed to be about two orders of magnitude smaller, as in the bacterial[20] and protozoal homologs.[16,21] They are approximately as efficient in hydroperoxide reduction as non-selenium GPx-type enzymes, be they natural [22] or tailored by site-directed mutagenesis.[23,24] Altogether the human organism is equipped with a minimum of 11 thiolperoxidases: the four efficient selenoenzymes of the GPx family, GPx-5 having the active site selenocysteine replaced by cysteine, and at least six distinct peroxiredoxins, the latter 7 likely being rather inefficient as peroxidases. Some of them are restricted to particular tissues or cellular compartments: GPx-3 is extracellular, GPx-5 has only been found in the epididymis, and Prx III is mitochondrial. The remaining eight, that obviously colocalize, would still appear redundant if they were not in charge of specific tasks. In fact, not even the selenium-containing ones appear to be able to substitute each other. GPx-1(-,-) mice, although without any obvious phenotype at first glance, proved to be highly susceptible to oxidative challenge by redox-cyclers, LPS exposure and viral infections. This sensitivity was largely unaffected by tritrating down the other selenoproteins by selenium deprivation (compiled in[15,25,26]). On the other hand, knock out of *gpx-4* proved to be lethal in early embryonic development for unknown reasons (M. Brielmeier, pers. communication).

The emerging scenario is that the most efficient and abundant GPx-1 is an antioxidant device that protects the organism against toxic levels of peroxides, whereas the remaining peroxidases make use of hydroperoxides to create specific (selena)disulfides bridges. A few examples may suffice to demonstrate that this distinction is not a semantic one: i) Activation of ASK-1 requires release of a tightly bound Trx(SH)$_2$ which is achieved by oxidation to TrxS$_2$;[27] the peroxidase involved is unknown, but Prx I or II would have the specificity and location to do the job. ii) GPx-4 in late spermatogenesis reacts with surface SH groups of its own and likely of other proteins to build-up the keratin-like material that has to embed the mitochondrial helix in the midpiece of spermatozoa.[28-30] iii) TrxII and IV were found to be associated with, and to specifically oxidize thiols of cyclophilin A.[31]

In other words, the products of the peroxidase reactions may be physiologically more important than the concommittant disappearance of H_2O_2. The kind of product formed depends on the thiol preferentially used that is unknown for many of these enzymes. The name-giving substrate of the GPx family, glutathione, is specifically used by the mammalian GPx-1, but is by no means the exclusive or preferred substrate of all members of the family. GPx-3 equally reacts with glutaredoxin and thioredoxin;[17] a GPx of *Plasmodium falciparum* hardly reacts with GSH but with thioredoxin instead;[22] a GPx homolog in *Trypanosoma cruzi* is specific for tryparedoxin, a remote relative of the thioredoxins[32] that had been discovered as the typical substrate for protozoal peroxiredoxins.[21] The only substrate common to all peroxiredoxins is an unphysiological one, DTT.[16] Physiological substrates have been identified in exceptional cases only: flavoproteins with vicinal SH groups (AhpF) in *Enterobacteria*,[33] thioredoxin in *Helicobacter pylori*,[20] yeast[34] and mammals (Prx I, II and III),[35] tryparedoxin in *Kinetoplastida*[16] and GSH for mammalian Prx VI.[36] Evidently, the steadily growing family offers chances for future surprises.

Chemical Principles and Lessons From Enzymology

Oxidation of thiols or selenols

Thiols may be oxidized via two-electron-transitions by hydroperoxides to sulfenic, sulfinic or sulfonic acids. For a productive attack by a hydroperoxide, e. g. H_2O_2, the thiol needs to be deprotonated.

$$R\text{-}S^- + H^+ + H_2O_2 \rightarrow R\text{-}SOH + H_2O \tag{3}$$

Oxidation of a thiol to a sulfenic acid is readily reversed through stepwise reduction by thiols:

$$R\text{-}SOH + R'\text{-}SH \rightarrow R\text{-}S\text{-}S\text{-}R' + H_2O \tag{4}$$

$$R\text{-}S\text{-}S\text{-}R' + R'\text{-}S^- \rightarrow R'\text{-}S\text{-}S\text{-}R' + R\text{-}S^- \tag{5}$$

In reaction (4) also undissociated thiols may react. At physiological pH, sulfenic acids are largely deprotonated and may themselves enforce dissociation of the co-reacting thiol accodring to (6):

$$\tag{6}$$

The reactions according to (3-6) appear most attractive for a reversible modification of protein thiols, since sulfinic or sulfonic acids would be less easily reduced by thiols again.   Thiol radicals, that may be formed by single-electron transitions with the aid of complexing transition metals, may also be disregarded as regulatory entities, because they react too unspecifically.   Like other sulfur-centered radicals[37] they appear more relevant to toxic phenomena.   Essentially the same considerations apply to the oxidation of selenols.

In contrast to thiols, selenols are largely dissociated at physiological pH values. The selenolate form being predominant anyway, oxidation by hydroperoxides may proceed in analogy to eq. (3), yet without vigorous catalytic proton abstraction. Another difference is the extreme reactivity of the resulting selenenic acid.   It instantly reacts with thiols and selenols. Exposure of selenols to H_2O_2 usually yields diselenides without any intermediate being detectable.   In fact, a single stable selenenic acid could so far be synthetized this way.[38]   In this particular case, the selenium was deeply hidden in a bowl like structure ("Bmt") that sterically prevented diselenide formation, while the ability of the selenenic acid to react with low molecular weight thiols was retained.

Another potential of selenenic acids became obvious from studies on the catalytic mechanism of the GPx mimic ebselen, a benzo-1,2-selenazole derivative (Fig. 1). By stepwise reduction with GSH (or other thiols) the selenazole ring is cleaved to yield an aromatic selenol with a substituted formamido group in ortho position.   Reaction with H_2O_2 would yield a selenenic acid. Apart from some diselenide, however, the original selenazole is detected in good yields, which reveals that selenenic acids readily react with amido groups.

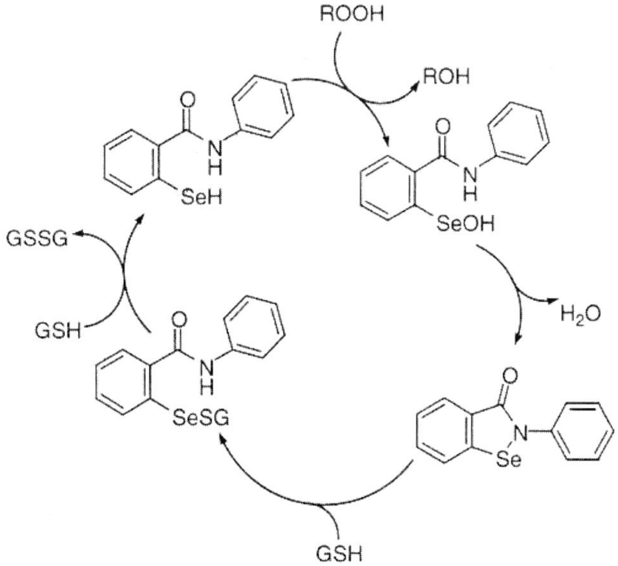

Figure 2-1. Catalytic cycle of the GPx mimic ebselen.

2.3.2 Catalytic triads of sulfur and selenium–containing peroxidases

Evidence from site-directed mutagenesis revealed that the peroxidase activity of peroxiredoxins depends on a cysteine that is strictly conserved in their N-terminal domain.[16] It is considered to be the residue which is attacked by the hydroperoxide. The presumed selenenic acid thereby formed according to eq. (3) was detected by X-ray crystallography in human PrxVI[39] and by derivatisation with 7-chloro-4-nitro-2-oxa-1,3-diazole followed by mass spectrometry of the adduct in AhpC of *Salmonella typhimurium*.[40]

As outlined in 3.1 the cysteine thiol has to be deprotonated to efficiently react with a hydroperoxide.   The first proposal how this might be achieved in peroxiredoxins was provided by the structure of PrxVI.[39] Here the thiol is in close proximity to the positively charged guanidino group of a strictly conserved arginine. Additional activation appeared feasible by a proton-donating histidine residue, which, however, is not conserved in most peroxiredoxins (Fig. 2A).   Extensive mutagenesis studies with tryparedoxin peroxidases of *Crithidia fasciculata*[41] and *Leishmania donovani*[42] revealed that the activation of Cys 52 is achieved by Arg 128 and hydrogen bonding from Thr 49 (Fig. 2B). These residues are actually conserved in most peroxiredoxins and thus may form the catalytic triad common to this family. Only in exceptional cases the threonine residue is replaced by serine[16] which, in mutants of TXNPx of *L. donovani*, functionally substituted for the threonine, while a valine did not.[42]

In the GPx-family the active site (seleno)cysteine is differently activated. In the X-ray structures of GPx-1[43] and GPx-3[44] the selenium is seen coordinated to the imino nitrogen of a tryptophan and the amido group of a glutamine (Fig. 3). These residues are strictly conserved in the whole family. Hydrogen bonds between the nitrogens and the selenium or sulfur are believed to assure quantitative dissociation of the selenol or thiol, respectively.   In complience with this assumption, substitution of Trp 136 and Gln 81 by acid residues, that should enforce association of the cysteine thiolate, inactivated the cysteine homolog of porcine GPx-4.[24] In selenocysteine GPx-type enzymes, however, the role of the residues coordinated to the selenocysteine is less clear, since the selenol should be dissociated anyway. In

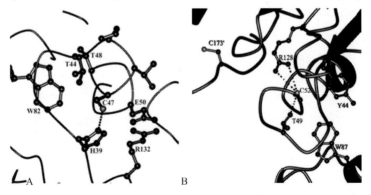

Figure 2-2. Catalytic centers of peroxiredoxins. A reaction center of human PrxVI according to Choi et al.[39] B: Model of the catalytic triad of tryparedoxin peroxidases[16].

fact, carboxymethylation of the (seleno)cysteine residue, which equally depends on dissociation, was as fast with authentic GPx-4 as with its cysteine homolog.[24] Nonetheless the reaction rate with ROOH was 3 orders of magnitude faster with authentic GPx-4. This reveals a superiority of selenol catalysis irrespective of the difference in thiol pKs versus selenol pKs.

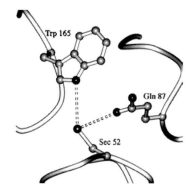

Figure 2-3. Catalytic triad common to selenium-containing glutathione peroxidases[15] based on the X-ray structure of bovine GPx-1[42] (Courtesy of Dr. S. Pilawa)

While in the peroxiredoxins, the primary oxidation product could be identified as an enzyme with a cysteine sulfenic residue, the precise chemical nature of the oxidized cytalytic intermediate in the GPx family remains puzzling. The hypothetical selenenic acid derivative could never be detected. In crystals of GPx-1 two oxygen atoms were seen bound to the selenium.[43] Obviously, overoxidation had happened, which leads to a sluggish enzyme form that requires long-term pre-incubation with thiols before full activity is re-established.[45] Recent attempts to identify the catalytic oxidation product of GPx-4 by LC-ESI-MS failed to detect any mass increment compatible with the addition of one ore more oxygen atoms. The molecular mass did not markedly increase when the donor substrate-free enzyme was exposed to excess ROOH. Instead, a decrease by two mass units was tentatively identified, which though is close to detection limit (F. Ursini *et al.*, unpublished data). The finding nevertheless points to a fast elimination of water upon the presumed formation of the selenenic acid. A straight forward possibility would be the formation of a selenadisulfide bridge within the protein according to eq. (7).

$$R\text{-}SeOH + R'\text{-}SH \rightarrow R\text{-}Se\text{-}S\text{-}R' \tag{7}$$

Yet according to molecular modelling, there is no cysteine in reach of the selenocysteine in GPx-4, nor could such bond by detected by ESI-MS of proteolytic digests (F. Ursini et al., unpublished data). An other explanation is suggested by the catalytic cycle of ebselen, where the selenenic acid form instantly reacts with the formamido group to rebuild the selenazole ring. Possibly, the oxidized selenium in GPx-4 reacts with the analogous glutamine amide to form a Se-N bond (eq. 8).

$$R\text{-}SeOH + R'\text{-}CO\text{-}NH_2 \rightarrow R\text{-}Se\text{-}NH\text{-}CO\text{-}R' \tag{8}$$

Such product would be equivalent in oxidation state to a selenenic acid. Although there is no direct analytical evidence for the existence of Se-N bonds in biological systems, we should consider their possible formation in catalytic processes and also in selenoprotein/protein interaction.

2.3.3 Alternate routes to (selena)disulfides

A ternary collision of two thiols with a peroxide to yield a disulfide and water does not seem to be the kind of reaction nature relies on. Rather the reactions according to eq. (3) and (4) should be considered as the predominant ways.

A formation of GSSG according to eq. (3) (R and R' being G) has been implicated in the GPx-like activity of certain glutathione-S-transferases which catalyse a nucleophilic attack by GSH on electrophiles, in this case a lipid hydroperoxide. The GSOH thus formed will instantly react the abundant GSH to form GSSG. A spontaneous formation of GSOH being slow due to the high pK of GSH, GSSG formation by hydroperoxides is overwhelmingly catalyzed. In most catalytic processes R in eq. (3) or (7) is the enzyme bearing the sulfenic or selenenic acid (or its equivalent) that reacts with GSH to form an S- or Se-glutathionylated enzyme, respectively. A second GSH will then cleave the (selena)disulfide bond to form the thermodynamically preferred symmetric disulfide GSSG and to regenerate the ground state enzyme (eq. 5; Fig. 4). This sequence of events is now commonly accepted as catalytic principle of GPx-type enzymes. It may similarly apply to one-cysteine peroxiredoxins such as PrxVI.

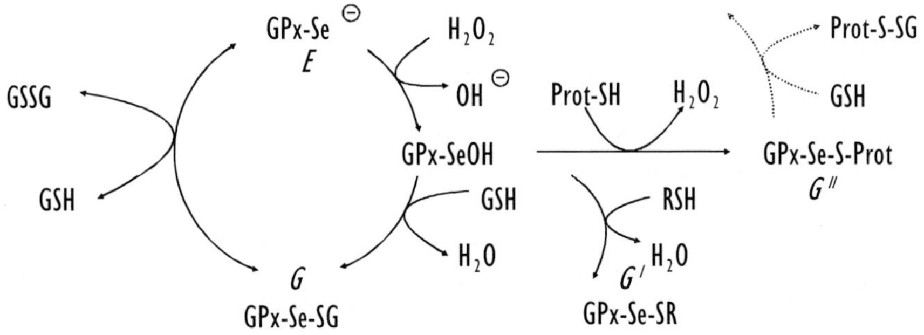

Figure 2-4. Catalytic cycle of glutathione peroxidases[15] with variations leading to dead-end intermediates

The catalytic cycle of glutathione peroxidases, as depicted in the left part of Fig. 4, may be taken as a model for protein glutathionylation and its reversal by GSH according to equations 3 – 5. The prerequisites for these reactions to actually take place are i) a highly activated selenol (or thiol) in E, ii) accessibility of the oxidized selenium (or sulfur) in F to GSH and iii) the acessibility of the (selena)disulfide bond in G to nucleophilic attack by a second GSH. In GPx-1, these conditions are ideally fulfilled. A selenolate is exposed in a flat well at the surface to react with the hydroperoxides.[43] The reaction center is surrounded by four arginines and a lysine that electrostatically direct the first GSH into a productive complex to react with F and thereafter the second one to cleave the Se-S bond in G. GSSG, once formed, is pushed away from the active site.[15,46] In consequence, the glutathionylated protein does not accumulate and, in fact, remains undetectable as long as traces of reduced GSH are present.[47] In order to understand how stable glutathionylations or thiylations in general are achieved at cellular GSH levels, we have therefore to consider the variations of the catalytic scheme. Clearly, the first GSH could be bound in a particular glutathionylated protein in a way that the Se-S bridge is shielded against the approach of the second one. Examples of this kind are unknown, but alternate substrates, e. g. mercaptosuccinate or penicillamine, create dead-end intermediates G' of GPx-1, from which E is not easily regenerated by GSH. An interesting example of another dead-end intermediate (G" in Fig. 4) formation is the polymerisation of GPx-4 that is observed during sperm maturation.[28] In this process Sec 46 specifically reacts with Cys 148 of another GPx-4 molecule (F. Ursini, unpublished), although at least four more SH groups are exposed to the surface. Clearly, GPx-4 itself here acts as a protein sulfhydryl modifying agent, and it appears attractive to look for analogous reactions that might explain the interference of GPx-4 with signaling cascades.[26]

Unlike in GPx-1, glutathionylation is easily detected in GPx-4. LC-ESI-MS revealed that GSH is specifically bound to Sec 46 as is presumed for the catalytic intermediate G (F. Ursini, unpublished). The intermediate is formed if the reduced enzyme is incubated with GSSG, thus confirming the reversibility of the G→E transition that had previously been deduced from electrochemical experiments.[48] This type of reaction is most commonly considered as the preferred mechanism of protein glutathionylation; although supporting experimental evidence in scarce. The

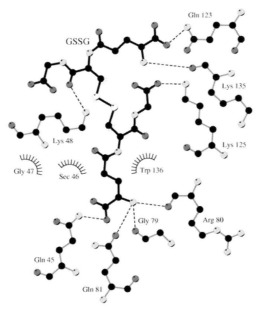

Figure 2-5. Model of interaction between GSSG and reduced GPx-4, as constructed by Autodock 3.0 followed by energy minimisation by AMBER6. (Courtesy of Dr. S. Pilawa)

GPx-4 example underscores the importance of structural peculiarities to allow such reactions. While oxidized mercaptoethanol promiscuously thiylated all surface exposed SH groups plus the selenol in GPx-4, the highly charged GSSG selectively modified Sec 46. Extensive molecular modeling suggests that GSSG is directed by electrostatic forces of Lys 48 and 125 and fixed by multiple hydrogen bonds to a position that permits an attack of the active site selenolate on the SS bridge (Fig. 5).

In short, protein thiylation may be achieved by i) reaction of GSH with sulfenic or selenenic acids of proteins, ii) reactions of the latter with suitable proteins or iii) reaction of GSSG with thiols or selenols of proteins. For the first two options, product formation is thermodynamically favoured. The third option, in view of the low cellular GSSG content, is thermodynamically unfavoured in principle but has to be considered for selenoproteins and also for protein SH groups if they are deprotonated by their microenvironment. In each of the cases, sterical fits and electrostatic compatibilities of the reactants are decisive criteria.

2.4 Disulfide reduction

The reduction equivalents for disulfide reduction are provided by NADPH in most species, less often by NADH. They are transferred to disulfides by a family of related flavoproteins such as glutathione reductases (GR), thioredoxin reductases (TR), trypanothione reductases, mycothiol reductase, and the bacterial AhpFs. These enzymes share a central redox center, in which a hydrid anion is transferred from the reduced pyridine nucleotide via enzyme-bound FAD to a disulfide bond that is coordinated to the isoalloxazine ring (Fig. 6). Thereby sulfhydrols are generated in the reaction center, one of which at least being dissociated due to a coordinated histidine residue. From there the reduction equivalents, by thiol/disulfide exchange, go to the substrate directly, as e. g. in GR, or mediated by selenium catalysis as in mammalian TRs. While the common initial steps of this type of catalysis have become textbook knowledge, the diversified downstream steps need to be discussed, since they determine substrate specificity.

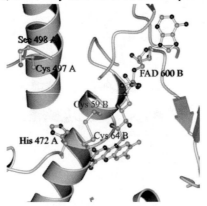

Figure 2-6. Intersubunit active site of mammalian thioredoxin reductase. The selenocysteine residue at the C-terminus of subunit A (Sec 498A) can interact with the central redox center

of subunit B (Cys 59 and Cys 64 and bound FAD). The model is based on the X-ray structure of rat TR (PDB ID 1H6V).[51] (Courtesy of Dr. S. Pilawa)

In most of the flavine-containing disulfide reductases the central redox center including the reactive thiolates is located at the bottom of a cleft and thus is accessible to ligands with optimum fit only. Accordingly, a GR does not reduce trypanothione, which is the oxidized bis-glutathionyl-derivative of GSH, nor can most of the TRs reduce any of the low molecular weight disulfides. They even tend to discriminate between homologous and foreign thioredoxins.[49] The exception to this rule are mammalian TRs. They posses a second redox center, a Cys-Sec-Gly motif at the end of a flexible C-terminal extension.[49] Within the dimeric enzyme, the flexible selenocysteine-containing arm can pick up the reduction equivalents from the central redox center of the opposite subunit and then swing out to reduce the disulfide bond of the bulky proteinaceous substrate, Trx (Fig. 6). The almost freely floating, solvent-exposed selenolate of such TRs can, of course, reduce a variety of disulfides including protein disulfides and even other oxidants such as hydroperoxides, alloxan, dehydroascorbate, vitamin K and nitrosoglutathione.[49] It may thus be assumed that mammalian TR is more generally involved in the reversal of oxidative protein modifications, documented examples being protein disulfide isomerase, GPx-3, and NK-lysine.[49]

The reduction products of TRs, the thioredoxins, are considered as broad-spectrum protein disulfide reductases.[50] They serve as reducing substrates for ribonucleotide reductase[49-51] and two-cysteine peroxiredoxins-type peroxidases,[16] but also reductively activate the transcription factor NFκB in the nucleus, are implicated in the regulation of cytosolic phosphorylation cascades and, extracellularly, display cytokine-like functions.[52] These pleiotropic roles should, however, not be taken to indicate lack of specificity. Specific binding to proteins evidenced by yeast two-hybrid analysis[52] as well as the multiplicity of Trxs and Trx-related molecules argue against promiscuity.

The reaction specificity of the Trx family is due to a CXXC motif, wherein the proximal cysteine is solvent exposed and largely dissociated for an attack on disulfide bridges, whereas the distal one is hidden in the protein core. The low pK of the exposed cysteine is likely due to proton shuttling between the vicinal SH groups that is facilitated by hydrogen bridges reaching the distal cysteine from OH groups in the protein core. Fig. 7 shows a close-up of a Trx-related protein, a tryparedoxin, to demonstrate how the reacting cysteine is activated by a network of hydrogen bonds.[53]

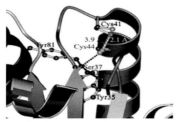

Figure 2-7. Active site of tryparedoxin: Oxidized and substrate-reduced tryparedoxin of C. fasciculata; close-up of the active site showing increase of S-S distance upon reduction (from 2.1 to 3.9 A) and network of hydrogen bridges activating Cys 41[53]

Figs. 8 A and B are to demonstrate how substrate specificities are achieved. Fig. 8A shows a close up of the intersubunit reaction center of an oxidized peroxiredoxin, tryparedoxin peroxidase, wherein the sulfenic acid form of Cys 52 has further reacted to form a disulfide bridge to Cys 173 of an inverted second subunit. Four charged groups of the tryparedoxin are attracted by complementing charges on the peroxidase surface to direct the Cys 41 of the tryparedoxin towards Cys 173 of its substrate (Fig. 8 A).[42]

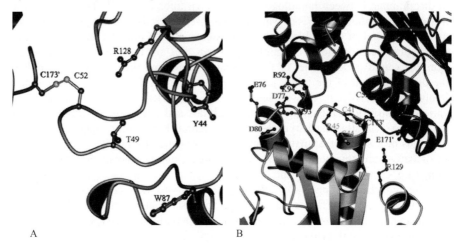

A   B

Figure 2-8. Interaction of tryparedoxin peroxidase with tryparedoxin. A: Intersubunit active site of oxidized tryparedoxin peroxidase showing a disulfide bridge between Cys 52 and Cys 173'. B: Catalytic intermediate of the reaction between reduced tryparedoxin and the peroxidase. Cys 41 of tryparedoxin (same view as in Figure 7) has reacted with Cys 173' of the peroxidase. Attraction of reactants is achieved by electrostatic interactions between E76, D77, D80 and R129 of tryparedoxin2 with R92, K93, K94 and E171' of tryparedoxin peroxidase.[16] (Courtesy of Dr. H.J. Hecht)

The tryparedoxin peroxidases are not equally fast reduced by the closely related Trxs and not at all by GSH or glutaredoxins.[54] Differences in the active site motifs, which are WCGPCK in Trx, WCPPCR in tryparedoxins and TCPYCR in glutaredoxins, may in part contribute to the specificity[55] but, again, the precise sterical fit and specific interactions of residues near the reaction centers appear to guarantee the appropriate selection of the reaction partners.

The route to disulfide reduction in target proteins are highly diversified in different living domains and even vary within one organism or cell. The common denominator appears to be the use of a reduced pyridine nucleotide by a member of the flavine-dependent disulfide reductases followed by thiol/disulfide exchange reactions. Staying with peroxidases as ultimate target proteins, the simplest system appears to be the enterobacterial one. Here the chain-initiating flavoprotein AhpF directly reduces AhpC, which is the peroxiredoxin-type peroxidase. In some bacteria, as in yeast and many other organisms the peroxiredoxin is fuelled by the TR/Trx system; trypanosomes need trypanothione as a redox mediator between the flavoprotein and the Trx-homologous tryparedoxins, and mammals mostly rely on glutathione to feed their selenoperoxidases. Whenever a redox mediator is needed to reduce a target peroxidase, ramifications of the pathways are observed such as the use of GSH for disulfide reshuffling or glutaredoxin-mediated ribonucleotide reduction or the reduction of alternate targets by Trxs or tryparedoxins. Despite the confusing scenario, it may safely be stated that nature has never left the realm of possible reactions to the chance of being simply driven by mass law.

2.5 Conclusions and Persepctives

The examples of sulfur and selenium catalysis presented in this article were selected to demonstrate that chemical reactions that are thermodynamically feasible and therefore might happen spontaneously are nevertheless catalyzed in nature. If spontaneous reaction rates are reasonably fast, as the oxidation of thiols by hydroperoxides, the catalytic power of selenium is preferentially used to further increase the turnover rates in order to prevent unspecific chemical reaction. Similarly, the fast selenium catalysis is used in mammals to guarantee the availability of reduced Trx that is needed for a wide range of tasks. This being obvious from the analyses of enzymatic systems, it would be hard to imagine that nature might try to solve the more delicate tasks of regulating metabolic processes without the principles known from enzyme catalysis, i. e. accelerating the reaction between metastable compounds with utmost specificity. This deduction may sound trivial but sharply conflicts with prevailing views of redox regulation that, e. g., implicate hydroxy radicals in regulatory processes, although they react at diffusion-limited rates with almost every component of a biological system, or expect well balanced conditions from ingesting whatever is called an "antioxidant".

Another aspect, that is relevant to regulation, is the observation that enzyme-catalyzed processes usually work unidirectionally. In our examples, thiol/disulfide exchange reactions are certainly reversible in principle, but nonetheless the formation of a particular disulfide and its cleavage is done by distinct catalytic pathways: The disulfide bond in oxidized Trx is formed by reaction with a peroxiredoxin, a ribonucleotide reductase or NFκB and it is reduced at the expense of NADPH by TR. The glutathione system works in an analogous way. Such separate forward and backwards routes are an obligatory requirement for the design of regulatory circuits. Only an enzyme-catalyzed interplay of hydroperoxides and particular thiols meets this requirement.

Engineering also tells us that regulatory circuits need sensors to open or close welves at remote sites of a system. In the context of redox regulation, thioredoxin can safely be rated to be a redox sensor. In the ASK-1 system, it triggers cellular responses in remote cellular compartments (apoptosis) depending on its redox state. Analogous functions may be envisaged for any kind of protein that, depending on a redox-dependent modification, interacts with an other proteins to modify its function in a meaningful way. Regulation by redox sensing may certainly be an efficient principle also working beyond the cellular context. It is tempting to speculate that some still orphan enzymes might fulfill this role. The extracellular GPx-3, for instance, is an efficent peroxidase in permanent shortage of reduction equivalents and thus is not in a position to cope with major hydroperoxide challenge. The enzyme could, however, by means of its oxidized selenol, communicate to a cell surface receptor to take adequate measures, because it had been bombarded with

hydroperoxides by an excited phagocyte upstream in the circulation.[26]   Another
candidate of this type would be TRANK, an extracellular peroxiredoxin that can
trigger responses reminding of those induced by inflammatory cytokines, while
intracellular peroxiredoxins tend to suppress cytokine responses when
overexpressed.[16,56]

In short, redox modifications of proteins and their reversal are suited for building
up regulatory circuits, if they are enzyme-catalyzed or at least share the specificity
with enzymatic processes.   A few examples of specific regulations by redox
mediators have been worked out at the molecular levels, while the mechanisms of
most related phenomena remain unclear.   These will be discussed in dedicated
chapters of this volume.   It appears advisable to discriminate between protein
modification for regulatory purposes and oxidative protein destruction by "reactive
oxygen species".   The first phenomenon is an ordered metabolic response, the
second one is a pathological event.   In borderline cases the metabolic response to an
oxidative challenge may be understood as an alarm reaction to prevent an oxidative
disaster.   A molecular understanding of the regulatory processes will certainly help
to discriminate between fine-tuning of the normal flux of redox equivalents in an
aerobic organism, the use of hydroperoxides for particular cellular functions, alarm
reactions in pre-pathological situations and an overt oxidative disaster, as it happens
in a septic crises, ischemia/reperfusion or suicide by redox cycling herbicides.

References

1.   Pontremoli, S., S. Traniello, M. Enser, S. Shapiro and B. L. Horecker. 1967. Regulation
     of fructose diphosphatase activity by disulfide exchange. *Proc Natl Acad Sci USA*
     58:286-293.

2.   Bonsignore, A., I. Lorenzoni, R. Cancedda, L. Silengo, D. Dina and A. De Flora. 1968.
     Metabolism of human erythrocyte glucose-6-phosphate dehydrogenase. IV) Reductive
     inactivation and autoinactivation of the enzyme. *Ital J Biochem* **17**:346-362.

3.   Eggleston, L. V. and H. A. Krebs. 1974. Regulation of the pentose phosphate cycle.
     Biochem J **138**:425-435.

4.   Kosower, E. M. and N. S. Kosower. 1969. Lest I forget thee, glutathione. *Nature*
     224:117-120.

5.   Sies, H. 1983. Reduced and oxidized glutathione efflux from liver. In *Glutathione:*
     Storage, Transport and Turnover in Mammals (Sakamoto, Y. ,T. Higashi, and N.
     Tateishi, eds.), pp. 63-88, Japan Sci Soc. Press/VNU Science Press, Tokyo/Utrecht.

6.   Brigelius, R. 1985. Mixed disulfides: biological functions and increase in oxidative
     stress. In *Oxidative stress* (Sies, H., ed.), pp. 243-272, Acad Press, New York.

7.   Hunter, T. 1995. Protein kinases and phosphatases: the yin and yang of protein
     phosphorylation and signaling. *Cell* **80**:225-236.

8.   Fischer, E. H. 1997. Cellular regulation by protein phosphorylation: a historical
     overview. *Biofactors* **6**:367-374.

9.   Finkel, T. 2000. Redox-dependent signal transduction. *FEBS Lett* **476**:52-54.

10. Flohé, L., R. Brigelius-Flohé, C. Saliou, M. G. Traber and L. Packer. 1997. Redox regulation of NF-κ B activation. *Free Radic Biol Med* **22**:1115-1126.

11. Babior, B. M. 1999. NADPH oxidase: an update. *Blood* **93**:1464-1476.

12. Bayraktutan, U., L. Blayney and A. M. Shah. 2000. Molecular characterization and localization of the NAD(P)H oxidase components gp91-phox and p22-phox in endothelial cells. *Arterioscler Thromb Vasc Biol* **20**:1903-1911.

13. Goossens, V., J. Grooten, K. De Vos and W. Fiers. 1995. Direct evidence for tumor necrosis factor-induced mitochondrial reactive oxygen intermediates and their involvement in cytotoxicity. *Proc Natl Acad Sci U S A* **92**:8115-8119.

14. Schewe, T. 2002. 15-lipoxygenase-1: a prooxidant enzyme. *Biol Chem* **383**:365-374.

15. Flohé, L. and R. Brigelius-Flohé. 2001. Selenoproteins of the glutathione system. In *Selenium. Its molecular biology and role in human health* (Hatfield, D. L., ed.), pp. 157-178, Kluwer Academic Publishers, Boston.

16. Hofmann, B., H. J. Hecht and L. Flohé. 2002. Peroxiredoxins. *Biol Chem* **383**:347-364.

17. Bjoernstedt, M., J. Xue, W. Huang, B. Akesson and A. Holmgren. 1994. The thioredoxin and glutaredoxin systems are efficient electron donors to human plasma glutathione peroxidase. *J Biol Chem* **269**:29382-29384.

18. Tamura, T. and T. C. Stadtman. 1996. A new selenoprotein from human lung adenocarcinoma cells: purification, properties, and thioredoxin reductase activity. *Proc Natl Acad Sci U S A* **93**:1006-1011.

19. Rubartelli, A., A. Bajetto, G. Allavena, E. Wollman and R. Sitia. 1992. Secretion of thioredoxin by normal and neoplastic cells through a leaderless secretory pathway. *J Biol Chem* **267**:24161-24164.

20. Baker, L. M., A. Raudonikiene, P. S. Hoffman and L. B. Poole. 2001. Essential thioredoxin-dependent peroxiredoxin system from Helicobacter pylori: genetic and kinetic characterization. *J Bacteriol* **183**:1961-1973.

21. Nogoceke, E., D. U. Gommel, M. Kiess, H. M. Kalisz and L. Flohé. 1997. A unique cascade of oxidoreductases catalyses trypanothione-mediated peroxide metabolism in Crithidia fasciculata. *Biol Chem* **378**:827-836.

22. Sztajer, H., B. Gamain, K. D. Aumann, C. Slomianny, K. Becker, R. Brigelius-Flohé and L. Flohé. 2001. The putative glutathione peroxidase gene of Plasmodium falciparum codes for a thioredoxin peroxidase. *J Biol Chem* **276**:7397-7403.

23. Rocher, C., J. L. Lalanne and J. Chaudière. 1992. Purification and properties of a recombinant sulfur analog of murine selenium-glutathione peroxidase. *Eur J Biochem* **205**:955-960.

24. Maiorino, M., K. D. Aumann, R. Brigelius-Flohé, D. Doria, J. van den Heuvel, J. McCarthy, A. Roveri, F. Ursini and L. Flohé. 1995. Probing the presumed catalytic triad of selenium-containing peroxidases by mutational analysis of phospholipid hydroperoxide glutathione peroxidase (PHGPx). *Biol Chem Hoppe Seyler* **376**:651-660.

25. Hatfield, D. L. 2001. *Selenium. Its molecular biology and role in human health*, Kluwer Academic Publishers,Boston, Dordrecht, London

26. Brigelius-Flohé, R. 1999. Tissue-specific functions of individual glutathione peroxidases. *Free Radic Biol Med* **27**:951-965.

27. Saitoh, M., H. Nishitoh, M. Fujii, K. Takeda, K. Tobiume, Y. Sawada, M. Kawabata, K. Miyazono and H. Ichijo. 1998. Mammalian thioredoxin is a direct inhibitor of apoptosis signal- regulating kinase (ASK) 1. *EMBO J* **17**:2596-2606.

28. Ursini, F., S. Heim, M. Kiess, M. Maiorino, A. Roveri, J. Wissing and L. Flohé. 1999. Dual function of the selenoprotein PHGPx during sperm maturation. *Science* **285**:1393-1396.

29. Maiorino, M., L. Flohé, A. Roveri, P. Steinert, J. B. Wissing and F. Ursini. 1999. Selenium and reproduction. *Biofactors* **10**:251-256.

30. Foresta, C., L. Flohé, A. Garolla, A. Roveri, F. Ursini and M. Maiorino. 2002. Male fertility is linked to the selenoprotein phospholipid hydroperoxide glutathione peroxidase. *Biol. Reprod.* **67**: in press.

31. Lee, S. P., Y. S. Hwang, Y. J. Kim, K. S. Kwon, H. J. Kim, K. Kim and H. Z. Chae. 2001. Cyclophilin a binds to peroxiredoxins and activates its peroxidase activity. *J Biol Chem* **276**:29826-29832.

32. Wilkinson, S. R., D. J. Meyer, M. C. Taylor, E. V. Bromley, M. A. Miles and J. M. Kelly. 2002. The *Trypanosoma cruzi* enzyme TcGPXI is a glycosomal peroxidase and can be linked to trypanothione reduction by glutathione or tryparedoxin. *J Biol Chem* **277**:17062.17071.

33. Storz, G., F. S. Jacobson, L. A. Tartaglia, R. W. Morgan, L. A. Silveira and B. N. Ames. 1989. An alkyl hydroperoxide reductase induced by oxidative stress in Salmonella typhimurium and Escherichia coli: genetic characterization and cloning of ahp. *J Bacteriol* **171**:2049-2055.

34. Chae, H. Z., S. J. Chung and S. G. Rhee. 1994. Thioredoxin-dependent peroxide reductase from yeast. *J Biol Chem* **269**:27670-27678.

35. Rhee, S. G., S. W. Kang, L. E. Netto, M. S. Seo and E. R. Stadtman. 1999. A family of novel peroxidases, peroxiredoxins. *Biofactors* **10**:207-209.

36. Fisher, A. B., C. Dodia, Y. Manevich, J. W. Chen and S. I. Feinstein. 1999. Phospholipid hydroperoxides are substrates for non-selenium glutathione peroxidase. *J Biol Chem* **274**:21326-21334.

37. Giles, G. I. and C. Jacob. 2002. Reactive sulfur species: an emerging concept in oxidative stress. *Biol Chem* **383**:375-388.

38. Goto, K., M. Nagahama, T. Mizushima, K. Shimada, T. Kawashima and R. Okazaki. 2001. The first direct oxidative conversion of a selenol to a stable selenenic acid: experimental demonstration of three processes included in the catalytic cycle of glutathione peroxidase. *Org. Lett.* **3**:3569-3572.

39. Choi, H. J., S. W. Kang, C. H. Yang, S. G. Rhee and S. E. Ryu. 1998. Crystallization and preliminary X-ray studies of hORF6, a novel human antioxidant enzyme. *Acta Crystallogr D Biol Crystallogr* **54**:436-437.

40. Poole, L. B. and H. R. Ellis. 2002. Identification of cysteine sulfenic acid in AhpC of alkyl hydroperoxide reductase. *Methods Enzymol* **348**:122-136.

41. Montemartini, M., H. M. Kalisz, H. J. Hecht, P. Steinert and L. Flohé. 1999. Activation of active-site cysteine residues in the peroxiredoxin-type tryparedoxin peroxidase of Crithidia fasciculata. *Eur J Biochem* **264**:516-524.

42. Flohé, L., H. Budde, K. Bruns, H. Castro, J. Clos, B. Hofmann, S. Kansal-Kalavar, D. Krumme, U. Menge, K. Plank-Schumacher, H. Sztajer, J. Wissing, C. Wylegalla and H. J. Hecht. 2002. Tryparedoxin peroxidase of Leishmania donovani: molecular cloning, heterologous expression, specificity, and catalytic mechanism. *Arch Biochem Biophys* **397**:324-335.

43. Epp, O., R. Ladenstein and A. Wendel. 1983. The refined structure of the selenoenzyme glutathione peroxidase at 0.2- nm resolution. *Eur J Biochem* **133**:51-69.

44. Ren, B., W. Huang, B. Akesson and R. Ladenstein. 1997. The crystal structure of seleno-glutathione peroxidase from human plasma at 2.9 A resolution. *J Mol Biol* **268**:869-885.

45. Günzler, W. A. and L. Flohé. 1985. Glutathione Peroxidase. In *CRC Handbook of Methods for Oxygen Radical Research* (Greenwald, R. A., ed.), pp. 285-290, CRC Press, Boca Raton, FL.

46. Aumann, K. D., N. Bedorf, R. Brigelius-Flohé, D. Schomburg and L. Flohé. 1997. Glutathione peroxidase revisited--simulation of the catalytic cycle by computer-assisted molecular modelling. *Biomed Environ Sci* **10**:136-155.

47. Flohé, L., W. Günzler, G. Jung, E. Schaich and F. Schneider. 1971. Glutathione peroxidase. II. Substrate specificity and inhibitory effects of substrate analogues. *Hoppe Seylers Z Physiol Chem* **352**:159-169.

48. Lehmann, C., U. Wollenberger, R. Brigelius-Flohé and F. W. Scheller. 2001. Modified gold electrodes for electrochemical studies of the reaction of phospholipid hydroperoxide glutathione peroxidase with glutathione and glutathione disulfide. *Electroanalysis* **13**:364-369.

49. Holmgren, A. 2001. Selenoproteins of the thioredoxin system. In *Selenium. Its molecular biology and role in human healt* (Hatfield, D. L., ed.), pp. 179-188, Kluwer Academic Publishers, Boston, Dordrecht, London.

50. Follmann, H. and I. Häberlein. 1995. Thioredoxins: universal, yet specific thiol-disulfide redox cofactors. *Biofactors* **5**:147-156.

51. Zhong, L., E. S. Arner and A. Holmgren. 2000. Structure and mechanism of mammalian thioredoxin reductase: the active site is a redox-active selenolthiol/selenenylsulfide formed from the conserved cysteine-selenocysteine sequence. *Proc Natl Acad Sci U S A* **97**:5854-5859.

52. Yodoi, J., H. Nakamura and H. Masutani. 2002. Redox regulation of stress signals: possible roles of dendritic stellate TRX producer cells (DST cell types). *Biol Chem* **383**:585-590.

53. Hofmann, B., H. Budde, K. Bruns, S. A. Guerrero, H. M. Kalisz, U. Menge, M. Montemartini, E. Nogoceke, P. Steinert, J. B. Wissing, L. Flohé and H. J. Hecht. 2001. Structures of tryparedoxins revealing interaction with trypanothione. *Biol Chem* **382**:459-471.

54. Castro, H., H. Budde, L. Flohé, B. Hoffmann, H. Lünsdorf, J. Wissing and A. M. Tomás. 2002. Specificity and kinetics of a mitochondrial peroxiredoxin of Leishmania infantum. *Free Rad Biol Med*, in press.

55. Steinert, P., K. Plank-Schumacher, M. Montemartini, H. J. Hecht and L. Flohé. 2000. Permutation of the active site motif of tryparedoxin 2. *Biol Chem* **381**:211-219.

56. Haridas, V., J. Ni, A. Meager, J. Su, G. L. Yu, Y. Zhai, H. Kyaw, K. T. Akama, J. Hu, L. J. Van Eldik and B. B. Aggarwal. 1998. TRANK, a novel cytokine that activates NF-κ B and c-Jun N-terminal kinase. *J Immunol* **161**:1-6

Chapter 3

THIOLS OF THIOREDOXIN AND GLUTAREDOXIN IN REDOX SIGNALING

Anne Holmgren

3.1 Introduction

The overall cytosolic redox environment in *Eshcerichia coli* and mammalian cells is reducing with a high level (1-10 mM) of free thiol, mainly glutathione (GSH) and proteins containing thiols whereas disulfides are rare. This is in contrast to the cell surface or the extracellular environment where oxidizing conditions prevail due to the presence of oxygen and proteins have many disulfides and no or few free thiols. The thioredoxin (thioredoxin reductase and thioredoxin) and the glutaredoxin (glutathione reductase, GSH and glutaredoxin) systems are responsible for maintaining the low intracellular redox potential using electrons from NADPH. Thioredoxin and glutaredoxin were originally disovered in the synthesis of deoxyribonucleotides for DNA replication by ribonucleotide reductase, one of several enzymes which requires disulfide reduction for each catalytic turnover. Other such enzymes are the methionine sulfoxide reductases and the family of thioredoxin peroxidases or peroxiredoxins, which use cysteine residues to reduce hydrogen peroxide with a mechanism-derived disulfide intermediate. Thioredoxin, via its classical active site Cys-Gly-Pro-Cys dithiol, is used to maintain protein SH-groups reduced, but can also make disulfides via its oxidized or disulfide form. Thioredoxins are today known to be regulating the activity of enzymes, transcription factors and receptors by reversible disulfide bond formation (thiol redox control). Oxidized thioredoxin may be made following rapid and temporal generation of superoxide and hydrogen peroxide. Analogous reactions for the glutaredoxins are to

33

H. J. Forman, J. Fukuto, and M. Torres (eds.), *Signal Transduction by Reactive Oxygen and Nitrogen Species: Pathways and Chemical Principles,* ©2003, Kluwer Academic Publishers. Printed in the Netherlands catalyze S-glutathionylation of protein SH-groups, from glutathione disulfide (GSSG). This is in itself another mechanism of thiol redox control of protein activity, which may lead to a stable glutathionylated protein or generation of a disulfide. Glutaredoxins catalyze the reduction of glutathionylated protein and mixed disulfides in general by GSH.

In this chapter the chemistry and structural basis for the reactivity of thiolates in thioredoxin and glutaredoxin will be described. This includes our current understanding of how the thioredoxin fold, redox potential, thiol pKa-values, reaction kinetics or conformational changes following redox reactions in these molecules can be used to regulate other proteins and thereby transmit signals in cells. Thioredoxin reductase utilizing NADPH is the only enzyme capable of reducing thioredoxin. The remarkable differences of the structure and mechanism of the enzymes from prokaryotes and mammalian cells will be described. The selenocysteine residue in thioredoxin reductases from mammalian cells is essential and this makes selenium a factor regulating the activity of thioredoxin reductase as well as of thioredoxin in mammalian cells. Furthermore, the enzyme is itself a reductase of hydrogen peroxide and other hydroperoxides. Recent data on how the drug ebselen operates as a substrate for thioredoxin and thioredoxin reductase and thereby interferes in redox signaling and the role of this in inflammation will be discussed.

3.2 Background

Thioredoxin was discovered by Peter Reichard and coworkers in 1964 as a small heatstable protein cofactor containing a dithiol required for the enzymatic synthesis of dCDP from CDP by a partially purified *Escherichia coli* ribonucleotide reductase.[1] This essential enzyme catalyzes the reduction of all four ribonucleoside disphosphates to the corresponding deoxyribonucleoside disphosphates by replacing the 2'-OH-group in the ribose moiety by a hydrogen using a free radical mechanism.[2]

The reduction of the ribose OH-group in CDP to a deoxyribose in dCDP required a hydrogen donor reductant and originally the dithiol of dihydrolipoic acid showed activity, whereas millimolar concentrations of monothiols like glutathione (GSH) or mercaptoethanol were inactive. Today we know that the ribonucleotide reductase subunit B1 contains catalytic cysteine residues and the target of thioredoxin as a disulfide reductase is another disulfide in the C-terminus formed from a conserved pair of cysteines.[3] Within the enzyme there is then a thiol-disulfide exchange leading to reduction of the active site disulfide. The concentration of ribonucleotide reductase is micromolar in the cytosol and the fact that GSH, the major monothiol in the cell present in up to 10 mM, is unable to work by itself, is due to kinetic reasons, since GSH is generally a weak reducant of disulfides unless glutaredoxin is present. NADPH was active as a hydrogen donor when coupled with an enzyme activity, thioredoxin reductase, required to regenerate a dithiol from the disulfide in the oxidized thioredoxin (Fig 1). *E. coli* thioredoxin (12 kDa) contained a single cystine disulfide group and after improvement of the purification procedure to get homogeneous protein, the amino acid sequence of *E. coli* thioredoxin with its now classical active site sequence: -Cys-Gly-Pro-Cys- was completed and published in 1968.[4] Useful single crystals were obtained in 1970 of the oxidized protein by co-crystallization with cupric ions,[5] which form part of the crystal lattice. By 1975, the three-dimensional structure of thioredoxin-S_2 was solved to 2.8 Å resolution.[6] The active site disulfide bridge was located in a protrusion of the thioredoxin, which consists of a central core of 5 β-strands surrounded by 4 α-helices with more than 75 per cent of the residues in well defined secondary structures explaining the high stability of the structure (the thioredoxin fold) (Fig. 2). The structure of reduced thioredoxin remained elusive for many years, although a localized conformational change was deduced from the three-fold increase in tryptophan fluorescence following chemical or enzymatic reduction of oxidized thioredoxin.[7] This large increase in tryptophan fluorescence quantum yield unique to the *E. coli* and other thioredoxins having a Trp-28 residue as well as the conserved Trp-31 has enabled direct measurements of the kinetics of thiol-disulfide exchange for thioredoxin.[8,9] One consequence of this was that the results clearly demonstrated that thioredoxin operates as an enzyme catalyzing dithiol-disulfide

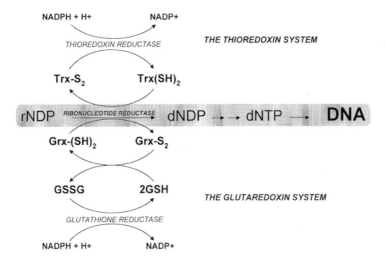

oxidoreductions.[9]

Figure. 1. Thioredoxin and glutaredoxin systems are hydrogen donors for ribonucleotide reductase in reducing ribonucleonucleoside diphosphates (rNDP) to deoxyribonucleoside diphosphates (dNDP). Deoxyribonucleotides (dNTP) are substrates for DNA-synthesis. Ribonucleotide reductase from *E. coli* and mammalian cells are dimeric enzymes where the RI subunit contains redox-active thiols, which participate in the catalytic mechanism. In exponentially growing *E. coli*, DNA replication requires reduction of more than 1000 protein disulfides per second.

For a rather long time up to 1975 thioredoxin was almost exclusively connected with ribonucleotide reductase and DNA synthesis as well as enzymatic sulfate or methionine sulfoxide reductions. The characterization of viable *E. coli* cells, which lacked thioredoxin, called into question its role in ribonucleotide reduction and lead to the discovery of glutaredoxin as a glutathione-dependent hydrogen donor for ribonucleotide reductase.[10,11] As seen from Fig. 1, glutaredoxin (Grx) enables the monothiol GSH to be active as a hydrogen donor for ribonucleotide reductase.[10] Glutaredoxin was initially characterized as a factor, which could stimulate dithiothreitol (DTT)-dependent activity of ribonucleotide reductase. Typically the strong dithiol reductant DTT (E'$_0$ -330 mV) at 1 mM only gave a very low activity since its K$_m$ is about 30 mM for ribonucleotide reductase, whereas that of reduced thioredoxin is 1.3 µM and that of reduced glutaredoxin is 0.13 µM.[3] DTT shows a fast reduction of the disulfide in glutaredoxin or thioredoxin. However, since there is no DTT in cells, the NADPH-dependent activity in extracts was not due to thioredoxin reductase but instead to glutathione reductase and glutathione.[10] Thioredoxin reductase was unable to reduce glutaredoxin nor was it reducing GSSG when thioredoxin is absent.[10] As seen from Fig. 1 the glutaredoxin system involves a disulfide reduction system with utilizing NADPH and the inherent but masked capacity of GSH to be a protein disulfide reductant. The name glutaredoxin refers to this fact and a binding site for glutathione in a redoxin with a redox active disulfide of the type originally discovered in thioredoxin (Cys-Xxx-Yyy-Cys).[10]

In many ways the lessons from studies of ribonucleotide reductase are very useful for understanding redox signaling involving protein disulfide reduction. There are no stable protein complexes formed and the reactions have bimolecular rates strongly influenced by the structure of the protein disulfide surroundings. Selective protein disulfide reductions can occur in the presence of a redox buffer of glutathione.

Thioredoxin and thioredoxin reductase in mammalian cells were purified initially based on use of the homologous ribonucleotide reductase to follow activity. This complicated assay and the fact that mammalian cytosolic thioredoxins now are known to contain three additional structural sulfhydryl groups, which upon air oxidation lead to additional disulfides, with protein aggregation and inactivation, made progress difficult. A major break-through was the realization that only reduced thioredoxin obtained after incubation with dithiothreithol could be purified as a single peak component by chromatography of liver or thymus extracts.[12] Since *E. coli* thioredoxin reductase shows no cross-reactivity with the human, rat or bovine cytosolic thioredoxins it was of no use for coupling to NADPH measuring reduction of 5,5'-dithiobis-(2-nitrobenzoic acid) (DTNB). Reduction of DTNB is used in the assay of thioredoxin from *E. coli* or yeast.[11] In contrast, the mammalian thioredoxin reductase of calf thymus showed completely different properties with a higher molecular weight and a wide substrate specificity, which involved direct NADPH-dependent reduction of DTNB.[13] To avoid ribonucleotide reductase as an assay system, insulin disulfide reduction was developed.[12,13] Furthermore, the general role of thioredoxin and thioredoxin reductase as the main disulfide reductase system of cells was shown e.g. by selective reduction of 5 out of 28 disulfide bonds in human fibrinogen by the thioredoxin system.[14] Human thioredoxin was discovered in platelets.[14] This was followed by disulfide reduction in trypsin, chymotrypsin, α2-macroglobulin, apocytochrome c and several other proteins and thioredoxin has been used as a tool to probe the exposure of disulfides and get selective reduction. The specificity is impressive since three of the five disulfides in trypsin react, but none of the 17 disulfides in albumin.[15] Since NADPH is used to drive the reaction and selectivity is given by thioredoxin, highly specific reductions can be obtained.

The stability of thioredoxin and thioredoxin reductase is impressive since in the experiments with trypsin, a 100:1 ratio of trypsin to thioredoxin was used and there was no sign of proteolytic inactivation of the thioredoxin system.[15]

The wide distribution of thioredoxin in mammalian cells and its presence irrespective of DNA synthesis and ribonucleotide reductase activity, which is only expressed in S-phase of proliferating cells, was proven by studies of the distribution of calf thioredoxin.[16] Results showed a localization in all cell compartments upon subcellular fractionation. Furthermore, immunohistochemical localization of thioredoxin and thioredoxin reductase in adult rats[17] demonstrated a general cytoplasmic staining with prominent expression particularly in epithelial cells including large amounts of thioredoxin in the terminally differentiated large nerve cells in the nervous system.[18] Axoplasmic transport of both thioredoxin and thioredoxin reductase in peripheral nerves and function-related changes in pancreatic B-cells with feeding starvation cycles as well as the affect of pharmacological agents in the gastric mucosa was observed.[19,20] All these findings pointed to a regulated expression of thioredoxin as well as movements within the cell and even out of the cell.

Thioredoxin cytokine work was initiated when Yodoi and coworkers identified adult T-cell leukemia derived factor (ADF) as human thioredoxin present in conditioned medium from HTLV-I infected lymphocytes with a role as a growth factor in upregulation of the IL-2 receptor.[21] This pointed to new functions in redox regulation of cellular activation.[22] Thioredoxin released from B-cells transformed with the EBV-virus was shown to be involved in lymphocyte immortalization.[23] Also from CD-4 positive T-cells a secreted factor, growth-promoting for normal and leukemic B-cells was identified as thioredoxin.[24] The mechanism has remained unclear since no classical receptor on the surface of cells has yet been identified. Today it appears possible that thioredoxin acts via its thiol-disulfide exchange activity or is transported into cells to bind to other proteins.

The concept of thiol redox control on redox regulation of cellular phenomena by changes in the structure of SH-groups on proteins was suggested to involve the thioredoxin system.[8-10] Only relatively recently it has been realized that there is an oxidizing mechanism controlling the disulfide status and redox regulation in cells, via the stimulus-mediated generation of superoxide and hydrogen peroxide by NADPH oxidase. Hydrogen peroxide is converted into a disulfide signal by glutathione peroxidases generating GSSG and oxidized glutaredoxins or by thioredoxin peroxidases (peroxiredoxins) generating disulfide forms of thioredoxins. Thioredoxin disulfides may be transmitted to generation of disulfides in other proteins or a reversal of thioredoxin-dependent reduction, as part of thiol redox control. Thus, the discovery of the abundant family of peroxiredoxins[25] is of major significance in understanding thioredoxin-dependent regulation of cellular activation in mammalian cells similar to the light-dependent regulation of photosynthetic enzymes in chloroplasts.[22,26,27]

E. coli glutaredoxin was purified to homogeneity[28] and its activity with ribonucleotide reductase, showed a higher turnover number than that of thioredoxin [29]. It was also discovered that pure glutaredoxin had inherent glutathione-disulfide oxidoreductase activity similar to the enzyme activity glutathione disulfide transhydrogenase from liver.[28,29] A simple spectrophotometric assay was developed which showed that *E. coli* bacteria were indeed a very rich source enzymes

catalyzing the GSH-dependent reduction of hydroethyldisulfide.[28,29]   In fact, the
activity in *E. coli* crude extracts is a 100-fold higher than what is specifically
measured as glutaredoxin assayed by the GSH-dependent activity with
ribonucleotide reductase.[28,29]   The additional activity was subsequently shown to be
caused by two new glutaredoxin species called glutaredoxin (Grx 2) and (Grx 3).[30]
Both Grx 2 and Grx 3 have the same Cys-Pro-Tyr-Cys active site as Grx1, but only
Grx 3 has a low activity with ribonucleotide reductase.[30] Grx 2, a 24.3 kDa protein
with structural similarity to glutathione S-transferases, represents a major protein in
E. coli and has other functions in defense against oxidative stress.[31,32]

Calf thymus glutaredoxin,[33] which acts as a species-specific electron donor for
calf thymus ribonucleotide reductase[34] contained the same conserved active site
sequence Cys-Pro-Tyr-Cys[35] as *E. coli* glutaredoxin.[36] However, it was not clear at
that time,[33] if glutaredoxin and a GSH-homocystine transhydrogenase from rat liver
renamed thioltransferase[37] were identical proteins.   The latter was reported to
contain 8.6 per cent carbohydrate.[37,38] However, the carbohydrate content was not
confirmed and sequencing of proteins including a revised sequence of calf thymus
glutaredoxin[39] showed the identity of the two proteins. Therefore, thioltransferase
and glutaredoxin are different names used for the same protein.

3.3 The thioredoxin system

Thioredoxin reductase (TrxR) reduces oxidized thioredoxin (Trx-S$_2$) at the
expense of NADPH (Reaction 1). Reduced thioredoxin &Trx-(SH)$_2$? is a general
disulfide reductase for disulfides in proteins generating sulfhydryl groups and
reforming Trx-S$_2$ (Reaction 2). The net reaction is that disulfides are reduced by
NADPH by the thioredoxin system acting as a protein disulfide reductase (Reaction
3):

$$\text{Trx-S}_2 + \text{NADPH} + \text{H}^+ \xrightarrow{\text{TrxR}} \text{Trx-(SH)}_2 + \text{NADP}^+ \qquad (1)$$

$$\text{Trx-(SH)}_2 + \text{Protein-S}_2 \rightarrow \text{Trx-S}_2 + \text{Protein-(SH)}_2 \qquad (2)$$

$$\text{Net:} \quad \text{NADPH} + \text{H}^+ + \text{Protein-S}_2 \rightarrow \text{NADP}^+ + \text{Protein-(SH)}_2 \qquad (3)$$

So far all living species have a thioredoxin system and this includes eubacteria,
archebacteria as well as eukaryotes. There are similarities but also some major
differences between the cytosolic thioredoxin systems in *E. coli* and that of
mammalian organisms. Thus, the *E. coli* and mammalian thioredoxins have the same
size and fold and conserved -Cys-Gly-Pro-Cys- active site.[40,41]   In contrast,
thioredoxin reductases from mammalian cells have very different properties when
compared with the enzymes from *E. coli*, yeast or plants.[42] The mammalian
cytosolic and mitochondrial isoenzymes are larger selenoproteins with a completely
different mechanism and a broad substrate specificity.[40,42]

3.4 Thioredoxin structure and mechanism

Reduced thioredoxin has only its N-terminally located Cys-residue (Cys-32)
exposed and the pKa-value of the thiolate is ~7.1 as known from NMR studies [43],

chemical modification[44] and titration using NMR[45] (Fig. 2). The C-terminally located Cys-residue is buried and has high pKa-value.[46]

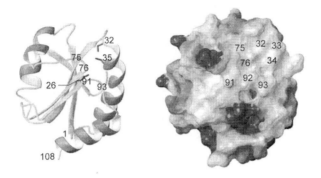

Figure 2. Three-dimensional NMR solution structure of reduced *E. coli* thioredoxin[46] illustrating the active site cysteines (Cys-32 and Cys-35), aspartic acid (Asp-26) and the hydrophobic surface area. Left, a cartoon of the polypeptide backbone with sidechain heavy atoms of residues 26, 32 and 35 displayed as sticks. Right, the molecular surface is colored according to electrostatic potential (dark is positive, medium is negative and light gray is uncharged). Atoms of residues of the proposed intermolecular hydrophobic surface interaction area are labeled at their approximate positions.

The mechanism of thioredoxin catalysis involves initial docking of reduced thioredoxin to the target protein using the hydrophobic surface area (Fig. 2) located around the active site[43] (see Fig. 3).

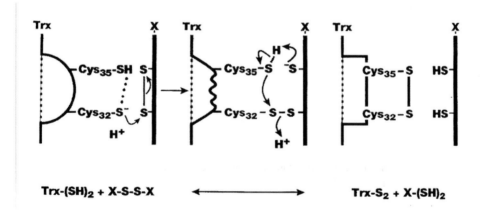

Figure 3. Mechanism of thioredoxin as a disulfide oxidoreductase. The shared hydrogen bond in the active site lowering the pKa value of Cys-32 is indicated. Note that Trx will catalyze the reaction in either direction depending on the redox potential of the X disulfide/dithiol pair.

This is followed by a nucleophilic attack on the disulfide by the thiolate to form a transient mixed disulfide with the substrate in the hydrophobic surface area Cys-35 thiolate is formed by deprotonation where the conserved buried Asp-26 with a high pKa-value accepts of the proton and assists in the catalyzis by general as a base chemistry.[47] The attack by the Cys-35 thiolate makes the disulfide in oxidized thioredoxin followed by a conformational change leading to dissociation of the reduced thiol substrate. Local increased structural mobility in the ns to μs time intervall in and around the active site in the reduced form of thioredoxin plays a major role in this process.[45,46] It is probably of major role in determining the binding of thioredoxin in the reduced form to other proteins.

Structural information supporting the mechanism has been obtained from the detailed NMR structures of oxidized and reduced thioredoxin from *E. coli*, by high resolution NMR.[46] Of particular interest is the structure by NMR of two complexes in the mixed disulfide form between human thioredoxin and a peptides from NF-kB[48] and Ref-1.[49] These structures have been generated by using a mutant of human thioredoxin where the C-terminally located Cys residue of the active site has been mutated to an Ala residue. Also the three structural Cys residues were mutated to Ala.[48] This enables preparation of the mixed disulfide and the detailed localization of the 13-residue peptide on the protein. Central to the redox regulatory function of the thioredoxin system is the ability of reduced thioredoxin to recognize targets for disulfide reduction. It was evident early that only peptides of proteins of a certain size could give high rates, whereas cystine or GSSG showed comparatively low rate constants.[8]

3.5 Structure of human thioredoxin in a mixed disulfide complex with a NFκB target peptide

Active NFκB is a hetero-dimer composed of 50 kDa and 65 kDa and the DNA-binding activity has been shown to be redox regulated by Trx which moves to the nucleus.[51] Oxidation of Cys-62 in P-50 leads to a disulfide linked dimer inhibiting DNA-binding. Thioredoxin restores binding by disulfide reduction.[51] An NMR solution structure of a mixed disulfide bonded complex between human thioredoxin and a 13-residue peptide comprising residues 58-68 of the P-50 subunit of NFκB[48] shows that the peptide is covalently linked by a disulfide bridge between Cys-32 of Trx and Cys-62 of the peptide. The peptide is located in a long boot-shaped conserved cleft on the surface of human and *E. coli* Trx (see Fig. 2) delineated by the active site loop, helices α-2, α-3 and α-4 and strands β-3 and β-4.[48,50] The peptide adopts a crescent-like conformation with a 110° bend centered around residue 60, that permits it to follow the path of the cleft. There are numerous hydrogen bonding electrostatic and hydrophobic interactions that involve residues 57 - 65 of the NFκB peptide and seem to confer substrate specificity.[48,50]

3.6 Structure of Trx in a mixed disulfide with a Ref-1 target peptide

DNA-binding of the transcription factor AP-1, which is a heterodimer of Fos and Jun is under redox control by Ref-1, a protein with both DNA-repair and redox activity.[52] The region essential for redox activity of Ref-1 is located in the aminoterminus of the protein and to a cysteine in position 65 with redox activity. Upon oxidation Ref-1 becomes inactive and it is known that Ref-1 and human thioredoxin form a complex as shown by cross-linking and analysis by yeast two hybrid systems [53]. Interestingly, the Ref-1 peptide contain Cys-65 is located in a similar crescent shaped groove on the surface of thioredoxin. The groove is formed by residues of the active site loop, helix α–3, β-strands 3 and 5, loops at β-strands 3 and 4.[53] Importantly the orientation of the Ref-1 peptide is opposite to that found in the complex of the Trx with the NFκB peptide,[53] but the groove is the same.[50]

The structures of the peptide complexes with thioredoxin has enabled deducing some preliminary rules determining the binding mode of a protein to thioredoxin. These include: the presence of an aromatic, or hydrophobic chain residue at P-2 or P+2 position, which is burried in a deep hydrophobic pocket.[50] The presence of an aliphatic residue at P-4 or P+4 position to bind to the aromatic ring of the conserved Tryp-31 of Trx, which is located on the protein surface. Potential charged residues at the P-5 and P+5 positions interacting with the side chain of carboxylates of Asp-58 and Asp-61 of hTrx. There are also conserved hydrogen bonds between the backbone of the peptide and the backbone of thioredoxin at the equivalent of Ile75 in *E. coli* Trx. This feature is conserved in glutaredoxin binding to glutathione and maybe therefore be a common theme in binding of ligands to the thioredoxin or glutaredoxin surface.

Obviously, the two peptides studied are locked into a position by the covalent disulfide bond the active site Cys-32. However, it appears that thioredoxin uses the long boot-shaped cleft to bind peptides in any orientation and thereby has a potential to target a wide range of proteins with disulfides suitable for reduction.

Another factor of interest is the role of thioredoxin as a molecular chaperone, which after non-covalent binding interactions changes the structure of the target protein and thereby get acess to sterically burried disulfides. Also conformational changes in the target protein may follow after reduction of one disulfide exposing new structures for reduction by thioredoxin as is obvious with coagulation factors like factor VIII.[15] Very stable vicinal disulfides with positive charges nearby like in the active site of bovine pancreatic trypsin inhibitor (BPTI) will not be substrates of thioredoxin despite the exposure of the disulfide bond (A. Holmgren unpublished results).

3.7 Redox potentials of thioredoxin family members

The redox potential (E'_0) of members of the thioredoxin fold superfamily of proteins (thioredoxins, glutaredoxins, protein disulfide isomerases DsbA etc)[54] span a wide range of values from strongly reducing like the cytosolic thioredoxin (-270 mV) to the oxidizing DsbA (-125 mV).[55] One factor determining the redox potential is the residues between the two active site half-cystines residues in thioredoxins and glutaredoxins. Thus, replacement of Pro-34 with His lowers the redox potential of thioredoxin by 35 mV.[55] The redox potentials are also linked to the stability of the proteins.[55] Thus, for a protein with a high redox potential ($E'_0 > -180$ mV) the reduced protein has a higher thermodynamic stability to denaturation compared with the disulfide form. In thioredoxin with E'_0 -270 mV the oxidized form is more stable.[55]

3.8 Thioredoxin reductase and selenium

The fact that administration of selenium compounds like selenite (SeO_3^{2-}) cause inhibition of tumor cell proliferation *in vivo* and the knowledge that thioredoxin reductase appeared to be more highly expressed in malignant cells prompted investigations on the reactivity of selenium compounds with pure mammalian thioredoxin reductase and thioredoxin.[56-58] Contrary to expectations selenite is a direct substrate for thioredoxin reductase as well as an efficient oxidant of Trx-$(SH)_2$.[57,59] With 200 μM NADPH and 50 nM calf thymus thioredoxin reductase, addition of 10 μM selenite caused oxidation of 40 μM NADPH in 12 min and 100 μM NADPH after 30 min demonstrating a direct reduction of selenite with redox cycling by oxygen.[58,59] This was demonstrated by running the reaction under anaerobic conditions where only 3 mol of NADPH was oxidized per mol of selenite according to Reaction 4:

$$SeO_3^{2-} + 3\,NADPH + 3\,H^+ \xrightarrow{\;TrxR\;} Se^{2-} + 3\,NADP^+ + 3\,H_2O \qquad (4)$$

Addition of thioredoxin stimulated the reaction further since selenite rapidly reacts with Trx-$(SH)_2$ to oxidize it to Trx-S_2.[56-59] Since glutathione reductase will not react with selenite, Reaction 4 should provide cells with selenide, a required precursor for selenophosphate and selenocysteine synthesis.[60] Selenite and glutathione may react to form selenodiglutathione (GS-Se-SG) which has been suggested to be a major metabolite of inorganic selenium salts in mammalian tissues,[60] a source of selenide as well as an inhibitor of neoplastic growth. Synthetic GS-Se-SG[58] is a direct efficient substrate for mammalian thioredoxin reductase and a highly efficient oxidant of reduced thioredoxin.[56,58] Since GSSG is not a substrate for mammalian thioredoxin reductase[61] the insertion of the selenium atom in the GSSG molecule to form GS-Se-SG makes this molecule highly reactive with the enzyme as well as thioredoxin.[58] Reduction of GS-Se-SG to yield selenide by glutathione reductase requires two mol of NADPH. We found only the first stoichiometric reduction to be fast with GS-Se⁻ as a product.[58] The second reaction was slow and inefficient. These results strongly suggest that the major selenide generation in cells is via thioredoxin reductase and thioredoxin. Thus, in mammalian cells the selenoenzyme thioredoxin reductase is also responsible for the synthesis of selenide required for its own synthesis. An oxygen dependent non-stoichiometric consumption of NADPH is given by the thioredoxin system in the presence of selenite, selenodiglutathione and selenocystine.[57-59] The latter is an efficient substrate for mammalian thioredoxin reductase with a K_m of 6 μM.[59] The mechanism may be that the XSe⁻ reacts as a charge transfer catalyst with a dithiol (or selenolthiol) to catalyze oxidation according to Reaction 5:

$$XSe^- + R\text{-}(SH)_2 + (O) \rightarrow XSe^- + R\text{-}S_2 + H_2O \qquad (5)$$

The overall effect will be O_2-dependent consumption of NADPH via the thioredoxin system. This provides an explanation for the lack of a potentially autooxidizable free pool of selenocysteine as well as the acute toxic effects of selenium compounds on cells, for example, leading to apoptosis.

Mammalian thioredoxin reductases display a surprisingly very wide substrate specificity as first observed during the purification to homogeneity an characterization.[61] This is in contrast to the smaller specific prokaryotic enzymes, which even do not react with mammalian thioredoxins despite the identical active sites and the closely related three-dimensional structures of the thioredoxins. A truly wide range of direct reductions are catalyzed by the mammalian cytosolic thioredoxin reductases[62-71] including reduction of lipid hydroperoxides and hydrogen peroxide as well as direct electron transfer to plasma glutathione peroxidase. The explanation is their content of a catalytically active selenocysteine residue.[65]

3.9 Structure and mechanism of mammalian thioredoxin reductase

Mammalian thioredoxin reductases are dimers with a subunit of 57 kDa or larger comprising a glutathione reductase scaffold elongated by 16-residues with the conserved C-terminal sequence -Gly-Cys-Sec-Gly[72-74] (Fig. 4).

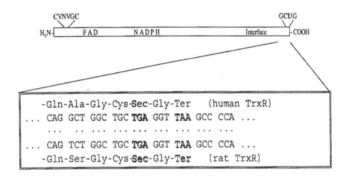

Figure 4: Structure of mammalian cytosolic thioredoxin reductase (TrxR) showing the overall homology to domains in glutathione reductase, the N-terminal redox active dithiol/disulfide and the C-terminal cysteine and selenocysteine (Sec or U) redox center.[74]

The selenocysteine (Sec) residue is encoded by a TGA codon, which is translated to selenocysteine by conversion of a serine using selenophosphate as a selenium donor. In the abscence of selenium, the TGA will act as a stop codon and a truncated, but inactive enzyme will be formed.[65] This is because the C-terminal - Cys-Sec- motif is the active site,[75] where thioredoxin is reduced (Fig. 5). In the oxidized form the active site is present as a selenenylsulfide and in the reduced form it is present as a selenolthiol.[75]

Electrons from NADPH go via FAD to the disulfide bridge in the N-terminus of one subunit and are then transferred from the dithiol across the subunits to the selenenylsulfide in the other subunit (Fig. 4 and Fig. 5). The Sec residue is also the basis for the hydroperoxidase activity of mammalian TrxR, which uses lipid hydroperoxides[66] or hydrogen peroxide as substrates.[65] Recently, the catalytically active Sec498 Cys mutant enzyme with a 100-fold lower k_{cat}[65] has been crystallized and the structure solved to 3Å resolution by X-ray crystallography.[76,77] The structure confirms the predicted close similarity to the three-dimensional structure of glutathione reductase (Fig. 6). The 16-residue C-terminal tail, which is unique to mammalian thioredoxin reductase and carries the Sec residue, folds in such a way that it can approach the active site disulfide of the other subunit in the dimer. The model of the complex of thioredoxin reductase with human thioredoxin-S_2 (Fig. 6) suggests that electron transfer from NADPH to the disulfide of the substrate is possible without large conformational changes.

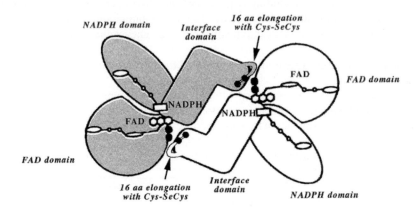

Thioredoxin Reductase

Figure 5. Schematic model of structure of mammalian thioredoxin reductase based on the known structure of glutathione reductase[75] see also Fig. 4. The Cys residues and the Sec are shown as black dots and the subunits are in different shades of gray.

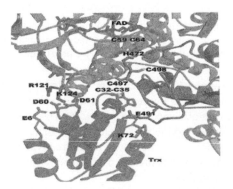

Figure 6. Model of a complex of rat thioredoxin reductase with human thioredoxin.[77] Residues at the Trx-TrxR interface are shown as stick models. The positions of the sulfur atoms of the catalytic disulfide Cys 59 - Cys 64, the Cys 497 and 498 as well as Cys 32 and Cys 35 are shown as spheres.

Thus, the C-terminal extension to glutathione reductase scaffold typical of mammalian thioredoxin reductase has two main functions. First, it extends the electron transport chain from the catalytic disulfide to the enzyme surface, where it can react with thioredoxin and a range of other substrates. Second, the C-terminal extension prevents the enzyme from acting as a glutathione reductase by blocking acess of GSSG to the redox active disulfide. The structure of the enzyme is also compatible with evolution of mammalian thioredoxin reductase from a glutathione reductase like protein rather than from the presumably much older prokaryotic counterpart. Such an evolutionary switch also rendered cell growth and redox regulation dependent upon selenium.[65] Why this happened is still not clear, but defense against oxidative stress and the evolution of multicellular organisms with oxygen dependent oxidative phosphorylation may be involved.

To the large group of substrates for mammalian thioredoxin reductase can also be added the selenazol drug ebselen (2-phenyl-1,2-benzoisoselenazol-3(2H)-one) an antioxidant and antiinflammatory agent, which has been known as a glutathione peroxidase mimick.[78] It was recently shown[79] that ebselen is an excellent substrate for mammalian thioredoxin reductase with a K_m of 2.5 μM and a K_{cat} of 600 per min giving rise to the ebselen selenol. Thereby ebselen stimulates the hydrogen peroxide reductase activity of the enzyme 15-fold. Ebselen is also an outstanding efficient oxidant of reduced thioredoxin to its disulfide form.[79] The mechanim of ebselen also involves ebselen diselenide and strongly suggest that reduction of hydrogen peroxide via the thioredoxin and thioredoxin reductase in the cytosol and mitochondria is the main effect or and target of this clinically promising drug,[80] which is used to treat stroke.

3.10 Redox regulation of cell signaling in stress

In mammalian cells it has recently been clarified that reduced thioredoxin in the cytosol is bound to apoptosis signaling kinase 1 (ASK1), a member of the mitogen-activated (MAP) kinase, kinase kinase family.[81] Trx is a negative regulator of ASK1-JNK/p38 pathway and only the Trx-$(SH)_2$ binds to ASK1. Oxidation leading to generation of Trx-S_2 leads to dissociation of the complex and signaling for apoptosis.[81] Also thioredoxin binding protein 2 (TBP2) or vitamin D upregulated protein (VDUP1) forms a complex with Trx-$(SH)_2$.[82] Upon oxidation, Trx-S_2 dissociates from the complex. The level of TBP2 controls the expression of Trx in cells and it was recently observed that the cancer drug SAHA which arrests cancer cell growth upregulates TBP2.[83] Also the human glutaredoxin in its reduced dithiol form binds to ASK1 and inhibits its signaling for apoptosis.[84] Oxidation of glutaredoxin during metabolic oxidative stress by glucose deprivation occurs in a H_2O_2 and GSSG-dependent way and leads to its dissociation from ASK1 resulting in cytotoxicity.[84] Apparently, the controlled expression of thioredoxin and glutaredoxin and the binding of only the dithiol forms to other proteins is a major mechanism of cellular signaling.

Acknowledgments

Supported by grants from the Swedish Cancer Society (961), the Swedish Medical Research Council and the Knut and Alice Wallenberg Foundation. The excellent secreterial assistance by Mrs Lena Ringdén is gratefully acknowledged.

References

1. Laurent, T. C., E. C. Moore and P. Reichard. 1964. Enzymatic synthesis of deoxyribonucleotides. IV. Isolation and characterization of thioredoxin, the hydrogen donor from *Escherichia coli* B. *J Biol Chem* **239**:3436-3444.

2. Jordan, A. and P. Reichard. 1998. Ribonucleotide reductases. *Annu Rev Biochem* **67**:71-98.

3. Sjöberg, B. M. and M. Sahlin. 2002. Thiols in redox mechanism of ribonucleotide reductase. *Methods Enzymol* **348**:1-21.

4. Holmgren, A. 1968. Thioredoxin 6. The amino acid sequence of the protein from *Escherichia coli* B. *Eur J Biochem* **6**:475-484.

5. Holmgren, A. and B. O. Söderberg. 1970. Crystallization and preliminary crystallographic data for thioredoxin from *Escherichia coli* B. *J Mol Biol* **54**:387-390.

6. Holmgren, A., B. O. Söderberg, H. Eklund and C. J. Brändén. 1975. Three-dimensional structure of *Escherichia coli* thioredoxin-S_2 to 2.8 Å resolution. *Proc Natl Acad Sci USA* **72**:2305-2309.

7. Stryer, L., A. Holmgren and P. Reichard. 1967. Thioredoxin. A localized conformational change accompanying reduction of the protein to the sulfhydryl form. *Biochemistry* **6**:1016-1020.

8. Holmgren, A. 1979. Reduction of disulfides by thioredoxin. Exceptional reactivity of insulin and suggested functions of thioredoxin in mechanism of hormone action. *J Biol Chem* **254**:9113-9119.

9. Holmgren, A. 1979. Thioredoxin catalyzes the reduction of insulin disulfides by dithiothreitol and dihydrolipoamide. *J Biol Chem* **254**:9627-9632.

10. Holmgren, A. 1976. Hydrogen donor system for *Escherichia coli* ribonucleoside-diphosphate reductase dependent upon glutathione. *Proc Natl Acad Sci USA* **73**:2275-2279.

11. Holmgren, A., I. Ohlsson and M. L. Grankvist. 1978. Thioredoxin from *Escherichia coli*. Radioimmunological and enzymatic determinations in wild type cells and mutants defective in phage T7 DNA replication. *J Biol Chem* **253**:430-436.

12. Engström, N. E., A. Holmgren, A. Larsson and S. Söderhäll. 1974. Isolation and characterization of calf liver thioredoxin. *J Biol Chem* **249**:205-210.

13. Holmgren, A. 1977. Bovine thioredoxin system. Purification of thioredoxin reductase from calf liver and thymus and studies of its function in disulfide reduction. *J Biol Chem* **252**:4600-4606.

14. Blombäck, B., M. Blombäck, W. Finkbeiner, A. Holmgren, B. Kowalska-Loth and G. Olovson. 1974. Enzymatic reduction of disulfide bonds in fibrinogen by the thioredoxin system. I. Identification of reduced bonds and studies on reoxidation process. *Thrombosis Res* **4**:55-75.

15. Holmgren, A. 1984. Enzymatic reduction-oxidation of protein disulfides by thioredoxin. *Meth Enzymol* **107**:295-300.

16. Holmgren, A. and M. Luthman. 1978. Tissue distribution and subcellular localization of bovine thioredoxin determined by radioimmunoassay. *Biochemistry* **17**:4071-4077.

17. Rozell, B., H. A. Hansson, M. Luthman and A. Holmgren. 1985. Immunohistochemical localization of thioredoxin and thioredoxin reductase in adult rats. *Eur J Cell Biol* **38**:79-86.

18. Stemme, S., H. A. Hansson, A. Holmgren and B. Rozell. 1985. Axoplasmic transport of thioredoxin and thioredoxin reductase in rat sciatic nerve. *Brain Res* **359**:140-146.

19. Hansson, H. A., A. Holmgren, B. Rozell and I. B. Täljedal. 1986. Immunohistochemical localization of thioredoxin and thioredoxin reductase in mouse exocrine and endocrine pancreas. *Cell Tissue Res* **245**:189-195.

20. Hansson, H. A., H. F. Helander, A. Holmgren and B. Rozell. 1988. Thioredoxin and thioredoxin reductase show function-related changes in the gastric mucosa: immunohistochemical evidence. *Acta Physiol Scand* **132**:313-320.

21. Tagaya, Y., Y. Maeda, A. Mitsui, N. Kondo, H. Matsui, J. Hamuro, N. Brown, K. I. Arai, T. Yokota, H. Wakasugi and J. Yodi. (1989) ATL-derived factor (ADF), an IL-2 receptor/Tac inducer homologous to thioredoxin: possible involvement of dithiol-reduction in the IL-2 receptor induction. *EMBO J* **8**:757-764.

22. Nakamura, H., K. Nakamura and J. Yodoi. 1997. Redox regulation of cellular activation. *Annu Rev Immunol* **15**:351-369.

23. Yodoi, J. and T. Tursz. 1991. ADF, a growth-promoting factor derived from adult T cell leukemia and homologous to thioredoxin: involvement in lymphocyte immortalization by HTLV-1 and EBV. *Adv Cancer Res* **57**:381-411.

24. Rosén, A., P. Lundman, M. Carlsson, K. Bhavani, B. R. Srinivasa, G. Kjellström, K. Nilsson and A. Holmgren. 1995. A CD4$^+$ T cell line-secreted factor, growth promoting for normal and leukemic B-cells, identified as thioredoxin. *Int Immunology* **7**:625-633.

25. Chae, H. Z., S. W. Kang and S. G. Rhee. 1999. Isoforms of mammalian peroxiredoxin that reduce peroxides in presence of thioredoxin. *Methods Enzymol* **300**:219-226.

26. Holmgren, A. 1985. Thioredoxin. *Annu Rev Biochem* **54**:237-271.

27. Holmgren, A. 1989. Thioredoxin and glutaredoxin systems. *J Biol Chem* **264**:13963-13966.

28. Holmgren, A. 1979. Glutathione-dependent synthesis of deoxyribonucleotides. Purification and characterization of glutaredoxin from *Escherichia coli*. *J Biol Chem* **254**:3664-3671.

29. Holmgren, A. 1979. Glutathione-dependent synthesis of deoxyribonucleotides. Characterization of the enzymatic mechanism of *Escherichia coli* glutaredoxin. *J Biol Chem* **254**:3672-3678.

30. Åslund, F., B. Ehn, A. Miranda-Vizuete, C. Pueyo and A. Holmgren. 1994. Two additional glutaredoxins exist in *Escherichia coli*: glutaredoxin-3 is a hydrogen donor for ribonucleotide reductase in a thioredoxin-glutaredoxin-1 double mutant. *Proc Natl Acad Sci USA* **91**: 9813-9817.

31. Potamitou, A., A. Holmgren and A. Vlamis-Gardikas. 2002. Protein levels of *Escherichia coli* thioredoxins and glutaredoxins and their relation to null mutants, growth phase and function. *J Biol Chem* **277**:18561-18567.

32. Vlamis-Gardikas, A., A. Potamitou, R. Zarivach, A. Hochman and A. Holmgren. 2002. Characterization of *Escherichia coli* null mutants for glutaredoxin 2. *J Biol Chem* **277**:10861-10868.

33. Luthman, M. and A. Holmgren. 1982. Glutaredoxin from calf thymus. I. Purification to homogeneity. *J Biol Chem* **257**:6686-6690.

34. Luthman, M., S. Eriksson, A. Holmgren and L. Thelander. 1979. Glutathione-dependent hydrogen donor system for calf thymus ribonucleoside diphosphate reductase. *Proc Natl Acad Sci USA* **76**:2158-2162.

35. Klintrot, I. M., J. O. Höög, H. Jörnvall, A. Holmgren and M. Luthman. 1984. The primary structure of calf thymus glutaredoxin. Homology with the corresponding *Escherichia coli* protein but elongation at both ends and with an additional half-cystine/cysteine pair. *Eur J Biochem* **144**:417-423.

36. Höög, J. O., H. Jörnvall, A. Holmgren, M. Carlquist and M. Persson. 1983. The primary structure of *Escherichia coli* glutaredoxin. Distant homology with thioredoxins in a superfamily of small proteins with a redox-active cystine disulfide/cysteine dithiol. *Eur J Biochem* **136**:223-232.

37. Axelsson, K., S. Eriksson and B. Mannervik. 1978. Purification and characterization of cytoplasmic thioltransferase (glutathione: disulfide oxidoreductase) from rat liver. *Biochemistry* **17**:2978-2984.

38. Gan, Z. R. and W. W. Wells. 1986. Purification and properties of thioltransferase. *J Biol Chem* **261**:996-1001.

39. Papayannopoulos, J. A., Z. R. Gan, W. W. Wells and K. Biemann. 1989. A revised sequence of calf thymus glutaredoxin. *Biochem Biophys Res Commun* **159**: 1448-1454.

40. Arnér, E. S. J. and A. Holmgren. 2000. Physiological functions of thioredoxin and thioredoxin reductase. *Eur J Biochem* **267**:6102-6109.

41. Powis, G. and W. R. Montfort. 2001. Properties and biological activities of thioredoxins. *Annu Rev Pharmacol Toxicol* **41**:261-295.

42. Williams Jr, C. H., L. D. Arscott, S. Muller, B. W. Lennon, M. L. Ludwig, P. F. Wang, D. M. Veine, K. Becker and R. H. Schirmer. 2000. Minireview: Thioredoxin reductase. Two modes of catalysis have evolved. *Eur J Biochem* **267**:6110-6117.

43. Holmgren, A. 1995. Thioredoxin structure and mechanism: conformational changes on oxidation of the active site sulfhydryls to a disulfide. *Structure* **3**:239-243.

44. Kallis, G. B. and A. Holmgren. 1980. Differential reactivity of the functional sulfhydryl groups of cysteine-32 and cysteine-35 present in the reduced form of thioredoxin from *Escherichia coli. J Biol Chem* **255**:10261-10265.

45. Dyson, H. J., M. F. Jeng, L. L. Tennant, I. Slaby, M. Lindell, D. S. Cui, S. Kuprin and A. Holmgren. 1997. Effects of buried charged groups on cysteine thiol ionization and reactivity in *Escherichia coli* thioredoxin: Structural and functional characterization of mutants of Asp 26 and Lys 57. *Biochemistry* **36**:2622-2636.

46. Jeng, M. F., A. P. Campbell, T. Begley, A. Holmgren, D. A. Case, P. E. Wright and H. J. Dyson. 1994. High-resolution solution structures of oxidized and reduced *Escherichia coli* thioredoxin. *Structure* **2**:853-868.

47. Chivers, P. T. and R. T. Raines. 1997. General acid/base catalysis in the active site of *Escherichia coli* thioredoxin. *Biochemistry* **36**:15810-15816.

48. Qin, J., G. M. Clore, W. M. P. Kennedy, J. Huth and A. M. Gronenborn. 1995. Solution structure of human thioredoxin in a mixed disulfide intermediate complex with its target peptide from the transcription factor NFkB. *Structure* **15**:289-297.

49. Qin, J., G. M. Clore, W. M. P. Kennedy and A. M. Gronenborn. 1996. The solution structure of human thioredoxin complexed with its target from Ref-1 reveals peptide chain reversal. *Structure* **4**:613-620.

50. Qin, J., Y. Yang, A. Velyvis and A. M. Gronenborn. 2000. Molecular views on redox regulation: Three-dimensional structures of redox regulatory proteins and protein complexes. *Antiox & Redox Signal* **2**:827-840.

51. Hirota, K., M. Murata, Y. Sachi, H. Nakamura, J. Takeuchi, K. Mori and J. Yodoi. 1999. Distinct roles of thioredoxin in the cytoplasm and in the nucleus. A two-step mechanism of redox regulation of transcription factor Nf-κB. *J Biol Chem* **274**:27891-27897.

52. Xanthoudakis, S.,G. G. Miao and T. Curran. 1994. The redox and DNA repair activities of Ref-1 are encoded by non-overlapping domains. *Proc Natl Acad Sci USA* **91**:23-27.

53. Hirota, K., M. Matsui, S. Iwata, A. Nishiyama, K. Mori and J. Yodoi. 1999. AP-1 transcriptional activity is regulated by a direct association between thioredoxin and Ref-1. *Proc Natl Acad Sci USA* **94**:3633-3638.

54. Martin, J. L. 1995. Thioredoxin - a fold for all reason. *Structure* **3**:245-250.

55. Åslund, F., K. D. Berndt and A. Holmgren. 1997. Redox potentials of glutaredoxins and other thiol-disulfide oxidoreductases of the thioredoxin superfamily determined by direct protein-protein redox equilibria. *J Biol Chem* **272**:30780-30786.

56. Ren, X., M. Björnstedt, B. Shen, M. Ericson and A. Holmgren. 1993. Mutagenesis of structural half-cystine residues in human thioredoxin and effects on regulation of activity by selenodiglutathione. *Biochemistry* **32**:9701-9708.

57. Kumar, S., M. Björnstedt and A. Holmgren. 1992. Selenite is a substrate for calf thymus thioredoxin reductase and thioredoxin and elicits a large non-stoichiometric oxidation of NADPH in the presence of oxygen. *Eur J Biochem* **207**:435-439.

58. Björnstedt, M., S. Kumar and A. Holmgren. 1992. Selenodiglutathione Is a Highly Efficient Oxidant of Reduced Thioredoxin and a Substrate for Mammalian Thioredoxin Reductase. *J Biol Chem* **267**:8030-8034.

59. Björnstedt, M., S. Kumar, L. Björkhem, G. Spyrou and A. Holmgren. 1997. Selenium and the thioredoxin and glutaredoxin systems. Proceedings of the Sixth International Symposium on Selenium in Biology and Medicine 1996. *Biomed Environ Sci* **10**:271-279.

60. Stadtman, T. C. 1996. Selenocysteine. *Annu Rev Biochem* **65**:83-100.

61. Luthman, M. and A. Holmgren. 1982. Rat liver thioredoxin and thioredoxin reductase: Purification and characterization. *Biochemistry* 21:6628-6633.

62. Lundström, J. and A. Holmgren. 1990. Protein disulfide-isomerase is a substrate for thioredoxin reductase and has thioredoxin-like activity. *J Biol Chem* **265**:9114-9120.

63. Nikitovic, D. and A. Holmgren. 1996. S-nitrosoglutathione is cleaved by the thioredoxin system with liberation of glutathione and redox regulating nitric oxide. *J Biol Chem* **271**:19180-19185.

64. Björnstedt, M., J. Xue, W. Huang, B. Åkesson and A. Holmgren. 1994. The thioredoxin and glutaredoxin systems are efficient electron donors to human plasma glutathione peroxidase. *J Biol Chem* **269**:29382-29384.

65. Zhong, L. and A. Holmgren. 2000. Essential role of selenium in the catalytic activities of mammalian thioredoxin reductase revealed by characterization of recombinant enzymes with selenocysteine mutations. *J Biol Chem* **275**:18121-18128.

66. Björnstedt, M., M. Hamberg, S. Kumar, J. Xue and A. Holmgren. 1995. Human thioredoxin reductase directly reduces lipid hydroperoxides by NADPH and selenocystine strongly stimulates the reaction via catalytically generated selenols. *J Biol Chem* **270**:11761-11764.

67. Holmgren, A. and C. Lyckeborg. 1980. Enzymatic reduction of alloxan by thioredoxin and NADPH-thioredoxin reductase. *Proc Natl Acad Sci USA* **77**:5149-5152.

68. Andersson, M., A. Holmgren and G. Spyrou. 1996. NK-lysin, a disulfide containing effector peptide of T-lymphocytes is reduced and inactivated by thioredoxin reductase. Implication for a protective mechanism against NK-lysin cytotoxicity. *J Biol Chem* **271**:10116-10120.

69. Arnér, E. S. J., J. Nordberg and A. Holmgren. 1996. Efficient reduction of lipoamide and lipoic acid by mammalian thioredoxin reductase. *Biochem Biophys Res Commun* **225**:268-274.

70. May, J. M., S. Mendiratta, K. E. Hill and R. F. Burk. 1997. Reduction of dehydroascorbate to ascorbate by the selenoenzyme thioredoxin reductase. *J Biol Chem* **272**:22607-22610.

71. May, J. M., C. E. Cobb, S. Mendiratta, K. E. Hill and R. F. Burk. 1998. Reduction of the ascorbyl free radical to ascorbate by thioredoxin reductase. *J Biol Chem* **273**: 23039-23045.

72. Tamura, T. and T. C. Stadtman. 1996. A new selenoprotein from human lung adenocarcinoma cells: Purification, properties and thioredoxin reductase activity. *Proc Natl Acad Sci USA* **93**:1006-1011.

73. Gladyshev, V. N., K. T. Jeang and T. C. Stadtman. 1996. Selenocysteine, identified as the penultimate C-terminal residue in human T-cell thioredoxin reductase, corresponds to TGA in the human placental gene. *Proc Natl Acad Sci USA* **93**: 6146-6151.

74. Zhong, L., E. S. J. Arnér, J. Ljung, F. Åslund and A. Holmgren. 1998. Rat and calf thioredoxin reductase are homologous to glutathione reductase with a carboxyterminal elongation containing a conserved catalytically active penultimate selenocysteine residue. *J Biol Chem* **273**:8581-8591.

75. Zhong, L., E. S. J. Arnér and A. Holmgren. 2000. Structure and mechanism of mammalian thioredoxin reductase: The active site is a redoxactive selenolthiol / selenenylsulfide formed from the conserved cysteine-selenocysteine sequence. *Proc Natl Acad Sci USA* **97**:5854-5859.

76. Zhong, L., K. Persson, T. Sandalova, G. Schneider and A. Holmgren. 2000. Purification, crystallization and preliminary crystallographic data for rat cytosolic selenocysteine-498 to cysteine mutant thioredoxin reductase. *Acta Cryst* **D56**: 1191-1193.

77. Sandalova, T., L. Zhong, Y. Lindqvist, A. Holmgren and G. Schneider. 2001. Three-dimensional structure of a mammalian thioredoxin reductase: implications for mechanism and evolution of a selenocysteine dependent enzyme. *Proc Natl Acad Sci USA* **98**:9533-9538.

78. Schewe, T. 1995. Molecular actions of ebselen - an antiinflammatory antioxidant. *Gen Pharmac* **26**:1153-1169.

79. Zhao, R., H. Masayasu and A. Holmgren. 2002. Ebselen: a substrate for human thioredoxin reductase strongly stimulating its hydroperoxide reductase activity and a superfast thioredoxin oxidant. *Proc Natl Acad Sci USA* **99**:8579-8584.

80. Zhao, R. and A. Holmgren. 2002. A novel antioxidant mechanism of ebselen involving ebselen diselenide, a substrate of mammalian thioredoxin and thioredoxin reductase. *J Biol Chem* **277**:in press.

81. Saitoh, M., H. Nishitoh, M. Fujii, K. Takeda, K. Tobiume, Y. Sawada, M. Kawabata, K. Miyazono and Ichijo, H. 1998. Mammalian thioredoxin is a direct inhibitor of apoptosis signal-regulating kinase (ASK1). *EMBO J* **17**:2596-2606.

82. Nishiyama, A., M. Matsui, S. Iwata, K. Hirota, H. Masutani, H. Nakamura, Y. Takagi, H. Sono, Y. Gon and J. Yodoi. 1999. Identification of thioredoxin-binding protein-

2/vitamin D(3) up-regulated protein 1 as a negative regulator of thioredoxin function and expression. *J Biol Chem* **274**:21645-21650.

83. Butler, L. M., X. Zhou, W. S. Xu, H. I. Scher, R. A. Rifkind, P. A. Marks and V. M. Richon. 2002.The histone deacetylase inhibitor SAHA arrests cancer cell growth, up-regulates thioredoxin-binding protein-2, and down-regulates thioredoxin. *Proc Natl Acad Sci USA* **99**:11700-11705.

84. Song, J. J., J. G. Rhee, M. Suntharalingam, S. A. Walsh, D. R. Spitz and Y. L. Lee. 2002. Role of glutaredoxin in metabolic oxidative stress: Glutaredoxin as sensor of oxidative stress mediated by H_2O_2. *J Biol Chem* in press.

Chapter 4

REACTIVITY AND DIFFUSIVITY OF NITROGEN OXIDES IN MAMMALIAN BIOLOGY

Jack R. Lancaster, Jr.

4.1   Introduction

The discovery of the synthesis of nitric oxide (NO, nitrogen monoxide)[1,2,3] and its role in vasodilation[4,5] and of its target, guanylate cyclase,[6] in mammals in the 1980's established a new paradigm in the remarkable history of cellular signaling mechanisms. Two unique properties of NO as a ubiquitous messenger are (1) its spontaneous disappearance *via* reaction with biological species and (2) its ready diffusibility over relatively great distances. The balance between the high diffusibility of NO (as a small, lipophilic molecule) and its relatively short lifetime determine its spatial range of action.[7] The biologically relevant chemistry of NO and its derivatives is complex, although the initial reactions of NO are straightforward.[8] These properties of NO and its derivatives have provided Nature with a remarkable opportunity for this molecule to serve multiple biological functions. Here I present an overview of these properties. The intent is to present a broad outline of general concepts, with an emphasis on interrelationships between these complex pathways, and also to provide a general appreciation for the importance of spatial diffusion in determining the biological chemistry of endogenously produced nitrogen oxides.

53

H. J. Forman, J. Fukuto, and M. Torres (eds.), Signal Transduction by Reactive Oxygen and Nitrogen Species: Pathways and Chemical Principles, ©2003, Kluwer Academic Publishers. Printed in the Netherlands

4.1.1 Reactions of Nitrogen Oxides

4.1.1 1   Reactions of Nitric Oxide

Progress in delineating the biological chemistry of the nitrogen oxides (*i.e.*, NO itself as well as all of its derived species) has proven exceedingly complex. Much of the reason for our relative ignorance is because most of these species are short-lived and thus difficult to measure. It is also true that the biochemistry of these species is intrinsically complex. This is because of three primary facts. First, there many possible chemical reactions of these species which can occur "in the test tube", and new candidate reactions continue to appear in the literature on a regular basis. Second, the precise conditions that determine the reactivities of the various nitrogen oxides in the biological setting (*e.g.*, concentrations of NO species and potential coreactants) are not known with accuracy. Finally, the biological milieu is extremely heterogeneous, both on a spatial and temporal scale. Nevertheless, it has proven possible to delineate the most important biological reactions of NO and its cogeners.

The chemistry of NO is dominated by the fact that it is a relatively stable free radical (possesses an unpaired electron). This phenomenon is a result of a simple fact: *the reactions of NO involve stabilization of the electron which is unpaired in NO.* In general, this can occur in two ways (1) association of NO with a transition metal ion and (2) combination with another radical species to pair the electron with another unpaired electron. In both cases the unpaired electron can no longer be considered lone and unpaired. These two basic reactions are dealt with below.

There are two major types of products of the reactions of nitrogen oxides, (1) other nitrogen oxides[4] (*e.g.*, nitrite (NO_2^-), nitrate (NO_3^-), peroxynitrite ($ONOO^-$) and (2) adducts containing nitrogen oxide functional groups. There are three major nitrogen oxide-containing adducts of biological interest, which can be formally considered as the product of the attack of (a) nitric oxide (NO: nitrosylation), (b) nitrosonium (NO^+: nitrosation), or (c) nitronium (NO_2^+: nitration)[5]. In all three cases, bonding (electron sharing) is conferred to the nitrogen oxide from the target atom (*e.g.*, transition metal, carbon) to the nitrogen atom of the nitrogen oxide. The type of bond formed, however, is fundamentally different for nitrosylation compared to nitrosation and nitration.

4.1.1.2   Reactions with Transition Metal Ions

Among the transition metal ions present in cells, iron is the most prevalent, being required for a variety of essential functions, most notably oxidation-reduction reactions.[10] With one apparent exception (certain strains of *lactobacilli*[11]) iron is essential for cellular Life and all mammalian cells possess a variety of mechanisms for the maintenance of intracellular iron homeostasis.[12] Because of the specific arrangements of the ligands around iron ions in its various forms in cells, the

---

[1] Excluding NO itself; dimerization of NO to form the species N_2O_2 is energetically unfavorable, primarily because there is no net increase in bonds resulting from dimer formation. Thus, the monomeric species is much favored on entropic grounds.[9]

[2] Among these three species, only NO exists free under biological conditions, as described below.

energetic separation between the five orbitals that house the 5 (Fe^{+3}) or 6 (Fe^{2+}) electrons on the iron atom (the d orbitals) is relatively small.[13] Because the pairing up of electrons requires a certain amount of energy, this means that the electrons can be considered to "reside" at least to some extent "spread out" amongst all five orbitals. The consequence is that iron ions when complexed to biological molecules can have appreciable paramagnetic "character" and so are sensitive targets for other paramagnetic molecules such as O_2, $O_2^{-\cdot}$, and NO^6. In addition, several iron-containing proteins such as hemoglobin, myoglobin, the cytochromes P450, and cytochrome oxidase are specifically designed to bind O_2, and since NO is so similar in structure it is highly reactive toward these particular proteins. Because of this ability of NO to interact with these O_2-carrying and O_2-metabolizing proteins, it has been used since as early as 1865 by biochemists as an "oxygen surrogate" to probe the functions of these proteins, long before there was any idea that this activity might have physiological relevance.[14]

A great deal of work has been done on the reaction of NO with hemoproteins, in particular with hemoglobin and myoglobin. Because of its close resemblance to O_2, when NO encounters hemoglobin in the reduced, deoxy state in the absence of O_2 it binds to the heme iron atom in a fashion very similar to that of O_2. Kinetic studies have revealed that the association constant for NO binding is 5-20 times faster than that for O_2, but the dissociation constant is very much slower.[15] This means that the equilibrium binding of NO is very much stronger than for O_2 (or even carbon monoxide, CO, as originally shown in 1865[14]). Indeed, in the absence of O_2 the half-time for the dissociation of NO from ferrohemoglobin, once formed, is very long, approximately 8 hours.[15] NO also forms a tight complex with another hemoprotein of particular biological interest (and with potentially major biological effects), mitochondrial cytochrome oxidase.[16] In marked contrast to the effects of NO on hemoglobin (where hemoglobin is in great excess and only under extreme conditions will the effects of NO cause significant inhibition of oxygen transport by hemoglobin), the inhibition of mitochondrial electron transfer by NO at physiological NO concentrations may have critical regulatory effects[16,17] and may act to extend the distance of diffusion of O_2 away from the surface of a blood vessel.[18] The precise chemistry of the reactions of NO with cytochrome oxidase under physiological conditions (*i.e.*, aerobically) are not clear, but appear to involve both metal-nitrosyl formation which is competitive with dioxygen and also a reaction involving oxidized forms of the enzyme where NO reduces the protein with concomitant hydration resulting in formation of nitrite (combination of hydroxide with the nitrosonium equivalent $OH^- + [NO^+] \rightarrow NO_2^-$).[17,19,20] The resultant nitrite dissociates from the enzyme very slowly, however, reduction (*via* electron flow from the respiratory chain under normal conditions) results in dissociation.

Under physiological conditions, most heme in hemoglobin is oxygenated. In this case, NO reacts extremely rapidly essentially "extracting" the equivalent of superoxide from the ferroheme, generating nitrate and the oxidized heme:[21,22,23]

---

[3] The bonding in transition metal complexes is determined by the interaction between the d orbital electrons of the metal with all the bonding and nonbonding electrons of the ligand,[13] as opposed to only the unpaired electron(s) in ligands such as NO or O_2. Thus, ligands such as CO can form very strong complexes with metals even though it is not paramagnetic.

$$\cdot NO + O_2Fe^{2+} \rightarrow NO_3^- + Fe^{3+} \tag{1}$$

Quantitatively, this reaction probably represents the major *in vivo* route for NO conversion into stable metabolites.[247] As described in more detail below, it noteworthy that this mechanism occurs mostly in the lumen of the blood vessel. NO also reacts with oxygen complexes of other metalloproteins, including lipoxygenase[26] and myeloperoxidase.[27,28]

The stimulation of soluble guanylate cyclase (GC) by NO is arguably the most physiologically important reaction of NO. Early work had demonstrated that GC as isolated contains heme which is a requirement for stimulation by NO, and that various heme species stimulate the activity of the enzyme.[29] Further work has led to the hypothesis that NO stimulates the enzyme by binding to the distal position and inducing formation of a pentacoordinate complex, thus resulting in breakage of the proximal histidine bond to the iron.[30,31] Although this model still holds generally, there is evidence that NO may also exert a modulatory effect on activity, presumably at a second site,[32] although this point is in contention.[33,34]

In addition to hemoproteins, the other major metabolically active pool of intracellular iron is the nonheme iron-sulfur proteins, which function primarily as electron carriers, most especially in the mitochondrial electron transfer chain.[35] Indeed, quantitatively much more iron is contained in iron-sulfur clusters than in cytochromes in the mitochondrion. As documented by numerous studies on the enzymology of iron sulfur-containing enzymes prior to the discovery of mammalian endogenous NO synthesis, these proteins are also targets of ·NO action[36,36,37,38,39,40] although the molecular mechanisms of this interaction *in vivo* are not known with nearly the detail as with hemoproteins. In early studies, treatment with NO (at albeit supraphysiological concentrations anaerobically) was shown to result in quantitative conversion of the iron in these clusters into dinitrosyliron complexes (DNIC) of thiol (added as excess cysteine)$(RS^-)_2 Fe(NO)_2$.[36] These complexes exhibit a characteristic electron paramagnetic resonance (EPR) signal, and so the observation of such signals in both NO-producing activated macrophages[41,42] as well as the tumor cell targets of these macrophages[43] supported such a mechanism. However, it has been reported that the intensity of these signals do not correlate well with a disappearance of the signals from the native mitochondrial iron-sulfur centers.[44] Although it may well be true that much or most of the iron which forms the DNIC complexes originates from non-iron-sulfur center cellular iron, the question of how much iron in these centers forms DNIC (or other iron-NO complexes), and which centers may be most sensitive to disruption, has not been resolved.

More attention has been directed to the precise identity of the nitrogen oxide species responsible for disruption of cellular iron-sulfur centers. As mentioned above, the disruption of these centers by NO played an important role in the initial identification by Hibbs of NO formation by mammalian cells (activated macrophages).[45] Cellular formation of NO results in loss in activity of the iron-sulfur soluble enzyme aconitase, as well as regions of the mitochondrial electron transfer chain which are rich in iron-sulfur centers.[45] Remarkably, a critical target of

---

[4] The possibility that a major species formed under these conditions is S-nitrosothiol hemoglobin[25] is dealt with below.

NO is the cytosolic form of aconitase, which in fact possesses either of two biological activities; in one form, which is dominant under conditions of cellular iron depletion, the protein functions as iron regulatory protein-1 (IRP-1) and is a key control in regulation of iron homeostasis by acting as a transcription factor increasing expression of transferrin receptor and ferritin.[46] This form possesses no aconitase activity, however, under conditions of iron repletion the protein acquires aconitase activity and loses IRP-1 activity. The mechanism of this conversion involves the presence (aconitase active, IRP-1 inactive) or absence (aconitase inactive, IRP-1 active) of an intact iron-sulfur center. Cellular NO formation, by disrupting the intactness of the iron-sulfur cluster, thus induces a cellular response (in terms of iron homeostasis) similar to iron depletion. Indeed, there is evidence that at least part of the effect of NO on the IRP system is due to its depletion of cellular iron pools.[47,48,49] NO also activates the iron sulfur-containing bacterial transcription factor SoxR, which is involved in regulating defensive responses to oxidative stress.[50] The precise mechanism of action of NO in protein-bound iron-sulfur centers is not clear, with the three possibilities being disruption of the cluster *via* direct attack of NO, as described above,[47,51] similar direct disruption *via* reactive nitrogen oxide products resulting from NO (*e.g.*, peroxynitrite),[51] and loss of activity from nitrogen oxide-induced thiol modification.[52] It is however difficult to demonstrate experimentally direct effects of NO (as compared to actions of higher nitrogen oxides) since such a demonstration would require anoxic, unphysiological conditions, which among other effects would dramatically increase the concentration of NO (also described below). In any event, it does appear that disruption of iron-sulfur center function can be a deleterious cellular injury since cells possess mechanisms for their repair.[53,54,55]

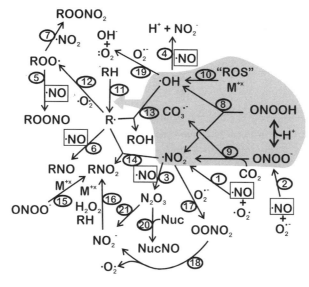

Scheme I. Overview of the Formation of Reactive Intermediates and Adducts from NO in the Biological Milieu. Reactions of free NO are denoted by boxes. For simplicity, all reactions are depicted as irreversible and are not necessarily completely balanced. For explanations of individual reactions (and also other reactions such as the formation of nitrate) see text.

4.1.1.2   Reactions with Other Radical Species

The major reactions of NO with other radical species of biological interest are its reactions with five species, dioxygen, superoxide, hydroxyl radical, nitrogen dioxide, and lipid peroxyl and alkoxyl radical species. In addition, there is recent evidence that nitrosative chemistry may be more widespread than previously appreciated *via* initial one-electron oxidation followed by nitrosylation.[56] These general mechanisms are presented in schematic overview form in Scheme I (Reactions 1-6) and are dealt with below. At the risk of instilling a sense of confusion in the reader, I present this scheme in order to emphasize the interrelationships between these complex reactions, which reveal an underlying network of interactions. Specific and more detailed information can be obtained from the literature references cited for each reaction.

Reaction with Dioxygen (Reaction 1 in Scheme I). NO reacts with dioxygen in aqueous solution *via* a fairly complex and not completely understood reaction to eventually produce nitrous acid (called autoxidation).[57,58,59] These reactions are listed below[8].

$$NO + O_2 \rightarrow [ONOO\cdot] \tag{2a}$$
$$[ONOO\cdot] + \cdot NO \rightarrow [N_2O_4] \tag{2b}$$
$$NO + \cdot NO \rightarrow [N_2O_2] \tag{3a}$$
$$[N_2O_2] + O_2 \rightarrow [N_2O_4] \tag{3b}$$
$$[N_2O_4] \rightarrow 2 \cdot NO_2 \tag{4}$$
$$NO + \cdot NO_2 \rightarrow N_2O_3 \tag{5}$$
$$N_2O_3 + H_2O \rightarrow 2 HNO_2 \tag{6}$$

with an overall stoichiometry

$$4 NO + O_2 + 2 H_2O \rightarrow 4 H^+ + 4 NO_2^- \tag{7}$$

The species in brackets are putative intermediates, and the sequence "a" compared to "b" for the first two steps correspond to the two possible sequences for the formation of the $[N_2O_4]$ species from 2 NO molecules and one O_2 molecule.[58] It is important to note that the equations above denote empirical formulas only. Thus, for example, there are several possible isomeric forms for $[N_2O_2]$ and $[N_2O_4]$.

It is instructive at this point to introduce the concept of the differentiation between the *chemical* importance of any reaction of nitrogen oxides and its *biological* importance. Perhaps the most direct example of this distinction is the reaction of NO with the heme of guanylate cyclase. In purely *chemical* terms, there is essentially no evidence that this reaction represents a significant portion of the

---

[5] Note that the unidirectionality of the arrows do not necessarily denote an irreversibility of these reactions; for example, the equilibrium for reactions 2a and 2b are highly likely to lie in the direction of reactants. Rate constants and equilibrium constants for these and other various reactions can be found in the cited references.

totality of reactions of NO in terms of overall flux of NO *in vivo*; in fact, the fate of the NO subsequent to heme nitrosyl formation is not known. However, this reaction is arguably the single most important reaction of NO *biologically*, since most of the signaling function of NO is accomplished through this mechanism. This point also applies to NO autoxidation. In particular, the kinetic expression for NO autoxidation is

$$\text{Rate} = 4k[\text{NO}]^2[\text{O}_2] \qquad (8)$$

with k = 2-2.9 x 10^6 $M^{-2}s^{-1}$.[57,58,59] The important point is that the rate of this reaction is exponential with NO concentration, and so will be very slow at low NO concentrations but very fast at high concentrations. For example, the "half-life" of NO at 10 nM concentration[9] is 30 hr. while it is 2 min. at 10 μM. Thus, for a range of concentrations in the physiological range (generally believed to be 1-50 nM) the consumption of NO *via* autoxidation will be exceedingly slow, and will not account for the loss of significant amounts. However, this reaction may have enormous *biological* consequences. In essence, although only minor amounts of total NO may proceed through this mechanism, it may result in the formation of species with potent biological activities. These reactions are considered below, when I discuss reactions of nitrogen oxides derived from NO (Section 1.2).

Reaction with Superoxide (Reaction 2 in Scheme I). The reaction of NO with superoxide is extremely rapid, and the product is peroxynitrite:[60,61]

$$\text{NO} + \text{O}_2{}^{\cdot -} \rightarrow \text{ONOO}^- \qquad (9)$$

The pK_a for the conjugate acid of peroxynitrite (peroxynitrous acid) is 6.8, and so ONOO⁻ will exist in virtually instantaneous equilibrium with ONOOH (since protonation is so rapid).

This reaction in particular has served as the focal point for the relationship between NO formation and cellular oxidative stress. As described in more detail below, both peroxynitrite and peroxynitrous acid are strong oxidants, much stronger than either NO or O_2^-, and so this extremely rapid reaction will serve to convert two mild oxidants into strong oxidants.[62] Another aspect of this reaction in terms of oxidative stress is that this reaction represents the formation of highly reactive oxidizing species without the participation of transition metal ions. Finally, production of both NO and O_2^- are deliberately upregulated in locations of inflammatory stimulation. These factors emphasize the potential importance of this reaction in oxidative stress in general, and in signaling in particular.[63,64]

Although the reaction of NO with O_2^- is exceedingly rapid, it is important to realize that the rates of appearance of these two reactants at any specific location will vary greatly, depending on many variables including the relative rates of synthesis. The rate of peroxynitrite formation will be a delicate balance between all these factors; specifically, for example, if the concentration or rate of appearance of one species at a particular location is greater than the other, then a simple treatment

---

[6] This is, strictly speaking, a misnomer since the term "half-life" can refer only to a first-order process, where its value is independent of concentration.

would predict that the rate at which peroxynitrite is formed will be determined entirely by the rate of appearance of the limiting reactant, and there will be an excess of the other reactant. Finally, another very important variable is the spatial distribution of NO and O_2^-, because both these species are diffusible (although to very different degrees, as described below) and peroxynitrite will be formed at distances away from the sources. This is illustrated in more detail below.

In terms of the rate of formation of peroxynitrite in the biological milieu, there are two most important determining and competing factors, the rates of reaction of superoxide with NO and with superoxide dismutase (SOD). In particular, the predominant fate of NO will obviously be determined by the faster reaction. Since the rate constants for these two reactions have been reported to be comparable in magnitude, then the rates will be determined by the relative concentrations of the two competing reactants (NO and SOD). As has been pointed out previously,[65] the intracellular concentrations of SOD are in general higher than NO concentrations, and so this raises a difficulty with regard to the relative occurrence of peroxynitrite formation compared to superoxide dismutation. Thus, the most important location for peroxynitrite formation may be in the extracellular space, where there is less (although not zero)[66] SOD. Finally, Koppenol and colleagues have shown that the rate of NO reaction with superoxide is in fact much faster than with SOD,[67,68] which would suggest that peroxynitrite would be formed at much lower concentrations of NO compared to SOD.

Reaction with Other Radical Species (Reactions 3-6 in Scheme I). As depicted in Scheme I, in addition to dioxygen and superoxide, NO also reacts with several other radical species. The reaction of NO with nitrogen dioxide (NO_2) (Reaction 3, Scheme I) is extremely rapid, and yields dinitrogen trioxide (N_2O_3, also known as nitrous anhydride).[66,69] As is clear from Scheme I, and dealt with in more detail below, NO_2 is a nitrogen oxide species which is central to the biological effects of NO formation. In addition to reaction with dioxygen, NO_2 is also formed as a result of the reaction of NO with superoxide. Another small radical species which will react with NO is hydroxyl radical (Reaction 4, Scheme I). This reaction, although very rapid,[70] has not received much attention in the biological setting until recently, and is discussed in more detail below regarding the oxidative actions of peroxynitrite. It is obvious however that this reaction could potentially serve a potent antioxidative function, by removing the highly reactive hydroxyl radical and forming the comparatively unreactive nitrite anion.

An exciting recent area of NO biology has been the demonstration, originally reported by Rubbo *et al.*,[71] that NO undergoes radical/radical recombination reactions with the initial radical products of lipid peroxidation (*e.g.* peroxyl species) to form several different nitrolipid species (illustrated by Reaction 5 in Scheme I). This action represents a potent antioxidative activity of NO, and there is also recent evidence that these species may possess specific regulatory and anti-inflammatory signaling activity as well.[72,73,74] Although not all shown in Scheme I for simplicity, other products of NO reaction (including $ONOO^-$, NO_2 (Reaction 7 in Scheme I),[75] and HNO_2) can also cause formation of nitrolipid species.[76] Finally, radical/radical recombination can occur between a one-electron oxidized species and NO (Reaction 6 in Scheme I); for example, NO reacts with protein-bound tyrosyl radical species in several proteins.[77] Recent results by Espey *et al.* reveal that such a mechanism could be a new previously unrecognized mechanism of nitrosation, as described in more detail below.[56]

4.1.2 Reactions of Nitrogen Oxide Derivatives of Nitric Oxide

4.1.2.2   Reactions of the Products of Reaction between NO and Superoxide (ONOO-, ONOOH, NO₂, OH, CO₃·)

The reaction of NO with superoxide produces the potent oxidant peroxynitrite (Reaction 2 in Scheme I), which gives rise to a collection of highly oxidizing species (denoted by the shaded area, Scheme I).[61] In the absence of reductants, peroxynitrite is stable at alkaline pH, but decomposes rapidly when protonated (Reaction 8 in Scheme I). The two decomposition routes for ONOOH are isomerization to form the unreactive nitrate anion (with release of H^+) and homolysis[10], yielding the radicals $\cdot NO_2$ and $\cdot OH$ (which occurs in approximately 25-35% yield)[78,79,80,81]

$$(65\text{-}75\%)\ NO_3^- + H^+ \leftarrow ONOOH \rightarrow \cdot NO_2 + \cdot OH\ (25\text{-}35\%) \qquad (10)$$

Hydroxyl radical is the prototypical oxidant in the biological setting[82] and NO_2 is also a good oxidant (forming nitrite anion).[83]

Under biologically relevant conditions, where the concentrations of CO_2 are high, the predominant reaction of peroxynitrite (dominating even the homolysis process) is reaction with CO_2 (Reaction 9 in Scheme I). Although the precise intermediate of this interaction is unclear (*i.e.*, whether a distinct molecular entity is formed *vs.* a "caged" intermediate), the association between $ONOO^-$ and CO_2 results in reactivity along two paths: CO_2-catalyzed isomerization to form nitrate (approximately 65% yield) and formation of NO_2 and the radical carbonate anion ($CO_3\cdot^-$, yet another strong oxidant):[83]

$$(65\%)\ NO_3^- + CO_2 \leftarrow ONOO^- + CO_2 \rightarrow \cdot NO_2 + CO_3\cdot^-\ (35\%) \qquad (11)$$

Thus, reaction of NO with superoxide produces five highly oxidizing species ($ONOO^-$, ONOOH, $\cdot NO_2$, $CO_3\cdot^-$, and $\cdot OH$). In addition, superoxide itself can give rise to highly oxidizing reactive oxygen species (especially in the presence of transition metal ions) (denoted "ROS" M^{+x} in Reaction 10 in Scheme I).[82] The overall effect is the oxidation of numerous biological targets to produce oxidized species (Reaction 11 in Scheme I) which are further targets for adduct formation by many of the same radical species which formed them. The formation of four of these adducts are presented below.

In the case of lipids, preferential oxidation occurs primarily *via* abstraction of an allylic hydrogen atom (*i.e.*, a carbon adjacent to a double bond) producing an alkyl radical, R· (Reaction 11 in Scheme I). These radicals are known to react at a diffusion-controlled rate with dioxygen, generating peroxyl radicals (Reaction 12 in Scheme I). These peroxyl radicals are capable of hydrogen extraction from another

---

[7] Cleavage of a bond such that the two electrons in the bond separate, thus generating two radical species.

lipid allylic carbon (forming the lipid peroxide LOOH), and so an initial hydrogen abstraction can result in a chain reaction whereby an initial event triggers multiple rounds of LOOH formation.[82] The only way this cycle can be terminated is *via* a reaction which results in extinguishing of the radical; this means another radical species must be involved (in order to pair up the electron) and so the reaction is a radical-radical recombination. As alluded to above, both NO (Reaction 5 in Scheme I) and NO_2 (Reaction 7 in Scheme I) can serve this antioxidative function, in the process forming a variety of nitrolipid adducts. These adducts are of varying stability, and in fact may serve as cellular messengers.[72,73,74]

The initial radical product of hydrogen abstraction (or, equivalently, one-electron oxidation followed by proton dissociation) R· will also react with other radicals to form adducts. In the case of the hydroxyl radical, recombination results in hydroxylation (Reaction 13 in Scheme I):

$$R· + ·OH \rightarrow ROH \qquad\qquad (12)$$

Hydroxylations resulting from peroxynitrite formation have been known by chemists previously[84] and its importance for its biological actions was pointed out by Beckman *et al.*[62] An additional recombination involves NO2, resulting in nitration of the molecule (Reaction 14 in Scheme I):

$$R· + ·NO_2 \rightarrow RNO_2 \qquad\qquad (13)$$

This is one mechanism whereby nitrotyrosine can be formed as a result of endogenous NO production, although contrary to common perception this is not a specific marker for peroxynitrite. Specifically, Reaction 14 in Scheme I (which was initially reported in the chemical literature)[85] requires only oxidation occurring in the presence of NO_2, which (as illustrated in Scheme I) is a central intermediate in multiple pathways involving a variety of nitrogen oxides.

The first demonstration of nitrotyrosine formation from NO_2 in aqueous solution was reported by Prutz *et al.*,[86] where it was shown that NO_2 performs two functions, first to oxidize tyrosine to generate the phenoxyl radical (Reaction 11 in Scheme I) and then to undergo radical-radical recombination with the tyrosyl radical resulting in nitrotyrosine (Reaction 14 in Scheme I). In addition, formation of dityrosine was detected, due to the recombination of two tyrosyl radicals (a reaction which is unfavored in general for tyrosine residues in protein). There exists selectivity in the nitration of protein tyrosine residues, both in terms of which tyrosines are nitrated and also in terms of susceptibility to different nitrating agents.[87] This effect may result from the requirement for a relatively aqueous microenvironment surrounding the phenolic ring for the initial oxidation and proton dissociation. This effect may explain for example the relative resistance of intracellularly expressed (as opposed to *in vitro* treated) green fluorescent protein to nitration in the face of significant peroxynitrite levels.[88]

There has been a great deal of controversy regarding the origin of nitrotyrosine *in vivo*, and its importance is due to the ubiquitous presence (detected by nitrotyrosine-specific antibody) of nitrated protein tyrosine in a remarkable array of pathophysiological and physiological conditions.[89] In addition to oxidation/recombination mechanisms (Reactions 11,14 in Scheme I), Beckman and colleagues demonstrated the formation of nitrotyrosine in several metalloproteins in the presence of peroxynitrite (requiring the presence of metals) suggesting a mechanism involving generation of nitronium (NO_2^+) equivalents[90,91] (Reaction 15 in Scheme I). The presence of CO_2, similar to metals, has been shown to enhance nitrative (as well as oxidative) chemistry of peroxynitrite due to its reaction as described above[92,93,94] (Reaction 9, Scheme I). Still another mechanism of nitration involves oxidation of nitrite by metalloprotein oxidases (in the presence of H_2O_2 as source of oxidizing equivalents)[95] (Reaction 16 in Scheme I), including myeloperoxidase and eosinohil peroxidase. Since the participants in this reaction (nitrite from NO, phagocytic peroxidases, and H_2O_2) are present under conditions of inflammation and immune stimulation, this mechanism is a good candidate for explaining nitrotyrosine formation detected *in vivo*. The primary product of this reaction is the one-electron oxidation of nitrite to $\cdot NO_2$, with minor formation of a two-electron oxidation product with $ONOO^-$-like properties.[96] This same study utilized knockout mice lacking eosinophil peroxidase or myeloperoxidase to demonstrate that these enzymes can indeed play a dominant role in nitrotyrosine formation.[96] Finally, Thomas *et al.* have recently shown that nitration can occur with nitrite and H_2O_2 in the presence of metals, such as heme or nohmeme iron,[97] which does not involve a specific enzyme and can be expected to occur under conditions of cellular injury. It thus seems reasonable to conclude that there is no one mechanism for nitrotyrosine formation, however, its presence does act as a good marker for endogenous NO formation, especially in the context of inflammation and/or oxidative stress. It must be stressed however that although chemical nitration of tyrosine residues in proteins may alter biological activity (analogous in principle to any covalent modification), there is no convincing evidence at present that this represents a mechanism of posttranslational modification which is of functional significance.

Oxidation and nitration from the synthesis of peroxynitrite formed by steady synthesis of NO and $O_2^{\cdot-}$ takes place only when the rates of formation of these two reactants are equal.[98,99,99,100] An excess of flux (formation per unit time) of either reactant eliminates these reactions. The simplest explanation for this result is that either reactant (NO or $O_2^{\cdot-}$) reacts with the active product(s) which cause oxidation/nitration, a result which has been verified experimentally, for both oxidation[99] and nitration.[97] The explanation lies in the fact that the reactions occur not with peroxynitrite itself, but the products of its homolysis, NO_2 and $\cdot OH$. Specifically, NO will react with both NO_2 (Reaction 3 in Scheme I) and with $\cdot OH$ (Reaction 4, Scheme I), as described above. Likewise, $O_2^{\cdot-}$ reacts with NO_2 to yield peroxynitrate[101] (Reaction 17, Scheme I), which at physiological pH (> 5) decomposes to nitrite and O_2[102] (Reaction 18, Scheme I), and with OH to yield OH^- and O_2[99] (Reaction 19, Scheme I). All of these reactions can be classified as radical-radical recombination reactions, and so proceed with rapid rate constants.[99] The final endproducts possess little or no oxidative/nitrative activities, and so these reactions imply that the oxidative/nitrative actions of peroxynitrite will be manifested *in vivo* only at locations where these fluxes are equal. Such conditions are likely to be quite rare, which has been used to argue against the physiological significance of these reactions of peroxynitrite. However, this conclusion entails several caveats, including the fact that these chemical studies were carried out in homogeneous solution where the only available reactants are those provided by chemical addition. *In vivo*, there exist a multitude of potential other targets for reaction of the immediate products of ONOO⁻ homolysis (NO_2 and OH), and at least in the case of ·OH the reaction rates with many targets are not greatly different the reaction with NO or $O_2^{·-}$. Thus, if these biological targets (nucleophilic groups in cellular components) are present at concentrations similar or higher than those of NO or $O_2^{·-}$ substantial reaction with these targets can occur. In addition, as described below, comprehensive consideration of the biological situation requires recognition that the locations of sources and targets are spatially heterogeneous, and there may be regions where the equal flux condition will hold and the locations of these regions, although small at any instant, may change with time.

4.1.2.3 Reactions of the Products of Reaction between NO and Oxygen (NO_2, N_2O_3)

As illustrated in Scheme I, NO_2 plays a central role in the biological chemistry of NO. NO_2 will readily dimerize with a rate constant not very different from the reaction of NO_2 with NO[66,69] (Reaction 3, Scheme I), however, under physiological conditions since all nitrogen oxide species are produced from NO there will be much more NO available for reaction than another molecule of NO_2. The product of the reaction between NO and NO_2 produces N_2O_3, a potent nitrosating agent which effectively displaces a proton on a nucleophile (Nuc; notably thiol, amine) (Reaction 20, Scheme I):

$$NucH + N_2O_3 \rightarrow NucNO + H^+ + NO_2^- \tag{14}$$

N_2O_3 also reacts with water, which, in addition to the nitrosation reaction (Eq. 14), generates nitrite (Reaction 21, Scheme I):

$$N_2O_3 + H_2O \rightarrow 2\,H^+ + 2\,NO_2^- \tag{15}$$

Nitrosative chemistry is important biologically. Historically, the consumption of nitrite and nitrate in the diet (most especially from cured foods) has been a major public health concern because of the formation of nitrosamine in the gastric contents because of the formation of nitrous acid.[103] These chemical conditions give rise to the same species responsible for nitrosative chemistry during NO autoxidation (Reaction 3, Scheme I), nitrous anhydride:

$$2\,HNO_2 \rightarrow H_2O + N_2O_3 \tag{16}$$

There is indeed evidence that chronic NO production resulting from inflammatory conditions is a risk factor for carcinogenesis [104]. In this regard, chemical modification of DNA (oxidation, nitrosation) by several reactive nitrogen oxides may be the mechanistic basis for this effect.[105,106]

Nitrosation can be classified as attack of the formal species nitrosonium (NO^+), although it is important to realize that this ion does not exist free under physiologically relevant conditions due to its extremely rapid reaction with hydroxide ($NO^+ + OH^- \rightarrow H^+ + NO_2^-$).[107] The formation of nitrosative intermediates in the aqueous autoxidation of NO is well documented,[108,109,110] although the precise identity of the isomeric nitrogen oxide species involved is unresolved;[108,110] in addition, the mechanism(s) for the formation of nitrosative products with aerobic NO in cells may be fundamentally different than the reaction in aqueous solution.[111] Indeed, the reaction of NO with O_2 in the biological milieu most probably occurs in hydrophobic phases (membranes, lipoproteins, etc.) and the exact chemistry may differ from that in the aqueous phase due to the difference in solvation effects of intermediate/transition states.[112] In any event, examination of the rate of NO consumption by intact non-erythroid cells reveals a process which, although O_2 is required, is too rapid for NO autoxidation.[18]

The two most important targets for nitrosation are amines and thiols. In the case of amines, the subsequent chemistry is different for each type of amine. For primary amines, the nitrosamine is the initial product which rapidly forms a diazonium compound:

$$RNH_2 + [NO^+] \rightarrow RN(H)NO + H^+ \rightarrow RN_2^+ + H_2O \qquad (17)$$

The diazonium is an excellent leaving group, which results in the formation of an alcohol:

$$RN_2^+ + OH^- \rightarrow ROH + N_2 \qquad (18)$$

The net result is the conversion of a primary amine into an alcohol. For secondary amines, the nitrosamine is stable:

$$(R, R')NH + [NO^+] \rightarrow (R,R')NNO + H^+ \qquad (19)$$

but is metabolized in cells to yield a variety of highly reactive and mutagenic species.[103] Although tertiary amines can also be nitrosated, relevance to biological systems is unclear since the reaction is much slower than for primary or secondary amines.[113]

Like amines, thiols also are excellent acceptors for nitrosation, yielding nitrosothiols:

$$RSH + [NO^+] \rightarrow RSNO + H^+ \qquad (20)$$

In reagent quantities NO can directly react with thiol resulting in oxidation and disulfide formation,[114] however this process is too slow to be of biological relevance with physiological NO concentrations. Most recently, Espey *et al.* demonstrated that under conditions of slight excess of NO formation over superoxide formation, nitrosative chemistry can occur.[56] Thus, peroxynitrite acts to oxidize a target which then reacts with NO (Reactions 11, 6 in Scheme I). This may represent a biological mechanism of nitrosation (e.g., nitrosothiol formation).

Nitrosation of thiol in both small (e.g., glutathione, cysteine) and large molecules (proteins) as a result of endogenous NO synthesis is well documented,[115] and like nitration not all cellular thiol groups are equally susceptible.[116,117,118] There is evidence that nitrosation of protein cysteine thiols may serve as a mechanism of regulation of protein function (see Chapter by Loscalzo),[119] however, a critical unresolved issue is the mechanism of this nitrosation; as described above, nitrosation requires the equivalent of one-electron oxidation of NO, and the cellular mechanism for this reaction is virtually completely unknown.[111] Nitrosothiols, in the presence of even trace amounts of transition metals (especially Cu^{1+}) will break down to liberate NO.[120]

A mechanism of protein thiol nitrosation of special interest is hemoglobin, wherein nitrosation of cysβ 93 has been proposed to serve a function as "preserver" of NO biological activity in the face of the very rapid irreversible consumption of NO *via* oxidation (Equation 1).[121] In addition, since changes in the accessibility of this thiol to covalent modification upon changes in the oxygenation state of hemoglobin has been known for many years,[122] a respiratory cycle has been proposed where uptake of O_2 and NO are proposed to exhibit positive cooperativity and so hemoglobin will deliver both O_2 and NO from the lung to O_2-deficient tissues.[119,123] However, important mechanistic details regarding formation, movement to the target (smooth muscle cell guanylate cyclase), and vasodilation from the nitrogen oxide species responsible are lacking, and more recent evidence has challenged the validity of this proposal.[124,125,126,127,128]

4.1.2.4 Reduction of NO

The biological transformations of NO described above are all oxidative, *i.e.*, NO is oxidized. One-electron reduction of NO yields the species nitroxyl (HNO) and its anion (NO⁻), which possesses biological properties distinctly different from NO.[8,129] Although chemical studies have shown that reduction of NO can occur *via* some biochemical mechanisms (*e.g.*, superoxide dismutase,[130] NO synthase,[131] mitochondrial metabolism,[132] and reaction of nitrosothiols with thiols),[133] little evidence has been presented to suggest that functionally relevant quantities of nitroxyl are produced *in vivo*.[134,135]

4.2 Diffusion of Nitrogen Oxides

4.2.1 Diffusion of NO in Biological Systems

The chemistry of NO at any specific biological location will be defined by the concentration of NO and of potential reactants at that specific location. Since NO is freely permeable through biological membranes, and exhibits one of the highest diffusion constants known, its actions are not confined to a specific cellular compartment or even a single cell producing it.[7,136] This property places NO in a unique category of cellular messengers, in that it is able to transmit its signal virtually instantaneously after it is produced by NO synthase. In addition, because of its reactivity, NO will disappear relatively rapidly and so there is no need for a specific mechanism of deactivation.

The rapid diffusibility of NO has critically important implications for its chemistry in the biological setting. I will examine two implications in particular.

4.2.2 Reaction vs. "Escape" of NO from the Cell Producing It

The speed with which NO moves by random diffusion can be illustrated by consideration of its root mean square distance of displacement, which describes the distance NO will move in any time interval based on its diffusion constant D (which is similar for aqueous solution and also tissue (brain): [137]

$$\langle \Delta \bar{x} \rangle^2 = 2Dt \tag{21}$$

An illustration of the extraordinarily rapid diffusibility of NO is illustrated by applying this relationship to ask how long, on average, it takes before NO will escape a cell which produces it. For a cell of 5 μm diameter, Equation 20 reveals that, on average, this time will be no more than:

$$t = \frac{(2.5\,\mu m)^2}{2 \times 3300\,\mu m^2 - s^{-1}} \approx 0.001\,\text{sec} \tag{21}$$

In terms of chemical reactivity, we can now ask how rapid an intracellular reaction must be in order to prevent this "escape". To a first approximation, the half-life of such a reaction should be in the range of this "escape" time, and so we calculate the rate for a reaction with this half-life:

$$rate = \frac{\ln 2}{0.001\ s} = 6.93 \times 10^2\ s^{-1} \tag{22}$$

Assuming a second-order reaction between NO and a reactant R (rate = k[NO][R]) and a very high intracellular R concentration (1mM), the value for the rate constant k is given by:

$$k = \frac{6.93 \times 10^2\ s^{-1}}{10^{-3}\ M} = 6.93 \times 10^5\ M^{-1}s^{-1} \tag{23}$$

This is the rate constant for a very rapid reaction. In addition, in order to provide continual inhibition of NO escape the reactant R must be continually maintained at this high concentration. Experimental support for the general concept that NO will rapidly escape a cell producing it is provided by the observation that coincubation of NO-producing isolated rat hepatocytes with intact erythrocytes eliminates the reaction of NO with targets within the NO-producing hepatocyte; it can be shown from first principles that the only explanation for this effect is if, on average, any individual NO molecule which acts within an NO-producing cell had at some time previously existed in the volume outside the cell.[138,139]

4.2.3 Spatial Heterogeneity Generates Chemical Heterogeneity

In vivo, different cell types produce different reactive oxygen and nitrogen species. Spatially, this means that the appearance (flux) of these reactive species at any point location will be determined by the distance from this point to the cells producing these species. Thus, the precise chemistry which occurs at different locations can be very different, especially when dealing with highly interactive networks of interreacting species such as depicted in Scheme I. I will present one illustration of this concept.

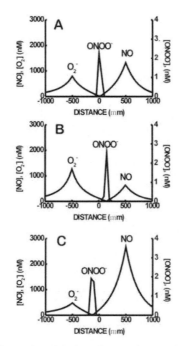

Figure 1. Theoretical effects of modulation of rates of NO and O_2^- Fluxes on ONOO$^-$ formation for separate sources located 1 mm apart. For explanation see text.

Using a reaction/diffusion simulation described previously,[136,139,140] Figure 1 presents steady-state profiles of concentrations of reactive species at various locations with a point source for NO separated from a point source for superoxide. For all three plots, the O_2^- source is located at position −500 μm, while the NO source is at +500 μm. The simulation utilizes experimentally determined rates (NO/O_2^- reaction, diffusion of NO, O_2^- , etc.) and will be reported in detail elsewhere. The essential point for this discussion is that the location of formation peroxynitrite will occur at locations in between these sources, as might be expected. In addition, the location of this zone of reaction will change in relation to these sources; in Fig. 1B the rate of O_2^- formation is increased relative to Fig. 1A and consequently the zone shifts closer to the NO source. Conversely, this zone shifts in the opposite direction if the NO formation rate increases (Fig. 1C) relative to the O_2^- rate (Fig. 1C). This illustrates the general concept that the spatial heterogeneity of different sources (and also sinks) for these species will generate a heterogeneity of different chemistries which will occur at each location. Thus, for example, while it may be true that in homogeneous solution maximal peroxynitrite formation will only occur with equal fluxes of NO and O_2^- (as described above), in the biological setting this condition may be met at some particular location, albeit in perhaps a relatively small volume at any instant. Additionally, with changing fluxes these specific locations may change with time, and thus "move around". The overall conclusion thus may be that the appropriate biologically relevant question may not <u>whether</u> a particular reaction takes place, but <u>where</u> a particular reaction takes place.

Acknowledgments

This work was supported, in part, by NIH grant DK46935.

Reference

1. Green, L. C., S. R. Tannenbaum and P. Goldman. 1981. Nitrate Synthesis in the Germfree and Conventional Rat. *Science 212:* 56-58.

2. Hibbs, J. B., Jr., R. R. Taintor and Z. Vavrin. 1987. Macrophage Cytotoxicity: Role for L-Arginine Deiminase and Imino Nitrogen Oxidation to Nitrite. *Science 235:* 473-476.

3. Stuehr, D. J., M. A. Marletta. 1985. Mammalian Nitrate Biosynthesis: Mouse Macrophages Produce Nitrite and Nitrate in Response to Escherichia Coli Lipopolysaccharide. *Proc. Natl. Acad. Sci. USA 82:* 7738-7742.

4. Ignarro, L. J., R. E. Byrns and K. S. Wood. **1988**. Biochemical and Pharmacological Properties of Endothelium- Derived Relaxing Factor and Its Similarity to Nitric Oxide Radical. In *Vasodilatation: Vascular Smooth Muscle, Peptides, Autonomic Nerves, and Endothelium;* (Vanhoutte, P. M., Ed.), pp. 427-435, Raven Press, New York, NY

5. Furchgott, R. F. **1988**. Studies on Relaxation of Rabbit Aorta by Sodium Nitrite: the Basis for the Proposal That the Acid-Activatable Factor From Bovine Retractor Penis Is Inorganic Nitrite and the Endothelium- Derived Relaxing Factor Is Nitric Oxide. In *Vasodilatation: Vascular Smooth Muscle, Peptides, Autonomic Nerves, and Endothelium;* (Vanhoutte, P. M., Ed.), pp. 401-414, Raven Press, New York, NY

6. Arnold, W. P., C. K. Mittal, S. Katsuki and F. Murad. 1977. Nitric Oxide Activates Guanylate Cyclase and Increases Guanosine 3':5'-Cyclic Monophosphate Levels in Various Tissue Preparations. *Proc. Natl. Acad. Sci. USA 74:* 3203-3207.

7. Lancaster, J. R., Jr. **2000**. The Physical Properties of Nitric Oxide. Determinants of the Dynamics of NO in Tissue. In *Nitric Oxide: Biology and Pathobiology;* (Ignarro, L. J., Ed.), pp. 209-224, Academic Press, San Diego, CA,

8. Fukuto, J. M., J. Y. Cho and C. H. Switzer. 2000. The Chemical Properties of Nitric Oxde and Related Nitrogen Oxides. In *Nitric Oxide: Biology and Pathobiology*; (Ignarro, L. J., Ed.), pp. 23-39, Academic Press, San Diego

9. Beckman, J. S. 1996. The Physiological and Pathological Chemistry of Nitric Oxide. In *Nitric Oxide. Principles and Actions*; (Lancaster, J. R., Jr., Ed.), pp. 1-82, Academic Press, San Diego, CA

10. Neilands, J. B. 1991. A Brief History of Iron Metabolism. *Biology of Metals 4:* 1-6.

11. Archibald, F. 1983. *Micrococcus Lysodiekticus*, an Organism Not Requiring Iron. *FEMS Microbiol. Lett. 19:* 29-32.

12. Weinberg, E. D. 1990. Cellular Iron Metabolism in Health and Disease. *Drug Metabolism Reviews 22:* 531-579.

13. Griffith, J. S. *The Theory of Transition-Metal Ions.* 1961. Cambridge University Press, Cambridge, UK

14. Hermann, L. 1865. Ueber Die Wirkungen Des Stickoxydulgases Auf Das Blut. *Arch. Anat. Physiol. ,Lpz.* 469-481.

15. Gibson, Q. H., F. J. W. Roughton. 1957. The Kinetics and Equilibria of the Reactions of Nitric Oxide With Sheep Hemoglobin. *J. Physiol. 136:* 507-526.

16. Brown, G. C. 2001. Regulation of Mitochondrial Respiration by Nitric Oxide Inhibition of Cytochrome c Oxidase. *Biochim. Biophys. Acta 1504:* 46-57.

17. Brookes, P. S., A.-L. Levonen, S. Shiva, P. Sarti and V. Darley-Usmar. 2002. Mitochondria: Regulators of Signal Transduction by Reactive Oxygen and Nitrogen Species. *Free Rad. Biol. Med. 33:* 755-764.

18. Thomas, D. D., X. Liu, S. P. Kantrow and J. R. Lancaster, Jr. 2001. The Biological Lifetime of Nitric Oxide: Implications for the Perivascular Dynamics of NO and O2. *Proc. Natl. Acad. Sci. U. S. A 98:* 355-360.

19. Giuffre, A., M. C. Barone, D. Mastronicola, E. D'Itri, P. Sarti and M. Brunori. 2000. Reaction of Nitric Oxide With the Turnover Intermediates of Cytochrome c Oxidase: Reaction Pathway and Functional Effects. *Biochemistry 39:* 15446-15453.

20. Sarti, P., A. Giuffre, E. Forte, D. Mastronicola, M. C. Barone and M. Brunori. 2000. Nitric Oxide and Cytochrome c Oxidase: Mechanisms of Inhibition and NO Degradation. *Biochem. Biophys. Res. Commun. 274:* 183-187.

21. Doyle, M. P., J. W. Hoekstra. 1981. Oxidation of Nitrogen Oxides by Bound Dioxygen in Hemoproteins. *J. Inorg. Biochem. 14:* 351-358.

22. Eich, R. F., T. Li, D. D. Lemon, D. H. Doherty, S. R. Curry, J. F. Aitken, A. J. Mathews, K. A. Johnson, R. D. Smith, G. N. J. Phillips and J. S. Olson. 1996. Mechanism of NO-Induced Oxidation of Myoglobin and Hemoglobin. *Biochemistry 35:* 6976-6983.

23. Herold, S. 1999. Kinetic and Spectroscopic Characterization of an Intermediate Peroxynitrite Complex in the Nitrogen Monoxide Induced Oxidation of Oxyhemoglobin

[Corrected and Republished Article Originally Printed in FEBS Lett 1998 Nov 13;439(1-2):85-8]. *FEBS Lett.* **443:** 81-84.

24. Wennmalm, A., G. Benthin, A. Edlund, N. Kieler-Jensen, S. Lundin, A. S. Petersson and F. Waagstein. 1994. Nitric Oxide Synthesis and Metabolism in Man. *Ann. N. Y. Acad. Sci.* **714:** 158-164.

25. McMahon, T. J., J. S. Stamler. 1999. Concerted Nitric Oxide/Oxygen Delivery by Hemoglobin. *Meth. Enz.* **301:** 99-114.

26. Coffey, M. J., R. Natarajan, P. H. Chumley, B. Coles, P. R. Thimmalapura, M. Nowell, H. Kuhn, M. J. Lewis, B. A. Freeman and V. B. O'Donnell. 2001. Catalytic Consumption of Nitric Oxide by 12/15- Lipoxygenase: Inhibition of Monocyte Soluble Guanylate Cyclase Activation. *Proc. Natl. Acad. Sci. U. S. A* **98:** 8006-8011.

27. Abu-Soud, H. M., S. L. Hazen. 2000. Nitric Oxide Is a Physiological Substrate for Mammalian Peroxidases. *J. Biol. Chem.* **275:** 37524-37532.

28. Eiserich, J. P., S. Baldus, M. L. Brennan, W. Ma, C. Zhang, A. Tousson, L. Castro, A. J. Lusis, W. M. Nauseef, C. R. White and B. A. Freeman. 2002. Myeloperoxidase, a Leukocyte-Derived Vascular NO Oxidase. *Science* **296:** 2391-2394.

29. Ignarro, L. J., K. S. Wood and M. S. Wolin. 1982. Activation of Purified Soluble Guanylate Cyclase by Protoporphyrin IX. *Proc. Natl. Acad. Sci. USA* **79:** 2870-2873.

30. Ignarro, L. J. 1994. Regulation of Cytosolic Guanylyl Cyclase by Porphyrins and Metalloporphyrins. *Adv. Pharmacol.* **26:** 35-65.

31. Reynolds, M. F., J. N. Burstyn. 2000. Mechanism of Activation of Souble Guanylyl Cyclase by NO. Allosteric Regulation Through Changes in Heme Coordination Geometry. In *Nitric Oxide: Biology and Pathobiology*; (Ignarro, L. J., Ed.), pp. 381-399, Academic Press, San Diego,

32. Zhao, Y., P. E. Brandish, D. P. Ballou and M. A. Marletta. 1999. A Molecular Basis for Nitric Oxide Sensing by Soluble Guanylate Cyclase. *Proc. Natl. Acad. Sci. U. S. A* **96:** 14753-14758.

33. Bellamy, T. C., J. Wood and J. Garthwaite. 2002. On the Activation of Soluble Guanylyl Cyclase by Nitric Oxide. *Proc. Natl. Acad. Sci. U. S. A* **99:** 507-510.

34. Ballou, D. P., Y. Zhao, P. E. Brandish and M. A. Marletta. 2002. Revisiting the Kinetics of Nitric Oxide (NO) Binding to Soluble Guanylate Cyclase: The Simple NO-Binding Model Is Incorrect. *Proc. Natl. Acad. Sci. U. S. A* **99:** 12097-12101.

35. Beinert, H. 2000. Iron-Sulfur Proteins: Ancient Structures, Still Full of Surprises. *J. Biol. Inorg. Chem.* **5:** 2-15.

36. Salerno, J. C., T. Ohnishi, J. Lim and T. E. King. 1976. Tetranuclear and Binuclear Iron-Sulfur Clusters in Succinate Dehydrogenase: a Method of Iron Quantitation by Formation of Paramagnetic Complexes. *Biochem. Biophys. Res. Comm.* **73:** 833-839.

37. Dervartanian, D. V., S. P. J. Albracht, J. A. Berden, B. F. Van Gelder and E. C. Slater. 1973. The EPR Spectrum of Isolated Complex III. *Biochim. Biophys. Acta* **292:** 496-501.

38. Drapier, J. C. 1997. Interplay Between NO and [Fe-S] Clusters: Relevance to Biological Systems. *Methods 11:* 319-329.

39. Reddy, D., J. R. Lancaster, Jr. and D. P. Cornforth. 1983. Nitrite Inhibition of Clostridium Botulinum: Electron Spin Resonance Detection of Iron-Nitric Oxide Complexes. *Science 221:* 769-770.

40. Bouton, C. 1999. Nitrosative and Oxidative Modulation of Iron Regulatory Proteins. [Review] [102 Refs]. *Cellular & Molecular Life Sciences 55:* 1043-1053.

41. Lancaster, J. R., Jr., J. B. Hibbs, Jr. 1990. EPR Demonstration of Iron-Nitrosyl Complex Formation by Cytotoxic Activated Macrophages. *Proc. Natl. Acad. Sci. USA 87:* 1223-1227.

42. Pellat, C., Y. Henry and J. C. Drapier. 1990. IFN-Gamma-Activated Macrophages: Detection by Electron Paramagnetic Resonance of Complexes Between L-Arginine-Derived Nitric Oxide and Non-Heme Iron Proteins. *Biochem. Biophys. Res. Comm. 166:* 119-125.

43. Drapier, J. C., C. Pellat and Y. Henry. 1991. Generation of EPR-Detectable Nitrosyl-Iron Complexes in Tumor Target Cells Cocultured With Activated Macrophages. *J. Biol. Chem. 266:* 10162-10167.

44. Vanin, A. F., G. B. Men'shikov, I. A. Moroz, P. I. Mordvintcev, V. A. Serezhenkov and D. S. Burbaev. 1992. The Source of Non-Heme Iron That Binds Nitric Oxide in Cultivated Macrophages. *Biochim. Biophys. Acta 1135:* 275-279.

45. Hibbs, J. B., Jr., R. R. Taintor, Z. Vavrin, D. L. Granger, J. C. Drapier, I. J. Amber and J. R. Lancaster, Jr. 1990. Synthesis of Nitric Oxide From a Terminal Guanidino Nitrogen Atom of L-Arginine: a Molecular Mechanism Regulating Cellular Proliferation That Targets Intracellular Iron. In *Nitric Oxide From L-Arginine: a Bioregulatory System*; (Moncada, S., Higgs, E. A., Eds.), pp. 189-223, Elsevier, Amsterdam,

46. Eisenstein, R. S. 2000. Iron Regulatory Proteins and the Molecular Control of Mammalian Iron Metabolism. *Annu. Rev. Nutr. 20:* 627-662.

47. Wardrop, S. L., R. N. Watts and D. R. Richardson. 2000. Nitrogen Monoxide Activates Iron Regulatory Protein 1 RNA-Binding Activity by Two Possible Mechanisms: Effect on the [4Fe-4S] Cluster and Iron Mobilization From Cells. *Biochemistry 39:* 2748-2758.

48. Oria, R., L. Sanchez, T. Houston, M. W. Hentze, F. Y. Liew and J. H. Brock. 1995. Effect of Nitric Oxide on Expression of Transferrin Receptor and Ferritin and on Cellular Iron Metabolism in K562 Human Erythroleukemia Cells. *Blood 85:* 2962-2966.

49. Pantopoulos, K., G. Weiss and M. W. Hentze. 1996. Nitric Oxide and Oxidative Stress (H_2O_2) Control Mammalian Iron Metabolism by Different Pathways. *Molec. Cell. Biol. 16:* 3781-3788.

50. Pomposiello, P. J., B. Demple. 2001. Redox-Operated Genetic Switches: the SoxR and OxyR Transcription Factors. *Trends Biotechnol. 19:* 109-114.

51. Cairo, G., R. Ronchi, S. Recalcati, A. Campanella and G. Minotti. 2002. Nitric Oxide and Peroxynitrite Activate the Iron Regulatory Protein-1 of J774A.1 Macrophages by Direct Disassembly of the Fe-S Cluster of Cytoplasmic Aconitase. *Biochemistry 41:* 7435-7442.

52. Oliveira, L., C. Bouton and J. C. Drapier. 1999. Thioredoxin Activation of Iron Regulatory Proteins. Redox Regulation of RNA Binding After Exposure to Nitric Oxide. *J. Biol. Chem. 274:* 516-521.

53. Hibbs, J. B., Jr., R. R. Taintor and Z. Vavrin. 1984. Iron Depletion: Possible Cause of Tumor Cell Cytotoxicity Induced by Activated Macrophages. *Biochem. Biophys. Res. Comm. 123:* 716-723.

54. Clementi, E., G. C. Brown, M. Feelisch and S. Moncada. 1998. Persistent Inhibition of Cell Respiration by Nitric Oxide: Crucial Role of S-Nitrosylation of Mitochondrial Complex I and Protective Action of Glutathione. *Proc. Natl. Acad. Sci. USA 95:* 7631-7636.

55. Yang, W., P. A. Rogers and H. Ding. 2002. Repair of Nitric Oxide-Modified Ferredoxin [2Fe-2S] Cluster by Cysteine Desulfurase (IscS). *J. Biol. Chem. 277:* 12868-12873.

56. Espey, M. G., D. D. Thomas, K. M. Miranda and D. A. Wink. 2002. Focusing of Nitric Oxide Mediated Nitrosation and Oxidative Nitrosylation As a Consequence of Reaction With Superoxide. *Proc. Natl. Acad. Sci. U. S. A 99:* 11127-11132.

57. Wink, D. A., R. W. Nims, J. F. Darbyshire, D. Christodoulou, I. Hanbauer, G. W. Cox, F. Laval, J. Laval, J. A. Cook and M. C. Krishna. 1994. Reaction Kinetics for Nitrosation of Cysteine and Glutathione in Aerobic Nitric Oxide Solutions at Neutral PH. Insights into the Fate and Physiological Effects of Intermediates Generated in the NO/O2 Reaction. *Chem. Res. Toxicol. 7:* 519-525.

58. Goldstein, S., G. Czapski. 1995. Kinetics of Nitric Oxide Autoxidation in Aqueous Solution in the Absence and Presence of Various Reductants. *J. Am. Chem. Soc. 117:* 12078-12084.

59. Ford, P. C., D. A. Wink and D. M. Stanbury. 1993. Autoxidation Kinetics of Aqueous Nitric Oxide. *FEBS Lett. 326:* 1-3.

60. Blough, N. V., O. C. Zafiriou. 1985. Reaction of Superoxide With Nitric Oxide to Form Peroxonitrite in Aqueous Solution. *Inorg. Chem. 24:* 3502-3504.

61. Radi, R., A. Denicola, B. Alvarez, G. Ferrer-Sueta and H. Rubbo. 2000. The Biological Chemistry of Peroxynitrite. In *Nitric Oxide: Biology and Pathobiology*; (Ignarro, L. J., Ed.), pp. 57-81, Academic Press, San Diego

62. Beckman, J. S., T. W. Beckman, J. Chen, P. A. Marshall and B. A. Freeman. 1990. Apparent Hydroxyl Radical Production by Peroxynitrite: Implications for Endothelial Injury From Nitric Oxide and Superoxide. *Proc. Natl. Acad. Sci. USA 87:* 1620-1624.

63. Beckman, J. S., W. H. Koppenol. 1996. Nitric Oxide, Superoxide, and Peroxynitrite: the Good, the Bad, and Ugly. [Review] [109 Refs]. *American Journal of Physiology 271:* C1424-C1437.

64. Radi, R., G. Peluffo, M. N. Alvarez, M. Naviliat and A. Cayota. 2001. Unraveling Peroxynitrite Formation in Biological Systems. *Free Radic. Biol. Med. 30:* 463-488.

65. Czapski, G., S. Goldstein. 1995. The Role of the Reactions of .NO With Superoxide and Oxygen in Biological Systems: a Kinetic Approach. *Free Rad. Biol. Med. 19:* 785-794.

66. Oury, T. D., B. J. Day and J. D. Crapo. 1996. Extracellular Superoxide Dismutase: a Regulator of Nitric Oxide Bioavailability. *Lab Invest 75:* 617-636.

67. Kissner, R., T. Nauser, P. Bugnon, P. G. Lye and W. H. Koppenol. 1997. Formation and Properties of Peroxynitrite As Studied by Laser Flash Photolysis, High-Pressure Stopped-Flow Technique, and Pulse Radiolysis. *Chem. Res. Toxicol. 10:* 1285-1292.

68. Nauser, T., W. H. Koppenol. 2002. The Rate Constant of the Reaction of Superoxide With Nitrogen Monoxide: Approaching the Diffusion Limit. *J. Phys. Chem. A 106:* 4084-4086.

69. Gratzel, M., A. Hanglein, J. Lilie and G. Beck. 1969. Pulsradiolytische Untersuchung Einiger Elementarprozesse Der Oxydation Und Reduktion Des Nitritions. *Ber. Bunsenges. Phys. Chem. 73:* 646-653.

70. Goldstein, S., G. Czapski. 2000. Reactivity of Peroxynitrite Versus Simultaneous Generation of (*)NO and O(2)(*)(-) Toward NADH. *Chem. Res. Toxicol. 13:* 736-741.

71. Rubbo, H., R. Radi, M. Trujillo, R. Telleri, B. Kalyanaraman, S. Barnes, M. Kirk and B. A. Freeman. 1994. Nitric Oxide Regulation of Superoxide and Peroxynitrite- Dependent Lipid Peroxidation. Formation of Novel Nitrogen- Containing Oxidized Lipid Derivatives. *J. Biol. Chem. 269:* 26066-26075.

72. O'Donnell, V. B., B. A. Freeman. 2001. Interactions Between Nitric Oxide and Lipid Oxidation Pathways: Implications for Vascular Disease. *Circ. Res. 88:* 12-21.

73. Coles, B., A. Bloodsworth, S. R. Clark, M. J. Lewis, A. R. Cross, B. A. Freeman and V. B. O'Donnell. 2002. Nitrolinoleate Inhibits Superoxide Generation, Degranulation, and Integrin Expression by Human Neutrophils: Novel Antiinflammatory Properties of Nitric Oxide-Derived Reactive Species in Vascular Cells. *Circ. Res. 91:* 375-381.

74. Balazy, M., T. Iesaki, J. L. Park, H. Jiang, P. M. Kaminski and M. S. Wolin. 2001. Vicinal Nitrohydroxyeicosatrienoic Acids: Vasodilator Lipids Formed by Reaction of Nitrogen Dioxide With Arachidonic Acid. *J. Pharmacol. Exp. Ther. 299:* 611-619.

75. Pryor, W. A., L. Castle and D. F. Church. 1985. Nitrosation of Organic Hydroperoxides by Nitrogen Dioxide/Dinitrogen Tetraoxide. *J. Am. Chem. Soc. 107:* 211-217.

76. O'Donnell, V. B., J. P. Eiserich, P. H. Chumley, M. J. Jablonsky, N. R. Krishna, M. Kirk, S. Barnes, V. M. Darley-Usmar and B. A. Freeman. 1999. Nitration of Unsaturated Fatty Acids by Nitric Oxide-Derived Reactive Nitrogen Species Peroxynitrite, Nitrous Acid, Nitrogen Dioxide, and Nitronium Ion. *Chem. Res. Toxicol. 12:* 83-92.

77. Gunther, M. R., B. E. Sturgeon and R. P. Mason. 2002. Nitric Oxide Trapping of the Tyrosyl Radical-Chemistry and Biochemistry. *Toxicology 177:* 1-9.

78. Merenyi, G., J. Lind, S. Goldstein and G. Czapski. 1998. Peroxynitrous Acid Homolyzes into *OH and *NO2 Radicals. *Chem. Res. Toxicol.* **11:** 712-713.

79. Richeson, C. E., P. Mulder, V. W. Bowry and K. U. Ingold. 1998. The Complex Chemistry of Peroxynitrite Decomposition: New Insights. *J. Am. Chem. Soc.* **120:** 7211-7219.

80. Coddington, J. W., J. K. Hurst and S. V. Lymar. 1999. Hydroxyl Radical Formation During Peroxynitrous Acid Decomposition. *J. Am. Chem. Soc.* **121:** 2443.

81. Kissner, R., W. H. Koppenol. 2002. Product Distribution of Peroxynitrite Decay As a Function of PH, Temperature, and Concentration. *J. Am. Chem. Soc.* **124:** 234-239.

82. Halliwell, B., J. M. Gutteridge. *Free Radicals in Biology and Medicine.* **2001.** Oxford University Press, Oxford

83. Augusto, O., M. G. Bonini, A. M. Amanso, E. Linares, C. C. Santos and S. L. De Menezes. 2002. Nitrogen Dioxide and Carbonate Radical Anion: Two Emerging Radicals in Biology. *Free Radic. Biol. Med.* **32:** 841-859.

84. Halfpenny, E. and P. L. Robinson. 1952. Pernitrous Acid. The Reaction Between Hydrogen Peroxide and Nitrous Acid, and the Properties of an Intermediate Product. *J. Chem. Soc.* 928-938.

85. Halfpenny, E., and P. L. Robinson. 1952. The Nitration and Hydroxylation of Aromatic Compounds by Pernitrous Acid. *J. Chem. Soc.* 939-946.

86. Prutz, W. A., H. Monig, J. Butler and E. J. Land. 1985. Reactions of Nitrogen Dioxide in Aqueous Model Systems: Oxidation of Tyrosine Units in Peptides and Proteins. *Arch. Biochem. Biophys* **243:** 125-134.

87. Souza, J. M., E. Daikhin, M. Yudkoff, C. S. Raman and H. Ischiropoulos. 1999. Factors Determining the Selectivity of Protein Tyrosine Nitration. *Arch. Biochem. Biophys.* **371:** 169-178.

88. Espey, M. G., S. Xavier, D. D. Thomas, K. M. Miranda and D. A. Wink. 2002. Direct Real-Time Evaluation of Nitration With Green Fluorescent Protein in Solution and Within Human Cells Reveals the Impact of Nitrogen Dioxide Vs. Peroxynitrite Mechanisms. *Proc. Natl. Acad. Sci. U. S. A* **99:** 3481-3486.

89. Greenacre, S. A., H. Ischiropoulos. 2001. Tyrosine Nitration: Localisation, Quantification, Consequences for Protein Function and Signal Transduction. *Free Radic. Res.* **34:** 541-581.

90. Ischiropoulos, H., L. Zhu, J. Chen, M. Tsai, J. C. Martin, C. D. Smith and J. S. Beckman. 1992. Peroxynitrite-Mediated Tyrosine Nitration Catalyzed by Superoxide Dismutase. *Arch. Biochem. Biophys.* **298:** 431-437.

91. Balavoine, G. G., Y. V. Geletti and D. Bejan. 1997. Catalysis of Peroxynitrite Reactions by Manganese and Iron Porphyrins. *Nitric. Oxide.* **1:** 507-521.

92. Gow, A., D. Duran, S. R. Thom and H. Ischiropoulos. 1996. Carbon Dioxide Enhancement of Peroxynitrite-Mediated Protein Tyrosine Nitration. *Arch. Biochem. Biophys.* **333:** 42-48.

93. Zhang, H., J. Joseph, M. Gurney, D. Becker and B. Kalyanaraman. 2002. Bicarbonate Enhances Peroxidase Activity of Cu,Zn-Superoxide Dismutase. Role of Carbonate Anion Radical and Scavenging of Carbonate Anion Radical by Metalloporphyrin Antioxidant Enzyme Mimetics. *J. Biol. Chem.* **277:** 1013-1020.

94. Uppu, R. M., G. L. Squadrito and W. A. Pryor. 1996. Acceleration of Peroxynitrite Oxidations by Carbon Dioxide. *Arch. Biochem. Biophys.* **327:** 335-343.

95. Eiserich, J. P., M. Hristova, C. E. Cross, A. D. Jones, B. A. Freeman, B. Halliwell and A. Van der Vliet. 1998. Formation of Nitric Oxide-Derived Inflammatory Oxidants by Myeloperoxidase in Neutrophils. *Nature* **391:** 393-397.

96. Brennan, M. L., W. Wu, X. Fu, Z. Shen, W. Song, H. Frost, C. Vadseth, L. Narine, E. Lenkiewicz, M. T. Borchers, A. J. Lusis, J. J. Lee, N. A. Lee, H. M. Abu-Soud, H. Ischiropoulos and S. L. Hazen. 2002. A Tale of Two Controversies: Defining Both the Role of Peroxidases in Nitrotyrosine Formation in Vivo Using Eosinophil Peroxidase and Myeloperoxidase-Deficient Mice, and the Nature of Peroxidase-Generated Reactive Nitrogen Species. *J. Biol. Chem.* **277:** 17415-17427.

97. Thomas, D. D., M. G. Espey, M. P. Vitek, K. M. Miranda and D. A. Wink. 2002. Protein Nitration Is Mediated by Heme and Free Metals Through Fenton-Type Chemistry: An Alternative to the NO/O2- Reaction. *Proc. Natl. Acad. Sci. U. S. A .*

98. Miles, A. M., D. S. Bohle, P. A. Glassbrenner, B. Hansert, D. A. Wink and M. B. Grisham. 1996. Modulation of Superoxide-Dependent Oxidation and Hydroxylation Reactions by Nitric Oxide. *J. Biol. Chem.* **271:** 40-47.

99. Jourd'heuil, D., F. L. Jourd'heuil, P. S. Kutchukian, R. A. Musah, D. A. Wink and M. B. Grisham. 2001. Reaction of Superoxide and Nitric Oxide With Peroxynitrite. Implications for Peroxynitrite-Mediated Oxidation Reactions in Vivo. *J. Biol. Chem.* **276:** 28799-28805.

100. Goldstein, S., G. Czapski, J. Lind and G. Merenyi. 2000. Tyrosine Nitration by Simultaneous Generation of (.)NO and O-(2) Under Physiological Conditions. How the Radicals Do the Job. *J. Biol. Chem.* **275:** 3031-3036.

101. Logager, T., K. Sehested. 1993. Formation and Decay of Peroxynitric Acid. *J. Phys. Chem.* **97:** 10047-10052.

102. Goldstein, S., G. Czapski, J. Lind and G. Merenyi. 1998. Mechanism of Decomposition of Peroxynitric Ion (O(2)NOO(-)): Evidence for the Formation of O(2)(*-) and (*)NO(2) Radicals. *Inorg. Chem.* **37:** 3943-3947.

103. Vermeer, I. T., and J. M. van Maanen. 2001. Nitrate Exposure and the Endogenous Formation of Carcinogenic Nitrosamines in Humans. *Rev. Environ. Health* **16:** 105-116.

104. Ohshima, H., and H. Bartsch. 1994. Chronic Infections and Inflammatory Processes As Cancer Risk Factors: Possible Role of Nitric Oxide in Carcinogenesis. *Mutat. Res.* **305:** 253-264.

105. Burney, S., J. L. Caulfield, J. C. Niles, J. S. Wishnok and S. R. Tannenbaum. 1999. The Chemistry of DNA Damage From Nitric Oxide and Peroxynitrite. *Mutat. Res.* **424:** 37-49.

106. Tamir, S., and S. R. Tannenbaum. 1996. The Role of Nitric Oxide (NO.) in the Carcinogenic Process. *Biochim. Biophys. Acta* **1288:** F31-F36.

107. Williams, D. L. H. *Nitrosation.* 1988. Cambridge University Press, Cambridge

108. Wink, D. A., J. F. Darbyshire, R. W. Nims, J. E. Saavedra and P. C. Ford. 1993. Reactions of the Bioregulatory Agent Nitric Oxide in Oxygenated Aqueous Media: Determination of the Kinetics for Oxidation and Nitrosation by Intermediates Generated in the NO/O2 Reaction. *Chem. Res. Toxicol.* **6:** 23-27.

109. Kharitonov, V. G., A. R. Sundquist and V. S. Sharma. 1994. Kinetics of Nitric Oxide Autoxidation in Aqueous Solution. *J. Biol. Chem.* **269:** 5881-5883.

110. Goldstein, S., G. Czapski. 1996. Mechanism of the Nitrosation of Thiols and Amines by Oxygenated NO Solutions: The Nature of the Nitrosating Intermediates. *J. Am. Chem. Soc.* **118:** 3425.

111. Espey, M. G., K. M. Miranda, D. D. Thomas and D. A. Wink. 2001. Distinction Between Nitrosating Mechanisms Within Human Cells and Aqueous Solution. *J. Biol. Chem.* **276:** 30085-30091.

112. Liu, X., M. S. Miller, M. S. Joshi, D. D. Thomas and J. R. Lancaster, Jr. 1998. Accelerated Reaction of Nitric Oxide With O_2 Within the Hydrophobic Interior of Biological Membranes. *Proc. Natl. Acad. Sci. USA* **95:** 2175-2179.

113. Lijinsky, W., L. Keefer, E. Conrad and B. R. Van de. 1972. Nitrosation of Tertiary Amines and Some Biologic Implications. *J. Natl. Cancer Inst.* **49:** 1239-1249.

114. Pryor, W. A., D. F. Church, C. K. Govindan and G. Crank. 1982. Oxidation of Thiols by Nitric Oxide and Nitrogen Dioxide: Synthetic Utility and Toxicological Implications. *J. Org. Chem.* **47:** 156-159.

115. Akaike, T. 2000. Mechanisms of Biological S-Nitrosation and Its Measurement. *Free Radic. Res.* **33:** 461-469.

116. Jaffrey, S. R., H. Erdjument-Bromage, C. D. Ferris, P. Tempst and S. H. Snyder. 2001. Protein S-Nitrosylation: a Physiological Signal for Neuronal Nitric Oxide. *Nat. Cell Biol.* **3:** 193-197.

117. Hess, D. T., A. Matsumoto, R. Nudelman and J. S. Stamler. 2001. S-Nitrosylation: Spectrum and Specificity. *Nat. Cell Biol.* **3:** E46-E49.

118. Gow, A. J., Q. Chen, D. T. Hess, B. J. Day, H. Ischiropoulos and J. S. Stamler. 2002. Basal and Stimulated Protein S-Nitrosylation in Multiple Cell Types and Tissues. *J. Biol. Chem.* **277:** 9637-9640.

119. Stamler, J. S., S. Lamas and F. C. Fang. 2001. Nitrosylation. the Prototypic Redox-Based Signaling Mechanism. *Cell* **106:** 675-683.

120. Williams, D. L. H. 1999. The Chemistry of S-Nitrosothiols. *Acc. Chem. Res.* **32:** 869-876.

121. Gow, A. J., B. P. Luchsinger, J. R. Pawloski, D. J. Singel and J. S. Stamler. 1999. The Oxyhemoglobin Reaction of Nitric Oxide. *Proc. Natl. Acad. Sci. USA* **96:** 9027-9032.

122. Riggs, A. F. 1961. The Binding of N-Ethylmaleimide by Human Hemoglobin and Its Effect Upon the Oxygen Equilibrium. *J. Biol. Chem.* **236:** 1948-1954.

123. Gross, S. S. 2001. Vascular Biology. Targeted Delivery of Nitric Oxide. *Nature* **409:** 577-578.

124. Huang, Z., J. G. Louderback, M. Goyal, F. Azizi, S. B. King and D. B. Kim-Shapiro. 2001. Nitric Oxide Binding to Oxygenated Hemoglobin Under Physiological Conditions. *Biochim. Biophys. Acta* **1568:** 252-260.

125. Zhang, Y., N. Hogg. 2002. Mixing Artifacts From the Bolus Addition of Nitric Oxide to Oxymyoglobin: Implications for S-Nitrosothiol Formation. *Free Radic. Biol. Med.* **32:** 1212-1219.

126. Deem, S., M. T. Gladwin, J. T. Berg, M. E. Kerr and E. R. Swenson. 2001. Effects of S-Nitrosation of Hemoglobin on Hypoxic Pulmonary Vasoconstriction and Nitric Oxide Flux. *Am. J. Respir. Crit Care Med.* **163:** 1164-1170.

127. Hobbs, A., M. Gladwin, R. Patel, D. Williams and A. Butler. 2002. Haemoglobin: NO Transporter, NO Inactivator or NOne of the Above? *Trends Pharmacol. Sci.* **23:** 406.

128. Joshi, M. S., T. B. Ferguson, Jr., T. H. Han, D. R. Hyduke, J. C. Liao, T. Rassaf, N. Bryan, M. Feelisch and J. R. Lancaster, Jr. 2002. Nitric Oxide Is Consumed, Rather Than Conserved, by Reaction With Oxyhemoglobin Under Physiological Conditions. *Proc. Natl. Acad. Sci. U. S. A* **99:** 10341-10346.

129. Hughes, M. N. 1999. Relationships Between Nitric Oxide, Nitroxyl Ion, Nitrosonium Cation and Peroxynitrite. *Biochim. Biophys. Acta* **1411:** 263-272.

130. Murphy, M. E. and H. Sies. 1991. Reversible Conversion of Nitroxyl Anion to Nitric Oxide by Superoxide Dismutase. *Proc. Natl. Acad. Sci. USA* **88:** 10860-10864.

131. Hobbs, A. J., J. M. Fukuto and L. J. Ignarro. 1994. Formation of Free Nitric Oxide From L-Arginine by Nitric Oxide Synthase: Direct Enhancement of Generation by Superoxide Dismutase. *Proc. Natl. Acad. Sci. USA* **91:** 10992-10996.

132. Cadenas, E., J. J. Poderoso, F. Antunes and A. Boveris. 2000. Analysis of the Pathways of Nitric Oxide Utilization in Mitochondria. *Free Radic. Res.* **33:** 747-756.

133. Arnelle, D. R. and J. S. Stamler. 1995. NO+, NO, and NO- Donation by S-Nitrosothiols: Implications for Regulation of Physiological Functions by S-Nitrosylation and Acceleration of Disulfide Formation. *Arch. Biochem. Biophys.* **318:** 279-285.

134. Espey, M., K. Miranda, D. Thomas and D. Wink. 2002. Ingress and Reactive Chemistry of Nitroxyl-Derived Species Within Human Cells. *Free Radic. Biol. Med.* **33:** 827.

135. Bartberger, M. D., W. Liu, E. Ford, K. M. Miranda, C. Switzer, J. M. Fukuto, P. J. Farmer, D. A. Wink and K. N. Houk. 2002. The Reduction Potential of Nitric Oxide (NO) and Its Importance to NO Biochemistry. *Proc. Natl. Acad. Sci. U. S. A* **99:** 10958-10963.

136. Lancaster, J. R., Jr. 1997. A Tutorial on the Diffusibility and Reactivity of Free Nitric Oxide. *Nitric Oxide 1:* 18-30.

137. Meulemans, A. 1994. Diffusion Coefficients and Half-Lives of Nitric Oxide and N-Nitroso-L-Arginine in Rat Cortex. *Neurosci. Lett. 171:* 89-93.

138. Stadler, J., H. A. Bergonia, M. Di Silvio, M. A. Sweetland, T. R. Billiar, R. L. Simmons and J. R. Lancaster, Jr. 1993. Nonheme Iron-Nitrosyl Complex Formation in Rat Hepatocytes: Detection by Electron Paramagnetic Resonance Spectroscopy. *Arch. Biochem. Biophys. 302:* 4-11.

139. Lancaster, J. R., Jr. 1996. Diffusion of Free Nitric Oxide. *Meth. Enz. 268:* 31-50.

140. Lancaster, J. R., Jr. 1994. Simulation of the Diffusion and Reaction of Endogenously Produced Nitric Oxide. *Proc. Natl. Acad. Sci. USA 91:* 8137-8141.

Chapter 5

BACTERIAL PEROXIREDOXINS

Leslie B Poole

5.1   Introduction

Defenses against oxidative damage include peroxide detoxifying enzymes as a critical component. Within this realm, heme-containing peroxidases and catalases have long been recognized to catalyze hydroperoxide metabolism utilizing the bound metal within the heme as the redox center.[1,2] The non-heme glutathione peroxidases, which catalyze the rapid reduction of hydrogen peroxide and organic hydroperoxides, have a covalently-incorporated selenium at the active site as part of a redox-active selenocysteine residue.[3,4] Only one exception to the requirement for metals in peroxidase active sites was known prior to 1985. This enzyme, a flavin-containing, cysteine-based peroxidase known as NADH peroxidase, is present in a subset of lactic acid bacteria and apparently evolved as an alternative to heme-based peroxide detoxification in these heme-deficient bacteria.[5,6]

In studies of bacterial defense responses evoked on treatment with hydrogen peroxide and cumene hydroperoxide, Bruce Ames's group identified mutant bacteria (*oxyR1* and *oxyR2* from *Salmonella typhimurium* and *Escherichia coli*, respectively) with constitutively upregulated expression of 9 genes apparently involved in protection against peroxide stress.[7] This finding led to the identification of two of the proteins, AhpF and AhpC, as a novel NAD(P)H-dependent peroxide reductase system with activity toward such bulky alkyl and aromatic hydroperoxides as *t*-butyl hydroperoxide and cumene hydroperoxide.[7-10] At around the same time, an antioxidant "protector" protein from yeast, identified in Earl Stadtman's laboratory,

H. J. Forman, J. Fukuto, and M. Torres (eds.), Signal Transduction by Reactive Oxygen and Nitrogen Species: Pathways and Chemical Principles, ©2003, Kluwer Academic Publishers. Printed in the Netherlands

was shown to protect glutamine synthetase from oxidative inactivation by a mixed-function oxidation system.[11] As sequence information for AhpC, the "protector" protein (subsequently designated thiol-specific antioxidant, or Tsa) and a number of other homologues of diverse or unknown functions became available, the recognition of these proteins as a family of cysteine-based peroxidases expressed in a wide range of organisms led to their designation as "peroxiredoxins" or Prxs.[12,13] The identification of Prx family members has greatly expanded over the last decade and includes not only representatives from every branch of the phylogenetic tree, but also multiple homologues (Prxs I through VI) in mammals with apparently diverse roles not only in oxidant stress protection, but also in differentiation, apoptosis and proliferation.[14,15] The recent linkage of these ubiquitous Prxs to the control of hydrogen peroxide levels, which directly mediate signaling pathways in higher organisms, has dramatically increased the attention given to this enzyme family among scientists in many fields.[15,16] This chapter focuses on the bacterial peroxiredoxins related to AhpC, and emphasizes the chemical, physical and biological properties, which underlie their important protective functions in these organisms. The chemical and structural attributes of bacterial Prxs generally extend to eukaryotic family members as well, although the multiple Prxs of mammals also appear to have evolved more specialized biological roles.[14,15,17]

5.2 Bacterial Representatives of the Peroxiredoxin Family

Essentially all eubacteria appear to express at least one, and in some cases two (or more), Prx family member(s) with greater than 35% amino acid sequence identity when compared with AhpC originally identified from *E. coli* and *S. typhimurium*. As discussed below, these proteins possess two conserved cysteinyl residues near their N- and C-termini, respectively (Cys46 and Cys165 in the *S. typhimurium* protein), which participate in catalysis; the surrounding sequences are also highly conserved. These two cysteine residues are linked in an unusual intersubunit, redox-active disulfide bond in the oxidized form of the active site. More distantly related are the so-called 1-Cys Prxs, with conservation of only the first active-site cysteine. Little information regarding the bacterial 1-Cys Prxs is available, although one member from *E. coli*, designated "BCP" for bacterioferritin co-migratory protein, was shown to have weak peroxidase activity in the presence of thioredoxin.[18] The "thiol peroxidases" (also called Tpx, p20 or scavengase) were originally characterized from *E. coli* as periplasmic enzymes distinct from AhpC-like peroxidases and utilizing thioredoxin as reductant.[19] Later work established the distant homology between these two groups of enzymes and supported a cysteine-based mechanism for peroxide reduction.[20,21] While most bacterial Prxs fall into one of the three divergent categories of *E. coli* Prxs (AhpC, BCP or Tpx), the two additional major clades of Prxs, represented by the "atypical" 2-Cys peroxiredoxin[17] PrxV (with an intrasubunit disulfide bond in the oxidized enzyme) and the 1-Cys peroxiredoxin PrxVI from mammals, also have as yet uncharacterized bacterial representatives (see Figs. 1 and 2 of Hofmann et al, 2002, for a review).[14]

While the majority of bacterial Prxs have been identified recently through searching of the genomic sequence information as it becomes available, functional studies confirming activities for at least some members have been conducted. The number of bacterial AhpC relatives characterized biochemically is relatively small (e.g. *Salmonella typhimurium* and *Escherichia coli*,[8,22] *Streptococcus mutans*,[23,24] *Amphibacillus xylanus*,[25-27] *Thermus aquaticus*,[28,29] *Helicobacter pylori*,[30] *Clostridium pasteurianum*,[31] *Chromatium gracile*,[32] and *Mycobacterium tuberculosis*[33-37]). Considerable information concerning the biological importance of Prxs has also been gained by genetic approaches, including disruption of the chromosomally-encoded structural genes (*Escherichia coli* and *Salmonella typhimurium*,[9,38-41] *Bacillus subtilis*,[42-44] *Helicobacter pylori*,[45,46] *Staphylococcus aureus*,[47] *Xanthomonas campestris*,[48-50] *Streptococcus mutans*,[23] *Streptococcus pyogenes*[51], *Bacteriodes fragilis*,[52,53] and *Mycobacterium* [including *M. tuberculosis*, *M. leprae*, *M. bovis*, *M. smegmatis*, *M. aurum* and *M. bovis*][54-60]). A recent survey of NADH-dependent *t*-butyl hydroperoxide reductase activity in a

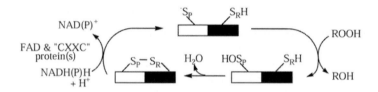

Figure 1. Catalytic cycle of 2-Cys peroxiredoxins. Scheme depicts catalysis at one of the two active sites of a dimer. S_P indicates the peroxidatic cysteine and S_R the resolving cysteine (Cys46 and Cys165, respectively, of S. typhimurium AhpC).

number of different bacterial strains was also conducted, leading to the conclusion that, while different levels of the two (or more) proteins involved in the peroxidase system were expressed by the various organisms, all possessed the ability to catalyze the reaction.[61] In one case, that of *Eubacterium acidaminophilum*,[62] the Prx has been reported to utilize a selenocysteine rather than a cysteine as the peroxidatic center, reminiscent of the glutathione peroxidases of higher organisms.

5.3 Catalytic Properties

By far the most prevalent, abundant and active Prxs in bacteria are the close relatives of AhpC (the "typical" 2-Cys Prxs). As mentioned above, these enzymes possess no metal or prosthetic group, but rely on a single peroxidatic cysteinyl residue for hydroperoxide reduction and a second cysteine for progression of the catalytic cycle to intersubunit disulfide bond formation (Fig. 1). For attack on the – O-O– bond of the hydroperoxide (ROOH, including peroxynitrite[34]), the strictly conserved peroxidatic cysteine (Cys46 in *S. typhimurium* and *E. coli* AhpC) is activated through electrostatic interaction with a conserved arginine residue (Arg119) and hydrogen bonding to a conserved threonine (Thr43, or serine in some Prxs),[17,63] lowering its pKa to a value which may be below 3.[34,64] Following attack by the thiolate anion, the first product, RO⁻, is likely protonated and released, and the catalytic cysteine is oxidized to a cysteine sulfenic acid (Cys-SOH) (Fig. 1).[64-66] This first step is common to all Prxs, and follows the same chemistry established previously for the enterococcal NADH peroxidase.[6,67] The analogous reaction, wherein the active site selenocysteine is oxidized to the selenenic acid (Cys-SeOH), represents the recognized pathway for hydroperoxide reduction by glutathione peroxidases, as well.[4]

Because of the high reactivity of cysteine sulfenic acid, chemical identification of this species within proteins has been quite challenging. Sulfenic acids are well known to react with any accessible thiols to form disulfide bonds, and are generally quite sensitive to further irreversible oxidation (to sulfinic, $-SO_2H$, and sulfonic, $-SO_3H$, acids) by such mild oxidants as peroxides and molecular oxygen.[68,69] Their presence can be inferred by their reactivity toward nucleophilic reagents such as cyanide or toward the sulfenic acid-specific reagent dimedone.[70] For quantitation, reactivity toward the chromophoric thiol-containing reagent 2-nitro-5-thiobenzoate (TNB), as well as subsequent release of this chromophore on dithiothreitol treatment of the isolated, labeled protein, can be analyzed.[64,66] Electrospray ionization mass spectrometry under "gentle" conditions can also be undertaken,[71] although overoxidation beyond the sulfenic acid state during the analysis is a common problem.[66] Fortunately, an electrophilic chemical modification agent, 7-chloro-4-nitrobenzo-2-oxa-1,3-diazole (NBD chloride), was identified which reacts with both thiols and sulfenic acids and preserves the oxygen in the product with the latter species.[64,65] Subsequent analyses of the isolated, labelled proteins then allow for identification of the respective thioether or sulfoxide adducts through UV-visible spectroscopic properties (λ_{max} at 420 or 347 nm, respectively) or mass spectroscopy (with the sulfenic acid-generated product having a mass 16 amu larger than that of the thiol adduct). To establish the formation of a sulfenic acid intermediate on reaction of the Cys46 thiolate with hydrogen peroxide or organic hydroperoxides, a mutant of *S. typhimurium* AhpC was studied wherein the second cysteine, termed the "resolving" cysteine (Cys165), was mutated to a serine residue (C165S). This mutation prevents progression to the disulfide bond-containing species and makes the enzyme much more susceptible to oxygen- and peroxide-mediated inactivation, but does not adversely affect catalytic turnover with hydroperoxide substrates in the presence of excess reductant.[66] Thus, a single equivalent of hydroperoxide was shown to rapidly and quantitatively convert the Cys46 thiolate to the sulfenic acid, a species that could be trapped with NBD chloride as described above or converted to a mixed disulfide bond on addition of TNB.[65,66]

Following generation of the sulfenic acid intermediate and release of the alcohol (or first water) product, Cys46-SOH of wild type AhpC reacts with Cys165-S⁻ in the other subunit to form the intersubunit disulfide bond of the oxidized enzyme, releasing a water molecule (Fig. 1). If disulfide bond formation were not rapid, the persistence of the sulfenic acid intermediate would provide more opportunity for inactivation of the enzyme through overoxidation of the peroxidatic cysteinyl center by oxygen or hydroperoxide substrate. While this appears to be only a rare occurrence for bacterial AhpC, the 2-Cys PrxII of higher organisms is clearly susceptible to such overoxidation as shown in the crystal structure obtained for the human enzyme, where the peroxidatic cysteine is present in the sulfinic acid (– SO_2H) form,[72] and by proteomics approaches detecting cysteine oxidation *in vivo*.[73] Formation of the disulfide bond clearly avoids such facile inactivation and stabilizes the protein as it awaits reduction and reactivation. In the final step of the catalytic cycle, an electron donor restores the intersubunit disulfide bond to the dithiol state, and the enzyme is ready for reaction with another molecule of hydroperoxide.

5.4 Electron Donors To Bacterial Prxs

As noted above, catalytic recycling of the 2-Cys Prxs requires a reducing substrate, which is nearly always a protein module containing a CXXC motif encoding the redox-active dithiol. The nucleophilic cysteine of this redox center attacks the intersubunit disulfide bond of the Prx and transfers two electrons for reestablishment of the reactive peroxidatic thiolate. In those bacterial systems where the AhpC homologues are greater than 55% identical in amino acid sequence with the *E. coli* and *S. typhimurium* proteins, there is invariably a specialized flavoprotein disulfide reductase encoded just downstream of the *ahpC* structural gene (~15 intervening base pairs for the gram positive sequences, and ~250 intervening base pairs for those from gram negative organisms).[12,74] This protein, designated AhpF or PrxR (for peroxiredoxin reductase), is homologous in its C-terminus to thioredoxin reductase (~35% identity) and possesses the tightly bound FAD within its binding domain and the redox-active disulfide center in the pyridine nucleotide binding domain that are characteristic of this group of flavoproteins. Interestingly, the additional ~200 amino acids at the N-terminus of this protein were suggested by prediction,[75] then proven by X-ray crystallography,[76] to consist of two thioredoxin (Trx)-like folds intimately associated to comprise a single domain. Only the second of the two Trx folds retains a CXXC motif characteristic of Trx (Cys129 and Cys132 in *S. typhimurium* AhpF numbering), and the active site acidic amino acid, rather than being an aspartate encoded a short distance upstream of the two active site cysteines, is a glutamate (Glu86) in a similar but mirrored position relative to the sulfurs and contributed by the first Trx fold.[76] That this domain is the direct reductant of AhpC was proven through several approaches, including separate expression of each fragment of the protein[75] and the appendage of this domain to *E. coli* Trx reductase to create a fusion protein now capable of AhpC reduction.[77]

For most 2-Cys Prxs of higher organisms, reductive recycling of the intersubunit disulfide bond reportedly occurs through interaction with reduced Trx. Most bacterial Prxs, which are more distant relatives of AhpC (less than 50% identity) appear to rely on Trx, as well, as their reducing substrate. Indeed, Trx is a well-known general reductant of protein disulfide bonds and seems to be capable of reducing nearly all bacterial Prxs, although those systems with dedicated AhpF/PrxR proteins are more efficiently recycled by the latter proteins. One consequence of Prx turnover with Trx rather than a PrxR is the resulting dependence on NADPH rather than NADH for their supply of electrons, as PrxR proteins are quite selective, if not completely specific for NADH.[74]

While PrxR and Trx proteins represent the reductants for the majority of Prxs, several cases of convergent evolution have given rise to new, specialized proteins for the recycling of these peroxidases. In two cases recently studied, glutaredoxin (Grx) homologues were the source of electrons for peroxidase activity. A three-protein system encoded in an operon upstream of rubredoxin in *Clostridium pasteurianum* includes an NADH-dependent Trx reductase homologue (Cp34) and a 75 amino acid Grx homologue (Cp9) in addition to the Prx (Cp20).[31] While *E. coli* Trx is capable of reducing Cp20, no activity of Cp34 with Trx is detectable. Cp9 is an efficient acceptor of electrons from Cp34 and in turn reduces Cp20, although the ability of Cp9 to reduce a more generic substrate such as insulin was also demonstrated. In a very unusual case of fusion of the electron donor to the Prx, the *Chromatium gracile* system includes a flavoprotein disulfide reductase (glutathione amide reductase), glutathione amide as mediator, and a Prx-Grx chimeric protein for catalysis of NADH-dependent peroxide reduction.[32] Finally, the electron donor to *Mycobacterium tuberculosis* AhpC (and those of other mycobacterial species) was unclear until recently as these organisms lack any homologue of PrxRs, yet neither Trx nor mycothiol, the small molecule sulfhydryl compound that replaces glutathione in these organisms, could act as reductants of the peroxidase. In a very recent paper, Carl Nathan's group reported not only the unique trimeric structure of a CXXC-containing protein, AhpD (as did a second group around the same time[78]), but also its participation, along with a lipoamide dehydrogenase (LpdC) and a lipoic acid- containing "E2" protein (SucB), in electron transfer to AhpC.[36] Thus, even with the diversity that has evolved in given organisms for the specific reductase systems utilized by Prxs, it is generally true that a pyridine nucleotide-dependent flavoprotein disulfide reductase and a Trx-like (or sometimes Grx-like) protein or module are required to sustain turnover of Prxs with their hydroperoxide substrates.

Figure 2. $(\alpha_2)_5$ decamer of S. typhimurium AhpC. Shown to the left is the X-ray crystal structure of decameric AhpC solved at 2.5 Å resolution (pdb code 1kyg). Shown to the right is the homodimer, with monomers in light and dark gray. The C-terminal arm containing Cys165' is seen reaching across the interface to form the disulfide bond with Cys46. Sulfurs are shown as large balls in both structures. These figures, converted to black and white, are

reproduced with permission from Wood, et al., Biochemistry **41**:5493-4403, copyright 2002, American Chemical Society.[63]

5.5 Redox-Sensitive Oligomerization And High-Resolution Structures

With the recent availability of six X-ray crystal structures of Prxs representing four typical 2-Cys Prxs[72,79,80] (including one bacterial representative, AhpC from *S. typhimurium*[63]), one atypical 2-Cys Prx (human PrxV which forms an intrasubunit disulfide bond[81]) and one 1-Cys Prx (human PrxVI[82]), several observations regarding these structures are striking.[17] Except for PrxV, a monomer, these enzymes form antiparallel homodimers and, in at least some (but probably all) of the 2-Cys Prxs, toroid-shaped $(\alpha_2)_5$ decamers with the active-site cysteines oriented around the outside of the ring (Fig. 2). Secondly, while the cysteinyl sulfurs of the oxidized Prxs must be within ~2 Å of one another to form the disulfide bond at the active site (as has been confirmed by the heme binding protein 23 [PrxI] structure[79]), the sulfurs of the reduced enzymes are ~13 Å apart, indicating that significant conformational rearrangements must take place during catalysis.

That these Prx proteins can form large aggregates has been recognized for some time, although the precise subunit composition and driving force for oligomerization was not clear. In 1999, Kitano *et al.*[26] reported that oligomerization of the *A. xylanus* AhpC to yield a decamer with 52 symmetry was promoted by high ionic strength and was accompanied by an increase in NADH-linked peroxidase activity (ionic strength effects on the flavoprotein reductase could not, however, be ruled out as causative of these activity changes).[25] AhpC from *Mycobacterium tuberculosis* was decameric as suggested by gel filtration studies and dissociated into dimers upon addition of high salt, conditions that also led to a decrease in activity.[35] Analytical ultracentrifugation studies of *S. typhimurium* AhpC were particularly informative; the reduced enzyme was decameric at all concentrations tested, while the oxidized enzyme was a mixture of species at most concentrations, but purely dimeric at 2.5 µM and purely decameric at 480 µM.[63] In both analytical ultracentrifugation and light scattering studies, addition of NaCl up to 1 M had little or no effect on the oligomerization state of either redox form of AhpC. Limited analytical ultracentrifugation studies were also carried out on more distantly related 2-Cys Prxs from *C. pasteurianum* and *H. pylori* which use small redox proteins (Grx-like Cp9 and Trx, respectively) rather than AhpF homologues for reductive recycling (see previous section). At concentrations from 6 to 50 µM, oxidized Cp20 was fully dimeric while the reduced protein was a mixture including larger species.[31] At similar concentrations, *H. pylori* AhpC was decameric in both redox forms, suggesting that either the enzyme does not break down into dimers on oxidation, or this effect requires further dilution of the enzyme.[30]

Like glutathione peroxidases, the structures of Prxs are built upon a Trx-like scaffold with a few secondary structural elements added as insertions.[17] Monomers are composed of a seven β-stranded core, which, in the α_2 dimers, is extended into a 14 strand β-sheet. Dimerization is supported, as well, by C-terminal tail swapping, and the second active site cysteine (Cys165 in *S. typhimurium* AhpC) resides within this tail. Residues beyond Cys165 (through Ile187) are "invisible" (disordered) in the AhpC structure as they are in most of the other Prx structures. It is likely that this mobile region accounts for the ~17 amide protons observed previously by high field NMR using ^{15}N-labelled protein and a ^{1}H-^{15}N heteronuclear multiple quantum correlation experiment (Poole, unpublished; see Ref. [83] for method).

A great deal of information can be gleaned by comparisons of three Prx structures,[17] particularly the oxidized dimer of PrxI,[79] the oxidized decamer of AhpC,[63] and the reduced-like decamer of PrxII[72] (overoxidized at the peroxidatic cysteine, but otherwise a faithful representation of the reduced structure of tryparedoxin peroxidase,[80] yet at much higher resolution). Based on these structures, the formation of the disulfide bond following sulfenic acid generation at the peroxidatic cysteine holds the sulfur of Cys46 out of the pocket formed by the residues of the loop preceding Cys46 (xDFTFVCPTE, with residues of the pocket, also known as region I, indicated by the underline). Disulfide bond formation in fact requires unraveling of about four residues of the helix containing the cysteine residue (VCPT) so that Cys46 can swing around and move closer toward the C-terminal tail of the other subunit of the homodimer containing Cys165. Some rearrangement in the C-terminus also reorients the Cys165 thiol for disulfide bond formation with Cys46. Destabilization of the region I loop vacated by the Cys46 sulfur also weakens its interaction with a second sequence (region II) in the adjacent homodimer, as observed directly in the AhpC structure. This explains the destabilization of the decamer on oxidation as observed by analytical ultracentrifugation studies of AhpC.[17,63] In the dimeric, oxidized Prx, the loop collapses, resulting in several new interactions among amino acids within this loop, and leaving no space for the Cys46 sulfur atom. Once the disulfide bond is reduced, the liberated thiol group of Cys46 swings back around with reformation of this segment of helix, reestablishes the opened-up form of the loop of region I, and again helps support strong interactions between regions I and II on adjacent homodimers (the dimer-dimer interface of the decamer). Disulfide bond formation therefore acts as a "molecular switch" in reorganizing the structure around Cys46 and promoting decamer dissociation. It is not clear why changes in the dimer-decamer equilibrium during different steps of catalysis would be advantageous to these enzymes, but one possibility could be that the diffusion of the smaller dimers away from the higher order complex following oxidation could be helpful in "finding" their reducing redox partners for reactivation. The need for the association of subunits into decamers is also not clear, although the arrangement of the active site surrounding the thiolate may depend on the region I-region II interaction between dimers. One might imagine a potential advantage to the cell, as well, in having ten active sites in close proximity if oxidative damage and formation of hydroperoxides occurs in localized regions. Still, the interconnections between redox-sensitive decamerization, enzymatic activity and biological function are far from clear in the peroxidase systems, an area ripe for future research.

As described above, the location of the catalytic cysteinyl sulfur groups changes during turnover of AhpC. In the reduced enzyme, the thiol(ate) of Cys46 is located within a narrow, solvent accessible pocket and able to react with even bulky hydroperoxide substrates. The sulfur of Cys165, on the other hand, is protected from solvent in the reduced enzyme. On oxidation, the sulfur of the peroxidatic cysteine is buried by the sulfenic acid oxygen, helping to prevent further oxidation of this species. Local unfolding of the active site then exposes the sulfenic acid for disulfide bond formation with Cys165, and it is likely that this unfolding event is highly coordinated with Cys165 thiol movement, again to avoid overoxidation (and inactivation) of the active site cysteine.[17] In the disulfide-bonded (oxidized) enzyme, the Cys165 sulfur is partially solvent accessible while the sulfur of Cys46 is not. This arrangement matches several lines of evidence supporting Cys165 as the "interchange" thiol during reduction of AhpC. That is, when the nucleophilic thiolate of the electron donor, AhpF, attacks the disulfide bond, the two proteins are transiently linked through a covalent disulfide bond, which then migrates, through another thiol-disulfide interchange reaction, to AhpF. If single cysteine mutants of AhpF and AhpC are tested for covalent complex formation (following pre-oxidation of one of the two proteins with DTNB), only the respective single mutants with free Cys165 (C46S AhpC) and Cys129 (C132S AhpF) rapidly and specifically form such a complex.[84] Furthermore, the thiol groups of single mutants of AhpC can be covalently linked, again through a disulfide bond, to a fluorescein derivative and subsequently serve as substrates for reduction by AhpF and its isolated, pre-reduced N-terminal domain.[75] In this case, only the fluorescein linked to Cys165 (in C46S) is released efficiently by the electron donor, while the corresponding fluorescein linked to Cys46 (in C165S) is only slowly released.[75,84] These experiments clearly point to Cys165 as the site of direct interaction with AhpF to bring the reducing equivalents into AhpC.

5.6 Reversible Switch in Ahpc Function Induced By Disulfide Stress: Ahpc*

In a remarkable discovery by Jon Beckwith and coworkers, AhpC in *E. coli* was found to interconvert between its relatively well-characterized peroxidase (wild type) form and a new, mutated species (designated AhpC*) without peroxidase activity, but with an apparent new biological function in supplying electrons to the glutathione/glutaredoxin redox system.[85] This notably high rate of spontaneous mutagenesis (0.5×10^{-3}) led to the expansion of a triplet repeat sequence 25 bp upstream of the codon for Cys46, and resulted from engineered deficiencies in both glutathione reductase and Trx reductase activities in the mutant bacteria which caused increased "disulfide stress". The effect of the resultant amino acid insertion, giving four rather than three phenylalanine residues in a row just prior to the region I loop, is to abrogate peroxidase activity at Cys46, yet support a disulfide reductase activity at Cys165, which can be demonstrated by catalytic reduction of 5,5'-dithiobis(2-nitrobenzoic acid) (DTNB).[85] This DTNB reductase activity was also observed for the wild type AhpC protein, indicating that the difference between the AhpC- and AhpC*-supported disulfide reductase activity *in vivo* is probably in their differing substrate specificities, although the identity of the physiological electron acceptor for AhpC* is as yet unclear. Evidence to date has shown that the ability of an AhpC*-expressing plasmid to complement the dithiothreitol requirement for the growth of *trxB gor* mutants requires expression of AhpF, Grx and a glutathione synthetic enzyme, but neither of the two Trx proteins.[85] Perhaps surprisingly, this stress-induced triplet expansion is reversible. On altering the environmental conditions from "disulfide stress" to "oxidative stress" through additional deletion of the *katG* gene in the *trxB⁻ gor⁻ ahpC** background, peroxidase activity was restored through reversion of the *ahpC* locus to its wild type sequence. These observations suggest a biological function for this switch between two functional forms of AhpC and a stress-response mechanism for the genetic transformation that is as yet poorly understood.

5.7 Regulation and Biological Functions of Bacterial Prxs

A great deal of the functional information about Prxs in bacteria has come from analyses of the effects of mutations in or around the genes themselves or in their regulators. In reviewing this information, it is important to first consider the deleterious effects that the Prx substrates themselves have on cells if left unchecked.

5.7.1 Sources and cellular effects of hydrogen peroxide, organic hydroperoxides and peroxynitrite

The control exerted by Prxs over H_2O_2 levels has been of particular interest in eukaryotic systems recently (reviewed by Hoffman et al.[14] and Rhee[15]), as H_2O_2 is now recognized to be an important mediator of cytokine- and growth factor-mediated signaling cascades.[16] In *E. coli*, H_2O_2 levels are maintained at a very low level during growth (~20 nM under the conditions examined), primarily due to the action of AhpC; endogenous levels around 2 μM and above are cytostatic.[41,86] Sources of H_2O_2 in *E. coli* and other organisms include "leakage" of electrons from endogenous reduced flavoproteins and exposure to redox-cycling antibiotics generated by plants and some microorganisms, or to the variety of reactive oxygen species generated on engulfment of bacteria by host phagocytes.[87] As concentrations rise, H_2O_2 acts as an oxidizing agent of cellular thiols, particularly where present in enzyme active sites. Other hallmarks of protein oxidation, the formation of methionine sulfoxide and protein carbonyls, may also result from the direct or indirect actions of hydrogen peroxide.

In addition to its effects on protein thiols, H_2O_2 adversely affects redox-sensitive metalloproteins and generates highly toxic hydroxyl radicals in the presence of free, reduced iron in cells.[87] For example, treatment of the reduced molecular chaperone Hsp33 with H_2O_2 results in the formation of two intramolecular disulfide bonds concomitant with the release of zinc.[88] H_2O_2 can also directly inactivate the exposed [4Fe-4S] clusters of both aconitase and fumarase.[41] All of these mechanisms of H_2O_2 modification appear to apply to peroxide-sensing transcriptional regulators, several of which are described below.[89] Thus, although H_2O_2 in bacteria is not a part of the elaborate signaling cascades present in higher organisms, it still has a clear role in redox regulation at many levels.

That bacteria must defend themselves against organic hydroperoxides is implied by the presence of systems in many bacteria, e.g. the organic hydroperoxide resistance protein (Ohr) and its regulator (OhrR), specifically tailored to respond to and detoxify them.[90-92]   At least some of these toxic agents are encountered through host-pathogen interactions between bacteria and their plant or animal hosts. Whether or not lipid hydroperoxides are major components of bacterial oxidative stress, given their lack of polyunsaturated fatty acids, is still a matter of debate.[41] Some bacteria (e.g. clostridia), however, produce significant levels of plasmalogens (alk-1-enyl phospholipids), oxidation-sensitive lipids that may act as "sinks" or "decoys" for oxidative damage in these obligate anaerobes.[31,93,94] Several reports have also linked AhpC homologues or mutants to resistance toward organic solvents (e.g. tetralin or toluene), implying the intermediacy of organic hydroperoxides in the toxicity of these solvents.[40,95]   Reactive nitrogen intermediates released by phagocytes or plant cells are the most likely source of bacterial peroxynitrite, another substrate of Prxs.[34] This species, formed by the reaction of nitric acid with superoxide, rapidly oxidizes iron-sulfur clusters, protein thiols and DNA.[87]

5.7.2 Bacterial peroxidase systems other than Prxs

A quick review of other enzymatic activities that to some extent overlap with Prx functions is included to put the biological roles of these antioxidants in perspective.   The most widely recognized detoxification enzyme for H_2O_2 is catalase, which in *E. coli* exists as two different gene products, HPI (the *katG* gene product) and HPII (the *katE* gene product).[87] Catalases contain heme and are very active toward hydrogen peroxide, albeit with quite a high K_m (k_{cat}/K_m ~10^6 M^{-1} s^{-1}, K_m ~ 6 mM), but cannot degrade organic hydroperoxides.[96,97] Two other cysteine-dependent peroxidase enzymes, which are distant members of the Prx family, Tpx and BCP, are described above. Tpx appears to be primarily found in the periplasm and preferentially reduces organic peroxides, with some activity toward H_2O_2.[19,98] BCP is likely cytoplasmic but exhibits very low peroxidase activity.[18]   Another even more distant relative of Prxs, a glutathione peroxidase homologue (BtuE), is encoded among genes in an operon associated with vitamin B12 transport (GenBank P06610). The function of the bacterial *btuE/gpx* gene product has not been well characterized, but several lines of evidence suggest that it does not confer glutathione peroxidase activity, yet may confer peroxide resistance to cells (Baker and Poole, unpublished).[51,99]

Two other antioxidant enzymes are also of interest.   The *dps* gene product of *E. coli* was previously shown to function as a DNA-binding protein in stationary phase.[87]   Recently, this protein and two of its homologues, NapA from *Helicobacter pylori* and Dpr from *Streptococcus mutans*, were shown to exhibit a ferritin-like capacity to bind large amounts of iron; data now indicate that these proteins catalyze the H_2O_2-dependent oxidation of Fe(II), preventing hydroxyl radical formation mediated by Fenton chemistry.[100-103]   The Ohr family of antioxidant proteins is found in many bacteria (although not in *E. coli*), but the enzymatic activity of these proteins is as yet not well characterized.[89-92] Nonetheless, the high degree of specificity of its regulator, OhrR, for organic hydroperoxides and not H_2O_2 or other stresses implies a specific function for this protein in organic hydroperoxide detoxification, as supported by knockout studies of *ohr*.[90,91] Both Ohr and Dpr were discovered by their peroxide detoxification activities in mutant bacteria lacking *ahpC* and *ahpF* homologues.[23,90,101]

5.7.3 Transcriptional regulation of bacterial Prx expression

As mentioned at the beginning of this chapter, the peroxide-sensitive transcriptional activator of AhpC expression in *E. coli* and *S. typhimurium* was identified through mutational studies to be OxyR. Early on, it was demonstrated that the deletion or mutation of this regulator led to large increases in spontaneous mutagenesis levels that could be supressed by overexpression of AhpC.[38,39] OxyR is exquisitely sensitive toward H_2O_2 levels as low as 100 nM and, on activation, is oxidized at Cys199 to form the sulfenic acid and/or a disulfide bond with Cys208.[87,89,104,105] Under normal reductive conditions in *E.coli*, the protein is then re-reduced by Grx-1 and reduced glutathione to deactivate it.[106] This system is autoregulatory, as H_2O_2-treated OxyR represses transcription of its own structural gene and activates transcription of *grxA* and *gorA* (the gene encoding glutathione reductase), leading to higher levels of its own reductants. Other activities induced by H_2O_2 through OxyR regulation in *E. coli* include AhpF, the primary reductant of AhpC, HPI catalase (*katG*), thioredoxin-2 (*trxC*), Dps, and the Fur repressor.[87,89] Homologues of OxyR have been identified in a large number of bacteria including most gram-negative and some gram-positive organisms. However, the arrangement of genes surrounding *oxyR* differs in various organisms.[87] While the *ahpCF* locus is quite distant from *oxyR* in *E. coli* and *S. typhimurium*, in *Mycobacterium*, *ahpC* is upstream of and on the opposite strand to *oxyR*. In *Xanthomonas campestris*, *ahpC*, *ahpF*, and *oxyR* are adjacent to one another and in the same orientation, although *ahpC* is transcribed separately from the other two genes.[49] Interestingly, exposure of this organism to H_2O_2 results in changes in both the level and activity of OxyR.

In *Bacillus subtilis*, a peroxide-sensitive transcriptional repressor known as PerR was shown to control expression of AhpC and AhpF, as well as of catalase (*katA*) and Dps (*mrgA*) homologues.[44,89] PerR is related to the ferric uptake repressor (Fur) superfamily of metalloproteins and binds two metals, including Zn(II) at one site and Fe(II) or Mn(II) at the other regulatory site.[89] Derepression by PerR on exposure to H_2O_2 likely involves oxidation of metal-coordinated thiol(s), disulfide bond formation, and metal dissociation similar to that seen for Hsp33 described above. PerR-like regulators have been found in both gram-positive organisms (including *Staphylococcus aureus*[107] and *Streptococcus pyogenes*[51]) and in gram-negative bacteria such as *Campylobacter jejuni*,[108] and in most cases were shown to regulate *ahpC* transcription.

Other conditions reported to result in increased expression of Prxs have included osmotic stress,[47] engulfment of bacteria by (or growth within) macrophages,[109-111] and lowering of pH,[45] stress conditions that would enhance the bacteria's ability to cope with increased peroxide levels. Interestingly, while co-regulation of catalase and AhpC implies coordinated responses of the two activities, where one of the activities is lacking, the other is often increased to compensate for this loss in peroxide detoxification capacity.

5.7.4 Biological roles of bacterial Prx expression

Most of this chapter has alluded to the detoxification of the reactive oxygen and nitrogen species known to be catalyzed by Prxs. In support of this biological role, knockouts of the respective *ahpC* genes have generally led to an increased sensitivity toward organic hydroperoxide treatment in particular.[9,22,43,50,51] Sensitivity toward H_2O_2 is more complex, and whether AhpC or catalases have the dominant role in its detoxification depends critically on the levels of H_2O_2 and the assay used. As clarified through an elegant set of experiments reported recently by Costa Seaver and Imlay,[41,86] AhpC is responsible for the degradation of essentially all of the endogenous H_2O_2 in *E. coli*. Only when H_2O_2 levels added exogenously exceed 5 to 10 μM does the expression of catalase contribute significantly to its removal. The typical disk inhibition assay used to assess cytotoxicity involves exposure to high levels of H_2O_2, which are more efficiently decomposed by catalases. As AhpC is apparently the only major contributor to organic peroxide resistance in *E. coli*, alterations in resistance toward these agents in response to increased or decreased AhpC activities are more readily discernable.

In some cases, disruption in *ahpC* genes has led to only modest or undetectable increases in sensitivity toward organic hydroperoxides due to the presence of other protective enzymes (as with Dpr in *S. mutans* and Ohr in *X. campestris*, above). On the other hand, intact *ahpC* was required for viability of *H. pylori* grown in 5% oxygen.[30] This organism is barely able to survive in its microaerophilic environment and actually requires many genes for viability. Interestingly, *ahpC* deletion mutants were viable when the growth conditions for the organism included only 2% oxygen.[46]

A similar importance of intact *ahpF* (or *nox1*) genes has not been observed. It should be noted that apparently all bacterial Prxs are capable of reductive recycling by Trxs; therefore, the NADPH/Trx reductase/Trx system can serve as a backup when AhpF homologues are unavailable.[74]

The AhpF-AhpC system expressed by many eubacteria seems optimal as a steady-state removal system for low levels of endogenous H_2O_2 or as an "emergency" system for rapid detoxification of a burst of peroxide, but not for sustained high levels of these substrates. AhpC in *E. coli* and *S. typhimurium* is present in high abundance.[8,12,98] AhpF, though less abundant,[8,22] is a highly efficient recycler of AhpC ($k_{cat}/K_m \sim 10^7$ M^{-1} s^{-1}).[24] Under conditions of oxidative stress, increases in glycolytic enzymes lead to augmented availability of NADH and would help sustain the peroxidatic activity of AhpC.[87] As pointed out by Costa Seaver and Imlay, however, long-term exposure to high μM concentrations of H_2O_2 would quickly exhaust the cells' ability to provide NADH; under these conditions, catalase is a far better choice given that H_2O_2 serves as its own reductant (and oxidant) during dismutation.[41]

Acknowledgments:

Thanks to many colleagues and lab members for constructive comments and shared information. Supported by research grants GM50389 from NIH and an Established Investigatorship from the American Heart Association.

References

1. Brill, A. S. 1966. Peroxidases and catalase. In *Comprehensive Biochemistry* (Florkin, M., and E. H. Stotz, eds.), pp 447-479, Elsevier, New York.

2. Chance, B., H. Sies and A. Boveris. 1979. Hydroperoxide metabolism in mammalian organs. *Physiol. Rev.* **59**:527-605.

3. Forstrom, J. W., J. J. Zakowski and A. L. Tappel. 1978. Identification of the catalytic site of rat liver glutathione peroxidase as selenocysteine. *Biochemistry* **17**:2639-2644.

4. Epp, O., R. Ladenstein and A. Wendel. 1983. The refined structure of the selenoenzyme glutathione peroxidase at 0.2-nm resolution. *Eur. J. Biochem.* **133**:51-69.

5. Dolin, M. I. 1957. The *Streptococcus faecalis* oxidases for reduced diphosphopyridine nucleotide. III. Isolation and properties of a flavin peroxidase for reduced diphosphopyridine nucleotide. *J. Biol. Chem.* **225**:557-573.

6. Poole, L. B. and A. Claiborne. 1989. The non-flavin redox center of the streptococcal NADH peroxidase. II. Evidence for a stabilized cysteine-sulfenic acid. *J. Biol. Chem.* **264**:12330-12338.

7. Christman, M. F., R. W. Morgan, F. S. Jacobson and B. N. Ames. 1985. Positive control of a regulon for defenses against oxidative stress and some heat-shock proteins in *Salmonella typhimurium*. *Cell* **41**:753-762.

8. Jacobson, F. S., R. W. Morgan, M. F. Christman and B. N. Ames. 1989. An alkyl hydroperoxide reductase from *Salmonella typhimurium* involved in the defense of DNA against oxidative damage. Purification and properties. *J. Biol. Chem.* **264**:1488-1496.

9. Storz, G., F. S. Jacobson, L. A. Tartaglia, R. W. Morgan, L. A. Silveira and B. N. Ames. 1989. An alkyl hydroperoxide reductase induced by oxidative stress in *Salmonella typhimurium* and *Escherichia coli*: Genetic characterization and cloning of *ahp*. *J. Bacteriol.* **171**:2049-2055.

10. Tartaglia, L. A., G. Storz, M. H. Brodsky, A. Lai and B. N. Ames. 1990. Alkyl hydroperoxide reductase from *Salmonella typhimurium*. Sequence and homology to

thioredoxin reductase and other flavoprotein disulfide oxidoreductases. *J. Biol. Chem.* **265**:10535-10540.

11. Kim, K., I. H. Kim, K. Y. Lee, S. G. Rhee and E. R. Stadtman. 1988. The isolation and purification of a specific "protector" protein which inhibits enzyme inactivation by a thiol/Fe(III)/O2 mixed-function oxidation system. *J. Biol. Chem.* **263**:4704-4711.

12. Chae, H. Z., K. Robison, L. B. Poole, G. Church, G. Storz and S. G. Rhee. 1994. Cloning and sequencing of thiol-specific antioxidant from mammalian brain: alkyl hydroperoxide reductase and thiol-specific antioxidant define a large family of antioxidant enzymes. *Proc. Natl. Acad. Sci. U.S.A.* **91**:7017-7021.

13. Chae, H. Z., S. J. Chung and S. G. Rhee. 1994. Thioredoxin-dependent peroxide reductase from yeast. *J. Biol. Chem.* **269**:27670-27678.

14. Hofmann, B., H.-J. Hecht and L. Flohé. 2002. Peroxiredoxins. *Biol. Chem.* **383**:347-364.

15. Rhee, S. G. 2002. Protein tyrosine phosphatases and peroxiredoxins. In *Signal Transduction by Reactive Oxygen and Nitrogen Species: Pathways and Chemical Principles* (Torres, M., J. M. Fukuto, and H. J. Forman, eds.), Kluwer Academic Publishers, Dordrecht, The Netherlands.

16. Forman, H. J. and E. Cadenas. 1997. *Oxidative Stress and Signal Transduction*, Chapman and Hall, New York.

17. Wood, Z. A., E. Schröder, J. R. Harris and L. B. Poole. 2003. Structure, Mechanism and Regulation of Peroxiredoxins. *Trends Biochem. Sci.*:in press.

18. Jeong, W., M. K. Cha and I. H. Kim. 2000. Thioredoxin-dependent hydroperoxide peroxidase activity of bacterioferritin comigratory protein (BCP) as a new member of the thiol-specific antioxidant protein (TSA)/alkyl hydroperoxide peroxidase C (AhpC) family. *J. Biol. Chem.* **275**:2924-2930.

19. Cha, M. K., H. K. Kim and I. H. Kim. 1995. Thioredoxin-linked 'thiol peroxidase' from periplasmic space of *Escherichia coli*. *J. Biol. Chem.* **270**:28635-28641.

20. Cha, M. K., H. K. Kim and I. H. Kim. 1996. Mutation and mutagenesis of thiol peroxidase of *Escherichia coli* and a new type of thiol peroxidase family. *J. Bacteriol.* **178**:5610-5614.

21. Zhou, Y., X. Y. Wan, H. L. Wang, Z. Y. Yan, Y. D. Hou and D. Y. Jin. 1997. Bacterial scavengase p20 is structurally and functionally related to peroxiredoxins. *Biochem. Biophys. Res. Commun.* **233**:848-852.

22. Poole, L. B. and H. R. Ellis. 1996. Flavin-dependent alkyl hydroperoxide reductase from *Salmonella typhimurium*. 1. Purification and enzymatic activities of overexpressed AhpF and AhpC proteins. *Biochemistry* **35**:56-64.

23. Higuchi, M., Y. Yamamoto, L. B. Poole, M. Shimada, Y. Sato, N. Takahashi and Y. Kamio. 1999. Functions for two types of NADH oxidases in energy metabolism and oxidative stress of *Streptococcus mutans*. *J. Bacteriol.* **181**:5940-5947.

24. Poole, L. B., M. Higuchi, M. Shimada, M. Li Calzi and Y. Kamio. 2000. *Streptococcus mutans* H2O2-forming NADH oxidase is an alkyl hydroperoxide reductase protein. *Free Radic. Biol. Med.* **28**:108-120.

25. Niimura, Y., L. B. Poole and V. Massey. 1995. *Amphibacillus xylanus* NADH oxidase and *Salmonella typhimurium* alkyl hydroperoxide reductase flavoprotein component show extremely high scavenging activity for both alkyl hydroperoxide and hydrogen peroxide in the presence of *S. typhimurium* alkyl hydroperoxide reductase 22-kDa protein component. *J. Biol. Chem.* **269**:25645-25650.

26. Kitano, K., Y. Niimura, Y. Nishiyama and K. Miki. 1999. Stimulation of peroxidase activity by decamerization related to ionic strength: AhpC protein from *Amphibacillus xylanus*. *J. Biochem. (Tokyo)* **126**:313-319.

27. Niimura, Y., Y. Nishiyama, D. Saito, H. Tsuji, M. Hidaka, T. Miyaji, T. Watanabe and V. Massey. 2000. A hydrogen peroxide-forming NADH oxidase that functions as an alkyl hydroperoxide reductase in *Amphibacillus xylanus*. *J. Bacteriol.* **182**:5046-5051.

28. Toomey, D. and S. G. Mayhew. 1998. Purification and characterisation of NADH oxidase from Thermus aquaticus YT-1 and evidence that it functions in a peroxide-reduction system. *Eur. J. Biochem.* **251**:935-945.

29. Logan, C. and S. G. Mayhew. 2000. Cloning, over-expression and characterization of peroxiredoxin and NADH-peroxiredoxin reductase from *Thermus aquaticus* YT-1. *J. Biol. Chem.* **275**:30019-30028.

30. Baker, L. M. S., A. Raudonikiene, P. H. Hoffman and L. B. Poole. 2001. Essential thioredoxin-dependent peroxiredoxin from *Helicobacter pylori*: genetic and kinetic characterization. *J. Bacteriol.* **183**:1961-1973.

31. Reynolds, C. M., J. Meyer and L. B. Poole. 2002. An NADH-dependent bacterial thioredoxin reductase-like protein, in conjunction with a glutaredoxin homologue, form a unique peroxiredoxin (AhpC) reducing system in *Clostridium pasteurianum*. *Biochemistry* **41**:1990-2001.

32. Vergauwen, B., F. Pauwels, F. Jacquemotte, T. E. Meyer, M. A. Cusanovich, R. G. Bartsch and J. J. Van Beeumen. 2001. Characterization of glutathione amide reductase from *Chromatium gracile*. Identification of a novel thiol peroxidase (Prx/Grx) fueled by glutathione amide redox cycling. *J. Biol. Chem.* **276**:20890-20897.

33. Hillas, P. J., F. S. del Alba, J. Oyarzabal, A. Wilks and P. R. Ortiz de Montellano. 2000. The AhpC and AhpD antioxidant defense system of *Mycobacterium tuberculosis*. *J. Biol. Chem.* **275**:18801-18809.

34. Bryk, R., P. Griffin and C. Nathan. 2000. Peroxynitrite reductase activity of bacterial peroxiredoxins. *Nature* **407**:211-215.

35. Chauhan, R. and S. C. Mande. 2001. Characterization of the *Mycobacterium tuberculosis* H37Rv alkyl hydroperoxidase AhpC points to the importance of ionic interactions in oligomerization and activity. *Biochem. J.* **354**:209-215.

36. Bryk, R., C. D. Lima, H. Erdjument-Bromage, P. Tempst and C. Nathan. 2002. Metabolic enzymes of mycobacteria linked to antioxidant defense by a thioredoxin-like protein. *Science* **295**:1073-1077.

37. Chauhan, R. and S. C. Mande. 2002. Site directed mutagenesis reveals a novel catalytic mechanism of *Mycobacterium tuberculosis* alkylhydroperoxidase C. *Biochem. J.* **in press**.

38. Storz, G., M. F. Christman, H. Sies and B. N. Ames. 1987. Spontaneous mutagenesis and oxidative damage to DNA in *Salmonella typhimurium*. *Proc Natl Acad Sci USA* **84**:8917-8921.

39. Greenberg, J. T. and B. Demple. 1988. Overproduction of peroxide-scavenging enzymes in *Escherichia coli* suppresses spontaneous mutagenesis and sensitivity to redox-cycling agents in *oxyR-* mutants. *EMBO J.* **7**:2611-2617.

40. Ferrante, A. A., J. Augliera, K. Lewis and A. M. Klibanov. 1995. Cloning of an organic solvent-resistance gene in Escherichia coli: the unexpected role of alkylhydroperoxide reductase. *Proc. Natl. Acad. Sci. USA* **92**:7617-7621.

41. Costa Seaver, L. and J. A. Imlay. 2001. Alkyl hydroperoxide reductase is the primary scavenger of endogenous hydrogen peroxide in *Escherichia coli*. *J. Bacteriol.* **183**:7173-7181.

42. Hartford, O. M. and B. C. Dowds. 1994. Isolation and characterization of a hydrogen peroxide resistant mutant of *Bacillus subtilis*. *Microbiology* **140**:297-304.

43. Antelmann, H., S. Engelmann, R. Schmid and M. Hecker. 1996. General and oxidative stress responses in *Bacillus subtilis*: cloning, expression and mutation of the alkyl hydroperoxide reductase operon. *J. Bacteriol.* **178**:6571-6578.

44. Bsat, N., L. Chen and J. D. Helmann. 1996. Mutation of the *Bacillus subtilis* alkyl hydroperoxide reductase (*ahpCF*) operon reveals compensatory interactions among hydrogen peroxide stress genes. *J. Bacteriol.* **178**:6579-6586.

45. Lundstrom, A. M. and L. Bolin. 2000. A 26 kDa protein of *Helicobacter pylori* shows alkyl hydroperoxide reductase (AhpC) activity and the mono-cistronic transcription of the gene is affected by pH. *Microb. Pathog.* **29**:257-266.

46. Olczak, A. A., J. W. Olson and R. J. Maier. 2002. Oxidative-stress resistance mutants of *Helicobacter pylori*. *J. Bacteriol.* **184**:3186-3193.

47. Armstrong-Buisseret, L., M. B. Cole and G. S. A. B. Stewart. 1995. A homologue to the *Escherichia coli* alkyl hydroperoxide reductase AhpC is induced by osmotic upshock in *Staphylococcus aureus*. *Microbiology* **141**:1655-1661.

48. Loprasert, S., S. Atichartpongkun, W. Whangsuk and S. Mongkolsuk. 1997. Isolation and analysis of the *Xanthomonas* alkyl hydroperoxide reductase gene and the peroxide sensor regulator genes *ahpC* and *ahpF-oxyR-orfX*. *J. Bacteriol.* **179**:3944-3949.

49. Mongkolsuk, S., S. Loprasert, W. Whangsuk, M. Fuangthong and S. Atichartpongkun. 1997. Characterization of transcription organization and analysis of unique expression patterns of an alkyl hydroperoxide reductase C gene (*ahpC*) and the peroxide regulator

operon *ahpF-oxyR-orfX* from *Xanthomonas campestris* pv. phaseoli. *J. Bacteriol.* **179**:3950-3955.

50. Mongkolsuk, S., W. Whangsuk, P. Vattanaviboon, S. Loprasert and M. Fuangthong. 2000. A *Xanthomonas* alkyl hydroperoxide reductase subunit C (*ahpC*) mutant showed an altered peroxide stress response and complex regulation of the compensatory response of peroxide detoxification enzymes. *J. Bacteriol.* **182**:6845-6849.

51. King, K. Y., J. A. Horenstein and M. G. Caparon. 2000. Aerotolerance and peroxide resistance in peroxidase and PerR mutants of *Streptococcus pyogenes*. *J. Bacteriol.* **182**:5290-5299.

52. Rocha, E. R. and C. J. Smith. 1999. Role of the alkyl hydroperoxide reductase (*ahpCF*) gene in oxidative stress defense of the obligate anaerobe *Bacteriodes fragilis*. *J. Bacteriol.* **181**:5701-5710.

53. Rocha, E. R., G. Owens and C. J. Smith. 2000. The redox-sensitive transcriptional activator OxyR regulates the peroxide response regulon in the obligate anaerobe *Bacteroides fragilis*. *J. Bacteriol.* **182**:5059-5069.

54. Deretic, V., W. Philipp, S. Dhandayuthapani, M. H. Mudd, R. Curcic, T. Garbe, B. Heym, L. E. Via and S. T. Cole. 1995. *Mycobacterium tuberculosis* is a natural mutant with an inactivated oxidative-stress regulatory gene: implications for sensitivity to isoniazid. *Mol. Microbiol.* **17**:889-900.

55. Wilson, T. M. and D. M. Collins. 1996. *ahpC*, a gene involved in isoniazid resistance of the *Mycobacterium tuberculosis* complex. *Mol. Microbiol.* **19**:1025-1034.

56. Dhandayuthapani, S., Y. Zhang, M. H. Mudd and V. Deretic. 1996. Oxidative stress response and its role in sensitivity to isoniazid in mycobacteria: characterization and inducibility of *ahpC* by peroxides in *Mycobacterium smegmatis* and lack of expression in *M. aurum* and *M. tuberculosis*. *J. Bacteriol.* **178**:3641-3649.

57. Sherman, D. R., K. Mdluli, M. J. Hickey, T. M. Arain, S. L. Morris, C. E. Barry and C. K. Stover. 1996. Compensatory *ahpC* gene expression in isoniazid-resistant *Mycobacterium tuberculosis*. *Science* **272**:1641-1643.

58. Heym, B., E. Stavropoulous, N. Honore, P. Domenech, B. Saint-Joanis, T. M. Wilson, D. M. Collins, M. J. Colston and S. T. Cole. 1997. Effects of overexpression of the alkyl hydroperoxide reductase AhpC on the virulence and isoniazid resistance of *Mycobacterium tuberculosis*. *Infect. Immun.* **65**:1395-1401.

59. Wilson, T., G. W. de Lisle, J. A. Marcinkeviciene, J. S. Blanchard and D. M. Collins. 1998. Antisense RNA to *ahpC*, an oxidative stress defence gene involved in isoniazid resistance, indicates that AhpC of *Mycobacterium bovis* has virulence properties. *Microbiology* **144**:2687-2695.

60. Springer, B., S. Master, P. Sander, T. Zahrt, M. McFalone, J. Song, K. G. Papavinasasundaram, M. J. Colston, E. Boettger and V. Deretic. 2001. Silencing of oxidative stress response in *Mycobacterium tuberculosis*: expression patterns of *ahpC* in virulent and avirulent strains and effect of *ahpC* inactivation. *Infect. Immun.* **69**:5967-5973.

61. Nishiyama, Y., V. Massey, K. Takeda, S. Kawasaki, J. Sato, T. Watanabe and Y. Niimura. 2001. Hydrogen peroxide-forming NADH oxidase belonging to the peroxiredoxin oxidoreductase family: existence and physiological role in bacteria. *J. Bacteriol.* **183**:2431-2438.

62. Sohling, B., T. Parther, K. P. Rucknagel, M. A. Wagner and J. R. Andreesen. 2001. A selenocysteine-containing peroxiredoxin from the strictly anaerobic organism *Eubacterium acidaminophilum. Biol. Chem.* **382**:979-986.

63. Wood, Z. A., L. B. Poole, R. R. Hantgan and P. A. Karplus. 2002. Dimers to doughnuts: redox-sensitive oligomerization of 2-cysteine peroxiredoxins. *Biochemistry* **41**:5493-4403.

64. Poole, L. B. and H. R. Ellis. 2002. Identification of cysteine sulfenic acid in AhpC of alkyl hydroperoxide reductase. *Methods Enzymol.* **348**:122-136.

65. Ellis, H. R. and L. B. Poole. 1997. Novel application of 7-chloro-4-nitrobenzo-2-oxa-1,3-diazole to identify cysteine sulfenic acid in the AhpC component of alkyl hydroperoxide reductase. *Biochemistry* **36**:15013-15018.

66. Ellis, H. R. and L. B. Poole. 1997. Roles for the two cysteine residues of AhpC in catalysis of peroxide reduction by alkyl hydroperoxide reductase from *Salmonella typhimurium. Biochemistry* **36**:13349-13356.

67. Yeh, J. I., A. Claiborne and W. G. J. Hol. 1996. Structure of the native cysteine-sulfenic acid redox center of enterococcal NADH peroxidase refined at 2.8 angstroms resolution. *Biochemistry* **35**:9951-9957.

68. Kice, J. L. 1980. Mechanisms and reactivity in reactions of organic oxyacids of sulfur and their anhydrides. *Adv. Phys. Org. Chem.* **17**:65-181.

69. Claiborne, A., J. I. Yeh, T. C. Mallett, J. Luba, E. J. Crane, 3rd, V. Charrier and D. Parsonage. 1999. Protein-sulfenic acids: diverse roles for an unlikely player in enzyme catalysis and redox regulation. *Biochemistry* **38**:15407-15416.

70. Allison, W. S. 1976. Formation and reactions of sulfenic acids in proteins. *Acc. Chem. Res.* **9**:293-299.

71. Fuangthong, M. and J. D. Helmann. 2002. The OhrR repressor senses organic hydroperoxides by reversible formation of a cysteine-sulfenic acid derivative. *Proc. Natl. Acad. Sci. USA* **99**:6690-6695.

72. Schröder, E., J. A. Littlechild, A. A. Lebedev, N. Errington, A. A. Vagin and M. N. Isupov. 2000. Crystal structure of decameric 2-Cys peroxiredoxin from human erythrocytes at 1.7 Å resolution. *Structure* **8**:605-615.

73. Wagner, E., S. Luche, L. Penna, M. Chevallet, A. V. Dorsselaer, E. Leize-Wagner and T. Rabilloud. 2002. A method for detection of overoxidation of cysteine: Peroxiredoxins are oxidised in vivo at the active site cysteine during oxidative stress. *Biochem. J.* **365**:in press.

74. Poole, L. B., C. M. Reynolds, Z. Wood, P. A. Karplus, H. R. Ellis and M. Li Calzi. 2000. AhpF and other NADH:peroxiredoxin oxidoreductases, homologues of low Mr thioredoxin reductase. *Eur. J. Biochem.* **267**:6126-6133.

75. Poole, L. B., A. Godzik, A. Nayeem and J. D. Schmitt. 2000. AhpF can be dissected into two functional units; tandem repeats of two thioredoxin-like folds in the N-terminus mediate electron transfer from the thioredoxin reductase-like C-terminus to AhpC. *Biochemistry* **39**:6602-6615.

76. Wood, Z. A., L. B. Poole and P. A. Karplus. 2001. Structure of intact AhpF reveals a mirrored thioredoxin-like active site and implies large domain rotations during catalysis. *Biochemistry* **40**:3900-3911.

77. Reynolds, C. M. and L. B. Poole. 2000. Attachment of the N-terminal domain of *Salmonella typhimurium* AhpF to *Escherichia coli* thioredoxin reductase confers AhpC reductase activity but does not affect thioredoxin reductase activity. *Biochemistry* **39**:8859-8869.

78. Nunn, C. M., S. Djordjevic, P. J. Hillas, C. R. Nishida and P. R. Ortiz de Montellano. 2002. The crystal structure of *Mycobacterium tuberculosis* alkylhydroperoxidase AhpD, a potential target for antitubercular drug design. *J. Biol. Chem.* **277**:20033-20040.

79. Hirotsu, S., Y. Abe, K. Okada, N. Nagahara, H. Hori, T. Nishino and T. Hakoshima. 1999. Crystal structure of a multifunctional 2-Cys peroxiredoxin heme-binding protein 23 kDa/proliferation-associated gene product. *Proc. Natl. Acad. Sci. USA* **96**:12333-12338.

80. Alphey, M. S., C. S. Bond, E. Tetaud, A. H. Fairlamb and W. N. Hunter. 2000. The structure of reduced tryparedoxin peroxidase reveals a decamer and insight into reactivity of 2Cys-peroxiredoxins. *J. Mol. Biol.* **300**:903-916.

81. Declercq, J. P., C. Evrard, A. Clippe, D. Vander Stricht, A. Bernard and B. Knoops. 2001. Crystal structure of human peroxiredoxin 5, a novel type of mammalian peroxiredoxin at 1.5 angstrom resolution. *J. Mol. Biol.* **311**:751-759.

82. Choi, H.-J., S. W. Kang, C.-H. Yang, S. G. Rhee and S.-E. Ryu. 1998. Crystal structure of a novel human peroxidase enzyme at 2.0 Å resolution. *Nature Struct. Biol.* **5**:400-406.

83. Hale, S. P., L. B. Poole and J. A. Gerlt. 1993. Mechanism of the reaction catalyzed by staphylococcal nuclease: identification of the rate-determining step. *Biochemistry* **32**:7479-7487.

84. Poole, L. B. 1999. Flavin-linked redox components required for AhpC reduction in alkyl hydroperoxide reductase systems. In *Flavins and Flavoproteins 1999* (Ghisla, S., P. Kroneck, P. Macheroux, and H. Sund, eds.), pp 691-694, Agency for Scientific Publications, Berlin.

85. Ritz, D., J. Lim, C. M. Reynolds, L. B. Poole and J. Beckwith. 2001. Conversion of a peroxiredoxin into a disulfide reductase by a triplet repeat expansion. *Science* **294**:158-160.

86. Costa Seaver, L. and J. A. Imlay. 2001. Hydrogen peroxide fluxes and compartmentalization inside growing *Escherichia coli*. *J. Bacteriol.* **183**:7182-7189.

87. Storz, G. and J. A. Imlay. 1999. Oxidative stress. *Curr. Opin. Microbiol.* **2**:188-194.

88. Graumann, J., H. Lilie, X. Tang, K. A. Tucker, J. H. Hoffmann, J. Vijayalakshmi, M. Saper, J. C. A. Bardwell and U. Jakob. 2001. Activation of the redox-regulated molecular chaperone Hsp33 -- A two-step mechanism. *Structure* **9**:377-387.

89. Mongkolsuk, S. and J. D. Helmann. 2002. Regulation of inducible peroxide stress responses. *Mol. Microbiol.* **45**:9-15.

90. Mongkolsuk, S., S. Loprasert, W. Whangsuk, M. Fuangthong and S. Atichartpongkun. 1997. Identification and characterization of a new organic hydroperoxide resistance (*ohr*) gene with a novel pattern of oxidative stress regulation from *Xanthomonas campestris* pv. phaseoli. *J. Bacteriol.* **180**:2636-2643.

91. Sukchawalit, R., S. Loprasert, S. Atichartpongkul and S. Mongkolsuk. 2001. Complex regulation of the organic hydroperoxide resistance gene (*ohr*) from *Xanthomonas* involves OhrR, a novel organic peroxide-inducible negative regulator, and posttranscriptional modifications. *J. Bacteriol.* **183**:4405-4412.

92. Atichartpongkul, S., S. Loprasert, P. Vattanaviboon, W. Whangsuk, J. D. Helmann and S. Mongkolsuk. 2001. Bacterial Ohr and OsmC paralogues define two protein families with distinct functions and patterns of expression. *Microbiology* **147**:1775-1782.

93. Johnston, N. C. and H. Goldfine. 1983. Lipid composition in the classification of the butyric acid-producing clostridia. *J. Gen. Microbiol.* **129**:1075-1081.

94. Morand, O. H., R. A. Zoeller and C. R. Raetz. 1988. Disappearance of plasmalogens from membranes of animal cells subjected to photosensitized oxidation. *J. Biol. Chem.* **263**:11597-11606.

95. Fukumori, F., H. Hirayama, H. Takami, A. Inoue and K. Horikoshi. 1998. Isolation and transposon mutagenesis of a *Pseudomonas putida* KT2442 toluene-resistant variant: involvement of an efflux system in solvent resistance. *Extremophiles* **2**:395-400.

96. Obinger, C., M. Maj, P. Nicholls and P. Loewen. 1997. Activity, peroxide compound formation, and heme d synthesis in *Escherichia coli* HPII catalase. *Arch. Biochem. Biophys.* **342**:58-67.

97. Hillar, A., B. Peters, R. Pauls, A. Loboda, H. Zhang, A. G. Mauk and P. C. Loewen. 2000. Modulation of the activities of catalase-peroxidase HPI of *Escherichia coli* by site-directed mutagenesis. *Biochemistry* **59**:5868-5875.

98. Link, A. J., K. Robison and G. M. Church. 1997. Comparing the predicted and observed properties of proteins encoded in the genome of *Escherichia coli* K-12. *Electrophoresis* **18**:1259-1313.

99. Moore, T. D. E. and P. F. Sparling. 1996. Interruption of the *gpxA* gene increases the sensitivity of *Neisseria meningitidis* to paraquat. *J. Bacteriol.* **178**:4301-4305.

100. Tonello, F., W. G. Dundon, B. Satin, M. Molinari, G. Tognon, G. Grandi, G. Del Giudice, R. Rappuoli and C. Montecucco. 1999. The *Helicobacter pylori* neutrophil-activating protein is an iron-binding protein with dodecameric structure. *Mol. Microbiol.* **34**:238-246.

101. Yamamoto, Y., M. Higuchi, L. B. Poole and Y. Kamio. 2000. Role of the *dpr* product in oxygen tolerance in *Streptococcus mutans. J. Bacteriol.* **82**:3740-3747.

102. Yamamoto, Y., L. B. Poole, R. R. Hantgan and Y. Kamio. 2002. An iron-binding protein, Dpr, from *Streptococcus mutans* prevents iron-dependent hydroxyl radical formation in vivo. *J. Bacteriol.* **184**:2932-2939.

103. Zhao, G., P. Ceci, A. Hari, L. Giangiacomo, T. M. Laue, E. Chiancone and N. D. Chasteen. 2002. Iron and hydrogen peroxide detoxification properties of DNA-binding protein from starved cells. A ferritin-like DNA-binding protein of *Escherichia coli. J. Biol. Chem.* **277**:27689-27696.

104. Zheng, M., F. Åslund and G. Storz. 1998. Activation of the OxyR transcription factor by reversible disulfide bond formation. *Science* **279**:1718-1721.

105. Kim, S. O., K. Merchant, R. Nudelman, W. F. Beyer, T. Keng, J. DeAngelo, A. Hausladen and J. S. Stamler. 2002. OxyR: a molecular code for redox-related signaling. *Cell* **109**:383-396.

106. Åslund, F., M. Zheng, J. Beckwith and G. Storz. 1999. Regulation of the OxyR transcription factor by hydrogen peroxide and the cellular thiol-disulfide status. *Proc. Natl. Acad. Sci. U.S.A.* **96**:6161-6165.

107. Horsburgh, M. J., M. O. Clements, H. Crossley, E. Ingham and S. J. Foster. 2001. PerR controls oxidative stress resistance and iron storage proteins and is required for virulence in *Staphylococcus aureus. Infect. Immun.* **69**:3744-3754.

108. van Vliet, A. H., M. L. Baillon, C. W. Penn and J. M. Ketley. 1999. *Campylobacter jejuni* contains two *fur* homologs: characterization of iron-responsive regulation of peroxide stress defense genes by the PerR repressor. *J. Bacteriol.* **181**:6371-6376.

109. Dhandayuthapani, S., L. E. Via, C. A. Thomas, P. M. Horowitz, D. Deretic and V. Deretic. 1995. Green fluorescent protein as a marker for gene expression and cell biology of mycobacterial interactions with macrophages. *Mol. Microbiol.* **17**:889-900.

110. Francis, K. P., P. D. Taylor, C. J. Inchley and M. P. Gallagher. 1997. Identification of the *ahp* operon of *Salmonella typhimurium* as a macrophage-induced locus. *J. Bacteriol.* **179**:4046-4048.

111. Rankin, S., Z. Li and R. R. Isberg. 2002. Macrophage-induced genes of *Legionella pneumophila*: protection from reactive intermediates and solute imbalance during intracellular growth. *Infect. Immun.* **70**:3637-3648.

Chapter 6

THE NOX ENZYMES AND THE REGULATED GENERATION OF REACTIVE OXYGEN SPECIES

Rebecca S. Arnold and J. David Lambeth

6.1 Introduction

6.1.1 Reactive Oxygen: Nature's Accident or Nature's Design

While its roots can be traced to earlier studies on catalase and peroxidases, the concept that reactive oxygen species (ROS) occur in nature in sufficient quantities to be of biological consequence began to blossom in the late 1960's with the discovery by McCord and Fridovich of superoxide dismutase.[1] The existence of catabolic enzymes that degrade superoxide, hydrogen peroxide and lipid hydroperoxides (Fig. 1), as well as the occurrence of DNA repair enzymes that deal with the downstream consequences of ROS, provided the intellectual milieu that quite logically lead to the concept that ROS are an unfortunate but unavoidable consequence of aerobic life. Accordingly, ROS are thought of as something deleterious for which biology has had to design defense mechanisms. This concept has dominated our thinking for the past 30 years, and a great deal of effort has gone into the study of both the catabolic mechanisms that are designed to rid the organism of ROS, and the mostly "accidental" sources of ROS. Among these sources are various electron transport processes including those occurring in mitochondria and in endoplasmic reticulum, xenobiotics (both redox-active drugs and toxins), a variety of metabolic enzymes that carry out oxidations, and irradiation (Fig. 1). The balance between this accidental generation of ROS and the capacity of the defense enzymes to rid us of

102

H. J. Forman, J. Fukuto, and M. Torres (eds.), Signal Transduction by Reactive Oxygen and Nitrogen Species: Pathways and Chemical Principles, ©2003, Kluwer Academic Publishers. Printed in the Netherlands

ROS has been thought to be critical for the genesis of a variety of diseases including Alzheimer's Disease, atherosclerosis, hypertension, cancer and aging itself.

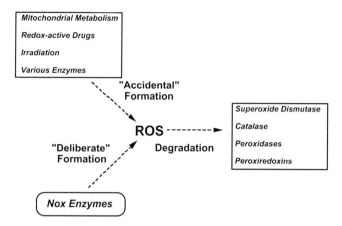

Figure 1. Production and Degradation of Reactive Oxygen Species (ROS). Shown are the unregulated generation of ROS by a variety of sources, the regulated production of ROS by Nox enzymes, and the degradation of ROS by catabolic enzymes.

However, this view of ROS as an unintended consequence of life on an aerobic planet is altered in a fundamental way by the recent discovery of a family of enzymes the sole function of which is to generate reactive oxygen (Fig. 1).[2-4] The first of these, the phagocyte NADPH oxidase,[5,6] had long been thought to be an exception to the rule. This oxidase exploits the deleterious properties of ROS as a mechanism to kill invading microbes. The recent description of a family of related NADPH oxidases, the Nox family,[2,3] and the finding of these enzymes in a variety of cell types[4,7] implies that the "deliberate" generation of ROS (Fig. 1) extends to a variety of tissues, including those not usually thought to participate in host defense. This finding also places the ROS catabolic enzymes in a new context, since the balance between generation and catabolism of ROS (as well as the localization of both classes of enzymes) will determine the steady state concentration of ROS in a given cell type. Understanding the function of ROS in cells, particularly in those cells that do not participate in host defense, represents a new frontier for research on ROS and free radicals in biology.

6.2 A Brief Visit to the Phagocyte Respiratory Burst Oxidase

Phagocytes such as neutrophils and macrophages initiate a massive consumption of oxygen upon exposure to microbes, microbial molecules (*e.g.,* formyl peptides) and inflammatory mediators (platelet activating factor, tumor necrosis factor, interleukins). Oxygen consumption is accompanied by the initial generation of superoxide, followed by the secondary generation of hydrogen peroxide, HOCl, hydroxyl radical and singlet oxygen. Together, these ROS participate along with non-oxidative mechanisms in killing the invading microbe, as evidenced by defective bacterial killing in the condition Chronic Granulomatous Disease in which phagocyte ROS generation is defective.[8]   The enzyme responsible for ROS generation is the phagocyte NADPH-oxidase, also known as the respiratory burst oxidase.[5] The catalytic moiety is a membrane glycoprotein known as gp91*phox*, which contains the NADPH binding site along with bound FAD and two hemes[9-12] (Fig. 2).

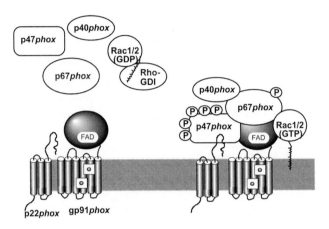

Figure 2. Regulation of the Phagocyte NADPH Oxidase by Protein Interactions. The left side of the figure shows the phagocyte NADPH oxidase components prior to cell activation, indicating the cytosolic location of many of the regulatory proteins including p47phox, p67phox, and Rac. The latter is in the GDP-associated form and is complexed to RhoGDI. Cell activation results in phosphorylation of p47phox and nucleotide exchange to generate the GTP complex with Rac, and these changes result in the assembly of an active multiprotein complex with the catalytic subunit, gp91phox.

This protein is complexed in the membrane with a regulatory protein, p22*phox,*[13] which is needed both to stabilize gp91*phox*[12] and to provide a docking site in the membrane for additional regulatory proteins.[14,15]   Exposure of phagocytes to activating stimuli initiates a complex series of events. The gp91*phox*-p22*phox* flavocytochrome heterodimer is located in the membranes of specific granules and secretory vesicles in resting cells, but moves to the plasma membrane upon cell activation.[16,17]   If the activating stimulus is a phagocytosed microbe, then these components move to the phagosomal membrane. In resting cells, the activator proteins p47*phox*, p67*phox* and Rac (Rac1 and Rac2) are located in the cytosol (Fig.2), along with p40*phox,*[18] which may function as a negative modulator of activity. Activation of the cell results in hyperphosphorylation of p47*phox*[19] as well as phosphorylation of p67*phox*[20] and of membrane components. These phosphorylations trigger assembly of p47*phox* with the flavocytochrome, probably by inducing a conformational change that exposes binding sites on p47*phox* for p22*phox* and gp91*phox.*[21] In resting cells, Rac contains bound GDP and resides in the cytosol in an inhibited complex with RhoGDI. Cell activation is accompanied by guanine nucleotide exchange, replacing bound GDP with GTP and resulting in dissociation of the inhibitory Rho GDI, exposure of the geranyl-geranyl lipid tail and translocation to the membrane (Fig.2). Both Rac and p47*phox* provide binding sites for p67*phox*. The latter regulatory protein contains an "activation domain" (see Fig. 2)[22] that interacts with the gp91*phox* catalytic subunit. Specifically, the activation domain regulates FAD reduction by stimulating the rate of hydride transfer from NADPH to FAD.[23] Sequential transfer of electrons from FAD through the two hemes results in transmembrane reduction of oxygen to form superoxide. Thus, the phagocyte NADPH oxidase is dormant in resting cells but becomes activated as a result of regulated protein assembly.

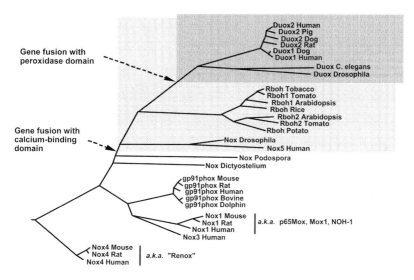

Figure 3. Evolutionary Relationships Among Nox Family Members. The dendrogram was generated from the regions of these proteins that are homologous to gp91phox. The Nox group (no background shading) contains only the gp91phox homology region. Those enclosed by the light-shaded box contain this domain plus a calcium binding domain, while those enclosed by a darker-shaded box contain the gp91phox domain, the calcium-binding domain and a domain showing homology with peroxidases.

6.3 The Nox/Duox Family of NAD(P)H Oxidases

Over the past decade it has become clear that ROS are produced in a variety of cell types in addition to inflammatory cells. In general, the amounts of ROS generated are considerably lower than that seen in activated neutrophils, but the availability of sensitive assays has permitted detection in a variety of cells. These include colonic epithelia, smooth muscle, fibroblasts, osteocytes, endothelial cells, keratinocytes, chondrocytes, adipocytes and a variety of cancer cells.[24] Frequently, ROS are generated in response to signals such as growth factors (Platelet Derived Growth Factor, Epidermal Growth Factor) or cytokines (Tumor Necrosis Factor, Interferon-γ). Although it was initially thought that ROS production was a byproduct of mitochondrial respiration, inhibition of ROS production by diphenyleneiodonium [*e.g.*],[25] which also inhibits the phagocyte NADPH oxidase, suggested that the ROS production might be due either to gp91*phox* itself, or to a related enzyme(s). By the late 1990's the increasing availability of EST and genomic sequence information provided confirmation of the existence of homologs of gp91*phox*, and the first of the human homologs, now termed Nox1, was described in 1999.[2] Other homologs had already been reported in plants (the Rboh enzymes), and additional human/mammalian homologs quickly followed.[3,4,7,26-29]

There are now a total of 7 Nox homologs in the human, and isologs in a variety of organisms ranging from nematodes to Drosophila to plants. These are shown in the family tree (Fig. 3). The Nox superfamily shows three distinct subgroupings, based on primary sequence and domain structure. The Nox subgroup itself contains gp91*phox* (an alternate name is Nox2) plus Nox1, Nox3, and Nox4. All of these proteins resemble gp91*phox* in possessing an N-terminal hydrophobic domain that contains 6 predicted transmembrane alpha helices (Fig. 4).

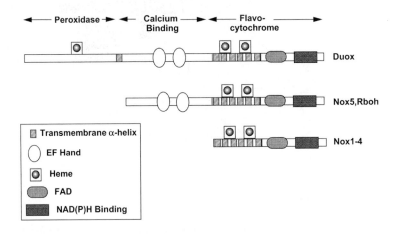

Figure 4. Domain Structure of Nox and Duox Enzymes. See text for details.

Based on studies with gp91*phox*, this domain contains both heme groups, which are buried more or less in the two leaflets of the bilayer (Fig. 2). The gp91*phox* group also contains a C-terminal flavoprotein domain, consisting of an FAD binding site and an NADPH binding site (Fig. 4). This group is indicated in Fig. 3 (proteins in the lower half of the figure, not highlighted by a box). The Nox5 group contains the gp91*phox* homology domain, but also possesses a calcium-binding domain that contains two EF-hand structures (Fig. 4). The latter are protein structures that bind calcium. The Duox group contains not only the gp91*phox*-like structure and the calcium-binding domain, but also a peroxidase homology domain (Fig. 4). The peroxidase domain is separated from the calcium-binding domain by an additional predicted transmembrane alpha helix. The family tree in Fig. 3 was generated based on a comparison of the amino acid sequences of the gp91*phox*-homology regions alone. The family tree thus generated fits with the idea that the evolution involved sequential gene fusion events, first with the calcium-binding domain, and then with the peroxidase homology domain. This can be visualized in Fig. 3, wherein every protein after the first gene fusion event (light grey background) possesses not only the gp91*phox* domain but also the calcium-binding domain. The subsequent gene fusion with the peroxidase homology domain results in proteins (dark grey background) that contain all three domains. Nox5 can also "revert" to the classical Nox family, since one of its splice forms contains only the gp91*phox*-homology domain and not the calcium-binding domain.[7]

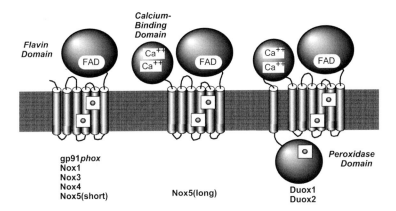

Figure 5. Predicted Topology of the Nox and Duox Enzymes. The transmembrane arrangement of the Nox1-4, Nox5 and Duox types of enzymes were predicted from the domain structures shown in Figure 4. The FAD domain resides on the cytosolic side of the membrane. For the Duox enzymes, the peroxidase domain is on the opposite side of the membrane from the FAD domain.

6.4 Predicted Membrane Topology of Nox/Duox Enzymes

The predicted topology of Nox1-4 enzymes (Fig. 5) is based largely on the existence of a cluster of 6 predicted transmembrane alpha helices, plus the above-described homologies with known proteins. Although no X-ray or NMR structure is available, many of the features of the topology have been experimentally verified for gp91*phox*. For example, the structure at left correctly places known glycosylation sites of gp91phox on the outside of the membrane[30] and a known binding site for p47*phox*[31] on the cytosolic side. In addition, this places the flavoprotein domain on the cytosolic side where it has access to NADPH. The transmembrane arrangement of alpha helices places 5 conserved histidines in these structures so that 4 of these are aligned in such a way as to bind hemes more or less in the outer and inner leaflets of the bilayer.[32] In effect, this structure provides a conduit for electrons to move from NADPH on the cytosolic side to flavin --> heme1 --> heme2 --> oxygen to generate superoxide transmembrane to the cytosol.

Nox5 builds on this basic structure by adding the calcium-binding domain at the N-terminus. This places the calcium-binding domain on the cytosolic side, where it can presumably respond to calcium fluxes, and where it may interact with the flavoprotein domain. Duox builds on the Nox5 structure by adding an additional transmembrane alpha helix and a peroxidase homology domain, both at the N-terminus. The presence of the additional transmembrane alpha helix places the peroxidase domain transmembrane to the cytosol (Fig. 5). Thus, while it remains to be proven, the Duox protein is expected to have one FAD plus three heme groups. The transmembrane arrangement of domains is functionally important, since reactive oxygen (specifically hydrogen peroxide) produced on the outside of the cell provides one of the substrates for peroxidative reactions that are catalyzed by the peroxidase domain.

6.6 Tissue Expression of Nox/Duox Enzymes

The tissue distribution of human Nox and Duox enzymes may provide clues as to their biological functions in mammals. In general, each isoform shows a fairly specific but somewhat overlapping range of tissue expression, often with a predominant location. For example, Nox1 is abundant in colonic epithelium,[2] while Nox4 is highly expressed in kidney epithelium,[28,29] hence the origin of the alternate name "Renox". However, by RT-PCR, Nox4 is also found in vascular endothelium, colonic cell lines, osteoclasts, placenta, ovary and pancreas, and Nox1 is found also in rat vascular smooth muscle, placenta and prostate.[7] Nox3 is seen in embryonic tissues, but has not been found in adult tissues. The long form Nox5 is expressed predominantly in spleen, sperm and mammary gland, with lesser amounts in cerebrum, while its short splice form is found in embryonic tissues. Both Duox1 and Duox2 are expressed in thyroid epithelium[33] and colon.[4] Duox1 is also found in cerebellum, while Duox2 is located also in pancreatic islets and prostate. While no generalized conclusions can be drawn from these expression patterns, the tissue expression patterns must be taken into account in any hypotheses regarding function.

6.7 Regulation of Nox Enzymatic Activity

Expression of Nox1 or Nox4 in fibroblasts results in increased generation of both superoxide and hydrogen peroxide. This implies that these enzymes, unlike the phagocyte enzyme, possess intrinsic basal activity. This is consistent with the finding that growth factors such as Keratinocyte Growth Factor (KGF) in colon and Angiotensin II and Platelet-Derived Growth Factor (PDGF) in vascular smooth muscle elevate both basal ROS generation and Nox1 gene expression. In addition, PDGF and Angiotensin II acutely elevate ROS in vascular smooth muscle in a response that peaks well before levels of Nox1 become elevated. Suppression of Nox1 expression eliminates this response, indicating that the acute elevation of ROS is mediated in part by Nox1. Thus, ROS production by Nox1 is mediated both at the level of gene expression and acute regulation. The mechanism of the acute regulation is not well understood, but might involve phosphorylation and/or cytosolic regulatory proteins identical to or homologous to the phox proteins found in phagocytes. In this regard, a homolog of p47*phox* termed p41*nox* (Fig. 6) has recently been discovered (Geiszt and Leto, unpublished, and Cheng and Lambeth, unpublished), and like Nox1, is expressed in colon epithelium. The similar domain structure of p47*phox* and p41*nox* strongly suggests an analogous role in the regulation of Nox enzymes. Both proteins contain the PX domain, which is involved in interaction with lipids, as well as two SH3 (Src Homology 3) domains. The latter participate in inter- and intra-molecular interactions with proline-rich domains such as those found in p22*phox*. Both p47*phox* and p41*nox* contain a C-terminal proline-rich domain (PRD in Fig. 6). Notably, p41*nox* lacks a polybasic region that contains most of the phosphorylation sites that participate in activating p47*phox*.

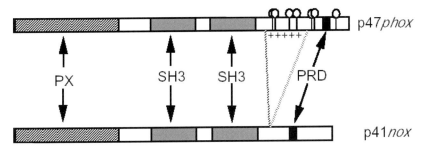

Figure 6. Domain Structure of p47phox and p41nox. A homolog of p47phox, one of the cytosolic proteins that regulates the phagocyte NADPH oxidase, has been identified in colon cells. p41nox and p47phox share the presence of a PX domain, two SH3 (Src-homology 3) domains, and a proline-rich domain (PRD). P41nox lacks a polybasic site that is present in p47phox, indicated by "+". Phosphorylation sites (circles) on p47phox that regulate its function are lacking in p41nox

As predicted from its structure, Nox5 is acutely regulated by calcium. In cells expressing Nox5, calcium-EGTA activates ROS generation,[26] presumably by altering protein interactions between this domain and the flavoprotein domain. While it remains to be proven, it can be speculated that the calcium-binding domain serves like the cytosolic factors for gp91*phox* to regulate electron flow. Duox1 and Duox2 are also likely to be regulated by cytosolic calcium, and calcium stimulation of a preparation containing Duox2 has been reported.[34]

6.8 Possible Functions of Nox Enzymes in Innate Immunity

Based on the known function of gp91*phox,* a role for other Nox and Duox enzymes in innate immunity seems reasonable. Several of the Nox enzymes are located in epithelial cells that provide a barrier between the organism and the microbe-rich outside world. For example, Nox1, Duox1 and Duox2 are expressed in relatively high levels in colonic epithelium, which serves as a barrier to the large quantities of bacteria present in the colon. Consistent with a role in innate immunity, Nox1 in colonic epithelial cells can be induced by the inflammatory mediator γ-interferon. In addition, Nox1 is present in gastric pit cells, and both Nox1 and p67*phox* were induced by LPS from *Helobacter pylori*, coinciding with induction of superoxide generation by these cells.[35] Nox4, which is highly expressed in kidney epithelium, might guard against retrograde bacterial infection. Duox1 is also located in bronchiolar and salivary duct epithelium, where it might be expected to provide reactive oxygen for microbicidal function. Thus, a role for Nox and Duox enzymes in innate immunity remains a likely possibility, although the case at present is largely circumstantial.

6.9 Functions of Nox Enzymes in Signal Transduction

Additional functions are strongly suggested by the localization of Nox and Duox enzymes in cells or tissues that do not function as a barrier to the outside world (*e.g.*, cerebellum, osteoclasts, vascular smooth muscle and endothelium). Other chapters in this volume document roles for reactive oxygen in signal transduction related to apoptosis and mitogenic regulation, and this seems a likely function for the Nox enzymes in some cell types. Functions in both innate immunity and in growth-related signal transduction are by no means mutually exclusive. In addition to microbial killing, clonal expansion of cells via mitogenic stimulation is a common feature of the immune response of several cell types as well as an essential feature of repair and replacement of disrupted barrier cells. Thus, the use by biology of non-toxic doses of otherwise toxic ROS molecules as intracellular or intercellular mediators may represent one of many examples of the opportunistic use in biology of the same molecule for different functions.

A role for Nox enzymes in signal transduction related to cell growth was first noted in NIH 3T3 cells that had been stably transfected with Nox1.[2] Surprisingly, these cells showed a transformed appearance and showed increased growth rates. The effect was general, because expression of Nox1 in an epithelial cell line also increased growth rates.[36] Nox1 functions as a potent oncogene, since injection of Nox1 expressing cells into athymic mice resulted in rapid tumor formation. Although the effect could be explained in part by increased mitogenic rate, this did not fully explain the marked tumorigenicity, which resulted also from Nox1-generated mediators serving as a potent angiogenic stimulus, causing marked stimulation of the growth and invasion of new blood vessels in the tumor. Nox1 expression resulted in the induction or repression of a large number of genes,[37] many related to growth, cell cycle, signal transduction or cancer. Unexpectedly, Nox1 failed to induce a number of proteins such as superoxide dismutases, catalase, glutathione peroxidase and others that are markers of an oxidative stress. In addition, the oxidation state of glutathione pools is not significantly altered, as would be expected during an oxidative stress. Thus, the low basal quantities of reactive oxygen that are generated in Nox1-expressing cells are not sufficient to cause a classical oxidative stress, and should rather be referred to as "oxidative signaling".

What then is (are) the signal molecule(s) and what are the pathways that are responsible for oxidative signaling, and how do these relate to Nox1-mediated gene expression and transformation? The mediator related to mitogenic signaling, gene expression, and angiogenesis has been identified as hydrogen peroxide, since catalase overexpression in Nox1-expressing cells reverses these effects.[37] Thus, while the enzyme generates superoxide, it is rapidly dismutated to form hydrogen peroxide, which functions either as the mediator itself or as a precursor. A large number of candidates exist for the direct target or targets of Nox1-generated hydrogen peroxide. Several amino acids can be oxidized by reactive oxygen including cysteine and methionine. Low pKa cysteine residues, which occur in a variety of enzymes, and transcription factors, are unusually reactive and in some cases their oxidation affects enzymatic function. For example, protein tyrosine phosphatases utilize an active site cysteine in their catalytic mechanisms, and this can be reversibly oxidized upon exposure to ROS, resulting in inhibition of enzyme activity. This mechanism markedly increases protein tyrosine phosphorylation following treatment of cells with growth factors such as PDGF and EGF.[38,39] In addition, a large number of transcription factors have low pKa cysteine groups, the oxidation of which affects their function.[40] Whether these mechanisms are relevant to Nox1-expressing fibroblasts, which generate rather low amounts of ROS, is unknown. However, MAP kinase pathways including the ERK1,2 pathway and the JNK pathway are activated by Nox1, and this activation is reversed by chemical antioxidants or catalase (Go, Y.M., Ritsick, D. and Lambeth, J.D., unpublished data). Both of these pathways regulate gene expression in these cells, and the ERK1,2 pathway is essential for the transformed phenotype. Thus, Nox1, via its generation of hydrogen peroxide, activates MAP kinase signaling cascades that are linked to gene expression and transformation.

6.10 Function of Duox Enzymes in Model Organisms and Humans

The predicted topology of Duox proteins (Fig. 5) suggests that this class of enzymes will both generate reactive oxygen and consume it. In preparations of Duox from thyroid membranes, only hydrogen peroxide is detected,[34] suggesting either that the enzyme directly generates hydrogen peroxide, or that superoxide is formed but is rapidly dismutated without release into the medium. The hydrogen peroxide that is produced is one substrate for peroxidases such as the thyroid peroxidase, lactoperoxidase or others. Alternatively, the hydrogen peroxide might be utilized directly by the peroxidase domain of Duox to catalyze the oxidation of an unknown substrate. The active site of the peroxidase domain of Duox is somewhat divergent in amino acid sequence from other peroxidases, giving rise to speculation that it would not have an enzymatic activity.[27] However, the peroxidase domain expressed in *E. coli* is highly active in carrying out peroxidative reactions,[4] indicating that in the presence of hydrogen peroxide and an appropriate substrate, the peroxidase domain will be active. Thus, it seems likely that Duox enzymes catalyze the coupled generation and peroxidative utilization of hydrogen peroxide.

Because its biological function in mammals is unknown, the genetically more tractable model organism *C. elegans* has been used to begin to understand the biological function(s) of Duox enzymes.[4] *C. elegans* Duox is expressed almost exclusively in hypodermal cells, the outermost layer of cells in the nematode. Disruption of Duox expression using RNA interference resulted in phenotypes reminiscent of those seen in collagen biosynthetic defects that are known to affect the structure of the cuticle, the exoskeletal structure overlying the hypodermal cells. Duox-impaired animals showed a severe disruption in the structure of the cuticle. This extracellular matrix structure is normally stabilized by inter-chain cross-linking between tyrosine residues, resulting from C-C bond formation between the tyrosine rings to generate dityrosine and trityrosine. In the RNA interference animals, di- and tri-tyrosine linkages were absent. The expressed peroxidase domain itself when supplied with hydrogen peroxide was capable of catalyzing tyrosine cross-linking *in vitro*. Thus, in nematodes, Duox functions in the peroxidative cross-linking of tyrosines in extracellular proteins to stabilize the structure of the cuticle.

The function of Duox1 and Duox2 in mammals is less clear. Both enzymes are expressed in thyroid,[4,27,34] where they are located in the plasma membrane.[34,41] There, they serve as a source of hydrogen peroxide that is utilized by thyroid peroxidase to carry out iodination of thyroglobulin, one of the steps in thyroid hormone biosynthesis. It is also speculated that these enzymes generate hydrogen peroxide in bronchioles and in salivary ducts, where it is utilized by lactoperoxidase as a microbicidal mechanism. It has been suggested that the peroxidase domain has superoxide dismutase activity, allowing rapid dismutation of superoxide so that the exclusive product of the enzyme is hydrogen peroxide. However, while peroxidases can react with superoxide, they do not generate hydrogen peroxide.[42] Thus, the role of the peroxidase domain of mammalian Duox enzymes is currently unexplored. Precedent from the *C. elegans* model suggests that the enzyme is designed to carry out peroxidative reactions on extracellular molecules such as extracellular matrix, small molecules, or microbes that are attempting to invade epithelial surfaces. Understanding of the function of Duox enzymes in higher organisms therefore represents an important area for future investigation.

6.11   Pathological Roles of Nox/Duox Enzymes

Excess reactive oxygen, often referred to as "oxidative stress", has been linked to a variety of diseases, including atherosclerosis, hypertension, cancer, neurodegenerative diseases (e.g. Alzheimer's, Parkinson's disease), diabetes, and aging. While these conditions may be linked to classical oxidative stress involving oxidative damage to biomolecules, the identification of ROS as signaling molecules indicates that, in some cases, overproduction of ROS may lead to diseases of signaling. Thus, chronic overexpression or inappropriate activation of Nox enzymes might be expected to be associated with such diseases.

Intriguingly, many human cancer cells overproduce hydrogen peroxide.[43] That this may be causally linked to the cancer phenotype is supported by the observation that low levels of hydrogen peroxide and superoxide stimulate cell growth [for review see[24]]. PDGF generates ROS in vascular smooth muscle cells, and elimination of ROS blocks the mitogenic effect of PDGF.[44] While the source of mitogenic ROS has not been clear, both Ras and Rac have been implicated in its generation.[45,46] This has lead to speculation that the origin is an oxidase similar to the phagocyte NADPH oxidase. That overexpression or inappropriate activation of Nox-like enzymes might plausibly contribute to tumorigenesis is supported by the ability of Nox1 to cause cell transformation and to render both fibroblasts and epithelial cells highly tumorigenic.[2,36] As described above, the mechanism for Nox1-induced tumorigenesis involves both increased mitotic rate and stimulation of angiogenesis in part via VEGF overproduction.

This leads to the question of whether Nox enzymes might account for the ROS generation in tumor cells. Nox isoforms have been detected in a variety of cancer cells lines, including the following: 1) Nox1 in colon and prostate cells; 2) Nox4 in ovarian, colon, and glioblastoma cells; 3) Nox5 in ovarian, colon and glioblastoma cells.[7] In addition, in recent unpublished studies from our laboratory, Nox1 mRNA was significantly elevated in 30-50% of tissue from both human colon and prostate cancers when compared with adjacent normal tissues. Thus, while cancer is a complex disease arising from multiple mutational and other events, Nox enzymes may be a contributing factor. These enzymes therefore represent an attractive target for development of novel anticancer drugs.

Overproduction of ROS has also been implicated in cardiovascular disease, including both hypertension and arteriosclerosis.[47,48] While the pathogenic mechanisms are complex and involve a variety of cell types, hypertrophy and/or hyperproliferation of smooth muscle cells contribute to both conditions. Excess Angiotensin II results in hypertension in part due to hyperproliferation and hypertrophy of vascular smooth muscle cells. This hormone stimulates reactive oxygen generation in these cells, and antisense to p22*phox* inhibits both ROS generation and growth stimulation.[49] This originally suggested the involvement of a gp91*phox*-like enzyme. Although it is still unclear whether p22*phox* binds to and regulates Nox1, subsequent studies showed that Nox1 is induced by Angiotensin II, PDGF and other growth-promoting factors.[50,51] Studies further showed that Nox1 mediates the acute agonist-stimulated generation of ROS in rat vascular smooth muscle, because elimination of Nox1 prevents cell proliferation in response to growth factors.[2,50] Similarly, prostaglandin F2α causes increased protein synthesis resulting in hypertrophy of vascular smooth muscle cells. Elimination of Nox1-dependent ROS production using either inhibitors or a ribozyme approach eliminated the increased protein synthesis seen in response to prostaglandin F2α.[52] Thus, Nox1 plays a critical role in rat in vascular smooth muscle cell growth related to hypertension. In addition, Nox1 is implicated in restenosis, or closure of the vessel due to proliferation of smooth muscle cells. In this model of atherosclerotic disease, physical removal of the vascular endothelial layer of cells using balloon catheter injury to the rat carotid vessel, resulted in induction of Nox1, gp91*phox* and p22*phox* in smooth muscle and fibroblasts, coinciding with increased generation of ROS and cell proliferation.[53] These effects may involve direct modulation by Nox1-generated ROS of signaling enzymes such as p38 kinase, Akt, c-Src and others[54,55] and may also be a result of the destruction by superoxide of nitric oxide (NO). As discussed in other chapters, NO plays an important role in vascular biology by inhibiting smooth muscle cell growth and migration, and by inducing relaxation of the vascular smooth muscle.

While ROS have been implicated in other diseases and indeed in aging itself, a role for Nox enzymes in these conditions remains unexplored. Aging is characterized by cumulative oxidative damage to DNA and proteins, the latter, for example, by oxidation of cysteine and methionine residues. Oxidized proteins can also become cross-linked and resist proteosomal degradation.[56] Oxidized proteins may lose function and accumulate, contributing to cellular senescence. Neurodegenerative diseases such as Alzheimer's and Parkinson's Diseases are similarly characterized by accumulation of oxidized proteins in the brain.[57,58] Similar changes occur as the normal brain ages, but are greatly accelerated in neurodegenerative diseases. ROS are also implicated in endothelial dysfunction seen in type II diabetes.[59,60] ROS levels increase when advanced glycation end products interact with their cell surface receptors. Elevated ROS in diabetes may contribute to the development of diabetic complications such as atherosclerosis and diabetic retinopathy. Thus, while a specific link has not yet been established between Nox homologs and aging, neurodegenerative diseases, or diabetes, these remain fertile areas for future exploration.

References

1. McCord, J. M., and Fridovich, I. 1969. Superoxide dismutase. An enzymic function for erythrocuprein (hemocuprein). *J. Biol. Chem.* **244:** 6049-55.

2. Suh, Y.-A., Arnold, R. S., Lassegue, B., Shi, J., Xu, X., Sorescu, D., Chung, A. B., Griendling, K. K., and Lambeth, J. D. 1999. Cell transformation by the superoxide-generating oxidase Mox1. *Nature* **401:** 79-82.

3. Lambeth, J. D., Cheng, G., Arnold, R. S., and Edens, W. E. 2000. Novel homologs of gp91phox. *TIBS* **25:** 459-461.

4. Edens, W. A., Sharling, L., Cheng, G., Shapira, R., Kinkade, J. M., Edens, H. A., Tang, X., Flaherty, D. B., Benian, G., and Lambeth, J. D. 2001. Tyrosine cross-linking of extracellullar matrix is catalyzed by Duox, a multidomain oxidase/peroxidase with homology to the phagocyte oxidase subunit gp91phox. *J. Cell Biol.* **154:** 879-891.

5. Sbarra, A. J., and Karnovsky, M. L. 1959. The biochemical basis of phagocytosis. Metabolic changes during the ingestion of particles by polymorphonuclear leukocytes. *J. Biol. Chem.* **234:** 1355-1368.

6. Babior, B. M. 1995. The respiratory burst oxidase. *Curr. Opin. Hematol.* **2:** 55-60.

7. Cheng, G., Cao, Z., Xu, X., Van meir, E. G., and Lambeth, J. D. 2001. Homologs of gp91^{phox}: cloning and tissue expression of Nox3, Nox4, and Nox5. *Gene* **131:** 140.

8. Tauber, A. I., Borregaard, N., Simons, E., and Wright, J. 1983. Chronic Granulomatous Disease: A Syndrome of Phagocytic Oxidase Deficiencies. *Medicine* **62:** 286-309.

9. Sumimoto, H., Sakamoto, N., Nozaki, M., Sakaki, Y., Takeshige, K., and Minakami, S. 1992. Cytochrome b558, a component of the phagocyte NADPH oxidase, is a flavoprotein. *Biochem. Biophys. Res. Commun.* **186:** 1368-1375.

10. Rotrosen, D., Yeung, C. L., Leto, T. L., Malech, H. L., and Kwong, C. H. 1992. Cytochrome b$_{558}$: The flavin-binding component of the phagocyte NADPH oxidase. *Science* **256:** 1459-1462.

11. Segal, A. W., West, I., Wientjes, F., Nugent, J. H. A., Chavan, A. J., Haley, B., Garcia, R. C., Rosen, H., and Scrace, G. 1992. Cytochrome b_{-245} is a flavocytochrome containing FAD and the NADPH-binding site of the microbicidal oxidase of phagoctyes. *Biochem. J.* **284:** 781-788.

12. Yu, L., Quinn, M. T., Cross, A. R., and Dinauer, M. C. 1998. Gp91(phox) is the heme binding subunit of the superoxide-generating NADPH oxidase. *Proc. Nat. Acad. Sci.* **95:** 7993-7998.

13. Parkos, C. A., Dinauer, M. C., Walker, L. E., Rodger, A. A., Jesaitis, A. J., and Orkin, S. H. 1988. Primary structure and unique expression of the 22-kilodalton light chain of human neutrophil cytochrome. *Proc. Nat. Acad. Sci.* **85:** 3319-3323.

14. de Mendez, I., Homayounpour, N., and Leto, T. 1997. Specificity of p47^{phox} SH3 domain interactions in NADPH oxidase assembly and activation. *Mol. Cell. Biol.* **17:** 2177-2185.

15. Sumimoto, H., Hata, K., Mizuki, K., Ito, T., Kage, Y., Sakaki, Y., Fukumaki, Y., Nakamura, M., and Takeshige, K. 1996. Assembly and activation of the phagocyte

NADPH oxidase: Specific Interaction of the N-terminal Src homology 3 domain of p47phox with p22phox is required for activation of the NADPH oxidase complex. *J. Biol. Chem.* **36:** 22152-22158.

16. Borregaard, N., Heiple, J. M., Simons, E. R., and Clark, R. A. 1983. Subcellular localization of the b-cytochrome component of the human neutrophil microbicidal oxidase: Translocation during activation. *J. Cell Biol.* **97:** 52-61.

17. Borregaard, N., Lollike, K., Kjeldsen, L., Sengelov, H., Bastholm, L., Nielsen, M. H., and Bainton, D. F. 1993. Human neutrophil granules and scretory vesicles. *Eur. J. Haematol.* **51:** 187-198.

18. Wientjes, F. B., Hsuan, J. J., Totty, N. F., and Segal, A. W. 1993. p40^{phox}, a third cytosolic component of the activation complex of the NADPH oxidase to contain *src* homology 3 domains. *Biochem. J.* **296:** 557-561.

19. El Benna, J., Faust, L., and Babior, B. 1994. The phosphorylation of the respiratory burst oxidase component p47^{phox} during neutrophil activation.   Phosphorylation of sites recognized by protein kinase C and by proline-directed kinases. *J. Biol. Chem.* **269:** 23431-23436.

20. El Benna, J., Dang, P. M-C., Gaudry, M., Fay, M., Morel, F., Hakim, J., and Gougerot-Pocidalo, M.   1997. Phosphorylation of the respiratory burst oxidase subunit p67^{phox} during human neutrophil activation. *J. Biol. Chem.* **272:** 17204-17208.

21. Lambeth, J. D. 2000. Regulation of the phagocyte respiratory burst oxidase by protein interactions. *J. Biochem. Mol. Biol.* **33:** 427-439.

22. Han, C.-H., Freeman, J. L. R., Lee, T., Motalebi, S. A., and Lambeth, J. D. 1998. Regulation of the neutrophil respiratory burst oxidase: Identification of an activation domain in p67^{phox}. *J. Biol. Chem.* **273,** 16663-16668.

23. Nisimoto, Y., Motalebi, S., Han, C.-H., and Lambeth, J. D. 1999. The p67^{phox} activation domain regulates electron transfer flow from NADPH to flavin in flavocytochrome b_{558}. *J. Biol. Chem.* **274:** 22999-23005.

24. Burdon, R.   1995. Superoxide and hydrogen peroxide in relation to mammalian cell proliferation. *Free Radic. Biol. Med.* **18:** 775-794.

25. Meier, B., Cross, A. R., Hancock, J. T., Kaup, F. J., and Jones, T. G. 1991. Identification of a superoxide-generating NADPH oxidase system in human fibroblasts. *B. J.* **275:** 241-245.

26. Banfi, B., Molinar, G., Maturana, A., Steger, K., Hegedus, B., Demaurex, N., and Krause, K.-H. 2001. A Ca^{2+}-activated NADPH oxidase in testis, spleen and lymph nodes. *J. Biol. Chem.* **276:** 37594-37601.

27. De Deken, X., Wang, D., Many, M. C., Costagliola, S., Libert, F., Vassart, G., Dumont, J. E., and Miot, F. 2000. Cloning of two human thyroid cDNAs encoding new members of the NADPH oxidase family. *J. Biol. Chem.* **275:** 23227-23233.

28. Geiszt, M., Kopp, J. B., Varnai, P., and Leto, T. L. 2000. Identification of Renox, an NAD(P)H oxidase in kidney. *Proc. Nat. Acad. Sci.* **97:** 8010-8014.

29. Shiose, A., Kuroda, J., Tsuruya, K., Hirai, M., Hirakata, H., Naito, S., Hattori, M., Sakaki, Y., and Sumimoto, H. 2001. A novel superoxide-producing NAD(P)H oxidase in kidney. *J. Biol. Chem.* **276,** 1417-1423.

30. Wallach, T. M., and Segal, A. W. 1997. Analysis of glycosylation sites on gp91phox, the flavocytochrome of the NADPH oxidase, by site-directed mutagenesis and translation in vitro. *Biochem. J.* **321:** 583-585.

31. Biberstine-Kinkade, B., Yu, L., and Dinauer, M. 1999. Mutagenesis of an arginine- and lysine-rich domain in the gp91^{phox} subunit of the phagocyte NADPH-oxidase flavocytochrome b_{558}. *J. Biol. Chem.* **274:** 10451-10457.

32. Bibersine-Kinkade, K. J., DeLeo, F. R., Epstein, R. I., LeRoy, B. A., Bauseef, W. M., and Dinauer, M. C. 2001. Heme-ligating histidines in flavocytochrome b558. Identification of specific histidines in gp91^{phox}. *J. Biol Chem.* **276:** 31105-31112.

33. Dupuy, C., Pomerance, M., Ohayon, R., Noel-Hudson, M.-S., Deme, D., Chaaraoui, M., Francon, J., and Virion, A. 2000. Thyroid oxidase (THOX2) gene expression in the rat thyroid cel line FRTL-5. *Biochem. Biophys. Res. Commun.* **277:** 287-292.

34. Dupuy, C., Ohayon, R., Valent, A., Noe-Hudson, M., Dee, D., and Virion, A. 1999. Purification of a novel flavoprotein involved in the thyroid NADPH oxidase. *J. Biol. Chem.* **274:** 37265-37269.

35. Kawahara, T., Teshima, S., Oka, A., Sugiyama, T., Kishi, K., and Rokutan, K. 2001. Type I Helicobacter pylori lipopolysaccharide stimulates toll-like receptor 4 and activates mitogen oxidase 1 in gastric pit cells. *Infect. Immun.* **69:** 4382-9.

36. Arbiser, J. L., Petros, J. A., Klafter, R., Govindajaran, B., McLaughlin, E. R., Brown, L. F., Cohen, C., Moses, M., Kilroy, S., Arnold, R. S., and Lambeth, J. D. 2002. Reactive oxygen generated by Nox1 triggers the angiogenic switch. *Proc. Nat. Acad. Sci.* **99:** 715-720.

37. Arnold, R. S., Shi, J., Murad, E., Whalen, A. M., Sun, C. Q., Polavarapu, R., Parthasarathy, S., Petros, J. A., and Lambeth, J. D. 2001. Hydrogen peroxide mediates the cell growth and transformation caused by the mitogenic oxidase Nox1. *Proc. Nat. Acad. Sci.* **98:** 5550-5555.

38. Lee, S.-R., Kwon, K.-S., Kim, S.-R., and Rhee, S. G. 1998. Reversible inactivation of protein tyrosine phosphatase 1B in A431 cells stimulated with epidermal growth factor. *J. Biol. Chem.* **273:** 15366-15372.

39. Barrett, W. C., DeGnore, J. P., Keng, Y-F., Zhang, Z.-Y., Yim, M. B., and Chock, P. B. 1999. Role of superoxide radical anion in signal transduction mediated by reversible regulation of protein-tyrosine phosphatase 1B. *J. Biol. Chem.* **274:** 34543-34546.

40. Schmidt, K., Amstad, K., Cerutti, P., and Baeuerle, P. 1995. The roles of hydrogen peroxide and superoxide as messengers in the activation of transcription factor NF-kB. *Chem. Biol.* **2:** 13-22.

41. Caillou, B., Dupuy, C., LaCroix, L., Nocera, M., Talbot, M., Ohayon, R., Deme, D., Bidart, J.-M., Schlumberger, M., and Virion, A. 2001. Expression of reduced

nicotinamide adenine dinucleotide phosphate oxidase (Thox1, LNOX, Duox) genes and proteins in human thyroid tissues. *J. Clin. Endocrin. Metabol.* **86:** 3351-3358.

42. Heinecke, J. W., and Shapiro, B. M. 1990. Superoxide peroxidase activity of ovoperoxidase, the cross-linking enzyme of fertilization. *J. Biol. Chem.* **265:** 9241-6.

43. Szatrowski, T., and Nathan, C. 1991. Production of large amounts of hydrogen peroxide by humor tumor cells. *Cancer Res.* **51:** 794-798.

44. Sundaresan, M., Yu, Z.-X., Ferrans, V. J., Irani, K., and Finkel, T. 1995. Requirement for generation of H_2O_2 for platelet-derived growth factor signal transduction. *Science* **270:** 296-299.

45. Sundaresan, M., Yu, Z.-X., Ferrans, V. J., Sulciner, D. J., Gutkind, J. S., Irani, K., Goldschmidt-Clermont, P. J., and Finkel, T. 1996. Regulation of reactive-oxygen-species generation in fibroblasts by Rac 1. *Biochem J.* **318:** 379-382.

46. Irani, K., Xia, Y., Zweier, J., Sollott, S., Der, C., Rearon, E., Sundaresan, M., Finkel, T., and Goldschmidt-Clermont, P. 1997. Mitogenic signaling mediated by oxidants in Ras-transformed fibroblasts. *Science* **275:** 1649-1652.

47. Griendling, K. K., and Alexander, R. W. 1997. Oxidative stress and cardiovascular disease. *Circulation* **96:** 3264-3265.

48. Griendling, K. K., Sorescu, D., and Ushio-Fukai, M. 2000. NAD(P)H oxidase: role in cardiovascular biology and disease. *Circ. Res* **86:** 494-501.

49. Ushio-Fukai, M., Zafari, A. M., Fukui, T., Ishizaka, N., and Griendling, K. 1996. p22[phox] is a critical component of the superoxide-generating NADPH/NADPH oxidase system and regulates angiotensin II-induced hypertrophy in vascular smooth muscle cells. *J. Biol. Chem.* **271:** 23317-23321.

50. Lassegue, B., Sorescu, D., Szocs, K., Yin, Q., Akers, M., Zhang, Y., Grant, S. L., Lambeth, J. D., and Griendling, K. K. 2001. Novel gp91(phox) homologues in vascular smooth muscle cells: Nox1 mediates angiotensin II-induced superoxide formation and redox-sensitive signaling pathways. *Circ. Res.* **88:** 888-94.

51. Wingler, K., Wunsch, S., Kreutz, R., Rothermund, L., Paul, M., and Schmidt, H. H. 2001. Upregulation of the vascular NAD(P)H-oxidase isoforms Nox1 and Nox4 by the renin-angiotensin system in vitro and in vivo. *Free Radi. Bio.l Med.* **31:** 1456-64.

52. Katsuyama, M., Fan, C., and Yabe-Nishimura, C. 2002. NADPH oxidase is involved in prostaglandin F2α -induced hypertrophy of vascular smooth muscle cells: induction of NOX1 by PGF2α . *J. Biol. Chem.* **277:** 13438-42.

53. Szocs, K., Lassegue, B., Sorescu, D., Hilenski, L. L., Valppu, L., Couse, T. L., Wilcox, J. N., Quinn, M. T., Lambeth, J. D., and Griendling, K. K. 2002. Upregulation of Nox-based NAD(P)H oxidases in restenosis after carotid injury. *Arterioscler, Thromb, Vas,c Biol.* **22:** 21-7.

54. Griendling, K. K., Sorescu, D., Lassegue, B., and Ushio-Fukai, M. 2000. Modulation of protein kinase activity and gene expression by reactive oxygen species and their role in vascular physiology and pathophysiology. *Arterioscle.r Thromb. Vasc. Biol.* **20:** 2175-83.

55. Griendling, K. K., and Ushio-Fukai, M. 2000. Reactive oxygen species as mediators of angiotensin II signaling. *Regul. Pept.* **91:** 21-7.

56. Shringarpure, R., and Davies, K. J. 2002. Protein turnover by the proteasome in aging and disease(1,2). *Free Radic. Biol. Med.***32:** 1084-9.

57. Butterfield, D. A., and Lauderback, C. M. 2002. Lipid peroxidation and protein oxidation in Alzheimer's disease brain: potential causes and consequences involving amyloid beta-peptide-associated free radical oxidative stress(1,2). *Free Radic. Bio. Med.* **32:** 1050-60.

58. Mattson, M. P. 2002. Involvement of superoxide in pathogenic action of mutations that cause Alzheimer's disease. *Meth. Enzymol.* **352:** 455-74.

59. Laight, D. W., Carrier, M. J., and Anggard, E. E. 2000. Antioxidants, diabetes and endothelial dysfunction. *Cardiovasc. Res.* **47:** 457-64.

60. Laight, D. W., Desai, K. M., Anggard, E. E., and Carrier, M. J. 2000. Endothelial dysfunction accompanies a pro-oxidant, pro-diabetic challenge in the insulin resistant, obese Zucker rat in vivo. *Eur. J. Pharmacol.* **402:** 95-9.

Chapter 7

NO SYNTHESIS AND NOS REGULATION

Ulrich Forstermann, Huige Li, Petra M. Schwarz and Hartmut Kleinert

7.1 Introduction

Nature has turned to nitric oxide (NO), a small bioactive gas, to mediate vital servoregulatory as well as cytotoxic functions. Among the important physiological functions of NO are neurotransmission, regulation of vascular tone and inhibition of platelet aggregation (via activation of soluble guanylyl cyclase), gene transcription and mRNA translation (via iron-responsive elements), and post-translational modifications of proteins (via ADP-ribosylation).[1] In higher concentrations, NO is capable of destroying parasites and tumor cells by inhibiting iron-containing enzymes or directly interacting with the DNA of these cells.[2] However, excess or inappropriate production of NO can be deleterious. This chapter will focus on the mechanisms by which cells accomplish and regulate the production of NO.

7.1.1 NO synthase isoforms (NOS)

In mammals, three distinct isoforms of NOS (L-arginine, NADPH:oxygen oxidoreductases, nitric oxide forming; EC 1.14.13.39) have been identified. These are products of different genes, with different localization, regulation, catalytic properties and inhibitor sensitivity, and with 51–57% homology between the human isoforms. All NOS isoforms catalyze a reaction of L-arginine, NADPH, and oxygen to the free radical NO, L-citrulline and $NADP^+$. nNOS (also referred to as NOS1 or NOS I) is a low output enzyme that is primarily expressed in neurons. iNOS (also referred to as NOS2 or NOS II) is a high output NOS whose expression can be

H. J. Forman, J. Fukuto, and M. Torres (eds.), Signal Transduction by Reactive Oxygen and Nitrogen Species: Pathways and Chemical Principles, ©2003, Kluwer Academic Publishers. Printed in the Netherlands

induced in a wide range of cells and tissues by cytokines and other agents. eNOS (also referred to as NOS3 or NOS III), a low output enzyme is the prototypical isoform being found in endothelial cells. The Ca^{2+}-dependence of NO synthesis distinguishes the NOS isoforms, with nNOS and eNOS having a much higher Ca^{2+} requirement than iNOS. The nature of nNOS and eNOS as low output enzymes and iNOS as a high output enzyme depends not so much on the conversion rate of the different isozymes, but rather reflects the short-lasting, pulsatile, Ca^{2+}-activated NO production of nNOS and eNOS versus the continuous, Ca^{2+}-independent NO production of iNOS. Reported K_m values for L-arginine and v_{max} values for NO- or L-citrulline formation differ between the isozymes. Reported half saturating L-arginine concentrations seem to be similar for nNOS and eNOS (1.5-3 µM),[3,4] and somewhat higher for iNOS (up to 30 µM).[5,6]

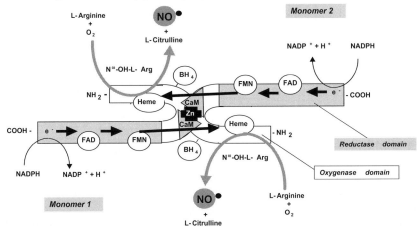

Figure 1: Principal structure of NOSs and scheme of NOS catalysis. As a common feature, all NOSs exhibit a bidomain structure, in which a C-terminal reductase domain of one monomer (that contains binding sites for FAD, FMN and NADPH) is linked by a CaM recognition site to an N-terminal oxygenase domain of the other monomer (containing binding sites for BH_4, heme and L–arginine).[9,11] A zinc tetrathiolate center seems to stabilize the NOS dimer without being essential for catalysis.[12] The electron flow is in the direction NADPH > FAD > FMN.[13] The electron flow in the NOS dimer goes from the flavins in the reductase domain of one monomer to the heme iron in the oxygenase domain of the other monomer.[14] For all three NOS isoforms, BH_4 as well as heme and L-arginine promote and/or stabilize the active dimeric form. The presence of heme appears to be mandatory, with BH_4 and L-arginine promoting dimer formation and stabilizing the dimer once formed. Recent studies with inhibitors of dimerization[15] on iNOS in intact cells suggest that once a dimer is formed there is little or no significant return to the monomer; it is unclear whether this will be true also for the other isoforms. It is generally accepted that L-arginine, NADPH and O_2 are substrates of all NOS isoforms and that $NADP^+$ and L-citrulline are products. It is also well accepted that N-hydroxy-L-arginine is an intermediate in the reaction.[16] Although NO is a likely product of this reaction, it is not totally clear whether NOSs directly synthesize NO (see below).

7.1.2 Structural aspects of NOS

The known NOS enzymes are dimeric in their active form and contain the bound cofactors (6R)-5,6,7,8-tetrahydrobiopterin (BH_4), FAD, FMN and iron protoporphyrin IX (heme) (Figure 1). Recent years have seen significant advances in structural studies of the NOS isoforms, with crystal structures being solved for oxygenase domains of murine and human iNOS[7-9] and bovine and human eNOS.[10] An unexpected finding in the crystal structure was a zinc tetrathiolate center located at the dimer interface, since the presence of zinc in NOS had not previously been detected[10] (Figure 1).

7.1.3 NOS Catalysis, Role of Cofactors

The principal scheme of NOS catalysis is depicted in Figure 1. Interestingly, there has been some debate as to whether NOS directly synthesizes NO. Other potential reactive nitrogen species, e.g. nitroxyl (NO⁻), peroxynitrite (ONOO⁻) or nitrosothiols could in principle be formed first, with subsequent reactions producing NO.[17] In experiments with purified nNOS, Schmidt *et al.*[18] were unable to detect authentic NO formation unless they added superoxide dismutase (SOD). They suggested that NO⁻ was being formed by nNOS, and that the NO⁻ was subsequently oxidized to NO by SOD.

Sub-saturating concentrations of BH_4 or L-arginine promote NADPH oxidase activity of NOS leading to superoxide formation at the expense of NO synthesis. It appears that nNOS and eNOS have a particular tendency to catalyze this uncoupled reaction;[19,20] iNOS has also been shown to produce detectable – yet lower – amounts of superoxide.[21] Occupation of the L-arginine binding site alone does not prevent superoxide formation. However, the concomitant addition of L-arginine and BH_4 restores NO production and abolishes superoxide generation by NOS.[20] Because of the very rapid reaction of superoxide with NO,[22] synthesis of both species by the same enzyme is likely to result in peroxynitrite formation.

The role of the reductase domain of the NOSs resembles that of the cytochrome P450 reductases. The FAD and FMN in the reductase domain accept electrons from NADPH and pass them on to the heme domain (Figure 1). Also the NOS heme domain has similarities to cytochrome P450 enzymes. However, there are functional differences between the hemes in the two types of enzymes. Unlike cytochrome P450 enzymes, NOS performs two separate oxidation steps, one to form N-hydroxy-L-arginine and the other to convert this intermediate to NO (or NO⁻). It has been known for more than 10 years that BH_4 is required for NOS activity, but there has been considerable controversy as to what its role(s) in catalysis may be (for review see[17]). Currently it seems likely that BH_4 plays a role in NOS catalysis as a redox agent. In addition to this catalytic role, BH_4 may play additional roles in promoting NO synthesis.

7.2 nNOS

7.2.1 Regulation of nNOS Activity

Neuronal-derived NO can be either beneficial or detrimental depending on the cellular context. nNOS activity must therefore be tightly regulated. nNOS is a Ca^{2+}/CaM-dependent enzyme.[23,24] Its activity is regulated by physiological changes in the intracellular Ca^{2+} concentration. CaM was the first protein shown to interact with nNOS[23,25] (and is necessary for the enzymatic activity of all three isoforms).

CaM binding increases the rate of electron transfer from NADPH to the reductase domain flavins[26] and also triggers electron transfer from the reductase domain to the heme center.[27] nNOS and eNOS, but not iNOS, contain 40–50 amino acid inserts in the middle of the FMN binding subdomain. These inserts have been described as auto-inhibitory loops.[28] Analysis of a mutant of nNOS with this loop deleted has shown that the insert acts by destabilizing CaM binding at low Ca^{2+} and by inhibiting electron transfer from FMN to the heme in the absence of Ca^{2+}/CaM.[29] Hormones and neurotransmitters leading to increases in intracellular Ca^{2+} will activate nNOS in neurons and probably other cells. As Ca^{2+} concentrations change rapidly in neuronal cells, the physiological NO production of these cells likely consists of short-lasting puffs of NO rather than a continuous NO formation.

In addition to this acute mechanism of regulation, other parameters are likely to be important for NO production by nNOS. An important one may be the subcellular localization of nNOS protein. In neurons, both soluble and particulate protein is found. Depending on the individual study, the particulate enzyme represents between 30 and 60% of the total neuronal nNOS protein.[30-32] In electron microscopy studies of kidney macula densa cells, the neuronal isoform has been seen associated mainly with small vesicles [33]. In skeletal muscle, nNOS protein is mostly particulate.[34,35] The particulate localization of part of the nNOS protein is most probably due to the PDZ/GLGF motif found in the NH_2-terminal sequence of the nNOS protein. This motif participates in protein-protein interactions with several other membrane-associated proteins.[36,37] In brain, synaptic association of nNOS is mediated by the binding of the PDZ/GLGF motif to the postsynaptic density protein PSD-95 and/or to the related PSD-93 protein.[38] NMDA receptors are also known to be associated with PSD-95.[39] Thus, the complexed nNOS may be the enzyme primarily activated during NMDA receptor-mediated Ca^{2+} influx into neuronal cells.

In skeletal muscle, the muscle-specific isoform (nNOSµ) is attached to the sarcolemma-dystrophin complex via the PDZ/GLGF motif and mainly interacts with α1-syntrophin.[40] High amounts of this enzyme are found at motor endplates.[35] nNOS splice variants lacking the PDZ/GLGF domain do not associate with PSD-95 in brain or with skeletal muscle sarcolemma.[38] Also, this membrane association of nNOS is likely to facilitate activation of the enzyme during sarcolemmal depolarization.

Recently, nNOS heterocomplexes with the molecular chaperone Hsp90 have been detected.[41] nNOS was not directly activated by Hsp90 *in vitro*, but this protein may help to incorporate heme into nNOS.[41]

The N-terminal extension of nNOS also contains a binding site for the 89-amino-acid protein PIN (protein inhibitor of nNOS).[41] PIN binds nNOS with a 1:2 stoichiometry.[42] The initial data suggested that only nNOS associated with PIN and that PIN was inhibitory by destabilizing nNOS.[42] However, more recent reports claim that PIN neither inhibits nNOS nor promotes monomerization.[43] Others have found that PIN has no effect on nNOS dimerization, but inhibits the activity of nNOS, eNOS and iNOS.[44] The identification of PIN as a light chain of dynein has led to the suggestion that PIN may be an axonal transport protein for nNOS rather than a regulator of nNOS activity.[43]

Finally, nNOS can be phosphorylated at serine and threonine residues by Ca^{2+}/CaM -dependent protein kinase II, and protein kinases A, C and G.[45,46] *In vitro*, phosphorylation by Ca^{2+}/CaM kinase II and protein kinases A and G reduces catalytic activity of the enzyme.[45,47]

7.2.2 Regulation of nNOS Expression

Although constitutively expressed in several cell types, the expression level of nNOS can be regulated dynamically. nNOS mRNA upregulation seems to represent a response of neuronal cells (and others) to stress or injury induced by physical, chemical and biological agents. Examples include pain induced by formalin,[48] brain injury caused by heat stress, axonal transection, colchicine treatment or experimental allergic encephalomyelitis,[49-51] and chronic electrical stimulation of skeletal muscle.[52] A similar response is observed in the rat paraventricular nucleus and adrenal cortex during immobilization stress,[53,54] and after mechanical or pathological lesions including spinal cord, axonal or nerve injuries,[55-57] hypophysectomie,[58] or middle cerebral artery occlusion leading to focal ischemia.[59] nNOS protein expression increased in olfactory bulb neurons during infections with vesicular stomatitis virus.[60] nNOS expression appears to be regulated also by changes in neuronal activity; its expression was higher when rat cerebellar granule cells were kept in the presence of 10 mM K^+ compared with 25 mM K^+.[61] In the same cell type, the inhibition of the glutamatergic transmission drastically increased nNOS expression.[62] nNOS expression can also be triggered by steroid hormones. It has been demonstrated that estradiol and pregnancy could induce nNOS expression in several tissues of the rat.[63-65] In male rats, testosterone treatment has been described to stimulate the expression of the neuronal isoform in the penis.[66] Corticosterone treatment resulted in an upregulation of heme oxygenase-2 and a concomitant decrease of nNOS transcription in rat brain.[67] Lithium and tacrine, a cholinesterase inhibitor currently used in the treatment of the symptoms of Alzheimer's disease, increased the expression of nNOS synergistically in the hippocampus of the rat. This effect could be inhibited by corticosterone.[68] Recent evidence obtained in our laboratory with murine N1E-115 neuroblastoma cells indicated that glucocorticoids inhibit nNOS expression by reducing the transcription of the gene.[69]

A downregulation of nNOS expression has been documented in guinea pig skeletal muscle and rat brain after *in vivo* treatment with bacterial lipopolysaccharide (LPS).[70,71] Treatment of rats with LPS or interferon-γ (IFN-γ decreased the expression of nNOS also in brain, stomach, rectum and spleen.[72]

7.3 iNOS

7.3.1 Regulation of iNOS Activity

It has been thought that iNOS is primarily regulated at the expressional level, by transcriptional and post-transcriptional mechanisms. Unlike nNOS and eNOS, little is known about the regulation of iNOS enzyme activity. However, recently, two proteins have been identified that interact with iNOS and regulate its activity. In the central nervous system, the protein kalirin appears to inhibit iNOS by preventing the formation of iNOS dimers and may play a neuroprotective role during inflammation.[73] In murine macrophages, a 110-kDa protein (named NAP110) has been identified that directly interacts with the amino terminus of iNOS, thereby preventing dimer formation and inhibiting NOS catalytic activity.[74]

7.3.2 Signaling Pathways Regulating iNOS Expression

Murine cells generally express iNOS in response to bacterial LPS, stimulatory cytokines such as IFN-γ, interleukin-1β (IL-1β, interleukin-6 (IL-6), tumor necrosis factor-α (TNF-α and/or other compounds. In contrast, most human cells require a complex cytokine combination including IFN-γ, IL-1β and TNF-α for iNOS induction.[75] The analysis of signal transduction pathways involved in the induction of iNOS expression revealed a marked heterogeneity and cell- and species specificity.[76] The different inducers of iNOS expression have been shown to activate different signal pathways in parallel. Also many compounds have been described that enhance or inhibit LPS/cytokine-induced iNOS expression in different cells and tissues by activating/blocking a wide variety a signal transduction pathways.[76]

7.3.3 Regulation of iNOS Transcription

The regulation of iNOS expression seems to be the main regulatory step to control iNOS activity. iNOS protein can synthesize NO continuously until the enzyme becomes degraded.[77] Regulation of the transcription of the iNOS gene has been believed to be the most important control mechanism for iNOS expression.

7.3.4 iNOS Promoter Sequences

The sequences of the avian, bovine, human, murine, rat and rainbow trout iNOS promoters exhibit homologies to binding sites for many transcription factors such as activating protein-1 (AP-1), CCAAT-enhancer box binding protein (C/EBP), camp-responsive element binding protein (CREB), interferon regulatory factor-1 (IRF-1), nuclear factor IL-6 (NF-IL6), nuclear factor-κB (NF-κB, octamer factor (oct) and signal transducer and activator of transcription-1α (STAT-1α).[77] All promoters contain a TATA box about 30 bp from the transcription start site. Near the TATA box, all mammalian promoters contain binding sites for the transcription factors NF-κB, octamer factor and for transcription factors induced by TNF-α. At position -900 bp, the rat, the murine and the human promoter display binding sites for IFN-γ-induced transcription factors, γ-interferon activation sequence (GAS), γ-interferon responsive element (ISRE), interferon-stimulated response element (ISRE). The murine and rat promoters also contain a NF-κB site at this position.

In transfection experiments, a murine 1000 bp iNOS promoter fragment showed full promoter functionality in homologous[78,79] and heterologous[80] cells. Fragments (1.7 and 4.5 kb) of the rat iNOS promoter showed cytokine-mediated inducibility in transfection experiments with rat mesangial cells. However, the inducibility of the transfected promoter was much lower (2 to 4-fold) than the inducibility of the endogenous iNOS mRNA expression.[81-83] In contrast, using primary rat hepatocytes[84,85] or rat insulin-producing RINm5F cells,[86] a much higher cytokine-mediated inducibility of a 1700 bp or 1000 bp rat iNOS promoter fragment was observed. Therefore, as with the human iNOS promoter (see below), the cellular background seems markedly to influence the inducibility of the rat iNOS promoter.

In transient transfection experiments with human A549 alveolar-, AKN bilary-, or DLD-1 colon epithelial cells, a human 1000-bp iNOS promoter fragment showed low, but significant basal promoter activity, but no induction with cytokines.[87] When this human 1000 bp promoter fragment was transfected into murine RAW 264.7 macrophages, it became markedly inducible with LPS, IFN-γ and IL-1β.[88] In human AKN-, A549- or DLD1 cells only iNOS promoter fragments larger than 3.8 kb showed significant induction with cytokines.[87,89] Thus, transcription factor binding sites relevant for cytokine induction of the human iNOS promoter in these cells seem to be located upstream of -3.8 kb of the 5'-flanking sequence of the human iNOS gene. Maximal cytokine-mediated induction (5 to 10-fold) was seen with a 16 kb human iNOS promoter fragment in transiently or stably transfected A549-, AKN- or DLD1 cells.[87,90,91] Interestingly, data by another group[92-94] showed much higher inducibility (up to 50-fold) of a human 8.3 kb iNOS promoter fragment in human A549 cells. The reason of this marked difference in promoter inducibility is not clear at this time.

7.3.4.1 *Nuclear Factor-κB (NF-κB)*

The transcription factor NF-κB[95] seems to be a central target for activators or inhibitors of iNOS expression. LPS, IL-1β, TNF-α and oxidative stress for instance have been shown to induce iNOS expression in different cell types by activating NF-κB. Also the inhibition of iNOS expression by numerous agents (e.g. glucocorticoids, antioxidants, TGF-β) has been shown to result from inhibition of NF-κB activation. This inhibition may result from direct capture of NF-κB by protein-protein interactions,[80,96] inhibition of nuclear translocation of NF-κB,[97] inhibition of NF̄κB transactivation activity[98] or from enhancement of the expression of I-κB, the specific inhibitor of NF-κB.[99]

Analyses using murine,[79,100] rat[82,101] and human[93,102] cells showed the important role of NF-κB binding sites for the induction of the iNOS promoter activity. In the murine iNOS promoter, the downstream NF-κB binding site (positions -76 bp to -85 bp) seems to be the most important one,[100] but also the upstream NF-κB site (positions -974 bp to -960 bp) seems to have some functionality and cooperativeness with the downstream site.[101] In rat mesangial cells a minimal promoter fragment ranging from the transcriptional start site up to –111 bp containing the downstream NF-κB site was sufficient to confer IL-1β-mediated iNOS promoter activation.[82] In the rat promoter, a third NF-κB site located at -901 to -892 bp with an opposite orientation ("a reverse NF-κB site") has been identified. Mutation of this "reverse NF-κB" reduced the IL-1β/IFN-γ-induced promoter activity of a 1.4-kb rat iNOS promoter fragment in transfection experiments using rat aortic smooth muscle cells.[103]

For the NF-κB binding sites in the human iNOS promoter, conflicting results have been published. Taylor *et al.*[102] reported interactions of multiple NF-κB binding sites between positions –5.2 kb and –6.5 kb in the induction of the iNOS promoter, one of which has been shown to be a mixed NF-κB/STAT-1α site.[104] In contrast, these authors attributed little functionality to the downstream NF-κB site of the human iNOS promoter (positions -115 bp to -106 bp, near the TATA-box).

Other groups, however, reported on the importance of this downstream NF-κB binding site for human iNOS promoter activity.[93,105]

7.3.4.2   Interferon Regulatory Factor-1 (IRF-1)

In murine RAW 264.7 macrophages, the essential role of the IRF-1 binding site (positions –913 bp to –923 bp) for the induction of the murine iNOS promoter has been shown.   Mutations of this binding site blocked the IFN-γ-mediated enhancement of the LPS-induced iNOS promoter activity in macrophages.[106] Supershift and *in vivo* footprint experiments showed the involvement of IRF-1 in the protein complexes bound to this binding site after LPS/IFN-γ incubation.[106,107]   Also, nuclear extracts from rat cardiomyocytes incubated with LPS and IFN-γ contained proteins binding to the homologous ISRE on the rat iNOS promoter.[108]   Finally, in macrophages and glial cells from IRF-1[-/-] mice, LPS/IFN-γ incubation resulted in a markedly reduced iNOS expression.[109-112]   However islets, chondrocytes and hepatocytes from IRF-1[-/-] mice showed normal iNOS induction in response to LPS/IFN-γ.[110,112]   In these cell types, IRF-1 does not seem to be essential for iNOS induction.   These data underline the cell specificity of the regulation of iNOS expression.   Induction of IRF-1 activity by IFN-γ depends on protein de novo synthesis and on STAT-1α activation.[113]

7.3.4.3   Signal Transducer and Activator of Transcription-1α (STAT-1α)

All mammalian iNOS promoters contain several homologies to STAT-1α binding sites (GAS). Gao *et al.*[114] reported that binding of STAT-1α to the GAS of the murine iNOS promoter (positions –934 bp to –942 bp) is required for optimal induction of the iNOS gene by IFN-γ and LPS. Also the IFN-γ-mediated enhancement of IL-1β-induced promoter activity in rat RINm5F cells was dependent on the GAS elements around position -900 bp of the rat iNOS promoter.[86]   In rat aortic smooth muscle cells, IFN-γ, although not able to induce iNOS expression, enhanced IL-1β-mediated iNOS induction. In a 1.4-kb rat iNOS promoter fragment, deletion of a GAS element increased IL-1β-induced activity, but inhibited IFN-γ-enhanced activity, suggesting a two-way effect of the GAS site on iNOS induction: enhancing induction through STAT-1α activation and inhibiting induction through a non-IFN-γ-mediated mechanism.[115]   In human A549 or DLD1 cells, inhibition of the IFN-γ-activated tyrosine kinase JAK2 by tyrphostin B42 (AG 490) reduced STAT-1α DNA binding activity and iNOS expression.[116]   STAT-1α is likely to be involved in the stimulation of iNOS induction, either directly by binding to the iNOS promoter, or indirectly by inducing IRF-1 activity.   In addition, positive cooperativeness between STAT-1α and IRF-1 is likely to occur for the induction of the iNOS promoter.[86]   Furthermore, iNOS induction was blocked in macrophages from mice with a disrupted STAT-1α gene.[117]

Interestingly in recent reports, a negative regulation of INF-γ-STAT-1α-mediated induction of iNOS expression by suppressor of cytokine signaling (SOCS) proteins have been described. In murine bone marrow-derived macrophages, overexpression of SOCS-1 or -3 protein inhibited induction of the mouse iNOS promoter by suppressing interactions between STAT-1α and GAS sites.[118] Cytokine-induced iNOS expression was enhanced in islets from SOCS-1 [-/-] mice.[119]

7.3.4.4 *cAMP-Induced Transcription Factors; cAMP-Responsive Element Binding Protein (CREB), CCAAT-Enhancer Box Binding Protein (C/EBP)*

Eberhardt *et al.*[82] reported that a CCAAT box (C/EBP binding site, positions – 155 bp to –163 bp) was essential for cAMP-mediated (but not IL-1β-mediated) induction of the rat iNOS promoter. Supershift experiments showed the involvement of C/EBP-β and C/EBP-δ in the cAMP regulation of the rat iNOS promoter. In rat C-6 glial cells transfection with dominant negative of C/EBP resulted in downregulation of the rat iNOS promoter activity. Also overexpression of wild-type activating transcription factor 2 (ATF2) enhanced, whereas a phosphorylation-defective form of ATF2 suppressed, rat iNOS promoter activity.[120] In a heterologous transfection system using rat pulmonary microvascular endothelial cells, IL-1β was able to induce the activity of a transient transfected human - 1034/+88 bp iNOS promoter fragment. C/EBP sites within the -205/+88 bp region of the human iNOS promoter (along with a NF-κB site at -115/-106 bp) were shown to be responsible for induction of iNOS promoter by IL-1β. Overexpression of C/EBP-α, C/EBP-δ, and liver-enriched activator protein (LAP) activated the cotransfected human -1034/+88 bp iNOS promoter fragment, whereas overexpression of liver-enriched inhibitory protein (LIP) strongly suppressed the activity.[121]

7.3.4.5 *Activating Protein-1 (AP-1)*

The role of the transcription factor AP-1 in the regulation of iNOS expression is controversial. Analysis of constructs containing murine iNOS promoter fragments with mutations in two AP1-like sites (U site, around -1125; L site, around -518) transiently transfected in J774A.1 cells showed a significant increase in the LPS-induced promoter activity when the L site was mutated, but not when the U site was mutated.[122] This suggests that the murine iNOS expression is negatively regulated, at least in part, through the L site in response to LPS. In human DLD1- and A549/8 cells, overexpression of AP-1 inhibited the promoter activity of a 7 kb human iNOS promoter fragment.[116] Similarly, agents like calyculin A, okadaic acid, phenylarsine oxide and anisomycin that markedly enhance c-jun and c-fos mRNA expression and AP-1 binding activity, inhibited cytokine-induced iNOS expression in human DLD1- and A549/8 cells.[116,123] In stably transfected A549/8 cells containing a 16-kb iNOS promoter-luciferase reporter gene construct, all of these compounds also reduced the cytokine-induced iNOS promoter activity.[123] Also in human DLD1 cells, the deletion of this repressor sequences enhanced the promoter activity of a 16-kb human iNOS promoter fragment (own unpublished results). In contrast to the above data, Marks-Konzalik *et al.*[93] described a marked inhibition (90%) of cytokine-induced activity of an 8.3-kb iNOS promoter fragment transfected into A549 cells after site-directed mutagenesis of an AP-1 binding sequence located 5301 bp upstream of the transcription start site. Furthermore, the addition of pharmacological or molecular MAP/ERK kinase-1 and p38 MAPK inhibitors, which markedly inhibited LPS/cytokine-induced 8.3-kb iNOS promoter activity, significantly diminished AP-1 binding. In supershift experiments, these authors detected Jun D and Fra-2 as components of the cytokine-induced AP-1-DNA-protein complexes.[93,94] At this time there is no explanation for the discrepancies between the above-mentioned findings.

7.3.4.6 Other Transcription Factors

Beside the important transcription factors describe above, a number of other transcription factors had been shown to regulate iNOS expression. Octamer factor (Oct), nonhistone high mobility group protein I(Y) (HMG-I(Y))[124,125] and hypoxia induced factor-1 (HIF-1)[126] have been shown to stimulate iNOS transcription. On the other hand, activation of peroxisome proliferator-activated receptor (PPAR)-γ decreased iNOS expression.[127] Also other members of the nuclear receptor superfamily such as the retinoic acid receptor (RAR)-α,[128] the estrogen receptor-(ER)-β,[129] the androstane receptor (CAR) and the pregnane X receptor (PXR)[130] have been implicated in iNOS regulation.

7.3.5 Regulation of iNOS mRNA Stability

In murine cells and tissues, iNOS expression has been explained mainly by iNOS promoter activity.[78,79] However, in murine peritoneal macrophages, TGF-β reduced IFN-γ-induced iNOS expression partly by destabilizing the iNOS mRNA.[131] In murine RAW 264.7 macrophages, a cycloheximide-sensitive iNOS mRNA degradation mechanism has been reported[132] and the potentiating effect of IFN-γ on LPS-induced iNOS mRNA expression has been attributed to an IFN-γ-mediated stabilization of the iNOS mRNA.[133] Recently, Korhonen *et al.*[134] described that dexamethasone inhibited LPS- (but not LPS and IFN-γ)-induced iNOS expression by destabilizing iNOS mRNA. Finally, Soderberg et al. described septic shock-dependent specific binding of the heterogeneous nuclear ribonucleoproteins hnRNP L and hnRNP I (PTB) to the 3′-UTR of the murine iNOS mRNA.[135] In rat vascular smooth muscle cells, BH_4 was shown to modulate iNOS mRNA stability.[136] The enhancing effect of PKC-δ activation on IL-1β-induced iNOS expression in rat INS β-cells was attributed to PKC-δ-mediated stabilization of the iNOS mRNA.[137]

In human A549-, AKN- or DLD1 cells nuclear run-on and transfection experiments revealed a significant basal activity of the human iNOS promoter that was only enhanced 2 to 5-fold by cytokines.[91,138,139] In contrast, iNOS mRNA expression could be detected only after cytokine induction. These findings suggest that regulation of iNOS mRNA stability plays an important role for iNOS induction.

Sequence analysis of the human iNOS mRNA reveals 4 sequence motifs (AUUUA) in the 3'-untranslated region (3'-UTR) that have been shown to confer mRNA destabilization to cytokine- and oncogene mRNAs.[140] The same sequence motifs are found twice in the 3'-UTR of the murine iNOS mRNA, and 4 times in the rat iNOS mRNA. In transfection experiments with human A549 or DLD1 cells, the 3'-UTR of the human iNOS mRNA destabilized the mRNA of a luciferase reporter gene. Gel retardation experiments showed high affinity interaction of the AUUUA binding protein HuR with the human iNOS 3′-UTR. HuR is known to positively regulate the stability of several inducible mRNAs[141]. Stable overexpression of HuR in human DLD1 cells resulted in an upregulation of cytokine-induced iNOS expression. Accordingly, downregulation of HuR expression in DLD1 cells resulted in reduction of cytokine-induced iNOS expression.[91] The exact mechanism by which HuR regulates iNOS mRNA expression in response to cytokine incubation remains to be analyzed.

Whereas HuR had been described to be a major positive regulator of the stability of AUUUA-containing mRNAs, the hnRNP D family of proteins (also named AUF1) is believed to destabilize AUUUA-containing mRNAs.[142,143] In gel retardation experiments, p37AUF1 protein showed high affinity binding to the human iNOS 3′-UTR interacting with AUUUA elements other than those binding HuR. Stable overexpression of p37AUF1 in DLD1 cells downregulated human iNOS mRNA expression.[144] Thus, severeal RNA binding proteins seem to be involved in the regulation of human iNOS mRNA stability.

7.3.6 Regulation of iNOS mRNA Translation and Protein Stability

Regulation of the iNOS mRNA translation and protein stability has also been described. Human primary cardiomyocytes express iNOS mRNA, but no iNOS protein in cell culture.[145] However, infection of these cells with a retroviral vector containing only the iNOS coding region (no 5'- and 3'-UTR) produced a marked iNOS protein expression. These data suggest that human cardiomyocytes express (protein) factors that inhibit iNOS mRNA translation by interacting with the 5'- and/or 3'-UTR sequences of the iNOS mRNA.[145] Also the inhibition of iNOS expression by different agents, e.g. by transforming growth factor-β1 (TGF-β1) in primary murine macrophages[131] and dexamethasone in rat mesangial cells[146] has been described to result from iNOS mRNA and protein destabilization.

By stable constitutive expression of iNOS in human epithelial kidney HEK293 cells, an involvement of the proteasome pathway in the degradation of iNOS protein has been demonstrated.[147] In human intestinal carcinoma HT29- and DLD-1 cells overexpression of caveolin-1 (cav-1) decreased cytokine-induced iNOS expression. Also a direct cav-1-iNOS protein-protein interaction has been shown. In the presence of proteasome inhibitors, the amount of iNOS protein was enhanced and cav-1 overexpression did not modify cytokine-induced of iNOS mRNA expression. These data suggest, that the interaction of cav-1 protein with iNOS-protein enhanced proteosomal degradation of iNOS.[148]

7.4 eNOS

eNOS was first identified and isolated from vascular endothelial cells,[149] but it is also expressed in several non-endothelial cell types and tissues, such as cardiac myocytes, blood platelets, some epithelium and neurons (for details see[150]). NO production from the endothelium is regulated at the level of enzyme activity and gene expression.

7.4.1 Regulation of eNOS Enzyme Activity

7.4.1.1 Posttranscriptional Modifications and Cellular Localization

Of the three NOS isoforms, only eNOS is acylated by both myristate and palmitate. eNOS is co-translationally and irreversibly myristoylated at an N-terminal glycine residue while palmitoylation occurs post-translationally and reversibly at cysteine residues Cys15 and Cys26. Dual acylation of eNOS is required for efficient localization to Golgi membranes and to plasmalemmal caveolae of endothelial cells.[151] The targeting of eNOS to caveolae is likely to facilitate the interactions of eNOS with other colocalized signaling and regulatory molecules.

7.4.1.2 Regulatory Proteins

The localization of eNOS within caveolae renders the enzyme inactive by its negative regulatory protein, caveolin, which is the major coat protein of caveolae.[152,153] Caveolin-1 interacts with eNOS via its so-called scaffolding domain.[153,154] Incubation of pure eNOS with peptides derived from the scaffolding domain of caveolin-1 results in inhibition of eNOS activity.[155] In cotransfection experiments, caveolin overexpression resulted in a reduction of eNOS activity.[154,155] The inhibitory effect of caveolin on eNOS has also been observed *in vivo*. Exposure of blood vessels to a membrane-permeable form of the caveolin-1 scaffolding domain results in uptake of the peptide into the endothelium and the adventitia, leading to a blockade of NO-mediated vasorelaxation.[156] In a rat model of cirrhosis, reduced NO production is associated with increased protein levels of caveolin-1 and increased binding of eNOS with caveolin.[157]

The Hsp90 family is a group of highly conserved stress proteins that are expressed in all eukaryotic cells. Hsp90 is associated with eNOS in resting endothelial cells. Stimuli that cause NO release, such as vascular endothelial growth factor (VEGF), histamine, fluid shear stress and estrogen, enhance the interaction between Hsp90 and eNOS in a time frame mirroring NO release.[158,159] Hsp90 can directly activate eNOS *in vitro*, and co-expression of eNOS with Hsp90 in COS cells increases NOS activity.[158] Hsp90 may act as an allosteric modulator of eNOS by inducing a conformational change in the enzyme that results in increased activity.[160]

7.4.1.3 Activation by Ca^{2+}/calmodulin

In unstimulated endothelial cells, eNOS remains inactive probably due to two mechanisms: (i) the binding of caveolin and (ii) the intramolecular "autoinhibitory" element (residues 594-645) within the FMN-binding domain, which inhibits eNOS activity and CaM binding.[28,161] Both hindrances can be overcome with Ca^{2+}/CaM.

In response to shear stress, as well as Ca^{2+}-mobilizing agents such as estradiol, bradykinin, acetylcholine and Ca^{2+} ionophore,[162] Ca^{2+}-bound CaM associates with eNOS, releases eNOS from caveolin-1,[163] and displaces the autoinhibitory element to allow CaM access to its binding site.[161] The activated eNOS-CaM complex generates NO until intracellular Ca^{2+} levels drop to the point where CaM dissociates and the inhibitory eNOS-caveolin complex reforms [162].

7.4.1.4 Regulation by Phosphorylation and Dephosphorylation

Ser^{1177} (human sequence) phosphorylation results in activation of eNOS. Kinases such as Akt/PKB,[164,165] PKA/PKG,[166] AMPK[167] and CaMK II[168] have been found to phosphorylate Ser^{1177} residue. Phosphorylation of this residue transforms eNOS into an electron-transfering enzyme by (i) imposing a negative charge at this residue, which increases electron flux at the reductase domain, and (ii) reducing CaM dissociation from eNOS.[169]

Stimuli for PI3K/Akt-mediated Ser^{1177} phosphorylation include shear stress, VEGF,[164,165] sphingosine 1-phosphate[170] and H_2O_2.[171] Phosphorylation of Ser^{1177} by CaMK II is stimulated by agonists of G-protein-coupled receptors, such as bradykinin/histamine.[168] Elevation of intracellular cAMP levels by stimulation of β_2-adrenoceptors results in eNOS activation.[172] PKA-dependent phosphorylation of Ser^{1177} can be the underlying mechanism.[166] Ischemic stress leads to Ser^{1177} phosphorylation via activation of the cardiac AMPK.[167]

Thr^{495} in the CaM-binding domain is phosphorylated in resting endothelial cells.[168] PKC and perhaps also AMPK [167] seem to be the kinases that maintain this phosphorylation.[168] (PKC-mediated) Thr^{495} phosphorylation inhibits eNOS activity.[168,173] In addition to Ser^{1177} phosphorylation, activation of eNOS by bradykinin, histamine and H_2O_2 is associated with Thr^{495} dephosphorylation (probably by protein phosphatase PP1) [168]. Ser^{1177} phosphorylation alone is only associated with a modest increase in enzyme activity.[165,169] A simultaneous Thr^{495} dephosphorylation is likely necessary to achieve maximal eNOS activation.[168] Thr^{495} phosphorylation reduces CaM binding,[167] and dephosphorylation of Thr^{495} enhances the association of CaM.[168]

Tyrosine phosphorylation of eNOS is associated with a decrease in the activity of the enzyme; however, the kinase(s) and the site(s) of the phosphorylation involved have not yet been identified.[174]

7.4.1.5 eNOS Activation Independent of Increases in $[Ca^{2+}]$

Although shear stress induces only a transient increase in intracellular Ca^{2+}, the resulting eNOS activation persists for hours. This long-lasting eNOS activation can also be observed in the absence of extracellular Ca^{2+}, and is not inhibited by the calmodulin antagonist calmidazolium [175].

Shear stress-induced eNOS activation is associated with enhanced levels of tyrosine phosphorylation and the stimulated NO synthesis can be blocked by tyrosine kinase inhibitors,[174,175] indicating that tyrosine phosphorylation of eNOS, or an associated regulatory protein is crucial for the Ca^{2+}-independent eNOS activation.

Phosphorylation of serine residues can also contribute to eNOS activation independent of increases in $[Ca^{2+}]$. Ser^{1177} phosphorylation enhances sensitivity of the enzyme to Ca^{2+}, and reduces CaM dissociation from eNOS, thereby allowing full eNOS activity at sub-physiological concentrations of Ca^{2+}.[165,169]

The lipid second messenger ceramide has also been shown to activate eNOS in a Ca^{2+}-independent manner.[176] This activation is associated with a translocation of eNOS from the endothelial membrane to intracellular sites.[176]   Also short-term exposure of endothelial cells to TNFα[177] or binding of high-density lipoprotein to scavenger receptors[178] have been shown to activate eNOS in endothelial cells; intracellular ceramide has been identified as the second messenger for both processes.

7.4.2 One Stimulus can Promote Multiple Mechanisms of eNOS Regulation

As discussed above, the activation of eNOS is a complex procedure. One stimulus may deal with many of the listed mechanisms. For example, an eNOS activation by receptor dependent agonists, e.g. bradykinin or histamine, involves Ca^{2+} elevation, Ser^{1177} phosphorylation, Thr^{495} dephosphorylation.[168]   Hsp90 binding,[158] and even a Ca^{2+}-independent component which is mediated by ceramide generation.[176] The shear stress-induced eNOS activation consists of an acute phase, which is mediated via a transient Ca^{2+} elevation, and a maintained phase, which is Ca^{2+}-independent and involves eNOS phosphorylation (on serine and tyrosine residues) and Hsp90 binding.

7.4.3 Regulation of eNOS Gene Expression

Although eNOS is a constitutively expressed enzyme, its expression is regulated by a number of biophysical, biochemical and hormonal stimuli, both under physiological conditions and in pathology.[150,179]

7.4.3.1 Shear Stress and Exercise

Shear stress produced by the flowing blood upregulates eNOS expression in cultured endothelial cells.[180] Also *in vivo*, increased levels of eNOS mRNA and protein are found in localized areas of blood vessels exposed to increased shear stress.[181]   Shear stress enhances eNOS gene transcription and stabilizes eNOS mRNA.[182]

The shear stress-induced eNOS transcription requires intracellular Ca^{2+} [183] and a pertussis toxin-sensitive G-protein.[184] The signal transduction is likely to involve a complex kinase cascade including Raf, Ras, MEK1/2, ERK1/2, and also the tyrosine kinase c-Src.[182]

Exercise training has assumed a major role in cardiac rehabilitation. Chronic exercise is associated with an increased eNOS expression in dogs.[185] Also patients with peripheral arterial occlusive disease show increased endogenous NO production after 14 days of physical exercise.[186]

7.4.3.2 Growth Status and Growth Factors

Proliferating endothelial cells possess higher eNOS mRNA and protein levels compared with confluent cells.[187] This results from a prolonged half-life of the eNOS mRNA.[188] In an *in vivo* model of endothelium denudation injury, as well as in a tissue culture wound model, eNOS protein and mRNA expression were significantly increased in regenerating and migrating endothelial cells,[189] which may be beneficial in the setting of arterial injury *in vivo*. The eNOS upregulation seems to be mediated, at least in part, via transforming growth factor -β_1 (TGF-β_1) which is highly expressed in injured arteries.[189] In bovine aortic endothelial cells, a putative nuclear factor 1 (NF-1) binding site in the bovine eNOS promoter proves to be responsible for the TGF-β_1-enhanced transcription.[190]

Fibroblast growth factors (aFGF and bFGF) increase eNOS expression. Transfection of the aFGF gene into spontaneously hypertensive rats leads to elevation of endothelial FGF levels and normalization of blood pressure, which is accompanied by an increased eNOS expression in the endothelium of thoracic aorta. Also in cultured bovine endothelial cells, bFGF enhances eNOS mRNA, protein and activity. The bFGF-induced eNOS expression seems to play an important role in angiogenesis.[179]

Also other growth factors, such as VEGF and platelet-derived growth factor (PDGF) enhance eNOS gene expression.[150]

7.4.3.3   Hormones

The vascular protective hormone estrogen upregulates eNOS mRNA and protein expression.[191] The increased eNOS expression resulted from an increased eNOS promoter activity with unchanged mRNA stability. In the absence of a *bona fide* estrogen responsive element in the human eNOS promoter, the increased eNOS promoter activity is likely to result from an enhanced binding activity of the transcription factor Sp1.[191]

In addition to its acute vasodilation effect, long-term treatment with insulin increases eNOS expression. The insulin-induced eNOS expression can be blocked by PI3K inhibitors. Chronic *in vivo* treatment of streptozotocin diabetic rats with insulin also upregulates aortic eNOS expression.[150]

In addition, angiotensin II and endothelin-1 upregulate eNOS expression, whereas thrombin downregulates eNOS gene expression by shortening the half-life of eNOS mRNA.[150]

7.4.3.4   Oxidative stress

Pathophysiological situations such as hypercholesterolemia, atherosclerosis, hypertension, smoking and diabetes are associated with increased oxidative stress. Recent evidence suggests that reactive oxygen species can increase eNOS expression. The mediator of eNOS upregulation seems to be H_2O_2 rather than superoxide.[192,193] The upregulation of eNOS mRNA expression results from increased transcription and stabilization.[193] The signaling cascade leading to H_2O_2-induced eNOS transcription is likely to involve CaM kinase II and janus kinase 2.[194]

7.4.3.5   Lysophosphatidylcholine (LPC)

LPC is generated from ox-LDL or from inflammatory cells and possesses pro-inflammatory and pro-atherogenic properties. In human[195] and bovine[196] endothelial cells, LPC has been shown to increase eNOS expression. It has been postulated that this LPC-induced eNOS expression represents an adaptive vasoprotective mechanism. The eNOS induction by LPC is a transcriptional event, dependent on transcription factor Sp1. The signal transduction involves a complex kinase cascade and protein phosphatase PP2A.[197]

7.4.3.6   *Statins, Rho GTPase, and the Actin Cytoskeleton*

Statins block HMG-CoA reductase, the rate-limiting step of cholesterol biosynthesis. In vascular endothelial cells, statins increase eNOS expression by stabilizing the eNOS mRNA, an effect that is unrelated to cholesterol reduction.[198,199] An eNOS upregulation by statins has also been documented *in vivo,* and this is often associated with protective effects, e.g. regression of atherosclerosis,[200,201] or augmentation of cerebral flow and reduction of cerebral infarct size.[202,203]

Rho GTPase has been identified as a negative regulator of eNOS expression. In human endothelial cells activation of Rho decreases eNOS expression, and, in contrast, inhibition of Rho increases eNOS expression.[204] Rho GTPases are subject to post-translational modification by geranylgeranylation. This targets them to the membrane and leads to their activation. By inhibiting HMG-CoA reductase, statins prevent the synthesis of L-mevalonate and geranylgeranylpyrophosphate (GGPP), and thus inhibit Rho geranylgeranylation and activation. Statin-induced eNOS expression can be bypassed and reversed with L-mevalonate and GGPP, indicating that inhibition of geranylgeranylation is involved.[198,204]

The actin cytoskeleton is a downstream sensor of Rho function. Endothelial cells overexpressing a dominant-negative RhoA mutant exhibit decreased actin stress fiber formation and increased eNOS expression.[205] Mice treated with the actin cytoskeleton disrupter cytochalasin D show increased vascular eNOS expression and activity.[205] Aortas from gelsolin$^{-/-}$ mice, which have genetic increases in actin stress fiber formation, have a decrease in eNOS expression, suggesting that the endothelial actin cytoskeleton and eNOS expression are inversely related.[205] Rho-mediated reduction of the endothelial actin cytoskeleton may represent the mechanic basis for the effect of statins on eNOS expression.[205]

7.4.3.7   *Angiotensin Converting Enzyme (ACE) Inhibitors and AT1-receptor Antagonists*

Beside their primary mode of action, inhibitors of ACE and AT1 angiotensin receptor blockers possess additional properties, including anti-atherosclerotic effects. An upregulation of eNOS expression may contribute to such effects.

ACE inhibitors upregulate eNOS in cultured endothelial cells,[206] as well as in experimental animals.[207,208] Spontaneously hypertensive rats show an upregulation of eNOS following ACE inhibition, which is associated with improved endothelium-dependent vasodilation, better cardiac function and an increased lifespan.[207] Also patients treated with the ACE inhibitor lisinopril show increased plasma levels of Nox.[209] Part of the eNOS-upregulating effect of ACE inhibitors is mediated through an increase in kinins and involves bradykinin B$_2$ receptors.[208]

Also the AT1-receptor antagonists (which have no effect on kinins) have been shown to normalize the diminished eNOS mRNA expression in stroke-prone spontaneously hypertensive rats, and in aortae from coarctation-hypertensive rats.[150]

7.4.3.8 Ca²⁺ Channel Blockers (CCB)

CCB are widely used for treatment of hypertension and ischemic heart disease. These drugs have also anti-atherosclerotic effects that are independent of the reduction of blood pressure. Incubation of vascular endothelial cells with felodipine increases eNOS protein expression and NO generation.[210] This effect has also been reported for other dihydropyridine CCBs like nifedipine and amlodipine,[211] for the benzothiazepine CCB diltiazem, but not for the phenylalkylamine CCB verapamil.[211] The upregulation of eNOS by dihydropyridine CCBs has also been observed in animals. Accordingly, benidipine inhibits intimal thickening of the carotid artery of mice by upregulating eNOS and increasing NO production.[212]

7.4.3.9 Alcohol and Wine

Individuals with moderate wine consumption enjoy significant reduction in cardiovascular mortality. This could, at least in part, be attributed to an increased expression or activity of eNOS, as shown in human endothelial cells.[213] No difference was detected between 'en barrique' and 'non-barrique' produced red wines, and also an equivalent amount of ethanol could not mimic the effect. These data indicate that the compounds stimulating eNOS expression derive from the grapes.[213] Interestingly, resveratrol, a polyphenolic phytoalexin found in grapes and wine, also increases eNOS expression.[214]

7.4.3.10 Glucocorticoids

Glucocorticoids are widely used as anti-inflammatory and immunosuppressant agents, with hypertension as a cardiovascular side effect. Downregulation of eNOS is likely to contribute to this phenomenon.[215] Glucocorticoids downregulate eNOS mRNA and protein expression in cultured endothelial cells, as well as in the aorta of glucocorticoid-treated rats. Treatment of rats with dexamethasone increases blood pressure progressively over 7 days without changing Na^+ and K^+ in plasma or urine, thereby excluding a mineralocorticoid-like mechanism. Glucocorticoids decrease eNOS mRNA stability and reduce the activity of the human eNOS promoter by decreasing the binding activity of the transcription factor GATA.[215] A decreased renal expression of eNOS (and iNOS) has also been reported in adrenocorticotropin (ACTH)-induced and corticosterone-induced hypertension.[216]

7.4.3.11 Staurosporine Analogs

Staurosporine and its analogs are widely used as protein kinase C (PKC) inhibitors. Independent of their PKC inhibitory effect, these compounds display an interesting structure-activity relationship in regulating eNOS expression.[217] Staurosporine and its glycosidic indolocarbazole analogs 7-hydroxystaurosporine (UCN-01) and 4'-N-benzoyl staurosporine (CGP 41251) enhance eNOS mRNA and protein expression in a concentration- and time-dependent manner. This upregulation is accompanied by an increased NO production in human endothelial cells. In contrast, the bisindolylmaleimide analogs GF 109203, Ro 31-8220 and Gö

6983 have no effect on eNOS expression. Gö 6976, a methyl- and cyanoalkyl-substituted nonglycosidic indolocarbazole derivative of staurosporine, even reduces eNOS expression. The upregulation of eNOS expression by staurosporine and glycosidic indolocarbazole analogs appears to be a transcriptional event with no effect on eNOS mRNA stability.[217]

7.4.4 Regulation of eNOS mRNA Stability

eNOS expression is regulated not only at the transcriptional level, but also post-transcriptionally. The stability of the eNOS mRNA can be altered by different compounds and stimuli. These include TNF-α, LPS, thrombin, statins, Rho GTPases, actin cytoskeleton, VEGF, ox-LDL, glucocorticoids, hypoxia and the proliferation status of the cells.[188,150]

TNF-α reduces eNOS mRNA half-life.[218] This destabilization seems to result from specific interaction of a TNF-α-induced cytosolic ~60 kDa protein with the 38 nt region (position +3785 nt to +3823 nt) of the eNOS 3'-UTR.[219,220] The eNOS mRNA and protein levels are higher in growing versus resting endothelial cells, resulting from an increased stability of eNOS mRNA.[188] The 43-nt region (position +3648 nt to +3690 nt) in the eNOS 3'-UTR seems to be responsible for the destabilization of eNOS mRNA.[188] Binding of a cytosolic protein (~51 kDa) to this region is 3-fold higher in confluent cells compared with proliferating cells.[188]

7.5 Conclusions

Of the three established NOS isozymes, nNOS and eNOS are low output, Ca^{2+}-activated enzymes whose main physiological function is signal transduction. They usually show a constitutive, basal expression, which however, can be modified by various agents and conditions. In addition to the intracellular Ca^{2+} level, post-translational modifications and the subcellular targeting of these enzymes determine their activity. iNOS is a high output enzyme that can produce cytotoxic amounts of NO. iNOS is mainly regulated at the level of expression, with transcriptional, post-transcriptional and translational mechanisms involved. The stimuli and conditions that determine iNOS expression are cell- and species-specific. Once expressed, iNOS does not seem to be subject to any major regulation of its enzymatic activity. Thus, nature has invented a large array of regulatory mechanisms controlling the production of the pluripotent molecule NO.

References

1. Förstermann, U., E. I. Closs, J. S. Pollock, M. Nakane, P. Schwarz, I. Gath and H. Kleinert. 1994. Nitric oxide synthase isozymes: Characterization, molecular cloning and functions. *Hypertension* **23**:1121-1131.

2. Nathan, C. F. and J. B. Hibbs. 1991. Role of nitric oxide synthesis in macrophage antimicrobial activity. *Curr Opin Immunol* **3**:65-70.

3. Mayer, B., M. John and E. Böhme. 1990. Purification of a calcium/calmodulin-dependent nitric oxide synthase from porcine cerebellum. Cofactor role of tetrahydrobiopterin. *FEBS Lett* **277**:215-219.

4. Pollock, J. S., U. Förstermann, J. A. Mitchell, T. D. Warner, H. H. H. W. Schmidt, M. Nakane and F. Murad. 1991. Purification and characterization of particulate endothelium-derived relaxing factor synthase from cultured and native bovine aortic endothelial cells. *Proc Natl Acad Sci USA* **88**:10480-10484.

5. Hevel, J. M., K. A. White and M. A. Marletta. 1991. Purification of the inducible murine macrophage nitric oxide synthase. Identification as a flavoprotein. *J Biol Chem* **266**:22789-22791.

6. Yui, Y., R. Hattori, K. Kosuga, H. Eizawa, K. Hiki and C. Kawai. 1991. Purification of nitric oxide synthase from rat macrophages. *J Biol Chem* **266**:12544-12547.

7. Li, H., C. S. Raman, C. B. Glaser, E. Blasko, T. A. Young, J. F. Parkinson, M. Whitlow and T. L. Poulos. 1999. Crystal structures of zinc-free and -bound heme domain of human inducible nitric-oxide synthase. Implications for dimer stability and comparison with endothelial nitric-oxide synthase. *J Biol Chem* **274**:21276-21284.

8. Fischmann, T. O., A. Hruza, X. D. Niu, J. D. Fossetta, C. A. Lunn, E. Dolphin, A. J. Prongay, P. Reichert, D. J. Lundell, S. K. Narula and P. C. Weber. 1999. Structural characterization of nitric oxide synthase isoforms reveals striking active-site conservation. *Nat Struct Biol* **6**:233-242.

9. Crane, B. R., A. S. Arvai, D. K. Ghosh, C. Wu, E. D. Getzoff, D. J. Stuehr and J. A. Tainer. 1998. Structure of nitric oxide synthase oxygenase dimer with pterin and substrate. *Science* **279**:2121-2126.

10. Raman, C. S., H. Li, P. Martasek, V. Kral, B. S. Masters and T. L. Poulos. 1998. Crystal structure of constitutive endothelial nitric oxide synthase: A paradigm for pterin function involving a novel metal center. *Cell* **95**:939-950.

11. Masters, B. S., K. McMillan, E. A. Sheta, J. S. Nishimura, L. J. Roman and P. Martasek. 1996. Neuronal nitric oxide synthase, a modular enzyme formed by convergent evolution: Structure studies of a cysteine thiolate-liganded heme protein that hydroxylates l-arginine to produce no. As a cellular signal. *Faseb J* **10**:552-558.

12. Hemmens, B., W. Goessler, K. Schmidt and B. Mayer. 2000. Role of bound zinc in dimer stabilization but not enzyme activity of neuronal nitric-oxide synthase. *J Biol Chem* **275**:35786-35791.

13. Adak, S., S. Ghosh, H. M. Abu-Soud and D. J. Stuehr. 1999. Role of reductase domain cluster 1 acidic residues in neuronal nitric- oxide synthase. Characterization of the fmn-free enzyme. *J Biol Chem* **274**:22313-22320.

14. Noble, M. A., A. W. Munro, S. L. Rivers, L. Robledo, S. N. Daff, L. J. Yellowlees, T. Shimizu, I. Sagami, J. G. Guillemette and S. K. Chapman. 1999. Potentiometric analysis of the flavin cofactors of neuronal nitric oxide synthase. *Biochemistry* **38**:16413-16418.

15. McMillan, K., M. Adler, D. S. Auld, J. J. Baldwin, E. Blasko, L. J. Browne, D. Chelsky, D. Davey, R. E. Dolle, K. A. Eagen, S. Erickson, R. I. Feldman, C. B. Glaser, C. Mallari, M. M. Morrissey, M. H. Ohlmeyer, G. Pan, J. F. Parkinson, G. B. Phillips, M. A. Polokoff, N. H. Sigal, R. Vergona, M. Whitlow, T. A. Young and J. J. Devlin. 2000. Allosteric inhibitors of inducible nitric oxide synthase dimerization discovered via combinatorial chemistry. *Proc Natl Acad Sci U S A* **97**:1506-1511.

16. Stuehr, D. J., N. S. Kwon, C. F. Nathan, O. W. Griffith, P. L. Feldman and J. Wiseman. 1991. N omega-hydroxy-l-arginine is an intermediate in the biosynthesis of nitric oxide from l-arginine. *J Biol Chem* **266**:6259-6263.

17. Alderton, W. K., C. E. Cooper and R. G. Knowles. 2001. Nitric oxide synthases: Structure, function and inhibition. *Biochem J* **357**:593-615.

18. Schmidt, H. H., H. Hofmann, U. Schindler, Z. S. Shutenko, D. D. Cunningham and M. Feelisch. 1996. No no from no synthase. *Proc Natl Acad Sci U S A* **93**:14492-14497.

19. Heinzel, B., M. John, P. Klatt, E. Bohme, and B. Mayer. 1992. Ca2+/calmodulin-dependent formation of hydrogen peroxide by brain nitric oxide synthase. *Biochem J* **281**:627-630.

20. Xia, Y., A. L. Tsai, V. Berka, and J. L. Zweier. 1998. Superoxide generation from endothelial nitric-oxide synthase. A ca2+/calmodulin-dependent and tetrahydrobiopterin regulatory process. *J Biol Chem* **273**:25804-25808.

21. Xia, Y., L. J. Roman, B. S. Masters and J. L. Zweier. 1998. Inducible nitric-oxide synthase generates superoxide from the reductase domain. *J Biol Chem* **273**:22635-22639.

22. Beckman, J. S., T. W. Beckman, J. Chen, P. A. Marshall and B. A. Freeman. 1990. Apparent hydroxyl radical production by peroxynitrite: Implications for endothelial injury from nitric oxide and superoxide. *Proc Natl Acad Sci U S A* **87**:1620-1624.

23. Förstermann, U., L. D. Gorsky, J. S. Pollock, K. Ishii, H. H. Schmidt, M. Heller and F. Murad. 1990. Hormone-induced biosynthesis of endothelium-derived relaxing factor/nitric oxide-like material in n1e-115 neuroblastoma cells requires calcium and calmodulin. *Mol Pharmacol* **38**:7-13.

24. Schmidt, H. H. H. W., J. S. Pollock, M. Nakane, L. D. Gorsky, U. Förstermann and F. Murad. 1991. Purification of a soluble isoform of guanylyl cyclase-activating-factor synthase. *Proc Natl Acad Sci U S A* **88**:365-369.

25. Bredt, D. S. and S. H. Snyder. 1990. Isolation of nitric oxide synthetase, a calmodulin-requiring enzyme. *Proc Natl Acad Sci USA* **87**:682-685.

26. Gachhui, R., H. M. Abu-Soud, D. K. Ghosha, A. Presta, M. A. Blazing, B. Mayer, S. E. George and D. J. Stuehr. 1998. Neuronal nitric-oxide synthase interaction with calmodulin-troponin c chimeras. *J Biol Chem* **273**:5451-5454.

27. Abu Soud, H. M., L. L. Yoho and D. J. Stuehr. 1994. Calmodulin controls neuronal nitric-oxide synthase by a dual mechanism. Activation of intra- and interdomain electron transfer. *J Biol Chem* **269**:32047-32050.

28. Salerno, J. C., D. E. Harris, K. Irizarry, B. Patel, A. J. Morales, S. M. Smith, P. Martasek, L. J. Roman, B. S. Masters, C. L. Jones, B. A. Weissman, P. Lane, Q. Liu and S. S. Gross. 1997. An autoinhibitory control element defines calcium-regulated isoforms of nitric oxide synthase. *J Biol Chem* **272**:29769-29777.

29. Daff, S., I. Sagami, and T. Shimizu. 1999. The 42-amino acid insert in the fmn domain of neuronal nitric-oxide synthase exerts control over ca(2+)/calmodulin-dependent electron transfer. *J Biol Chem* **274**:30589-30595.

30. Arbones, M. L., J. Ribera, L. Agullo, M. A. Baltrons, A. Casanovas, V. Riveros Moreno and A. Garcia. 1996. Characteristics of nitric oxide synthase type i of rat cerebellar astrocytes. *Glia* **18**:224-232.

31. Hecker, M., A. Mülsch and R. Busse. 1994. Subcellular localization and characterization of neuronal nitric oxide synthase. *J Neurochem* **62**:1524-1529.

32. Rodrigo, J., V. Riveros-Moreno, M. L. Bentura, L. O. Uttenthal, E. A. Higgs, A. P. Fernandez, J. M. Polak, S. Moncada and R. Martinez-Murillo. 1997. Subcellular localization of nitric oxide synthase in the cerebral ventricular system, subfornical organ, area postrema, and blood vessels of the rat brain. *J Comp Neurol* **378**:522-534.

33. Tojo, A., S. S. Gross, L. Zhang, C. C. Tisher, H. H. Schmidt, C. S. Wilcox and K. M. Madsen. 1994. Immunocytochemical localization of distinct isoforms of nitric oxide synthase in the juxtaglomerular apparatus of normal rat kidney. *J Am Soc Nephrol* **4**:1438-1447.

34. Chang, W. J., S. T. Iannaccone, K. S. Lau, B. S. Masters, T. J. McCabe, K. McMillan, R. C. Padre, M. J. Spencer, J. G. Tidball and J. T. Stull. 1996. Neuronal nitric oxide synthase and dystrophin-deficient muscular dystrophy. *Proc Natl Acad Sci USA* **93**:9142-9147.

35. Gath, I., E. I. Closs, U. Gödtel-Armbrust, S. Schmitt, M. Nakane, I. Wessler and U. Förstermann. 1996. Inducible no synthase ii and neuronal no synthase i are constitutively expressed in different structures of guinea pig skeletal muscle: Implications for contractile function. *FASEB J* **10**:1614-1620.

36. Ponting, C. P., C. Phillips, K. E. Davies and D. J. Blake. 1997. Pdz domains: Targeting signalling molecules to sub-membranous sites. *Bioessays* **19**:469-479.

37. Schepens, J., E. Cuppen, B. Wieringa and W. Hendriks. 1997. The neuronal nitric oxide synthase pdz motif binds to -g(d,e)xv* carboxyterminal sequences. *FEBS Lett* **409**:53-56.

38. Brenman, J. E., D. S. Chao, S. H. Gee, A. W. McGee, S. E. Craven, D. R. Santillano, Z. Wu, F. Huang, H. Xia, M. F. Peters, S. C. Froehner and D. S. Bredt. 1996. Interaction of nitric oxide synthase with the postsynaptic density protein psd-95 and α 1-syntrophin mediated by pdz domains. *Cell* **84**:757-767.

39. Kornau, H.-C., L. T. Schenker, M. B. Kennedy and P. H. Seeburg. 1995. Domain interaction between nmda receptor subunits and the postsynaptic density protein psd-95. *Science* **269**:1737-1740.

40. Brenman, J. E., D. S. Chao, H. Xia, K. Aldape and D. S. Bredt. 1995. Nitric oxide synthase complexed with dystrophin and absent from skeletal muscle sarcolemma in duchenne muscular dystrophy. *Cell* **82**:743-752.

41. Bender, A. T., A. M. Silverstein, D. R. Demady, K. C. Kanelakis, S. Noguchi, W. B. Pratt and Y. Osawa. 1999. Neuronal nitric-oxide synthase is regulated by the hsp90-based chaperone system in vivo. *J Biol Chem* **274**:1472-1478.

42. Jaffrey, S. R. and S. H. Snyder. 1996. Pin: An associated protein inhibitor of neuronal nitric oxide synthase. *Science* **274**:774-777.

43. Rodriguez-Crespo, I., W. Straub, F. Gavilanes and P. R. Ortiz de Montellano. 1998. Binding of dynein light chain (pin) to neuronal nitric oxide synthase in the absence of inhibition. *Arch Biochem Biophys* **359**:297-304.

44. Hemmens, B. and B. Mayer. 1998. Enzymology of nitric oxide synthases. *Methods Mol Biol* **100**:1-32.

45. Nakane, M., J. Mitchell, U. Förstermann and F. Murad. 1991. Phosphorylation by calcium calmodulin-dependent protein kinase ii and protein kinase c modulates the activity of nitric oxide synthase. *Biochem Biophys Res Commun* **180**:1396-1402.

46. Dinerman, J. L., J. P. Steiner, T. M. Dawson, V. Dawson and S. H. Snyder. 1994. Cyclic nucleotide dependent phosphorylation of neuronal nitric oxide synthase inhibits catalytic activity. *Neuropharmacology* **33**:1245-1251.

47. Komeima, K., Y. Hayashi, Y. Naito and Y. Watanabe. 2000. Inhibition of neuronal nitric-oxide synthase by calcium/ calmodulin- dependent protein kinase iiα through ser847 phosphorylation in ng108-15 neuronal cells. *J Biol Chem* **275**:28139-28143.

48. Lam, H. H., D. F. Hanley, B. D. Trapp, S. Saito, S. Raja, T. M. Dawson and H. Yamaguchi. 1996. Induction of spinal cord neuronal nitric oxide synthase (nos) after formalin injection in the rat hind paw. *Neurosci Lett* **210**:201-204.

49. Sharma, H. S., J. Westman, P. Alm, P. O. Sjoquist, J. Cervos Navarro and F. Nyberg. 1997. Involvement of nitric oxide in the pathophysiology of acute heat stress in the rat. Influence of a new antioxidant compound h-290/51. *Ann N Y Acad Sci* **813**:581-590.

50. Lumme, A., S. Vanhatalo, M. Sadeniemi and S. Soinila. 1997. Expression of nitric oxide synthase in hypothalamic nuclei following axonal injury or colchicine treatment. *Exp Neurol* **144**:248-257.

51. Calza, L., L. Giardino, M. Pozza, A. Micera and L. Aloe. 1997. Time-course changes of nerve growth factor, corticotropin-releasing hormone, and nitric oxide synthase isoforms and their possible role in the development of inflammatory response in experimental allergic encephalomyelitis. *Proc Natl Acad Sci USA* **94**:3368-3373.

52. Reiser, P. J., W. O. Kline and P. L. Vaghy. 1997. Induction of neuronal type nitric oxide synthase in skeletal muscle by chronic electrical stimulation in vivo. *J Appl Physiol* **82**:1250-1255.

53. Calza, L., L. Giardino and S. Ceccatelli. 1993. Nos mrna in the paraventricular nucleus of young and old rats after immobilization stress. *Neuroreport* **4**:627-630.

54. Tsuchiya, T., J. Kishimoto and Y. Nakayama. 1996. Marked increases in neuronal nitric oxide synthase (nnos) mrna and nadph-diaphorase histostaining in adrenal cortex after immobilization stress in rats. *Psychoneuroendocrinology* **21**:287-293.

55. Herdegen, T., S. Brecht, B. Mayer, J. Leah, W. Kummer, R. Bravo and M. Zimmermann. 1993. Long-lasting expression of jun and krox transcription factors and nitric oxide synthase in intrinsic neurons of the rat brain following axotomy. *J Neurosci* **13**:4130-4145.

56. Vizzard, M. A. 1997. Increased expression of neuronal nitric oxide synthase in bladder afferent and spinal neurons following spinal cord injury. *Dev Neurosci* **19**:232-246.

57. Lin, L. H., A. Sandra, S. Boutelle and W. T. Talman. 1997. Up-regulation of nitric oxide synthase and its mrna in vagal motor nuclei following axotomy in rat. *Neurosci Lett* **221**:97-100.

58. Villar, M. J., S. Ceccatelli, K. Bedecs, T. Bartfai, D. Bredt, S. H. Synder and T. Hokfelt. 1994. Upregulation of nitric oxide synthase and galanin message-associated peptide in hypothalamic magnocellular neurons after hypophysectomy. Immunohistochemical and in situ hybridization studies. *Brain Res* **650**:219-228.

59. Zhang, Z. G., M. Chopp, S. Gautam, C. Zaloga, R. L. Zhang, H. H. H. W. Schmidt, J. S. Pollock and U. Förstermann. 1994. Upregulation of neuronal nitric oxide synthase and mrna, and selective sparing of nitric oxide synthase-containing neurons after focal cerebral ischemia in rat. *Brain Res* **654**:85-95.

60. Komatsu, T. and C. S. Reiss. 1997. Ifn-gamma is not required in the il-12 response to vesicular stomatitis virus infection of the olfactory bulb. *J Immunol* **159**:3444-3452.

61. Tascedda, F., R. Molteni, G. Racagni and M. A. Riva. 1996. Acute and chronic changes in k(+)-induced depolarization alter nmda and nnos gene expression in cultured cerebellar granule cells. *Mol Brain Res* **40**:171-174.

62. Baader, S. L. and K. Schilling. 1996. Glutamate receptors mediate dynamic regulation of nitric oxide synthase expression in cerebellar granule cells. *J Neurosci* **16**:1440-1449.

63. Ceccatelli, S., L. Grandison, R. E. Scott, D. W. Pfaff and L. M. Kow. 1996. Estradiol regulation of nitric oxide synthase mrnas in rat hypothalamus. *Neuroendocrinology* **64**:357-363.

64. Weiner, C. P., R. G. Knowles and S. Moncada. 1994. Induction of nitric oxide synthases early in pregnancy. *Am J Obstet Gynecol* **171**:838-843.

65. Xu, D. L., P. Y. Martin, J. St. John, P. Tsai, S. N. Summer, M. Ohara, J. K. Kim and R. W. Schrier. 1996. Upregulation of endothelial and neuronal constitutive nitric oxide synthase in pregnant rats. *Am J Physiol* **271**:R1739-R1745.

66. Reilly, C. M., P. Zamorano, V. S. Stopper and T. M. Mills. 1997. Androgenic regulation of no availability in rat penile erection. *J Androl* **18**:110-115.

67. Weber, C. M., B. C. Eke and M. D. Maines. 1994. Corticosterone regulates heme oxygenase-2 and no synthase transcription and protein expression in rat brain. *J Neurochem* **63**:953-962.

68. Bagetta, G., M. T. Corasaniti, G. Melino, A. M. Paoletti, A. Finazzi Agro and G. Nistico. 1993. Lithium and tacrine increase the expression of nitric oxide synthase mrna in the hippocampus of rat. *Biochem Biophys Res Commun* **197**:1132-1139.

69. Schwarz, P. M., B. Gierten, J.-P. Boissel and U. Förstermann. 1998. Expressional down-regulation of neuronal-type nitric oxide synthase i by glucocorticoids in n1e-115 neuroblastoma cells. *Mol Pharmacol* **54**:258-263.

70. Gath, I., U. Gödtel-Armbrust and U. Förstermann. 1997. Expressional downregulation of neuronal-type no synthase i in guinea pig skeletal muscle in response to bacterial lipopolysaccharide. *FEBS Lett.* **410**:319-323.

71. Liu, S. F., I. M. Adcock, R. W. Old, P. J. Barnes and T. W. Evans. 1996. Differential regulation of the constitutive and inducible nitric oxide synthase mrna by lipopolysaccharide treatment in vivo in the rat. *Crit Care Med* **24**:1219-1225.

72. Bandyopadhyay, A., S. Chakder and S. Rattan. 1997. Regulation of inducible and neuronal nitric oxide synthase gene expression by interferon-gamma and vip. *Am J Physiol* **272**:C1790-C1797.

73. Ratovitski, E. A., M. R. Alam, R. A. Quick, A. McMillan, C. Bao, C. Kozlovsky, T. A. Hand, R. C. Johnson, R. E. Mains, B. A. Eipper and C. J. Lowenstein. 1999. Kalirin inhibition of inducible nitric-oxide synthase. *J Biol Chem* **274**:993-999.

74. Ratovitski, E. A., C. Bao, R. A. Quick, A. McMillan, C. Kozlovsky and C. J. Lowenstein. 1999. An inducible nitric-oxide synthase (nos)-associated protein inhibits nos dimerization and activity. *J Biol Chem* **274**:30250-30257.

75. Geller, D. A. and T. R. Billiar. 1998. Molecular biology of nitric oxide synthases. *Cancer Metastasis Rev* **17**:7-23.

76. Kleinert, H., J. P. Boissel, P. M. Schwarz and U. Förstermann. 2000.Regulation of the expression of nitric oxide synthase isoforms. In *Nitric oxide: Biology and pathobiology* (Ignarro, L. J., ed.), pp. 105-128 (Academic Press, New York).

77. MacMicking, J., Q. W. Xie and C. Nathan. 1997. Nitric oxide and macrophage function. *Annu Rev Immunol* **15**:323-350.

78. Xie, Q. W., R. Whisnant and C. Nathan. 1993. Promoter of the mouse gene encoding calcium-independent nitric oxide synthase confers inducibility by interferon g and bacterial lipopolysaccharide. *J Exp Med* **177**:1779-1784.

79. Lowenstein, C. J., E. W. Alley, P. Raval, A. M. Snowman, S. H. Snyder, S. W. Russell and W. J. Murphy. 1993. Macrophage nitric oxide synthase gene - two upstream regions mediate induction by interferon-g and lipopolysaccharide. *Proc Natl Acad Sci USA* **90**:9730-9734.

80. Kleinert, H., C. Euchenhofer, I. Ihrig Biedert and U. Forstermann. 1996. Glucocorticoids inhibit the induction of nitric oxide synthase ii by down-regulating cytokine-induced activity of transcription factor nuclear factor-kb. *Mol Pharmacol* **49**:15-21.

81. Eberhardt, W., D. Kunz, R. Hummel and J. Pfeilschifter. 1996. Molecular cloning of the rat inducible nitric oxide synthase gene promoter. *Biochem Biophys Res Commun* **223**:752-756.

82. Eberhardt, W., C. Pluss, R. Hummel and J. Pfeilschifter. 1998. Molecular mechanisms of inducible nitric oxide synthase gene expression by il-1b and camp in rat mesangial cells. *J Immunol* **160**:4961-4969.

83. Beck, K. F., W. Eberhardt, S. Walpen, M. Apel and J. Pfeilschifter. 1998. Potentiation of nitric oxide synthase expression by superoxide in interleukin-1b-stimulated rat mesangial cells. *FEBS Lett* **435**:35-38.

84. Kuo, P. C., K. Y. Abe and R. A. Schroeder. 1997. Oxidative stress increases hepatocyte iNOS gene transcription and promoter activity. *Biochem Biophys Res Commun* **234**:289-292.

85. Guo, H., C. Q. Cai and P. C. Kuo. 2002. Hepatocyte nuclear factor-4α mediates redox sensitivity of inducible nitric-oxide synthase gene transcription. *J Biol Chem* **277**:5054-5060.

86. Darville, M. I. and D. L. Eizirik. 1998. Regulation by cytokines of the inducible nitric oxide synthase promoter in insulin-producing cells. *Diabetologia* **41**:1101-1108.

87. de Vera, M. E., J. M. Wong, J. Y. Zhou, E. Tzeng, H. R. Wong, T. R. Billiar and D. A. Geller. 1996. Cytokine-induced nitric oxide synthase gene transcription is blocked by the heat shock response in human liver cells. *Surgery* **120**:144-149.

88. Kolyada, A. Y., N. Savikovsky and N. E. Madias. 1996. Transcriptional regulation of the human inos gene in vascular-smooth-muscle cells and macrophages: Evidence for tissue specificity. *Biochem Biophys Res Commun* **220**:600-605.

89. Taylor, B. S. and D. A. Geller. 2000. Molecular regulation of the human inducible nitric oxide synthase (inos) gene. *Shock* **13**:413-424.

90. Hausding, M., A. Witteck, F. Rodriguez-Pascual, C. von Eichel-Streiber, U. Förstermann and H. Kleinert. 2000. Inhibition of small g proteins of the rho family by statins or clostridium difficile toxin b enhances cytokine-mediated induction of no synthase ii. *Br J Pharmacol* in press.

91. Rodriguez-Pascual, F., M. Hausding, I. Ihrig-Biedert, H. Furneaux, A. P. Levy, U. Forstermann and H. Kleinert. 2000. Complex contribution of the 3'-untranslated region to the expressional regulation of the human inducible nitric-oxide synthase gene. Involvement of the rna-binding protein hur. *J Biol Chem* **275**:26040-26049.

92. Chu, S. C., J. Marks Konczalik, H. P. Wu, T. C. Banks and J. Moss. 1998. Analysis of the cytokine-stimulated human inducible nitric oxide synthase (inos) gene: Characterization of differences between human and mouse inos promoters. *Biochem Biophys Res Commun* **248**:871-878.

93. Marks-Konczalik, J., S. C. Chu and J. Moss. 1998. Cytokine-mediated transcriptional induction of the human inducible nitric oxide synthase gene requires both activator protein 1 and nuclear factor-kb-binding sites. *J Biol Chem* **273**:22201-22208.

94. Kristof, A. S., J. Marks-Konczalik and J. Moss. 2001. Mitogen-activated protein kinases mediate activator protein-1-dependent human inducible nitric-oxide synthase promoter activation. *J Biol Chem* **276**:8445-8452.

95. Ghosh, S., M. J. May and E. B. Kopp. 1998. Nf-kb and rel proteins: Evolutionarily conserved mediators of immune responses. *Annu Rev Immunol* **16**:225-260.

96. Mukaida, N., M. Morita, Y. Ishikawa, N. Rice, S. Okamoto, T. Kasahara and K. Matsushima. 1994. Novel mechanism of glucocorticoid-mediated gene repression. Nuclear factor-kb is target for glucocorticoid-mediated interleukin 8 gene repression. *J Biol Chem* **269**:13289-13295.

97. Jeon, Y. J., S. H. Han, Y. W. Lee, S. S. Yea and K. H. Yang. 1998. Inhibition of nf-κ b/rel nuclear translocation by dexamethasone: Mechanism for the inhibition of inos gene expression. *Biochem Mol Biol Int* **45**:435-441.

98. Yu, Z., W. Zhang and B. C. Kone. 2002. Stat3 inhibits transcription of the inducible nitric oxide synthase gene by interacting with nf-κb. *Biochem J* **11**:

99. de Vera, M. E., B. S. Taylor, Q. Wang, R. A. Shapiro, T. R. Billiar and D. A. Geller. 1997. Dexamethasone suppresses inos gene expression by upregulating i-kb-a and inhibiting nf-kb. *Am J Physiol* **273**:G1290-G1296.

100. Xie, Q. W., Y. Kashiwabara and C. Nathan. 1994. Role of transcription factor nf-kb/rel in induction of nitric oxide synthase. *J Biol Chem* **269**:4705-4708.

101. Spink, J. M., J. Cohen and T. Evans. 1995. Delineation of the vsmc inos enhancer and its interaction with the trasncription factor, rela. *Endothelium* **3**:S-50.

102. Taylor, B. S., M. E. de Vera, R. W. Ganster, Q. Wang, R. A. Shapiro, S. M. Morris, Jr., T. R. Billiar and D. A. Geller. 1998. Multiple nf-kb enhancer elements regulate cytokine induction of the human inducible nitric oxide synthase gene. *J Biol Chem* **273**:15148-15156.

103. Teng, X., H. Zhang, C. Snead and J. D. Catravas. 2000. A reverse nuclear factor-κb element in the rat type ii nitric oxide synthase promoter mediates the induction by interleukin-1beta and interferon-gamma in rat aortic smooth muscle cells. *Gen Pharmacol* **34**:9-16.

104. Ganster, R. W., B. S. Taylor, L. Shao and D. A. Geller. 2001. Complex regulation of human inducible nitric oxide synthase gene transcription by stat 1 and nf-κ b. *Proc Natl Acad Sci U S A* **98**:8638-8643.

105. Nunokawa, Y., S. Oikawa and S. Tanaka. 1996. Human inducible nitric oxide synthase gene is transcriptionally regulated by nuclear factor-kb dependent mechanism. *Biochem Biophys Res Commun* **223**:347-352.

106. Martin, E., C. Nathan and Q. W. Xie. 1994. Role of interferon regulatory factor 1 in induction of nitric oxide synthase. *J Exp Med* **180**:977-984.

107. Goldring, C., S. Reveneau, M. Algarte and J. F. Jeannin. 1996. In vivo footprinting of the mouse inducible nitric oxide synthase gene: Inducible protein occupation of numerous sites including oct and nf-il6. *Nucleic Acids Res* **24**:1682-1687.

108. Kinugawa, K., T. Shimizu, A. Yao, O. Kohmoto, T. Serizawa and T. Takahashi. 1997. Transcriptional regulation of inducible nitric oxide synthase in cultured neonatal rat cardiac myocytes. *Circ Res* **81**:911-921.

109. Kamijo, R., H. Harada, T. Matsuyama, M. Bosland, J. Gerecitano, D. Shapiro, J. Le, S. I. Koh, T. Kimura, S. J. Green, T. W. Mak, T. Taniguchi and J. Vilcek. 1994. Requirement for transcription factor irf-1 in no synthase induction in macrophages. *Science* **263**:1612-1615.

110. Shiraishi, A., J. Dudler and M. Lotz. 1997. Role of ifn regulatory factor-1 in synovitis and nitric oxide production. *J Immunol* **159**:3549-3554.

111. Fujimura, M., T. Tominaga, I. Kato, S. Takasawa, M. Kawase, T. Taniguchi, H. Okamoto and T. Yoshimoto. 1997. Attenuation of nitric oxide synthase induction in irf-1-deficient glial cells. *Brain Res* **759**:247-250.

112. Blair, L. A., L. B. Maggi, Jr., A. L. Scarim and J. A. Corbett. 2002. Role of interferon regulatory factor-1 in double-stranded rna-induced inos expression by mouse islets. *J Biol Chem* **277**:359-365.

113. Boehm, U., T. Klamp, M. Groot and J. C. Howard. 1997. Cellular responses to interferon-g. *Annu Rev Immunol* **15**:749-795.

114. Gao, J., D. C. Morrison, T. J. Parmely, S. W. Russell and W. J. Murphy. 1997. An interferon-g-activated site (gas) is necessary for full expression of the mouse inos gene in response to interferon-gamma and lipopolysaccharide. *J Biol Chem* **272**:1226-1230.

115. Teng, X., H. Zhang, C. Snead and J. D. Catravas. 2002. Molecular mechanisms of inos induction by il-1 beta and ifn-gamma in rat aortic smooth muscle cells. *Am J Physiol Cell Physiol* **282**:C144-152.

116. Kleinert, H., T. Wallerath, G. Fritz, I. Ihrig-Biedert, F. Rodriguez-Pascual, D. A. Geller and U. Förstermann. 1998. Cytokine induction of no synthase ii in human dld-1 cells: Roles of the jak-stat, ap-1 and nf-kb-signaling pathways. *Br J Pharmacol* **125**:193-201.

117. Meraz, M. A., J. M. White, K. C. F. Sheehan, E. A. Bach, S. J. Rodig, A. S. Dighe, D. H. Kaplan, J. K. Riley, A. C. Greenlund, D. Campbell, K. Carver-Moore, R. N. DuBois, R. Clark, M. Aguet and R. D. Schreiber. 1996. Trageted disruption of the stat1 gene in mice reveals unexpected physiologic specificity in the jak-sat signaling pathway. *Cell* **84**:431-442.

118. Crespo, A., M. B. Filla and W. J. Murphy. 2002. Low responsiveness to ifn-gamma, after pretreatment of mouse macrophages with lipopolysaccharides, develops via diverse regulatory pathways. *Eur J Immunol* **32**:710-719.

119. Chong, M. M., H. E. Thomas and T. W. Kay. 2002. Suppressor of cytokine signaling-1 regulates the sensitivity of pancreatic beta cells to tumor necrosis factor. *J Biol Chem* **277**:27945-27952.

120. Bhat, N. R., D. L. Feinstein, Q. Shen and A. N. Bhat. 2002. P38 mapk-mediated transcriptional activation of inducible nitric-oxide synthase in glial cells. *J Biol Chem* **277**:29584-29592.

121. Kolyada, A. Y. and N. E. Madias. 2001. Transcriptional regulation of the human inos gene by il-1beta in endothelial cells. *Mol Med* **7**:329-343.

122. Kizaki, T., K. Suzuki, Y. Hitomi, K. Iwabuchi, K. Onoe, S. Haga, H. Ishida, T. Ookawara and H. Ohno. 2001. Negative regulation of lps-stimulated expression of inducible nitric oxide synthase by ap-1 in macrophage cell line j774a.1. *Biochem Biophys Res Commun* **289**:1031-1038.

123. Kleinert, H., D. A. Geller and U. Förstermann. 1999. Analysis of the molecular mechanism regulating the induction of the human no synthase ii promoter in human a549/8 alveolar carcinoma cells. *Naunyn-Schmiedeb Arch Pharmacol* **359**:R 37.

124. Xie, Q. 1997. A novel lipopolysaccharide-response element contributes to induction of nitric oxide synthase. *J Biol Chem* **272**:14867-14872.

125. Pellacani, A., M. T. Chin, P. Wiesel, M. Ibanez, A. Patel, S. F. Yet, C. M. Hsieh, J. D. Paulauskis, R. Reeves, M. E. Lee and M. A. Perrella. 1999. Induction of high mobility group-i(y) protein by endotoxin and interleukin-1b in vascular smooth muscle cells. Role in activation of inducible nitric oxide synthase. *J Biol Chem* **274**:1525-1532.

126. Melillo, G., T. Musso, A. Sica, L. S. Taylor, G. W. Cox and L. Varesio. 1995. A hypoxia-responsive element mediates a novel pathway of activation of the inducible nitric oxide synthase promoter. *J Exp Med* **182**:1683-1693.

127. Ricote, M., A. C. Li, T. M. Willson, C. J. Kelly and C. K. Glass. 1998. The peroxisome proliferator-activated receptor-g is a negative regulator of macrophage activation. *Nature* **391**:79-82.

128. Sirsjo, A., A. C. Gidlof, A. Olsson, H. Torma, M. Ares, H. Kleinert, U. Förstermann and G. K. Hansson. 2000. Retinoic acid inhibits nitric oxide synthase-2 expression through the retinoic acid receptor-α . *Biochem Biophys Res Commun* **270**:846-851.

129. Nuedling, S., R. H. Karas, M. E. Mendelsohn, J. A. Katzenellenbogen, B. S. Katzenellenbogen, R. Meyer, H. Vetter and C. Grohe. 2001. Activation of estrogen receptor beta is a prerequisite for estrogen-dependent upregulation of nitric oxide synthases in neonatal rat cardiac myocytes. *FEBS Lett* **502**:103-108.

130. Toell, A., K. D. Kroncke, H. Kleinert and C. Carlberg. 2002. Orphan nuclear receptor binding site in the human inducible nitric oxide synthase promoter mediates responsiveness to steroid and xenobiotic ligands. *J Cell Biochem* **85**:72-82.

131. Vodovotz, Y., C. Bogdan, J. Paik, Q. W. Xie and C. Nathan. 1993. Mechanisms of suppression of macrophage nitric oxide release by transforming growth factor-b. *J Exp Med* **178**:605-613.

132. Evans, T., A. Carpenter and J. Cohen. 1994. Inducible nitric-oxide-synthase mrna is transiently expressed and destroyed by a cycloheximide-sensitive process. *Eur J Biochem* **219**:563-569.

133. Weisz, A., S. Oguchi, L. Cicatiello and H. Esumi. 1994. Dual mechanism for the control of inducible-type no synthase gene expression in macrophages during activation by interferon-gamma and bacterial lipopolysaccharide - transcriptional and post-transcriptional regulation. *J Biol Chem* **269**:8324-8333.

134. Korhonen, R., A. Lahti, M. Hamalainen, H. Kankaanranta and E. Moilanen. 2002. Dexamethasone inhibits inducible nitric-oxide synthase expression and nitric oxide production by destabilizing mrna in lipopolysaccharide-treated macrophages. *Mol Pharmacol* **62**:698-704.

135. Soderberg, M., F. Raffalli-Mathieu and M. A. Lang. 2002. Inflammation modulates the interaction of heterogeneous nuclear ribonucleoprotein (hnrnp) i/polypyrimidine tract binding protein and hnrnp l with the 3'untranslated region of the murine inducible nitric-oxide synthase mrna. *Mol Pharmacol* **62**:423-431.

136. Linscheid, P., A. Schaffner and G. Schoedon. 1998. Modulation of inducible nitric oxide synthase mrna stability by tetrahydrobiopterin in vascular smooth muscle cells. *Biochem Biophys Res Commun* **243**:137-141.

137. Carpenter, L., D. Cordery and T. J. Biden. 2001. Protein kinase cdelta activation by interleukin-1beta stabilizes inducible nitric-oxide synthase mrna in pancreatic beta-cells. *J Biol Chem* **276**:5368-5374.

139. Linn, S. C., P. J. Morelli, I. Edry, S. E. Cottongim, C. Szabo and A. L. Salzman. 1997. Transcriptional regulation of human inducible nitric oxide synthase gene in an intestinal epithelial cell line. *Am J Physiol* **272**:G1499-G1508.

140. Caput, D., B. Beutler, K. Hartog, R. Thayer, S. Brown-Shimer and A. Cerami. 1986. Identification of a common nucleotide sequence in the 3´-untranslated region of mrna molecules specifying inflammatory mediators. *Proc Natl Acad Sci USA* **83**:1670-1674.

141. Brennan, C. M. and J. A. Steitz. 2001. Hur and mrna stability. *Cell Mol Life Sci* **58**:266-277.

142. Misquitta, C. M., V. R. Iyer, E. S. Werstiuk and A. K. Grover. 2001. The role of 3'-untranslated region (3'-utr) mediated mrna stability in cardiovascular pathophysiology. *Mol Cell Biochem* **224**:53-67.

143. Mitchell, P. and D. Tollervey. 2000. Mrna stability in eukaryotes. *Curr Opin Genet Dev* **10**:193-198.

144. Kleinert, H., K. Mangasser-Stephan, Y. Yao, M. Fechir, A. Bouazzaoui, F. Rodriguez-Pascual and U. Förstermann. 2002. Post-transcriptional regulation of the human inos expression by rna-binding proteins. *Nitric Oxide* **6**:412.

145. Lüß, H., R. K. Li, R. A. Shapiro, E. Tzeng, F. X. McGowan, T. Yoneyama, K. Hatakeyama, D. A. Geller, D. A. Mickle, R. L. Simmons and T. R. Billiar. 1997. Dedifferentiated human ventricular cardiac myocytes express inducible nitric oxide

synthase mrna but not protein in response to il-1, tnf, ifn-g, and lps. *J Mol Cell Cardiol* **29**:1153-1165.

146. Kunz, D., G. Walker, W. Eberhardt and J. Pfeilschifter. 1996. Molecular mechanisms of dexamethasone inhibition of nitric oxide synthase expression in interleukin-1b-stimulated mesangial cells: Evidence for the involvement of transcriptional and posttranscriptional regulation. *Proc Natl Acad Sci U S A* **93**:255-259.

147. Musial, A. and N. T. Eissa. 2001. Inducible nitric-oxide synthase is regulated by the proteasome degradation pathway. *J Biol Chem* **276**:24268-24273.

148. Felley-Bosco, E., F. C. Bender, F. Courjault-Gautier, C. Bron and A. F. Quest. 2000. Caveolin-1 down-regulates inducible nitric oxide synthase via the proteasome pathway in human colon carcinoma cells. *Proc Natl Acad Sci U S A* **97**:14334-14339.

149. Förstermann, U., J. S. Pollock, H. H. H. W. Schmidt, M. Heller and F. Murad. 1991. Calmodulin-dependent endothelium-derived relaxing factor/nitric oxide synthase activity is present in the particulate and cytosolic fractions of bovine aortic endothelial cells. *Proc Natl Acad Sci USA* **88**:1788-1792.

150. Li, H., T. Wallerath and U. Förstermann. 2002. Physiological mechanisms regulating the expression of endothelial-type no synthase. *Nitric Oxide Biol Chem* **7**:103-118.

151. Shaul, P. W. 2002. Regulation of endothelial nitric oxide synthase: Location, location, location. *Annu Rev Physiol* **64**:749-774.

152. Feron, O., L. Belhassen, L. Kobzik, T. W. Smith, R. A. Kelly and T. Michel. 1996. Endothelial nitric oxide synthase targeting to caveolae. Specific interactions with caveolin isoforms in cardiac myocytes and endothelial cells. *J Biol Chem* **271**:22810-22814.

153. Ju, H., R. Zou, V. J. Venema, and R. C. Venema. 1997. Direct interaction of endothelial nitric-oxide synthase and caveolin-1 inhibits synthase activity. *J Biol Chem* **272**:18522-18525.

154. Michel, J. B., O. Feron, K. Sase, P. Prabhakar and T. Michel. 1997. Caveolin versus calmodulin. Counterbalancing allosteric modulators of endothelial nitric oxide synthase. *J Biol Chem* **272**:25907-25912.

155. Garcia-Cardena, G., P. Martasek, B. S. Masters, P. M. Skidd, J. Couet, S. Li, M. P. Lisanti and W. C. Sessa. 1997. Dissecting the interaction between nitric oxide synthase (nos) and caveolin. Functional significance of the nos caveolin binding domain in vivo. *J Biol Chem* **272**:25437-25440.

156. Bucci, M., J. P. Gratton, R. D. Rudic, L. Acevedo, F. Roviezzo, G. Cirino and W. C. Sessa. 2000. In vivo delivery of the caveolin-1 scaffolding domain inhibits nitric oxide synthesis and reduces inflammation. *Nat Med* **6**:1362-1367.

157. Shah, V., M. Toruner, F. Haddad, G. Cadelina, A. Papapetropoulos, K. Choo, W. C. Sessa and R. J. Groszmann. 1999. Impaired endothelial nitric oxide synthase activity associated with enhanced caveolin binding in experimental cirrhosis in the rat. *Gastroenterology* **117**:1222-1228.

158. Garcia-Cardena, G., R. Fan, V. Shah, R. Sorrentino, G. Cirino, A. Papapetropoulos and W. C. Sessa. 1998. Dynamic activation of endothelial nitric oxide synthase by hsp90. *Nature* **392**:821-824.

159. Russell, K. S., M. P. Haynes, T. Caulin-Glaser, J. Rosneck, W. C. Sessa and J. R. Bender. 2000. Estrogen stimulates heat shock protein 90 binding to endothelial nitric oxide synthase in human vascular endothelial cells. Effects on calcium sensitivity and no release. *J Biol Chem* **275**:5026-5030.

160. Fulton, D., J. P. Gratton and W. C. Sessa. 2001. Post-translational control of endothelial nitric oxide synthase: Why isn't calcium/calmodulin enough? *J Pharmacol Exp Ther* **299**:818-824.

161. Chen, P. F. and K. K. Wu. 2000. Characterization of the roles of the 594-645 region in human endothelial nitric-oxide synthase in regulating calmodulin binding and electron transfer. *J Biol Chem* **275**:13155-13163.

162. Feron, O., F. Saldana, J. B. Michel and T. Michel. 1998. The endothelial nitric-oxide synthase-caveolin regulatory cycle. *J Biol Chem* **273**:3125-3128.

163. Michel, J. B., O. Feron, D. Sacks and T. Michel. 1997. Reciprocal regulation of endothelial nitric-oxide synthase by ca2+-calmodulin and caveolin. *J Biol Chem* **272**:15583-15586.

164. Fulton, D., J. P. Gratton, T. J. McCabe, J. Fontana, Y. Fujio, K. Walsh, T. F. Franke, A. Papapetropoulos and W. C. Sessa. 1999. Regulation of endothelium-derived nitric oxide production by the protein kinase akt [published correction appears in *nature* 1999;400:792]. *Nature* **399**:597-601.

165. Dimmler, S., I. Fleming, B. Fisslthaler, C. Hermann, R. Busse and A. M. Zeiher. 1999. Activation of nitric oxide synthase in endothelial cells by akt-dependent phosphorylation. *Nature* **399**:601-605.

166. Butt, E., M. Bernhardt, A. Smolenski, P. Kotsonis, L. G. Frohlich, A. Sickmann, H. E. Meyer, S. M. Lohmann and H. H. Schmidt. 2000. Endothelial nitric-oxide synthase (type iii) is activated and becomes calcium independent upon phosphorylation by cyclic nucleotide-dependent protein kinases. *J Biol Chem* **275**:5179-5187.

167. Chen, Z. P., K. I. Mitchelhill, B. J. Michell, D. Stapleton, I. Rodriguez-Crespo, L. A. Witters, D. A. Power, P. R. Ortiz de Montellano and B. E. Kemp. 1999. Amp-activated protein kinase phosphorylation of endothelial no synthase. *FEBS Lett* **443**:285-289.

168. Fleming, I., B. Fisslthaler, S. Dimmler, B. E. Kemp and R. Busse. 2001. Phosphorylation of thr(495) regulates ca(2+)/calmodulin-dependent endothelial nitric oxide synthase activity. *Circ Res* **88**:E68-75.

169. McCabe, T. J., D. Fulton, L. J. Roman and W. C. Sessa. 2000. Enhanced electron flux and reduced calmodulin dissociation may explain "calcium-independent" enos activation by phosphorylation. *J Biol Chem* **275**:6123-6128.

170. Igarashi, J. and T. Michel. 2001. Sphingosine 1-phosphate and isoform-specific activation of phosphoinositide 3-kinase beta. Evidence for divergence and convergence

of receptor-regulated endothelial nitric-oxide synthase signaling pathways. *J Biol Chem* **276**:36281-36288.

171. Thomas, S. R., K. Chen and J. F. Keaney, Jr. 2002. Hydrogen peroxide activates endothelial nitric-oxide synthase through coordinated phosphorylation and dephosphorylation via a phosphoinositide 3-kinase-dependent signaling pathway. *J Biol Chem* **277**:6017-6024.

172. Queen, L. R., B. Xu, K. Horinouchi, I. Fisher, and A. Ferro. 2000. Beta(2)-adrenoceptors activate nitric oxide synthase in human platelets. *Circ Res* **87**:39-44.

173. Hirata, K., R. Kuroda, T. Sakoda, M. Katayama, N. Inoue, M. Suematsu, S. Kawashima and M. Yokoyama. 1995. Inhibition of endothelial nitric oxide synthase activity by protein kinase c. *Hypertension* **25**:180-185.

174. Fleming, I., J. Bauersachs, B. Fisslthaler and R. Busse. 1998. Ca2+-independent activation of the endothelial nitric oxide synthase in response to tyrosine phosphatase inhibitors and fluid shear stress. *Circ Res* **82**:686-695.

175. Ayajiki, K., M. Kindermann, M. Hecker, I. Fleming and R. Busse. 1996. Intracellular ph and tyrosine phosphorylation but not calcium determine shear stress-induced nitric oxide production in native endothelial cells. *Circ Res* **78**:750-758.

176. Igarashi, J., H. S. Thatte, P. Prabhakar, D. E. Golan and T. Michel. 1999. Calcium-independent activation of endothelial nitric oxide synthase by ceramide. *Proc Natl Acad Sci U S A* **96**:12583-12588.

177. Bulotta, S., R. Barsacchi, D. Rotiroti, N. Borgese and E. Clementi. 2001. Activation of the endothelial nitric-oxide synthase by tumor necrosis factor-α . A novel feedback mechanism regulating cell death. *J Biol Chem* **276**:6529-6536.

178. Li, X. A., W. B. Titlow, B. A. Jackson, N. Giltiay, M. Nikolova-Karakashian, A. Uittenbogaard and E. J. Smart. 2002. High density lipoprotein binding to scavenger receptor, class b, type i activates endothelial nitric-oxide synthase in a ceramide-dependent manner. *J Biol Chem* **277**:11058-11063.

179. Li, H., T. Wallerath, T. Münzel, and U. Förstermann. 2002. Regulation of endothelial-type no synthase expression in pathophysiology and in response to drugs. *Nitric Oxide Biol Chem* **6**: in press.

180. Nishida, K., D. G. Harrison, J. P. Navas, A. A. Fisher, S. P. Dockery, M. Uematsu, R. M. Nerem, R. W. Alexander and T. J. Murphy. 1992. Molecular cloning and characterization of the constitutive bovine aortic endothelial cell nitric oxide synthase. *J Clin Invest* **90**:2092-2096.

181. Nadaud, S., M. Philippe, J. F. Arnal, J. B. Michel and F. Soubrier. 1996. Sustained increase in aortic endothelial nitric oxide synthase expression in vivo in a model of chronic high blood flow. *Circ Res* **79**:857-863.

182. Davis, M. E., H. Cai, G. R. Drummond and D. G. Harrison. 2001. Shear stress regulates endothelial nitric oxide synthase expression through c-src by divergent signaling pathways. *Circ Res* **89**:1073-1080.

183. Xiao, Z., Z. Zhang and S. L. Diamond. 1997. Shear stress induction of the endothelial nitric oxide synthase gene is calcium-dependent but not calcium-activated. *J Cell Physiol* **171**:205-211.

184. Malek, A. M., L. W. Jiang, I. Lee, W. C. Sessa, S. Izumo and S. L. Alpert. 1999. Induction of nitric oxide synthase mrna by shear stress requires intracellular calcium and g-protein signals and is modulated by pi 3 kinase. *Biochem Biophys Res Commun* **254**:231-242.

185. Sessa, W. C., K. Pritchard, N. Seyedi, J. Wang and T. H. Hintze. 1994. Chronic exercise in dogs increases coronary vascular nitric oxide production and endothelial cell nitric oxide synthase gene expression. *Circ Res* **74**:349-353.

186. Arosio, E., L. Cuzzolin, S. De Marchi, P. Minuz, M. Degan, F. Crivellente, M. Zannoni and G. Benoni. 1999. Increased endogenous nitric oxide production induced by physical exercise in peripheral arterial occlusive disease patients. *Life Sci* **65**:2815-2822.

187. Arnal, J. F., J. Yamin, S. Dockery and D. G. Harrison. 1994. Regulation of endothelial nitric oxide synthase mrna, protein, and activity during cell growth. *Am J Physiol* **267**:C1381-C1388.

188. Searles, C. D., Y. Miwa, D. G. Harrison and S. Ramasamy. 1999. Posttranscriptional regulation of endothelial nitric oxide synthase during cell growth. *Circ Res* **85**:588-595.

189. Poppa, V., J. K. Miyashiro, M. A. Corson and B. C. Berk. 1998. Endothelial no synthase is increased in regenerating endothelium after denuding injury of the rat aorta. *Arterioscler Thromb Vasc Biol* **18**:1312-1321.

190. Inoue, N., R. C. Venema, H. S. Sayegh, Y. Ohara, T. J. Murphy and D. G. Harrison. 1995. Molecular regulation of the bovine endothelial cell nitric oxide synthase by transforming growth factor-b1. *Arterioscler Thromb Vasc Biol* **15**:1255-1261.

191. Kleinert, H., T. Wallerath, C. E. Euchenhofer, I. Ihrig-Biedert, H. Li and U. Förstermann. 1998. Estrogens increase transcription of the human endothelial no synthase gene: Analysis of the transcription factors involved. *Hypertension* **31**:582-588.

192. Lopez-Ongil, S., O. Hernandez-Perera, J. Navarro-Antolin, G. Perez de Lema, M. Rodriguez-Puyol, S. Lamas and D. Rodriguez-Puyol. 1998. Role of reactive oxygen species in the signalling cascade of cyclosporine a-mediated up-regulation of enos in vascular endothelial cells. *Br J Pharmacol* **124**:447-454.

193. Drummond, G. R., H. Cai, M. E. Davis, S. Ramasamy and D. G. Harrison. 2000. Transcriptional and posttranscriptional regulation of endothelial nitric oxide synthase expression by hydrogen peroxide. *Circ Res* **86**:347-354.

194. Cai, H., M. E. Davis, G. R. Drummond and D. G. Harrison. 2001. Induction of endothelial no synthase by hydrogen peroxide via a ca(2+)/calmodulin-dependent protein kinase ii/janus kinase 2-dependent pathway. *Arterioscler Thromb Vasc Biol* **21**:1571-1576.

195. Zembowicz, A., J. L. Tang and K. K. Wu. 1995. Transcriptional induction of endothelial nitric oxide synthase type iii by lysophosphatidylcholine. *J Biol Chem* **270**:17006-17010.

196. Hirata, K., N. Miki, Y. Kuroda, T. Sakoda, S. Kawashima and M. Yokoyama. 1995. Low concentration of oxidized low-density lipoprotein and lysophosphatidylcholine upregulate constitutive nitric oxide synthase mrna expression in bovine aortic endothelial cells. *Circ Res* **76**:958-962.

197. Cieslik, K., A. Zembowicz, J. L. Tang and K. K. Wu. 1998. Transcriptional regulation of endothelial nitric-oxide synthase by lysophosphatidylcholine. *J Biol Chem* **273**:14885-14890.

198. Laufs, U., V. L. Fata and J. K. Liao. 1997. Inhibition of 3-hydroxy-3-methylglutaryl (hmg)-coa reductase blocks hypoxia-mediated down-regulation of endothelial nitric oxide synthase. *J Biol Chem* **272**:31725-31729.

199. Laufs, U., V. La Fata, J. Plutzky and J. K. Liao. 1998. Upregulation of endothelial nitric oxide synthase by hmg coa reductase inhibitors. *Circulation* **97**:1129-1135.

200. Sumi, D., T. Hayashi, N. K. Thakur, M. Jayachandran, Y. Asai, H. Kano, H. Matsui and A. Iguchi. 2001. A hmg-coa reductase inhibitor possesses a potent anti-atherosclerotic effect other than serum lipid lowering effects--the relevance of endothelial nitric oxide synthase and superoxide anion scavenging action. *Atherosclerosis* **155**:347-357.

201. Kano, H., T. Hayashi, D. Sumi, T. Esaki, Y. Asai, N. K. Thakur, M. Jayachandran and A. Iguchi. 1999. A hmg-coa reductase inhibitor improved regression of atherosclerosis in the rabbit aorta without affecting serum lipid levels: Possible relevance of up-regulation of endothelial no synthase mrna. *Biochem Biophys Res Commun* **259**:414-419.

202. Endres, M., U. Laufs, Z. Huang, T. Nakamura, P. Huang, M. A. Moskowitz and J. K. Liao. 1998. Stroke protection by 3-hydroxy-3-methylglutaryl (hmg)-coa reductase inhibitors mediated by endothelial nitric oxide synthase. *Proc Natl Acad Sci U S A* **95**:8880-8885.

203. Laufs, U., K. Gertz, P. Huang, G. Nickenig, M. Bohm, U. Dirnagl and M. Endres. 2000. Atorvastatin upregulates type iii nitric oxide synthase in thrombocytes, decreases platelet activation, and protects from cerebral ischemia in normocholesterolemic mice. *Stroke* **31**:2442-2449.

204. Laufs, U. and J. K. Liao. 1998. Post-transcriptional regulation of endothelial nitric oxide synthase mrna stability by rho gtpase. *J Biol Chem* **273**:24266-24271.

205. Laufs, U., M. Endres, N. Stagliano, S. Amin Hanjani, D. S. Chui, S. X. Yang, T. Simoncini, M. Yamada, E. Rabkin, P. G. Allen, P. L. Huang, M. Bohm, F. J. Schoen, M. A. Moskowitz and J. K. Liao. 2000. Neuroprotection mediated by changes in the endothelial actin cytoskeleton. *J Clin Invest* **106**:15-24.

206. Linz, W., P. Wohlfart, B. A. Scholkens, T. Malinski and G. Wiemer. 1999. Interactions among ace, kinins and no. *Cardiovasc Res* **43**:549-561.

207. Linz, W., P. Wohlfart, B. A. Schoelkens, R. H. Becker, T. Malinski and G. Wiemer. 1999. Late treatment with ramipril increases survival in old spontaneously hypertensive rats. *Hypertension* **34**:291-295.

208. Bachetti, T., L. Comini, E. Pasini, A. Cargnoni, S. Curello and R. Ferrari. 2001. Ace-inhibition with quinapril modulates the nitric oxide pathway in normotensive rats. *J Mol Cell Cardiol* **33**:395-403.

209. Kohno, M., K. Yokokawa, M. Minami, K. Yasunari, K. Maeda, H. Kano, T. Hanehira and J. Yoshikawa. 1999. Plasma levels of nitric oxide and related vasoactive factors following long-term treatment with angiotensin-converting enzyme inhibitor in patients with essential hypertension. *Metabolism* **48**:1256-1259.

210. Ding, Y. and N. D. Vaziri. 1998. Calcium channel blockade enhances nitric oxide synthase expression by cultured endothelial cells. *Hypertension* **32**:718-723.

211. Ding, Y. and N. D. Vaziri. 2000. Nifedipine and diltiazem but not verapamil up-regulate endothelial nitric-oxide synthase expression. *J Pharmacol Exp Ther* **292**:606-609.

212. Yamashita, T., S. Kawashima, M. Ozaki, Y. Rikitake, T. Hirase, N. Inoue, K. Hirata and M. Yokoyama. 2001. A calcium channel blocker, benidipine, inhibits intimal thickening in the carotid artery of mice by increasing nitric oxide production. *J Hypertens* **19**:451-458.

213. Wallerath, T., D. Poleo, H. Li and U. Förstermann. 2002. Red wine increases the expression of human endothelial no synthase (enos): A mechanism that may contribute to its beneficial cardiovascular effects. *J Am Coll Cardiol* **in press**:

214. Wallerath, T., G. Deckert, T. Ternes, H. Anderson, H. Li, K. Witte and U. Förstermann. 2002. Resveratrol, a polyphenolic phytoalexin present in red wine, enhances expression and activity of endothelial nitric oxide synthase. *Circulation* **106**: in press.

215. Wallerath, T., K. Witte, S. C. Schafer, P. M. Schwarz, W. Prellwitz, P. Wohlfart, H. Kleinert, H. A. Lehr, B. Lemmer and U. Förstermann. 1999. Down-regulation of the expression of endothelial no synthase is likely to contribute to glucocorticoid-mediated hypertension. *Proc Natl Acad Sci U S A* **96**:13357-13362.

216. Lou, Y. K., C. Wen, M. Li, D. J. Adams, M. X. Wang, F. Yang, B. J. Morris and J. A. Whitworth. 2001. Decreased renal expression of nitric oxide synthase isoforms in adrenocorticotropin-induced and corticosterone-induced hypertension. *Hypertension* **37**:1164-1170.

217. Li, H. and U. Förstermann. 2000. Structure-activity relationship of staurosporine analogs in regulating expression of endothelial nitric-oxide synthase gene. *Mol Pharmacol* **57**:427-435.

218. Yoshizumi, M., M. A. Perrella, J. C. Burnett, Jr. and M. E. Lee. 1993. Tumor necrosis factor downregulates an endothelial nitric oxide synthase mrna by shortening its half-life. *Circ Res* **73**:205-209.

219. Alonso, J., L. Sanchez de Miguel, M. Monton, S. Casado and A. Lopez-Farre. 1997. Endothelial cytosolic proteins bind to the 3' untranslated region of endothelial nitric oxide synthase mrna: Regulation by tumor necrosis factor-a. *Mol Cell Biol* **17**:5719-5726.

220. Sanchez de Miguel, L., J. Alonso, F. Gonzalez Fernandez, J. de la Osada, M. Monton, J. A. Rodriguez-Feo, J. I. Guerra, M. M. Arriero, L. Rico, S. Casado and A. Lopez-Farre. 1999. Evidence that an endothelial cytosolic protein binds to the 3'-untranslated region of endothelial nitric oxide synthase mrna. *J Vasc Res* **36**:201-208.

Chapter 8

S-NITROSOTHIOLS IN CELL SIGNALING

Joseph Loscalzo

8.1 Introduction

S-nitrosothiols are simple organic thioesters of nitrite that were first synthesized by Tasker and Jones in 1909.[1,2] These compounds are analogues of nitrite esters of alcohols and, as such, have been recognized to have nitrovasodilator properties for over 20 years. In 1973, Needleman demonstrated that the pharmacological effects of nitrovasodilators could be potentiated by thiols, and postulated that thiols were important in the regulation of a receptor for nitrovasodilators.[3] Ignarro and colleagues suggested that S-nitrosothiols were the pharmacological intermediates in the thiol-dependent activation of guanylyl cyclase by nitrosoguanidine and sodium nitroprusside,[4,5] and we first demonstrated that the S-nitroso-derivative of N-acetyl-L-cysteine directly accounted for the antiplatelet and guanylyl cyclase-activating effects of nitroglycerin.[6] Other S-nitrosothiols were subsequently shown to have nitrovasodilator effects, including S-nitroso-glutathione[7] and S-nitroso-captopril.[8,9,10]

With the demonstration that the endothelium produces nitric oxide,[11,12] the possibility that S-nitrosothiols formed endogenously was proposed. We first showed that thiols potentiate the action of endothelial nitric oxide[13,14] and, subsequently, that S-nitrosothiols form *in vivo*.[15] Furthermore, we showed that the S-nitrosothiol pool comprises both low-molecular-weight and S-nitroso-protein species, and postulated that S-nitrosation of proteins represents a form of posttranslational modification that modulates protein function and cell phenotype.[16] This hypothesis has been borne out in recent years with improved molecular tools with which to identify S-nitrosothiols, and an increasing range of proteins that have been shown to undergo S-nitrosation.[17] In this chapter, I will review the biological

H. J. Forman, J. Fukuto, and M. Torres (eds.), Signal Transduction by Reactive Oxygen and Nitrogen Species: Pathways and Chemical Principles, ©2003, *Kluwer Academic Publishers. Printed in the Netherlands*

chemistry of S-nitrosation, and the cell and molecular biology of S-nitrosothiol elaboration, including the role of S-nitrosothiols in cell signaling.

8.2 Biological Chemistry

As early as 1839, von Löwig reported his work with Weidmann in which they detected the transient development of a red color in a reaction solution of a mercaptan with sodium nitrite.[18] S-nitrosothiols or thionitrites were first identified by Tasker and Jones (1,2), who reported the formation of the unstable intermediate phenylthionitrite from the reaction of benzenethiol with nitrosyl chloride; this intermediate decomposed rapidly to yield the disulfide and NO according to the following reaction scheme:

$$ \text{(1)} $$

In general, S-nitrosothiols can be readily synthesized by the reaction of a thiol with acidified nitrous acid (which generates nitrosonium, NO^+) or with an alkyl nitrite, according to the following reactions:

$$ RSH + NO^+ \rightarrow RSNO + H^+ \qquad (2) $$

$$ RS^- + RONO \rightarrow RSNO + RO^- \qquad (3) $$

Both aryl and alkyl thionitrites can also by synthesized quantitatively by the reaction of the parent thiol with dinitrogen tetraoxide.[19] It is important to point out that the reaction of nitric oxide with thiols does not yield an S-nitrosothiol; rather, it leads to the synthesis of the disulfide and nitrous oxide by the following reaction scheme:

$$ 2\,RSH + 2\,NO \rightarrow RSSR + N_2O + H^+ \qquad (4) $$

The likely nitrosating species in biological systems include dinitrogen trioxide or nitrous anhydride (N_2O_3), the formation of which from molecular oxygen and nitric oxide is kinetically facilitated in the microenvironment of a biological membrane:[20]

$$ 4\,NO + O_2 + 2\,N_2O_3 \xrightarrow{\ 2\,RSH\ } 2\,RSNO + 2\,NO_2^- + 2\,H^+ \qquad (5) $$

and reversible trans-S-nitrosation reactions:[22]

$$ RSH + R'SNO \rightleftharpoons RSNO + R'SH \qquad (6) $$

Trans-S-nitrosation reactions will be discussed in more detail below. Peroxynitrite is also capable of reacting with thiols to yield S-nitrosothiols,[23] although the physiological relevance of this reaction is uncertain.

The physical properties of S-nitrosothiols have been well characterized and summarized in a comprehensive review by Oae and Shinhama.[24] Primary and secondary S-nitrosothiols are typically red or red-orange in color, while the tertiary S-nitrosothiols tend to be green or red-green. They all have nearly equivalent dipole moments, i.e., ~2-3 D; infrared absorption bands of the NO bond of somewhat longer wavelengths than those of the corresponding alkyl nitrites (~1500-1700 cm^{-1}); ultraviolet and visible absorption maxima at ~250 nm and ~340 nm; and characteristic chemical shifts in the [^{15}N]-nuclear magnetic resonance spectrum (785.2 ppm in CDCl$_3$ relative to [^{15}N]H$_4$NO$_3$). The X-ray crystal structure of the stable S-nitrosothiol, 2-(acetylamino) -2-carboxy-1,1-dimethylethylthionitrite (S-nitroso-N-acetyl-D,L-penicillamine), has been determined and shown to have a C-S bond that is longer than the calculated covalent bond length (1.841 Δ) and an S-N bond that is approximately equal to the predicted covalent bond length (1.771 Δ).[25]

The chemical stability of S-nitrosothiols and the mode of release of the nitric oxide moiety are the subjects of some debate. Homolytic cleavage of the S-NO bond photochemically was first shown in 1966 with the reaction yielding the homodisulfide and nitric oxide radical:[26]

$$2\ RSNO \xrightarrow{\ h\nu\ } 2\ RSNO^{\bullet} \rightarrow 2\ RS^{\bullet} + 2\ NO \ \rightarrow\ RSSR + 2\ NO \qquad (7)$$

This photolytic cleavage of the S-NO bond is likely responsible, at least in part, for the action spectrum of vascular tissue, and supports the importance of S-nitrosothiol formation *in vivo*. Furchgott and colleagues[27] first showed that light can induce relaxation of blood vessels. This initial observation was subsequently shown to be a likely consequence of the release of nitric oxide from a photolabile store,[28] and to require adequate glutathione stores.[29] Some reports are, however, inconsistent with this hypothesis, and suggest alternative sources for the photo-inducible vascular relaxation factor.[30,31]

Homolytic cleavage of the S-NO bond by thermal decomposition has been reported *in vitro*, and appears to involve an autocatalytic mechanism in oxygenated solution that depends on the generation of N$_2$O$_3$;[32] the bulkiness of the alkyl side chains of the model compounds studied in this report decreases the rate of decomposition by this mechanism. Recent data, however, suggest that this mechanism of decomposition is not relevant under physiological conditions;[33] heterolytic cleavage mechanisms are likely to be the principal determinants of S-nitrosothiol stability in biological systems.

Heterolytic cleavage determines the more common fate of the S-NO bond, and is principally a consequence of electrophilic attack of the partially negatively charged sulfur. This process is catalyzed by transition metals, especially copper (I), which can greatly catalyze the decomposition of S-nitrosothiols in aqueous solution.[34,35] In the absence of Cu(I), S-nitrosothiols are quite stable in aqueous solution; however, trace Cu(I) promotes decomposition of S-nitrosothiols to nitric oxide, thiols, and disulfides, and reducing agents, including thiols and ascorbate, facilitate this Cu(I)-dependent decomposition.[36] Reductive heterolysis has also been considered a mechanism by which superoxide anion reacts with S-nitrosothiols to facilitate their decomposition. Recent evidence, however, does not support the relevance of this reaction in biological systems; rather, carbon dioxide radical anion can react with S-nitrosothiols to yield nitric oxide and the corresponding thiol:[37]

$$GSNO + CO_2^{\bullet -}(+ H^+) \rightarrow GSH + NO^* + CO_2 \qquad (8)$$

Thiols are also important determinants of S-nitrosothiol stability, and can engage in two types of reactions, trans-S-nitrosation (Eqn. 6) and S-thiolation:

$$RSH + R'SNO \rightarrow RSSR' + NO^- + H^+ \qquad (9)$$

Trans-S-nitrosation reactions are reversible with rate constants of 1-100 $M^{-1}s^{-1}$ and equilibrium constants not unlike those of the analogous thiol-disulfide exchange reactions, i.e., near unity. S-thiolation reactions yield nitroxyl anion, which can react with oxygen and thiols; disulfide formation represents an important byproduct of this reaction. Although trans-S-nitrosation reactions are more rapid than S-thiolation reactions, the latter are thermodynamically favored owing to the rapid consumption of nitroxyl anion under physiological conditions.

8.3 Cell and Molecular Biology

The cellular metabolism of S-nitrosothiols is complex and has not yet been fully elucidated. S-nitrosothiols can both enter cells by passive diffusion and be metabolized at the cell surface to facilitate entry into the cell by trans-S-nitrosation, likely in the membrane micro-environment. We have shown that cell-surface protein disulfide isomerase (PDI) catalyzes the trans-S-nitrosative decomposition of extracellular low-molecular-weight S-nitrosothiols,[38] which appear to undergo denitrosation with subsequent reaction with molecular oxygen in the cell membrane to form the nitrosating species, nitrous anhydride (N_2O_3), which subsequently leads to S-nitrosation of intracellular thiols.[39]

These observations are based on studies with S-nitrosoglutathione, which is a naturally occurring S-nitrosothiol.[40] The intracellular fate of S-nitrosoglutathione is controversial, and appears to depend on the concentration of intracellular glutathione (~ mM concentration), the local pH, and oxygen tension.

Under anaerobic conditions, the major product of the reaction of GSNO with GSH is NH_3 and GSSG with nitric oxide and nitrous oxide as minor products; under aerobic conditions, the major products are nitrite and glutathione disulfide.[41] More recent data suggest that in endothelial cells the metabolism of S-nitrosoglutathione has an absolute requirement for cysteine in the cell culture medium, and that thiols in general, rather than glutathione exclusively, modulate S-nitrosoglutathione decomposition.[42] Other factors that may be involved in the intracellular metabolism of S-nitrosoglutathione include a partly characterized AGSNO lyase."[43,44] In addition, both cellular glutathione peroxidase (GP x-1)[45] and formaldehyde dehydrogenase[46] may serve as denitrosating enzymes, the former providing trans-S-nitrosating capability, as well.[45]

Trans-S-nitrosation reactions likely account for the cellular effects of exogenous low-molecular-weight S-nitrosothiols.[22] There are many examples of S-nitrosation reactions of protein cysteines yielding changes in protein function, which led to our original suggestion that protein S-nitrosation is a form of posttranslational modification akin to O-phosphorylation of serine, threonine, or tyrosine residues.[16] Since our original report, there have been over 80 examples of protein S-nitrosation that illustrate this principle. Until recently, however, only a few proteins have been shown to undergo S-nitrosation *in vivo*, including serum albumin,[16,21] the N-methyl-D-aspartate receptor,[47] the ryanodine-sensitive calcium release channel,[48] methionine adenosyl transferase,[49] and caspase-3.[50] Owing to severe technical limitations, the detection of S-nitrosated proteins *in vivo* has generally proven elusive. Recently, however, Jaffrey and colleagues[51,52] have developed an indirect, but sensitive and specific, method for detecting S-nitrosated proteins *in vivo*. Under conditions that do not alter S-nitrosated thiols or disulfides, cysteine thiols are blocked with methanethiolsulfonate, after which S-nitrosothiols are selectively reduced with ascorbate, then biotinylated with the sulfhydryl-specific reagent, N-[6-(biotinamido)hexyl]-3'-(2'-pyridyldithio)proprionamide. Using this methodology in mouse brain, these investigators identified sixteen proteins that are specifically S-nitrosated, and showed that S-nitrosation of these proteins is dependent upon neuronal nitric oxide synthase expression.[51] The identity of certain of these proteins confirmed the observations of other investigators using less rigorous methods (H-ras and the N-methyl-D-aspartate receptor), and suggests that there is specificity to S-nitrosation that may be a consequence of proximity of the protein cysteinyl functionality to the source of nitric oxide flux or S-nitrosothiol intermediate, the pK of the cysteinyl functionality (see below), and the microenvironment of the cysteinyl side chain.

It is important to point out that S-nitrosothiols can modulate protein function and cell phenotype by S-thiolation reactions, as well, as has been shown for glyceraldehyde-3-phosphate dehydrogenase.[53] The relative contribution of S-nitrosation and S-thiolation by S-nitrosothiols in specific cell types has not yet been ascertained.

In the extracellular space, serum albumin represents a principal S-nitrosated protein and serves a carrier and buffer function for the transport of nitric oxide.[16,17] S-nitroso-albumin has a half-life in heparinized plasma of ~5.5 hours,[54] undergoes trans-S-nitrosation reactions with S-nitrosoglutathione,[17,55-56] and in the steady-state exists in concentrations of ~80 nM.[57] One group of investigators has proposed the interesting hypothesis that S-nitrosation of the β-chain of hemoglobin occurs *in vivo* and serves as an allosteric modulator of oxygen release from the heme group;[58] however, recent data convincingly refute this hypothesis, showing that S-nitroso-hemoglobin is unstable in the reductive environment of erythrocytic cytosol and lacks allosteric function.[59]

A principal determinants of protein S-nitrosation is the pK of the sulfhydryl group. Typically ~8.4 in aqueous environments, The pK of a cysteinyl sulfhydryl functionality can become anomalously low within the constraints of the tertiary structure of a protein and proximity to charged side-chains, especially histidine and aspartate, rendering its primary form as thiolate anion. For example, the pK of serum albumin's single cysteinyl group (cys34) is ~4.5,[60] rendering it susceptible to electrophilic attack by the partially positively charged nitric oxide moiety of a low-molecular-weight S-nitrosothiol such as S-nitrosoglutathione.[16-17,61]

The role of S-nitrosothiols in cell signaling have been studied principally as activators of guanylyl cyclase, which was first demonstrated in 1980.[4] In addition, S-nitrosothiols appear to inhibit phosphoinositide-3-kinase activation by suppressing phosphatase-dependent dephosphorylation of the src-family kinase member lyn at tyr508.[62] S-nitrosation of p21[ras] at cys118 induces guanidine nucleotide exchange, leading to S-nitrosation-dependent modification of downstream signaling activities, including mitogen-activated protein kinase signaling.[63,64] Palmitate turnover is increased by S-nitrosocysteine and S-nitrosoglutathione via H-ras in NIH 3T3 cells,[65] and S-nitrosation of cys130 in the transcription factor c-Myb impaired its binding to DNA.[66] Both p53 and the p50 subunit of nuclear factor κ-B can be S-nitrosated, leading to the induction of apoptosis[67] and inhibition of nuclear factor κ-B redox activation, respectively.[68] The transcription factor AP-1also undergoes S-

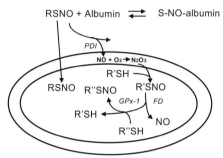

nitrosation, leading to its inactivation.[69] Caspase S-nitrosation is believed to have antiapoptotic effects.[47,70]

Figure 1. Metabolic fate of S-nitrosothiols in a cell system. RSNO, R'SNO, S-nitrosothiols; PDI, protein disulfide isomerase; GPx-1, cellular glutathione peroxidase; FD, formaldehyde dehydrogenase

8.4 Conclusions

The chemistry and biology of S-nitrosothiols have evolved considerably since their first identification over 150 years ago. In the context of nitric oxide biology, these nitrosated thiol derivatives serve several functions: they represent a stable pool of bioactive nitric oxide that is relatively resistant to oxidant stress, they offer a mechanism by which to buffer the steady-state flux of nitric oxide in biological systems, and, perhaps most importantly, they represent a unique mechanism of cell signaling through posttranslational modification of protein thiol targets. The structural and biochemical determinants of cysteinyl functionalities that render them susceptible to S-nitrosation vs. thiolation remain to be defined, and enzyme catalysts that govern their formation, cellular transport, and catabolism have begun to be identified. Clearly, this field suffers from a deficiency of specific and sensitive technologies for the direct measurement of S-nitrosated species, but offers extraordinary potential for understanding of the complex biological chemistry and molecular biology of nitric oxide.

References

1. Tasker, H. S. and H. O. Jones. 1909. Action of mercaptans on acid chlorides. I. Oxalylchloride; the mono- and dithiooxalates. *J Chem Soc* **95**:1904-1909.

2. Tasker, H. S. and H. O. Jones. 1909. Action of mercaptans on acid chlorides. II. Acid chlorides of phosphorus, sulphur and nitrogen. *J Chem Soc* **95**:1910-1918.

3. Needleman, P., B. Jakschik and E.M. Johnson, Jr. 1973. Sulfhydryl requirement for relaxation of vascular smooth muscle. *J Pharmacol Therap* **187**:324-331.

4. Ignarro, L. J., J. D. Edwards, D. Y. Gruetter, B. K. Barry and C. A. Gruetter. 1980. Possible involvement of S-nitrosothiols in the activation of guanylate cyclase by nitroso compounds. *FEBS Lett* **110**:275-278.

5. Ignarro, L. J. and C. A. Gruetter. 1980. Requirement of thiols for activation of coronary arterial guanylate cyclase by glyceryltrinitrite and sodium nitrite. *Biochim Biophys Acta* **631**:221-231.

6. Loscalzo, J. 1985. N-Acetylcysteine potentiates inhibition of platelet aggregation by nitroglycerin. *J Clin Invest* **76**:703-708.

7. Park, J. W., G. E. Billman and G. E. Means. 1993. Transnitrosation as a predominant mechanism in the hypotensive effect of S-nitrosoglutathione. *Biochem Mol Biol Int* **30**:885-891.

8. Loscalzo, J., D. Smick, N. Andon and J. Cooke. 1989. S-Nitrosocaptopril. I. Molecular characterization and effects on the vasculature and on platelets. *J Pharmacol Exp Therap* **249**:726-729.

9. Cooke, J. P., N. Andon and J. Loscalzo, J. 1989. S-nitrosocaptopril. II. Effects on vascular reactivity. *J Pharmacol Exp Therap* **249**:730-734.

10. Shaffer, J. E., F. L. Thomson, B. J. Han, J. P. Cooke and J. Loscalzo. 1991. The hemodynamic effects of S-nitrosocaptopril in anesthetized dogs. *J Pharmacol Exp Therap* **256**:704-709.

11. Furchgott, R. F. and J. V. Zawadzki. 1980. The obligatory role of endothelial cells in the relaxation of arterial smooth muscle by acetylcholine. *Nature* **288**:373-376.

12. Ignarro, L. J., R. E. Byrns, G. M. Buga and K. S. Wood. 1987. Endothelium-derived relaxing factor from pulmonary artery and vein possesses pharmacologic and chemical properties identical to those of nitric oxide radical. *Circ Res* **61**:866-879.

13. Stamler, J. S., M. E. Mendelsohn, P. Amarante, D. Smick, N. Andon, P. F. Davies, J. P. Cooke, J. P. and J. Loscalzo. 1989. N-Acetylcysteine potentiates platelet inhibition by endothelium-derived relaxing factor. *Circ Res* **65**:789-795.

14. Cooke, J. P., J. Stamler, N. Andon, P. F. Davies, G. McKinley and J. Loscalzo. 1990. Flow stimulates endothelial cells to release a nitrovasodilator that is potentiated by reduced thiol. *Am J Physiol* **28**:H804-H812.

15. Stamler, J. S., O. Jaraki, J. Osborne, D. I. Simon, J. F. Keaney Jr., J. Vita, D. Singel, C. R. Valeri and J. Loscalzo. 1992. Nitric oxide circulates in mammalian plasma primarily as an S-nitroso adduct of serum albumin. *Proc Nat'l Acad Sci (USA)* **89**:7674-7677.

16. Stamler, J. S., D. I. Simon, J. A. Osborne, M. E. Mullins, O. Jaraki, T. Michel, D. J. Singel and J. Loscalzo. 1992. S-Nitrosylation of proteins with nitric oxide: synthesis and characterization of biologically active compounds. *Proc Nat'l Acad Sci (USA)* **89**:444-448.

17. Lane, P., G. Hao and S. S. Gross. 2001. S-Nitrosylation is emerging as a specific and fundamental posttranslational protein modification head-to-head comparison with O-phosphorylation. *Science* STKE:RE1-RE7. von Löwig, C. 1836. Ueber Sulfaethylschwefelsaeure *Ann Physik Chemie* **47**:153-160.

18. Oae, S., Y. H. Kim, D. Fukushima and K. Shinhama. 1978. New syntheses of thionitrites and their chemical reactivities. *J Chem Soc Perkin Trans* **1**:913-917.

19. Liu, X., M. J. S. Miller, M. S. Joshi, D. D. Thomas and J. R. Lancaster Jr. 1998. Accelerated reaction of nitric oxide with O_2 within the hydrophobic interior of biological membranes. *Proc Nat'l Acad Sci (USA)* **95**:2175-2179.

20. Scharfstein, J. S., J. F. Keaney Jr., A. Slivka, G. N. Welch, J. A. Vita, J. S. Stamler and J. Loscalzo 1994. In vivo transfer of nitric oxide between a plasma protein-bound reservoir and low molecular weight thiols. *J Clin Invest* **94**:1432-1439.

21. Liu, Z., M. A. Rudd, J. E. Freedman and J. Loscalzo. 1998. S-transnitrosation reactions are involved in the metabolic fate and biological actions of nitric oxide. *J Pharmacol Exp Therap* **284**:526-534.

22. Viner, R. L., T. D. Williams and C. Schoneich. 1999. Peroxynitrite modification of protein thiols: oxidation, nitrosylation, and glutathiolation of functionally important cysteine residue(s) in the sarcoplasmic reticulum Ca-ATPase. *Biochemistry* **38**:12408-12415.

23. Oae, S. and K. Shinhama. 1983. Organic thionitrites and related substances: a review. *Org Prep Proc Int* **15**:165-198.

24. Field, L., R. V. Dilts, R. Ravichandran, P. G. Lenhert and G.E. Carnahan. 1978. An unusually stable thionitrite from N-acetyl-D,L-penicillamine; X-ray crystal and molecular structure of 2-(acetylamino)-2-carboxy-1,1-dimethylethyl thionitrite. *J C S Chem Comm* pp.249-250.

25. Barrett, J., L. J. Fitzgibbones, J. Glauser, R. H. Still and P.N.W. Young. 1966. Photochemistry of the S-nitroso derivatives of hexane-1-thiol and hexane-1,6-dithiol. *Nature* 1966;**211**:848.

26. Furchgott, R. F., S. J. Ehrreich and E. Greenblatt. 1961. The photoactivated relaxation of rabbit aorta. *J Gen Physiol* **44**:499-519.

27. Matsunaga, K. and R.F. Furchgott. 1989. Interactions of light and sodium nitrite in producing relaxation of rabbit aorta. *J Pharmacol ExpTherap* **248**:687-695.

28. Megson, L. L., S. A. Holmes, K. S. Magid, R. J. Pritchard, R. J. and F. W. Flitney. 2000. Selective modifiers of glutathione biosynthesis and 'repriming' of vascular smooth muscle photorelaxation. *Br J Pharmacol* **130**:1575-1580.

29. Wolin, M. S., H. A. Omar, M. P. Mortelliti, M. P. and P.D. Cherry. 1991. Association of pulmonary artery photorelaxation with hydrogen peroxide metabolism by catalase. *Am J Physiol* **261**:H1141- H1147.

30. Chaudhry, H., M. Lynch, K. Schomacker, R. Birngruber, K. Gregory and I. Kochevar. 1993. Relaxation of vascular smooth muscle induced by low-power laser radiation. *Photochem Photobiol* **58**:661-669.

31. Grossi, L., P. C. Montevecchi and S. Strazzari. 2001. Decomposition of S-nitrosothiols: unimolecular versus autocatalytic mechanism. *J Am Chem Soc* **123**:4853-4854.

32. Bartberger, M. D., J. D. Mannion, S. C. Powell, J. S. Stamler, K. N. Houk and E.J. Toone. 2001. Dissociation energies of S-nitrosothiols: on the origins of nitrosothiol decomposition rates. *J Am Chem Soc* **123**:8868-8869.

33. McAninly J., D., L. H. Williams, S. C. Askew, A. R. Butler and C. Russell. 1993. Metal ion catalysis in nitrosothiol (RSNO) decomposition. *J Chem Soc Chem Comm* 1758-1759.

34. Dicks, A. P., H. R. Swift, D. L. H. Williams, A. R. Butler, H. H. Al-Sa'doni and B. G. Cox. 1996. Identification of Cu$^+$ as the effective reagent in nitric oxide formation from S-nitrosothiols (RSNO). *J Chem Soc Perkin Trans* 2:481-487.

35. Scorza, G., D. Pietraforte and M. Minetti. 1997. Role of ascorbate and protein thiols in the release of nitric oxide from S-nitroso-albumin and S-nitroso-glutathione in human plasma. *Free Rad Biol Med* **22**:633-642.

36. Ford, E., M. N. Hughes and P. Wardman. 2002. The reaction of superoxide radicals with -nitrosoglutathione and the products of its reductive heterolysis. *J Biol Chem* **277**:2430-2436.

37. Zai, A., M. A. Rudd, A. W. Scribner and J. Loscalzo. 1999. Cell-surface protein disulfide isomerase catalyzes transnitrosation and regulates intracellular transfer of nitric oxide. *J Clin Invest* **103**:393-399.

38. Ramachandran, N., P. Root, X. M. Jiang, P. J. Hogg and B. Mutus. 2001. Mechanism of transfer of NO from extracellular S-nitrosothiols into the cytosol by cell-surface protein disulfide isomerase. *Proc Nat'l Acad Sci (USA)* **98**:9539-9544.

39. Gaston, B., J. Riley, J. M. Drazen, J. Fackler, P. Ramdev, D. Arnelle, M. E. Mullins, J. D. Sugarbaker, C. Chee, D. J. Singel, J. Loscalzo and J.S. Stamler. 1993. Endogenous nitrogen oxides and bronchodilator S-nitrosothiols in human airways. *Proc Nat'l Acad Sci (USA)* **90**:10957-10961.

40. Singh, S. P., J. S. Wishnok, M. Keshive, W. M. Deen, and S. R. Tannenbaum. 1997. The chemistry of the S-nitrosoglutathione/glutathione system. *Proc Nat'l Acad Sci (USA)* **93**:14428-14433.

41. Zeng, H., N. Y. Spencer and N. Hogg. 2001. Metabolism of S-nitrosoglutathione by endothelial cells. *Am J Physiol Heart Circ Physiol* **281**:H432-H439.

42. Gordge, M. P., P. Addis, A. A. Noronha-Dutra and J. S. Hothersall. 1998. Cell-mediated biotransformation of S-nitrosoglutathione. *Biochem Pharmacol* **55**:657-665.

43. Misiti, F, M. Castagnola, C. Zuppi, B. Giardina and I. Messana. 2001. I. Role of ergothioneine on S-nitrosoglutathione catabolism. *Biochem J* **356**:799-804.

44. Freedman, J. E., B. Frei, G. N. Welch and J. Loscalzo. 1995; Glutathione peroxidase potentiates the inhibition of platelet function by S-nitrosothiols. *J Clin Invest* **96**:394-400.

45. Liu, L., A. Hausladen, M. Zeng, L. Que, J. Hettman and J. S. Stamler. 2001. A metabolic enzyme for S-nitrosothiols conserved from bacteria to humans. *Nature* **410**:480-484.

46. Lipton, S. A., Y. B. Choi, Z. H. Pan, S. Z. Lei, H. S. V. Chen, N. J. Sucher, J. Loscalzo, D. J. Singel and J.S. Stamler. 1993. A redox-based mechanism for the neuroprotective and neurodestructive effects of nitric oxide and related nitroso-compounds. *Nature* **364**:626-632.

47. Xu, L, J. P. Eu, G. Meissner and J.S. Stamler. 1998. Activation of the cardiac calcium release channel (ryanodine receptor) by poly-S-nitrosylation. *Science* **279**:234-237.

48. Perez-Mato, L., C. Castro, F. A. Rutz, F. J. Corrales and J. M. Mato. Methionine adenosyltrasnferase S-nitrosylation is required for the basic and acidic amino acids surrounding the target thiol. *J Biol Chem* **274**:7075-7079.

49. Mannick, J. B., A. Hausladen, L. Liu, D. T. Hess, M. Zeng, G. X. Miao, L. S. Kane, A. J. Gow and J. S. Stamler. 1999. Fas-induced caspase denitrosylation. *Science* **284**:651-654.

50. Jaffrey, S. R., H. Erdjument-Bromage, C. D. Ferris, P. Tempst and S. H. Snyder. 2001. Protein S-nitrosylation: a physiological signal for neuronal nitric oxide. *Nature Cell Biol* **3**:193-197.

51. Jaffrey, S. R. and S. H. Snyder. 2001. The biotin switch method for the detection of S-nitrosylated proteins. *Science STKE* **86**:PL1.

52. Mohr, S., H. Hallak, A. De Boitte, E. G. Lapetina and B. Brune. 1999. Nitric oxide-induced S-glutathionylation and inactivation of glyceraldehyde-3-phosphate dehydrogenase. *J Biol Chem* **274**:9427-9430.

53. Tsikas, D., J. Sandmann, S. Rossa, F. M. Gutzki and J. S. Frolich. 1999. Measurement of S-nitrosoalbumin by gas chromatography-mass spectrometry. I. Preparation, purification, isolation, characterization and metabolism of S-[^{15}N]-nitrosoalbumin in human blood in vitro. *J Chromatogr B Biomed Sci Appl* **726**:1-12.

54. Tsikas, D., J. Sandmann, P. Lueszen, A. Savva, S. Rossa, D. O. Stichtenoth and J.C. Frolich. 2001. S-Transnitrosylation of albumin in human plasma and blood in vitro and in vivo in the rat. *Biochim Biophys Acta* **1546**:422-434.

55. Jourd'heuil, D., K. Hallen, M. Feelisch and M. B. Grisham. 2000. Dynamic state of S-nitrosothiols in human plasma and whole blood. *Free Rad Biol Med* **28**:409-417.

56. Marley, R., R. P. Patel, N. Orie, E. Ceaser, V. Darley-Usmar and K. Moore. 2001. Formation of nanomolar concentrations of S-nitroso-albumin in human plasma by nitric oxide. *Free Rad Biol Med* **31**:688-696.

57. Jia, L., C. Bonaventura, J. Bonaventura and J. S. Stamler. 1996. S-Nitrosohaemoglobin: a dynamic activity of blood involved in vascular control. *Nature* **380**:221-226.

58. Gladwin, M. T., X. Wang, C. D. Reiter, B. K. Yang, E. X. Vivas, C. Bonaventura and A. N. Schechter. 2002. S-Nitrosohemoglobin is unstable in .the reductive red cell environment and lacks O_2/NO-linked allosteric function. *J Biol Chem,* e-publication.

59. Lewis, S. D., D. C. Misra and J. A. Shafer. 1980. Determination of interactive thiol inoizations in bovine serum albumin, glutathione, and other thiols by potentiometric difference titration. *Biochemistry* **23**:6129-6137.

60. Keaney, J. F., Jr., D. I. Simon, J. S. Stamler, O. Jaraki, J. Scharfstein, J. A. Vita and J. Loscalzo. 1993. NO forms an adduct with serum albumin that has endothelium-derived relaxing factor-like properties. *J Clin Invest* **91**:1582-1589.

61. Pigazzi, A., S. Heydrick, F. Folli, S. Benoit, A. Michelson and J. Loscalzo. 1999. Nitric oxide inhibits thrombin-receptor-activating peptide-induced phosphoinositide 3-kinase activity in human platelets. *J Biol Chem* **274**:14368-14375.

62. Lander, H. M., D. P. Hajjar, B. L. Hempstead, U. A. Mizra and B. T. Chait. 1995. A molecular redox switch on p21[ras]. *J Biol Chem* **272**:4323-4326.

63. Lander, H. M., J. M. Ogiste, S. F. Pearce, R. Levi and A. Novogrodski. 1995. Nitric oxide-stimulated guanine nucleotide exchange on p21[ras]. *J Biol Chem* **270**:7017-7020.

64. Baker, T. L., M. A. Booden and J. E. Buss. 2000. S-nitrosocysteine increases palmitate turnover on H-Ras in NIH 3T3 cells. *J Biol Chem* **275**:467-473.

65. Brendeford, E. M., K. B. Andersson and O. S. Gabrielsen. 1998. Nitric oxide (NO) disrupts specific DNA binding of the transcription factor c-Myb in vitro. *FEBS Lett* **425**:52-56.

66. Calmels, S., P. Hainaut and H. Ohshima. 1997. Nitric oxide induces conformational and functional modifications of wild-type p53 tumor suppressor protein. *Cancer Res* **57**:3365-3369.

67. DellaTorre, A., R. A. Schroeder, C. Punzalan and P.C. Kuo. 1999. Endotoxin-mediated S-nitrosylation of p50 alters NK-κ B-dependent gene transcription in ANA-1 murine macrophages. *J Immunol* **162**:4101-4108.

68. Tabuchi, A., K. Sano, E. Oh, T. Tsuchiya and M. Tsuda. 1994. Modulation of AP-1 activity by nitric oxide (NO) in vitro: NO-mediated modulation of AP-1. *FEBS Lett* **351**:123-127.

69. Dimmeler, S., J. Haendeler, A. Sause and A. M. Zeiher. 1998. Nitric oxide inhibits APO-1/Fas-mediated cell death. *Cell Growth Differ* **9**:415-422.

Chapter 9

HYDROGEN PEROXIDE AS INTRACELLULAR MESSENGER: IDENTIFCATION OF PROTEIN TYROSINE PHOSPHATASES AND PTEN AS H_2O_2 TARGET

Sue Goo Rhee, Seung-Rock Lee, Kap-Seok Yang, Jaeyul Kwon, and Sang Won Kang

9.1 Introduction

Hydrogen peroxide (H_2O_2) is generated in all aerobic organisms as a result of normal cellular metabolism. Thus, electrons that leak from the electron transport chain of mitochondria cause the univalent reduction of molecular oxygen to superoxide anion ($O_2^{\bullet-}$), which is then spontaneously or enzymatically dismutated to H_2O_2 (1). H_2O_2 is readily converted to hydroxyl radicals ($^{\bullet}OH$) via the Fenton reaction in the presence of iron (or copper) and cellular reductants, and these radicals induce irreversible oxidative damage to various cellular components.[2] Hydrogen peroxide is also generated by arachidonic acid-metabolizing enzymes,[3] xanthine oxidase,[4] nitric oxide synthase,[4] and cytochrome P450,[5] as well as in the cellular response to ultraviolet radiation.

Furthermore, many cell types produce H_2O_2 in response to a variety of extracellular stimuli including cytokines, neurotransmitters, peptide growth factors, hormones, and phorbol myristate acetate (PMA).[6-13] H_2O_2 thus produced in response to receptor stimulation is known to mediate the activation of such crucial protein kinases as the members of MAPK family, Src family, receptor protein

167

H. J. Forman, J. Fukuto, and M. Torres (eds.), Signal Transduction by Reactive Oxygen and Nitrogen Species: Pathways and Chemical Principles, ©2003, Kluwer Academic Publishers. Printed in the Netherlands

tyrosine kinase family, and PKC family.[13] The addition of exogenous H_2O_2 or the intracellular production in response to receptor stimulation affect the function of various proteins including transcription factors, phospholipases, protein phosphatases, ion channels, and G proteins.[13] Accordingly, H_2O_2 is now recognized as a ubiquitous intracellular messenger under subtoxic conditions.[14-20] Understandably, there has long been skepticism about the messenger role of H_2O_2, partly because it seemed illogical for nature to employ a dangerous molecule like H_2O_2 for such a crucial function. However, if one considers that NO, itself a reactive radical, is easily converted to the more reactive OONO while also functioning as an established intracellular messenger, the proposed H_2O_2 function is not implausible, Moreover, recent evidence that specific inhibition of H_2O_2 generation results in a complete blockage of signaling by PDGF, EGF, and angiotensin II is a strong indication that H_2O_2 serves in a messenger role.[10,11,21] However, the mechanism by which H_2O_2 mediates receptor signaling has not been well characterized.

The mechanism of the receptor-mediated generation of reactive oxygen species (ROS) has been studied extensively in phagocytic cells, in which $O_2^{\cdot-}$ (and thus H_2O_2) is produced via the reduction of O_2 by an intricate NADPH oxidase complex.[22] Several new types of NADPH oxidase were identified in non-phagocytic cells, and some of these enzymes were shown to be responsible for H_2O_2 production induced by various agonists.[23] The mechanisms by which these non-phagocytic NADPH oxidases are activated by various receptors are described in the chapter by Arnold and Lambeth. Identifying the target proteins for H_2O_2 interaction is also key to understanding how H_2O_2 mediates signaling processes. The goal of this report is to review our current knowledge of the molecules on which H_2O_2 acts directly to propagate the signal.

9.2 Identification of Proteins Containing Cysteine Residues that are Sensitive to Oxidation by Hydrogen Peroxide

Since H_2O_2 is readily converted to $^{\cdot}OH$, it is considered a cytotoxic agent that causes damage to many cellular components. Oxidative damage has been implicated in numerous disorders and the general process of aging. Accordingly, the target for H_2O_2 action has been studied mainly in relation to these disorders, leading to the identification of irreversibly damaged cellular components, including lipid-derived malondialdehyde,[24] carbonyl group-containing proteins,[25] 4-hydroxynonenal-conjugated protein,[26] and 8-hydroxy-2'-deoxyguanosine derived from DNA.[27] The intracellular messenger role of H_2O_2, however, requires reversible modification by or binding of H_2O_2. H_2O_2 itself is a mild oxidant and is relatively inert to most biomolecules. However, H_2O_2 is able to oxidize cysteine (Cys-SH) residues in proteins to Cys sulfenic acid or disulfide, both of which are readily reduced back to Cys-SH by various cellular reductants. Because unaltered sulfhydryl groups are crucial for the catalytic and structural functions of many proteins, their oxidation in normal cells is likely related to the physiological consequences of H_2O_2 production.

We developed a procedure for detecting proteins that contain H_2O_2-sensitive cysteine (or selenocysteine) residues as a means by which to study protein oxidation by H_2O_2 in cells.[28] The procedure is based on the facts that H_2O_2 and biotin-conjugated iodoacetamide (BIAM) selectively and competitively react with cysteine residues that exhibit a low pK_a, and that the decrease in the labeling of cell lysate proteins with BIAM caused by prior exposure of cells to H_2O_2 or to an agent that induces H_2O_2 production can be monitored by streptavidin blot analysis. The cysteine thiolate anion (Cys-S$^-$) is more readily oxidized by H_2O_2 than is the Cys sulfhydryl group (Cys-SH). Because the pK_a values of most protein Cys-SH residues are ~8.5,[29] few proteins would be expected to possess a Cys-SH residue that is readily susceptible to oxidation by H_2O_2 in cells. Protein Cys residues exist as thiolate anions at neutral pH often because the negatively charged thiolate is stabilized by salt bridges to positively charged amino acid residues. Proteins with Cys-S$^-$ residues include protein disulfide isomerase,[30] protein tyrosine phosphatase,[31] creatine kinase,[32] glyceraldehyde-3-phosphate dehydrogenase,[33] and peroxiredoxin.[34] In addition, a few proteins contain selenocysteine (Cys-SeH) residues, which are readily oxidized at neutral pH because their pK_a values are 5.7.[29]

Given that oxidized Cys residues do not react with sulfhydryl agents such as BIAM, proteins that contain an H_2O_2-sensitive Cys, but not those containing Cys residues with a normal pK_a, would be expected to exhibit a decreased extent of labeling with BIAM after exposure to H_2O_2. The extent of BIAM labeling can be readily measured by SDS-PAGE and blot analysis of the biotinylated proteins with HRP-conjugated streptavidin and ECL (Fig 1A). However, because most protein Cys residues are not sensitive to oxidation by H_2O_2 at neutral pH, the decrease in BIAM labeling due to Cys oxidation would be difficult to detect if all cysteines of cellular proteins were labeled. This problem was circumvented by performing the labeling reaction at pH 6.5, at which proteins with low-pK_a Cys residues are predominantly carboxymethylated by BIAM because Cys-S$^-$ is more susceptible to carboxymethylation than is Cys-SH (Fig 1A).

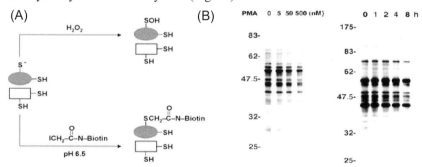

Figure 1. Identification of proteins containing cysteine residues that are sensitive to oxidation by H_2O_2 at neutral pH. (A) Use of BIAM labeling to identify H_2O_2-senstive proteins (B) Effects of exposure of K562 (left panel) and HT22 (right panel) cells to PMA and glutamate, respectively, on the labeling of proteins in cell lysates with BIAM. K562 cells were incubated for 30 min with the indicated concentrations of PMA, and HT22 cells were incubated for the indicated times with 5 mM glutamate. The cells were then lysed in a pH 6.5 buffer containing BIAM, and the resulting lysates (10 µg of protein) were subjected to blot analysis with HRP-conjugated streptavidin. The positions of molecular size standards are shown on the left. [This figure was modified from ref.[28]].

This BIAM labeling procedure was applied to mouse hippocampal HT22 cells in which H_2O_2 production was induced by glutamate, and human erythroleukemia K562 cells in which H_2O_2 production was induced by phorbol myristate acetate (Fig 1 B). It revealed that several cellular proteins contain cysteine or selenocysteine residues that are selectively oxidized by H_2O_2. Three of these H_2O_2-sensitive proteins were identified as a member of the protein disulfide isomerase family, thioredoxin reductase, and creatine kinase, all of which were previously known to contain at least one reactive cysteine or selenocysteine at their catalytic sites.

It was also investigated whether, at pH 6.5, BIAM specifically labels low-pK_a Cys residues within a protein molecule, and whether the labeling is decreased by prior exposure to H_2O_2. Mammalian muscle creatine kinase contains a reactive Cys at position 283 and three additional Cys residues. When purified creatine kinase was incubated in the presence of H_2O_2 at pH 7.0, and then subjected the protein to BIAM labeling at pH 6.5, Cys^{283} was the major site of modification of creatine kinase by BIAM at pH 6.5 as well as the site of oxidation by H_2O_2.[28] This procedure should thus prove useful for the identification of proteins that are oxidized by H_2O_2 generated in response to a variety of extracellular agents.

One of the limitations for the BIAM method is that the bulky alkylating reagent cannot react with cysteine residue located inside of the narrow pocket. For example, protein tyrosine phosphatases contain a low pK_a cysteine at their active site, but they could not be efficiently labeled at pH 6.5, probably because the reactive cysteine is not accessible to the bulky BIAM.

9.3 Reversible Inactivation of Protein Tyrosine Phosphatases in Cells Stimulated with Growth Factors.

Proteins with low-pK_a cysteine residues include protein tyrosine phosphatases (PTPs). All PTPs contain an essential cysteine residue (pK_a, 4.7 to 5.4) in the signature active site motif, His-Cys-X-X-Gly-X-X-Arg-Ser/Thr (where X is any amino acid), that exists as a thiolate anion at neutral pH.[35] This thiolate anion contributes to formation of a thiol-phosphate intermediate in the catalytic mechanism of PTPs. The active site cysteine is the target of specific oxidation by various oxidants, including H_2O_2, and this modification can be reversed by incubation with thiol compounds such as dithiothreitol (DTT) and reduced glutathione. These observations suggested that PTPs might undergo H_2O_2-dependent inactivation in cells. However, such evidence was not available until we demonstrated the ability of intracellularly produced H_2O_2 to oxidize PTP1B in EGF-stimulated A431 cells.[36] Because the active site Cys of PTP1B could not be labeled with BIAM, active PTP1B was labeled with $[^{14}C]$iodoacetic acid (IAA) in cell lysates and the decrease in $[^{14}C]$radioactivity associated with immunoprecipitated PTP1B was taken as the measure of its oxidation (Fig. 2).

The amount of oxidatively inactivated PTP1B was maximal (~40%) 10 min after exposure of cells to EGF and returned to baseline values by 40 min, suggesting that the oxidation of this phosphatase by H_2O_2 is reversible in cells. These results, together with the observation that increased levels of PDGF- or EGF-induced protein tyrosine phosphorylation requires H_2O_2 production, indicate that the activation of receptor protein tyrosine kinase (RTK) per se by binding of the corresponding growth factor may not be sufficient to increase the steady state level of protein tyrosine phosphorylation in cells. Rather, the concurrent inhibition of PTPs by H_2O_2 may also be required. This suggests that the extent of autophosphorylation of RTKs and their substrates would return to basal values after degradation of H_2O_2 and the subsequent reactivation of PTPs by electron donors. Experiments with purified PTP1B suggest that the oxidized enzyme is reactivated more effectively by thioredoxin (Trx) than by glutaredoxin or glutathione at the physiological concentrations of these reductants. Thus, Trx might be a physiological electron donor for PTP1B, as well as for other protein tyrosine phosphatases.

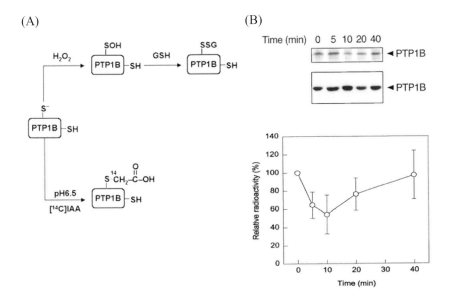

Figure 2. Detection of reversible oxidation of PTP1B with the use of [14C]iodoacetic acid (IAA) labeling. (A) The catalytic site Cys-SH of PTP1B is selectively oxidized by H_2O_2 to Cys-SOH and then glutathionylated, whereas those PTP1B molecules that had not been oxidized in the cell are alkylated by [14C]IAA. (B) Effect of EGF stimulation on [14C]IAA labeling of PTP1B from A431 cells. A431 cells were stimulated with EGF (200 ng/ml) for the indicated times and then lysed in a buffer containing 2 mM [14C]IAA. PTP1B was precipitated from the cell lysates with a specific mAb and subjected to SDS-PAGE on a 10% gel. The dried gel was exposed to x-ray film (Kodak) for 4 days to yield an autoradiogram (upper panel), after which the separated proteins were probed by immunoblot analysis with rabbit antibodies to PTP1B and immune complexes were visualized with horseradish peroxidase-conjugated antibodies to rabbit immunoglobulin G and ECL reagents (Amersham) (middle panel). The intensity of the PTP1B bands on the autoradiograms was quantitated with a Phosphorimager. Data are expressed as a percentage of the value for immunoprecipitates derived from unstimulated A431 cells and are means ± SEM from three independent experiments (bottom panel). [This figure was modified from ref.[36]].

With the use of the recombinant 37-kDa form of PTP1B, we showed that the site of oxidation by H_2O_2 was the essential residue Cys[215]. The oxidized products of cysteine include sulfenic acid (Cys-SOH), disulfide (Cys-S-S-Cys), sulfinic acid (Cys-SO₂H), and sulfonic acid (Cys-SO₃H). The oxidative product of PTP1B was proposed to be sulfenic acid. The disulfide intermediate was excluded as the H_2O_2-modified form of PTP1B on the basis of the observation that only one out of six DTNB (5,5'-dithiobis-2-nitorbenzoic acid)-sensitive residues was lost after H_2O_2 oxidation, and the sulfinic and sulfonic acid intermediates on the basis of the observation that the oxidized PTP1B could be reduced back to its original state by DTT. Cysteine sulfenic acid is highly unstable and readily undergoes condensation with a thiol. However, the sulfenic acid intermediate of PTP1B is likely stabilized by the fact that, according to the x-ray structure of the 37-kDa form of PTP1B,[37] no cysteine residues are located near Cys^{215}. Furthermore, the sulfenate anion (Cys-SO[-]) is also likely stabilized by a salt bridge to Arg^{221} which was shown to stabilize the thiolate anion of Cys^{215} and consequently to reduce its pK_a. Alternatively, Cys^{215}-SOH might react with glutathione as Chock's laboratory detected glutathionylated PTP-1B in A431 cells treated with EGF.[38]

Insulin stimulation also induces the production of intracellular H_2O_2.[39] Goldstein's laboratory demonstrated the reversibly oxidation of PTP1B and possibly other PTPs in insulin-stimulated cells by directly measuring the catalytic activity of PTPase activity in cell homogenates under strictly anaerobic conditions.[40] About 62% of total cellular PTPase activity was found to be reversibly inactivated in 3T3-L1 adipocytes and hepatoma cells stimulated with insulin. PTP1B, selectively immunoprecipitated from cell homogenates was inhibited up to 88% following insulin stimulation. These results suggested that H_2O_2 produced in response to insulin contributes to the insulin-stimulated cascade of protein tyrosine phosphorylation by oxidatively inactivating PTP1B and other PTPs.

Tonks' laboratory developed another method to reveal reversible oxidation of PTPs in cells.[41] This method is based on the fact that those PTPs with the oxidized Cys-SOH at their active site are resistant to alkylation by IAA and can be reactivated by treatment with DTT, whereas any PTPs that had not been oxidized by H_2O_2 in the cell became irreversibly inactivated by alkylation with IAA (Fig. 3A). To visualize active PTPs, an aliquot of cell lysate that had been prepared under anaerobic conditions was subjected to SDS-PAGE containing a radioactively labeled substrate, and proteins in the gel were renatured in the presence of DTT. Under these conditions, the activity of the PTPs in which the active site Cys had been subjected to H_2O_2 oxidation to sulfenic acid was recovered, whereas those that were not oxidized in response to the initial stimulus and were irreversibly alkylated in the lysis step remained inactive. The bands corresponding to activated PTPs were then detected by dephosphorylation of a [^{32}P]phosphate-labeled poly Glu-Tyr. Using this in-gel assay, Tonks' laboratory showed that the SH2- domain-containing PTP, SHP-2, was reversibly oxidized in Rat-1 cells treated with PDGF (Fig. 3B).

SHP-2 is a ubiquitously expressed cytosolic PTP that contains two SH2 domains and a PTP catalytic domain. The crystal structure revealed that SHP-2 exists in an inactive conformation in which the active site is occluded by the N-terminal SH2 domain.[42] Binding of autophosphorylated PDGF receptor promotes adoption of an open, active conformation in which the catalytic site of SHP-2 is free to interact with substrates.[42] Thus, one would anticipate that following binding of SHP-2 to autophosphorylated receptor, tyrosine phosphorylation of the PDGF receptor would be decreased. However, the peak of PDGF receptor autophosphorylation occurred during the time at which SHP-2 was associated with the receptor. These seemingly contradictory observations can be explained if the PDGF receptor-bound SHP-2 is temporarily inhibited. It is likely that the temporary inhibition is achieved through H_2O_2 produced in response to PDGF stimulation and SHP-2 becomes reactivated

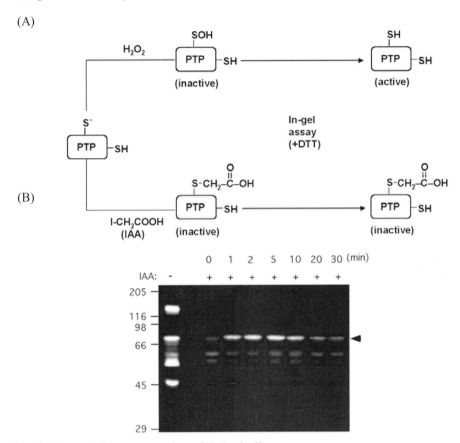

after the intracellular concentration of H_2O_2 declines.

Figure 3. Detection of reversible oxidation of PTPs with the use of in-gel assay. (A) Scheme for the in-gel assay to identify PTP that is sensitive to oxidation by H_2O_2. (B) PDGF-induced oxidation of SHP-2 in Rat-1 cells. Serum-starved Rat-1 cells were exposed to 50 ng/ml PDGF-BB for the times indicated. Lysates were prepared in the presence of 10 mM IAA, and an in-gel phosphatase assay conducted using SDS-PAGE containing [^{32}P]phosphate-labeled poly Glu-Tyr. The reaction was then terminated by fixing, and the gel was exposed to film. The presence of a PTP was visualized by dephosphorylation as the appearance of a clear, white area on the black background of labeled substrate. The arrowhead indicates the 70-kDa SHP-2 that was transiently oxidized following stimulation of Rat-1 cells with PDGF. [This figure was modified from ref.[41]].

9.4 Reversible Inactivation of the 3'-Phosphatase of Phosphoinositides PTEN by H₂O₂.

PTEN is a member of the PTP family and reverses the action of phosphoinositide (PI) 3-kinase by catalyzing the removal of the phosphate attached to the 3'-hydroxyl group of the PI inositol ring.[43] By negatively modulating the PI 3-kinase–Akt signaling pathway, PTEN functions as an important tumor suppressor.[44] The essential Cys[124] residue of PTEN is surrounded by three basic amino acid residues in the active site pocket.[45] We showed that Cys[124] of purified PTEN is readily oxidized by H₂O₂ to form a disulfide with Cys[71].[46] Oxidized PTEN migrated faster on SDS-PAGE under nonreducing conditions than did the unoxidized protein, probably because it was in a more compact protein structure.[46]

PTEN was also oxidized in cells exposed to H₂O₂. Exposure of NIH3T3 cells to H₂O₂ at concentrations as low as 50 μM resulted in the appearance of the higher-mobility (oxidized) form of PTEN, and the intensity of this band increased as the concentration of H₂O₂ increased (Fig. 4A). The amount of the oxidized form of PTEN also increased with time of incubation of cells with 0.5 mM H₂O₂, reaching a maximum at 10 to 40 min and decreasing gradually thereafter (Fig. 4B). The amount of the reduced form of the protein decreased as the amount of the oxidized form increased and vice versa, suggesting that PTEN was inactivated by H₂O₂ and then reactivated by cellular reductants as the concentration of H₂O₂ declined. When extracts derived from NIH 3T3 cells treated with H₂O₂ were incubated with DTT before gel electrophoresis, only the lower-mobility PTEN band was detected (Fig. 4A, B), consistent with the notion that the higher-mobility form of the protein contains a disulfide. To demonstrate that Cys[124] contributes to the disulfide responsible for the increase in PTEN mobility, we transfected NIH 3T3 cells separately with vectors encoding either the Cys124→Ser (C124S) mutant of human PTEN tagged with HA or the HA-tagged wild-type protein. Exposure of the transfected cells to H₂O₂ revealed that the HA-tagged wild-type protein, but not the HA-tagged C124S mutant, underwent reversible oxidation (Fig. 4C), confirming that Cys[124] forms the disulfide bond responsible for the mobility shift of PTEN.

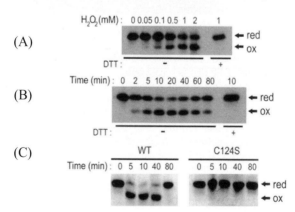

Figure 4. Oxidation of PTEN in NIH 3T3 cells exposed to H₂O₂. (A and B) Oxidation of endogenous PTEN in NIH 3T3 cells exposed to H₂O₂. Cells were incubated either for 5 min with the indicated concentrations of H₂O₂ (A) or for the indicated times with 0.5 mM H₂O₂ (B). Cellular protein extracts were then alkylated with N-ethylmaleimide and subjected to

nonreducing SDS-PAGE followed by immunoblot analysis with antibodies to PTEN. One protein sample (rightmost lane) was treated with 100 mM DTT for 5 min before electrophoresis. (C) Oxidation of wild-type PTEN, but not of the C124S mutant, in NIH 3T3 cells exposed to H_2O_2. Transfected cells expressing either HA-tagged wild-type PTEN or the HA-tagged C124S mutant were incubated with 0.5 mM H_2O_2 for the indicated times, after which cellular protein extracts were exposed to N-ethylmaleimide and subjected to nonreducing SDS-PAGE and immunoblot analysis with an antibody to HA. [This figure was modified from ref.[46]].

We also demonstrated that PTEN is oxidized in NIH3T3 cells stimulated with PDGF. Nevertheless, the extent of PTEN oxidation observed in PDGF-stimulated cells was much smaller than that apparent in the same cells treated with 50 μM H_2O_2 probably because the total amount of H_2O_2 produced in response to PDGF is much less than 50 μM. It is also likely that H_2O_2 is produced and accumulates only locally at sites near activated receptors, and that the concentration of H_2O_2 is therefore sufficiently high to induce and maintain PTEN oxidation only at these sites. Only a subset of PTEN molecules is thus likely to undergo oxidation in stimulated cells.

As in the case of PTP1B, the reduction of H_2O_2-oxidized PTEN in cells appears to be mediated predominantly by thioredoxin. Thus, thioredoxin was more efficient than was glutaredoxin or glutathione with regard to the reduction of oxidized PTEN in vitro; thioredoxin co-immunoprecipitated with PTEN from cell lysates; and incubation of cells with 2,4-dinitro-1-chlorobenzene (an inhibitor of thioredoxin reductase) delayed the reduction of oxidized PTEN, whereas incubation with buthionine sulfoximine (an inhibitor of glutathione biosynthesis) did not.

9.5 Conclusion

The proposed role of H_2O_2 in growth factor signaling is depicted in Fig. 5. In the model, H_2O_2 induces reversible inactivation of PTPs and PTEN through oxidation of their essential Cys residues. This model suggests that the receptor-mediated activation of RTK and PI 3-kinase may not be sufficient for the accumulation of tyrosine phosphorylated proteins and 3'-phosphorylated PIs because of the opposing activity of PTPs and PTEN, respectively. The concomitant inactivation of PTPs and PTEN by H_2O_2 produced in response to receptor stimulation might also be necessary for these effects. This model is consistent with the previous observations that H_2O_2 generation and accumulation are necessary for downstream actions of PDGF and EGF that are mediated by RTK and PI 3-kinase. In the absence of the proposed function of H_2O_2, the activation of RTK and PI 3-kinase would result in futile cycles of phosphorylation / dephosphorylation of proteins and phosphoinositide.

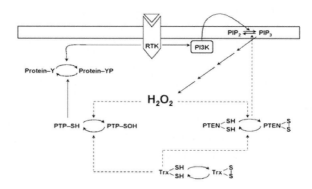

Figure 5. Scheme depicting the role of H_2O_2 as a modulator of protein tyrosine phosphorylation and PtdIns[3,4,5] P_3 (PIP$_3$) production. The essential role of PIP$_3$ in H_2O_2 production[47] as well as the electron-donating role of thioredoxin (Trx) in reduction of inactivated PTP and PTEN are also indicated. RTK, receptor tyrosine kinase; PI3K, PI 3-kinase; Y and YP, tyrosine and tyrosine phosphate, respectively.

References

1. Fridovich, I. (1997). Superoxide anion radical (O2-.), superoxide dismutases, and related matters. *J Biol Chem* **272:**18515-18517

2. Stadtman, E. R. (1992). Protein oxidation and aging. *Science* **257:**1220-1224

3. Funk, C. D. (2001). Prostaglandins and leukotrienes: advances in eicosanoid biology. *Science* **294:**1871-1875.

4. Saugstad, O. D. (1996). Role of xanthine oxidase and its inhibitor in hypoxia: reoxygenation injury. *Pediatrics* **98:**103-107.

5. Bernhardt, R. (1996). Cytochrome P450: structure, function, and generation of reactive oxygen species. *Rev Physiol Biochem Pharmacol* **127:**137-221

6. Krieger-Brauer, H. I. and H. Kather. (1995). The stimulus-sensitive H_2O_2-generating system present in human fat-cell plasma membranes is multireceptor-linked and under antagonistic control by hormones and cytokines. *Biochem J* **307:**543-548

7. Ohba, M., M. Shibanuma, T. Kuroki and K. Nose. (1994). Production of hydrogen peroxide by transforming growth factor-beta 1 and its involvement in induction of egr-1 in mouse osteoblastic cells. *J Cell Biol* **126:**1079-1088

8. Thannickal, V. J., K. D. Aldweib and B. L. Fanburg. (1998). Tyrosine phosphorylation regulates H_2O_2 production in lung fibroblasts stimulated by transforming growth factor beta1. *J Biol Chem* **273:**23611-23615

9. Tan, S., Y. Sagara, Y. Liu, P. Maher and D. Schubert. (1998). The regulation of reactive oxygen species production during programmed cell death. *J Cell Biol* **141:**1423-1432

10. Sundaresan, M., Z. X. Yu, V. J. Ferrans, K. Irani and T. Finkel. (1995). Requirement for generation of H_2O_2 for platelet-derived growth factor signal transduction. *Science* **270**:296-299

11. Bae, Y. S., S. W. Kang, M. S. Seo, I. C. Baines, E. Tekle, P. B. Chock and S. G. Rhee. (1997). Epidermal growth factor (EGF)-induced generation of hydrogen peroxide. Role in EGF receptor-mediated tyrosine phosphorylation. *J Biol Chem* **272**:217-221

12. Robertson, F. M., A. J. Beavis, T. M. Oberyszyn, S. M. O'Connell, A. Dokidos, D. L. Laskin, J. D. Laskin and J. J. Reiners Jr. (1990). Production of hydrogen peroxide by murine epidermal keratinocytes following treatment with the tumor promoter 12-O-tetradecanoylphorbol- 13-acetate. *Cancer Res* **50**:6062-6067

13. Rhee, S. G., Y. S. Bae, S. R. Lee and J. Kwon. (2000). Hydrogen peroxide: A key messenger that modulates protein phosphorylation through cysteine oxidation. *Science's stke* www.stke.org/cgi/contentfull/OC_sigtrans;2000/53/pe1

14. Rhee, S. G. (1999). Redox signaling: hydrogen peroxide as intracellular messenger. *Exp Mol Med* **31**:53-59

15. Finkel, T. (1998). Oxygen radicals and signaling. *Curr Opin Cell Biol* **10**:248-253

16. Suzuki, Y. J. and G. D. Ford. (1999). Redox regulation of signal transduction in cardiac and smooth muscle. *J Mol Cell Cardiol* **31**:345-353.

17. Thannickal, V. J and B. L. Fanburg. (2000). Reactive oxygen species in cell signaling. *Am J Physiol Lung Cell Mol Physiol* **279**:L1005-1028.

18. Griendling, K. K. and M. Ushio-Fukai. (2000). Reactive oxygen species as mediators of angiotensin II signaling. *Regul Pept* **91**:21-27.

19. Patel, R. P., D. Moellering, J. Murphy-Ullrich, H. Jo, J. S. Beckman and V. M. Darley-Usmar. (2000). Cell signaling by reactive nitrogen and oxygen species in atherosclerosis. *Free Radic Biol Med* **28**:1780-1794.

20. Forman, H. J. and M. Torres. (2001). Signaling by the respiratory burst in macrophages. *IUBMB Life* **51**:365-371.

21. Ushio-Fukai, M., R. W. Alexander, M. Akers, Q. Yin, Y. Fujio, K. Walsh and K. K. Griendling. (1999). Reactive oxygen species mediate the activation of Akt/protein kinase B by angiotensin II in vascular smooth muscle cells. *J Biol Chem* **274**:22699-22704

22. Babior, B. M. (1999). NADPH oxidase: an update. *Blood* **93**:1464-1476.

23. Lambeth, J. D. (2002). Nox/Duox family of nicotinamide adenine dinucleotide (phosphate) oxidases. *Curr Opin Hematol* **9**:11-17.

24. Luo, X. P., M. Yazdanpanah, N. Bhooi and D. Lehotay. (1995). Determination of aldehydes and other lipid peroxidation products in biological samples by gas chromatography-mass spectrometry. *Anal Biochem* **228**:294-298

25. Levine, R. L., J. A. Williams, E. R. Stadtman and E. Shacter. (1994). Carbonyl assays for determination of oxidatively modified proteins. *Methods Enzymol* **233**:346-357

26. Yoritaka, A., N. Hattori, K. Uchida, M. Tanaka, E. R. Stadtman and Y. Mizuno. (1996). Immunohistochemical detection of 4-hydroxynonenal protein adducts in Parkinson disease. *Proc Natl Acad Sci U S A* **93**:2696-2701

27. Teixeira, A. J., M. R. Ferreira, W. J. van Dijk, G. van de Werken and A. P. de Jong. (1995). Analysis of 8-hydroxy-2'-deoxyguanosine in rat urine and liver DNA by stable isotope dilution gas chromatography/mass spectrometry. *Anal Biochem* **226**:307-319

28. Kim, J. R., H. W. Yoon, K. S. Kwon, S. R. Lee and S. G. Rhee. (2000). Identification of proteins containing cysteine residues that are sensitive to oxidation by hydrogen peroxide at neutral pH [In Process Citation]. *Anal Biochem* **283**:214-221

29. Besse, D., F. Siedler, T. Diercks, H. Kessler and L. Moroder. (1997). The redox potential of Selenocysteine in unconstrained cyclic peptides. *Angew. Chem. Int. Ed. Engl.* **36**:883-885

30. Kemmink, J., N. J. Darby, K. Dijkstra, M. Nilges and T. E. Creighton. (1996). Structure determination of the N-terminal thioredoxin-like domain of protein disulfide isomerase using multidimensional heteronuclear 13C/15N NMR spectroscopy. *Biochemistry* **35**:7684-7691

31. Stone, R. L. and J. E. Dixon. (1994). Protein-tyrosine phosphatases. *J Biol Chem* **269**:31323-31326

32. Furter, R., E. M. Furter-Graves and T. Wallimann. (1993). Creatine kinase: the reactive cysteine is required for synergism but is nonessential for catalysis. *Biochemistry* **32**:7022-7029

33. Mercer, W. D., S. I. Winn and H. C. Watson. (1976). Twinning in crystals of human skeletal muscle D-glyceraldehyde-3- phosphate dehydrogenase. *J Mol Biol* **104**:277-283

34. Choi, H. J., S. W. Kang, C. H. Yang, S. G. Rhee and S. E. Ryu. (1998). Crystallization and preliminary X-ray studies of hORF6, a novel human antioxidant enzyme. *Acta Crystallogr D Biol Crystallogr* **54**:436-437

35. Lohse, D. L., J. M. Denu, N. Santoro and J. E. Dixon. (1997). Roles of aspartic acid-181 and serine-222 in intermediate formation and hydrolysis of the mammalian protein-tyrosine-phosphatase PTP1. *Biochemistry* **36**:4568-4575

36. Lee, S. R., K. S. Kwon, S. R. Kim and S. G. Rhee. (1998). Reversible inactivation of protein-tyrosine phosphatase 1B in A431 cells stimulated with epidermal growth factor. *J Biol Chem* **273**:15366-15372

37. Barford, D., A. J. Flint and N. K. Tonks. (1994). Crystal structure of human protein tyrosine phosphatase 1B. *Science* **263**:1397-1404.

38. Barrett, W. C., J. P. DeGnore, Y. F. Keng, Z. Y. Zhang, M. B. Yim and P. B. Chock. (1999). Roles of superoxide radical anion in signal transduction mediated by reversible regulation of protein-tyrosine phosphatase 1B. *J Biol Chem* **274**:34543-34546.

39. May, J. M. and C. de Haen. (1979). Insulin-stimulated intracellular hydrogen peroxide production in rat epididymal fat cells. *J Biol Chem* **254**:2214-2220

40. Mahadev, K., A. Zilbering, L. Zhu and B. J. Goldstein. (2001). Insulin-stimulated hydrogen peroxide reversibly inhibits protein- tyrosine phosphatase 1b in vivo and enhances the early insulin action cascade. *J Biol Chem* 276:21938-21942.

41. Meng, T. C., T. Fukada and N. K. Tonks. (2002). Reversible oxidation and inactivation of protein tyrosine phosphatases in vivo. *Mol Cell* **9:**387-399.

42. Hof, P., S. Pluskey, S. Dhe-Paganon, M. J. Eck and S. E. Shoelson. (1998). Crystal structure of the tyrosine phosphatase SHP-2. *Cell* **92:**441-450.

43. Maehama, T. and J. E. Dixon. (1998). The tumor suppressor, PTEN/MMAC1, dephosphorylates the lipid second messenger, phosphatidylinositol 3,4,5-trisphosphate. *J Biol Chem* **273:**13375-13378.

44. Maehama, T., G. E. Taylor and Dixon. (2001). PTEN and Myotubularin:Novel Phosphoinositide Phosphatases. *Annu. Rev. Biochem.* **70:**247-279

45. Lee, J. O., H. Yang, M. M. Georgescu, A. Di Cristofano, T. Maehama, Y. Shi, J. E. Dixon, P. Pandolfi and N. P. Pavletich. (1999). Crystal structure of the PTEN tumor suppressor: implications for its phosphoinositide phosphatase activity and membrane association. *Cell* **99:**323-334.

46. Lee, S. R., K. S. Yang, J. Kwon, C. Lee, W. Jeong and S. G. Rhee. (2002). Regulation of PTEN by superoxide and H_2O_2 through the reversible formation of a disulfide between Cys124 and Cys71. *J. Biol. Chem.* **277:**20336-20342

47. Bae, Y. S., J. Y. Sung, O. S. Kim, Y. J. Kim, K. C. Hur, A. Kazlauskas and S. G. Rhee. (2000). Platelet-derived growth factor-induced H_2O_2 production requires the activation of phosphatidylinositol 3-kinase. *J Biol Chem* **275:**10527-10531

Chapter 10

4-HYDROXYNONENAL SIGNALING

Giuseppe Poli, Gabriella Leonarduzzi and Elena Charpotto

10.1 Introduction

The lipid-derived aldehyde, 4-hydroxy-trans-2-nonenal (HNE), at the time of its discovery in natural fats by Hermann Esterbauer,[1] was regarded as a mere byproduct of auto-oxidation of unsaturated fatty acids. Years later, it turned out that HNE is a normal constituent of mammalian tissue membranes[2] and nowadays the detection of its increased steady-state level is often taken as a specific marker of oxidative stress,[3] i.e. the prevalence within the cell of oxidizing species over the cellular antioxidant potential.

The hydroxyalkenal HNE derives from the oxidation of n-6 polyunsaturated fatty acids, among which are the two most represented fatty acids in biomembranes, namely arachidonic and linoleic acids. HNE is an unusual compound containing three functional groups that in many cases act in concert and help to explain its high reactivity. There is, first of all, a conjugated system consisting of a C=C double bond and a C=O carbonyl group in HNE. The hydroxyl group at carbon 4 contributes to the reactivity both by polarizing the C=C bond and by facilitating internal cyclization reactions such as thio-acetal formation.

Virtually all of the biochemical effects of HNE can be explained by its high reactivity towards thiol and amino groups. Primary reactants for HNE are the amino acids cysteine, histidine and lysine, which – either free or protein-bound - undergo readily Michael addition reactions to the C=C double bond (Fig.1). Besides this type of reaction, which confers rotational freedom to the C2-C3 bond, secondary

H. J. Forman, J. Fukuto, and M. Torres (eds.), Signal Transduction by Reactive Oxygen and Nitrogen Species: Pathways and Chemical Principles, ©2003, *Kluwer Academic Publishers. Printed in the Netherlands*

reactions may occur involving the carbonyl and the hydroxyl group. Amino groups may alternatively react with the carbonyl group to form Schiff bases.

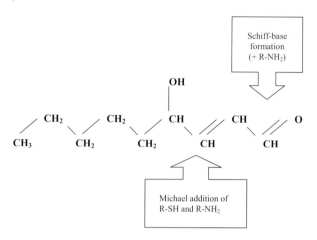

Figure 1. Structure and main reactions of 4-hydroxynonenal

An increasing number of reports provides evidence of a causative involvement of this biogenic aldehyde in the pathogenesis of a great number of inflammatory and degenerative processes, including atherosclerosis, Alzheimer's and Parkinson's diseases, liver fibrosis, glomerulosclerosis, chronic obstructive bronco-pulmonary diseases (see for a review [4, 5]).

HNE may actually interfere, at the molecular level, with cellular and tissue functions and significantly contribute to the pathogenesis and/or the progression of a variety of human disease processes. However, as regards HNE signaling to the nucleus, the relevant literature presently appears to accumulate in a rather heterogeneous way. Here we aim to overview the most recent achievements on this subject, possibly providing a more systematic analysis of the modulation of the various signal transduction processes by the aldehyde.

10.2 HNE Reaction at the Cell Plasma Membrane Level and/or Nuclear Localisation

The availability of fluorescent monoclonal antibodies raised against HNE-histidine (HNE-His) adducts allowed us to semi-quantitatively follow the intracellular fate of micromolar amounts of the aldehyde, added to the incubation medium of a large variety of cells. Most of the cell types tested thus far show an active aldehyde metabolism,[6,7] however, the relatively high reactivity of HNE for -SH and -NH$_2$ groups allows a certain amount of the aldehyde to escape from metabolic disposition and then influence the function of certain biomolecules. As regards the experimental incubation of cell lines with exogenously added HNE, it has been calculated that enzymatic metabolism may consume about 55-70% of the original amount,[6] while the remaining amount can react at the plasma membrane level and be taken up by the cells.

Cell uptake and intracellular tropism of 4-hydroxynonenal has so far been successfully monitored by using fluorescent antibodies against HNE-His adducts in hepatic stellate cells,[8] cells of the macrophage lineage,[9] hepatocytes in suspension or in primary culture, neuroblastoma cells, colon carcinoma cells (Poli *et al.*, unpublished data). In all these cells, after 5-10 min of prevailing aldehyde consumption, one can consistently recover HNE in the cytoplasmic space. Further, of great interest, a nuclear tropism of the aldehyde appears more evident with time.[8,9] Of interest, while in the majority of tested cell types HNE nuclear localisation was recorded as a relatively early event, that was not the case for hepatocytes. In these cells, fluorescent HNE-His adducts were evident in nuclei only after about one hour treatment and not in all cells (Figure 2). This finding could be related to the high specialisation of hepatocytes, maybe favouring the use of the EGF-dependent signaling pathway by HNE rather then direct internalisation. Also the high metabolic potential of these cells combined to high thiol content could contribute to a delayed and quenched nuclear localisation of the aldehyde.

To evaluate cell uptake and intracellular distribution of HNE, as well as its potential ability to modulate gene expression, the effect of proper thiol and carbonyl reagents has been investigated. As regards HNE possible interaction with plasma membrane, relevant appear the result we obtained on cultured macrophages treated with the thiol reagent 4-(chloro-mercuri)-benzene-sulfonic acid (PCMBS), or the carbonyl reagent hydroxylamine (HYD), 5 min before addition of the aldehyde (10 µM). Due to its molecular weight, PCMBS does not enter the cell plasma membrane but reacts exclusively with extracellular thiols;[10] macrophage pretreatment with 100 µM PCMBS induced about a 50% decrease of total thiols without affecting cell viability. But, more important, the drug showed a significant while transient quenching of HNE intracellular uptake, which turned out in a net inhibition of a HNE sensitive gene, that coding for transforming growth factor-β1.[11]

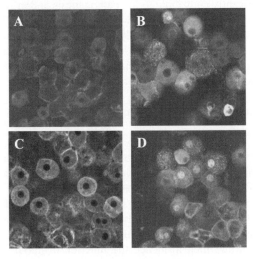

Figure 2. Localization of fluorescent HNE-Histidine adducts in rat hepatocyte suspensions. Cell suspensions (50,000 cells) were incubated at 37°C with 10 µM HNE for 5 min (A), 15 min (B), 30 min (C), 60 min (D). Then hepatocyte suspensions were cyto-centrifuged and the

slides were fixed and stained for indirect immunofluorescence using a monoclonal HNE-histidine 1g4h7 primary antibody and a FITC-conjugated secondary antibody.fluorescence was detected by a laser scanning confocal microscope (Zeiss, LSCM 510). The lens used was a plan Neofluar 40X/0.75; zoom: 1.4.

On the contrary, pretreatment of the same macrophagic cells with the carbonyl reagent HYD (1 mM) did not afford any significant inhibition of HNE cell uptake.[11] Taken together, these findings point to a key role of -SH groups (Michael addition reactions) on the outer face of the plasma membrane in regulating the uptake and consequent intracellular distribution of aldehydes like HNE.

However, it cannot be excluded, at least in defined types of cells, a surface reaction between –NH$_2$ groups and HNE. Indeed, it has recently been demonstrated that HNE is able to bind to EGF receptors[12,13] or PDGF receptors.[14] In particular, Nègre-Salvayre and colleagues demonstrated that 4-hydroxynonenal mimicked oxidised LDL as regards the ability, in a cell-free system, to derivatize free reactive amino groups of EGF receptor. Moreover, HNE (0.1 µM) was shown to trigger phosphorylation and activation of the receptor on human endothelial cell line CRL-1998.[12] Liu and colleagues confirmed the ability of HNE (500 µM) to bind EGF receptor and activate it by immunoblot analysis using human epidermoid carcinoma A431 cells.[13] Again Nègre-Salvayre's group proved that the aldehyde was able to undergo addition reaction with amino groups present in the PDGF β receptor, actually leading to its activation as clearly shown in rabbit arterial smooth muscle cells treated with 0.5 µM HNE.[14]

Thus, the lipid peroxidation product 4-hydroxynonenal, which is both lipophilic and highly diffusible, certainly may migrate from the site of origin within the cytosolic environment and eventually self-concentrate in intact form in the nucleus. Moreover, when it reaches the extracellular environment, or it is produced at the plasma membrane level, HNE is also able to signal to the nucleus of cells from the plasma membrane surface. Both ways allow modulation of gene expression.

10.3 HNE Modulation of Protein Kinase C Isoforms

In all cell types where 4-hydroxynonenal has shown intracellular migration, a consistent modulation of protein kinase C superfamily isoenzymes has been demonstrated. Analytical studies on this specific molecular event have been carried out by Chiarpotto et al.[15] on rat isolated hepatocytes in single cell suspension. Of note, the effect of HNE on the different PKC isoforms varied with the final concentration used. In fact, when rat hepatocytes were incubated in a medium where the level of aldehyde was maintained steadily in the high nanomolar range (100 nM), strong activation of classic isoforms β$_1$ and β$_2$ was observed. This event was proved causally related with a significant increase of intracellular traffic and secretion of the protein cathepsin D.[15] When an identical concentration was steadily applied to NT2 differentiated neuronal cells in culture, the aldehyde led again to a selective activation of β$_1$ and β$_2$ PKC isoforms, and, concomitantly, to a significant increase of intracellular amyloid β production.[16]

On the contrary, HNE in the low micromolar range (1-10 µM) was able to markedly activate hepatocyte novel isoforms of protein kinase C (nPKCs), in particular the δ isoform. The latter type of PKC modulation appears related with the well recognised pro-apoptotic effect of HNE, as clearly proved in cells of the macrophage lineage, by the full prevention of apoptosis by cell pretreatment with rottlerin, a selective inhibitor of nPKCs (Chiarpotto *et al.*, manuscript submitted for publication).

10.4 Mitogen Activated Protein Kinase Pathway and 4-Hydroxynonenal

Whether directly interacting with PKC or through activation of EGF receptor, 4-hydroxynonenal signaling has been demonstrated by a number of laboratories to consistently involve the mitogen-activated protein kinase pathway (MAPK).

The first demonstration of a net activation of c-Jun amino terminal kinases (JNKs) by HNE, in a concentration (1 µM) definitely compatible with those found in vivo, was provided by Parola and colleagues on primary cultures of human hepatic stellate cells. By using monoclonal antibodies specific for HNE-His adducts, authors demonstrated that externally added HNE was able to reach the nucleus of these cells and co-localise with translocated JNK, pointing to the involvement of the aldehyde in the nuclear translocation of the kinase. Of note, in the same experimental model, 1 µM HNE was unable to affect the enzymatic activity of extracellular signal-regulated protein kinases ERK1 and ERK2, another key element of MAPK pathway.[8]

Few months later, the group of Uchida also reported on the strong while transient activation of JNKs by HNE (25 µM final concentration) in this case using a rat liver epithelial RL34 cancer cell line. JNK activation was preceded by phosphorylation of the kinase, apparently mediated by HNE.[17] By the way, in the experiments by Parola et al. on hepatic stellate cells in primary culture, such a phosphorylating effect of the aldehyde was not observed.[8] More important in the report by Uchida et al. was the first demonstration that HNE was able to strongly phosphorylate and activate another key path of the MAPK system, namely the p38 mitogen-activated protein kinase, while activity of ERKs was on the contrary scarcely affected.[17]

Other reports then followed which stressed again a primary involvement of JNK activation in the MAPK modulation by 4-hydroxynonenal (5-10 µM), in particular with regard to the pro-apoptotic effect of the aldehyde as investigated in PC12 human neuroblastoma cell line.[18-20] Song and colleagues showed that JNK activation was an early but transient event also in PC12 cells and that it was preceded by activation of an upstream kinase of JNK, i.e. stress-activated protein kinase 1 (SEK1). Again the aldehyde did not show any significant effect on ERK activity.[18] Camandola et al., besides confirming HNE induced activation of JNK by confocal microscopy technique, showed this event as largely dependent upon the intracellular calcium derangement[20] that actually takes place in cells challenged by low micromolar doses of the aldehyde.[21]

Reports on HNE-dependent up-regulation of JNKs in other cell types, possibly with further insight in the signaling mechanism, are expected to increase in number in the near future. One of the last available in the literature was that showing that transfection of human myeloid HL-60 cells with glutathione-S-transferase isozyme 4 significantly quenched the pro-apoptotic effect of 20 µM HNE, mainly through a marked retardation of JNK involvement.[22]

The major involvement of JNK in the modulation of MAPK by 4-hydroxynonenal, already confirmed in a variety of cell types, does not necessarily mean that other key elements of this phosphorylation path may not be involved. Uchida's group, after having shown a marked p38 kinase activation by HNE in RL34 epithelial cell line,[17] very recently provided clear proofs in support of a crucial role of this kinase in HNE-induced expression of cyclooxygenase type 2. In fact, RL34 pretreatment with a p38 selective inhibitor (SB203580) allowed to suppress such gene induction.[23]

As regards extracellular signal-regulated protein kinases ERK1 and ERK2, most of the so far performed in vitro studies did not provide evidence of modulation by HNE. However, the cell type may significantly influence the effect of the aldehyde on the enzyme activity. Using rat aortic smooth muscle cell line, Ruef and colleagues observed a strong but transient activation of these two kinases just at 5 min incubation with 2.5 μM HNE. Such activation then rapidly declined until disappearance within half an hour.[24]

Finally, in relation to HNE-dependent modulation of MAPKs, it appears important to draw the attention on the observation of JNK activation in primary cultured astrocytes by hydrogen peroxide through a mechanism that apparently implies arachidonic acid metabolism.[25] But arachidonic acid, when undergoes to oxidative breakdown, represents a major source of HNE; then, JNK activation by hydrogen peroxide could actually involve HNE as one of the chemical mediators.

10.5 Up-Regulation of Redox-Sensitive Transcription Factors by HNE

As comprehensively analysed by a quite recent review related to free radical research, several transcription factors have shown to be modulated by intracellular redox reactions,[26] in particular those belonging to two families of transcriptionally active peptides, called activator protein-1 (AP-1) and nuclear factor κB (NF-κB).

AP-1 is a homo- or heterodimeric factor mainly composed of the c-Jun and c-Fos proteins. Other subunits are Jun-B, Jun-D, Fos-B, Fra-1, Fra-2 (see[27] for a review). If one examine the large number of mitogenic and pro-inflammatory molecules able to induce c-jun and c-fos genes, often he will find these factors accompanied by or inducing a net increase of ROS steady-state levels.[28-29] Indeed, the alteration of thiol redox status of the cells, as achieved by agents stimulating intracellular ROS generation, definitely appears a strong stimulus for AP-1 activation.

The evidence of a marked increase of AP-1 nuclear binding by HNE was among the earlier observations of a cell signaling ability expressed by the aldehyde. Such effect was first demonstrated by Camandola et al. in cultivated cells of the macrophage lineage, as exerted by HNE in the low micromolar range (1-10 μM). A significant up-regulation of the transcription factor was evident already 10-15 min after cell challenge with the aldehyde with a maximum increase between 30 and 60 min incubation and a later slow decrease at least within two hours of observation.[30] Still in our laboratory, HNE-induced up-regulation of AP-1 nuclear binding was found reproducible in a variety of cell types, like hepatic stellate cells,[8] hepatocytes (submitted for publication), neuronal cells.[20]   The stimulating effect of 4-hydroxynonenal on the transcription factor was also confirmed by other laboratories, for example, still in the low micromolar concentration range, in rat aortic smooth muscle cells[24] and rat epithelial cancer cells.[17]   Such strong activation of AP-1 binding and activity by HNE, actually found in all cell types so far considered, resulted to be consistently based on a net induction of c-Jun gene expression and synthesis.

As in the case of AP-1, also NF-κB is a redox-sensitive transcription factor of primary interest in the pathogenesis of various human diseases. NF-κB consists of homodimers and heterodimers of structurally related DNA-binding subunits. The more frequent form of NF-κB is a heterodimer complex consisting of the p50 and p65 peptides. The dipeptide is present within the cell in an inactive form when bound to the inhibitory peptide IκB. Also in the case of NF-κB, most of its inducers rely on the production of ROS and related reactions with cell macromolecules.[31]

Interestingly, NF-κB does not respond to lipid peroxidation-derived aldehydes (HNE) in the same consistent way showed by AP-1. In our hands, addition of HNE (1-10 μM) to the culture medium of both macrophages and stellate cells strongly up-regulated AP-1 nuclear binding, as above reported, but never significantly modified the extent of NF-κB nuclear translocation in these cells.[8,30] However, using cultured rat cortical neurons, Camandola et al. recently showed that 10μM HNE suppressed constitutive activation of the transcription factor.[32] Further, the observed ability of HNE to strongly inhibit also NF-κB activation by phosphatase inhibitors like okadaic acid and pervanadate, together with the lack of any effect both on p50 and p65 levels and on the binding of the factor to DNA, pointed to the inhibition of IκB phosphorylation as a likely crucial target of the aldehyde.[32] Those findings and such interpretation were actually confirming the previous demonstration of HNE ability to inhibit IκB phosphorylation and proteolysis, and by this way to counteract the consequent liberation of the active form of NF-κB.[33]

The majority of reports on the possible modulation of NF-κB by 4-hydroxynonenal showed either inhibition or lack of effect. However, a stimulatory effect was very recently reported as exerted by 1 μM HNE at least on vascular smooth muscle cells.[34] Although this is the only evidence of potential up-regulation of NF-κB activity by HNE so far, it still serves to demonstrate how the aldehyde's action on this factor can differ in the different cell types.

10.6 HNE-Dependent Modulation of Gene Expression

Modulation of AP-1 and maybe NF-κB nuclear binding by HNE will likely interfere with the regulation of expression of a number of genes that have consensus sequences for these peptides in their promoter regions. At present, clear evidence of HNE-dependent overexpression is available for a number of genes whose optimal transcription mainly requires AP-1 activation.

The first two reports on the ability of HNE (1-10μM) to induce gene expression were related to pathophysiology of fibrogenesis. In relation to this, seminal experiments have been those carried out on the profibrogenic effect of the prooxidant hepatotoxin carbon tetrachloride (CCl$_4$). Following rat chronic intoxication, CCl$_4$ showed a strong stimulatory effect on the expression of the key fibrogenic cytokine transforming growth factor β1 (TGFβ1) and of procollagen type I, which was fully prevented when animals were suitably supplemented with the antioxidant α-tocopherol.[35] Using human hepatic stellate cells in primary culture, Parola et al. showed that HNE was able to markedly up-regulate both expression and synthesis of procollagen type I, the main type of collagen which accumulates in liver fibrosis.[36] Moreover, defined other hydroxyalkenals of biological interest showed similar ability to significantly stimulate procollagen type I gene expression, i.e. 4-hydroxyhexenal, 4-hydroxyoctenal, 4-hydroxyundecenal,[36] while parental saturated compounds did not exert such effect. Using different cell cultures of the macrophage lineage, Leonarduzzi *et al.* demonstrated that 4-hydroxynonenal, but not the parental compounds 2-nonenal and nonanal, markedly up-regulated both expression and synthesis of TGFβ1.[7] This cytokine shows a great variety of effects on cell function and proliferation, including strong enhancement of fibrogenesis.[37]

In primary cultures of rat aortic smooth muscle cells three hours incubation with 1 μM HNE led to strongly up-regulate the synthesis of platelet-derived growth factor-AA (PDGF-AA). Simultaneous treatment with N-acetyl cysteine was able to prevent the overexpression of this cytokine, indirectly indicating that this other evidence relevant to atherosclerosis was likely oxidant-mediated.[24] Still related to inflammation is the reported effect of HNE on the expression and synthesis of monocyte chemotactic peptide-1 (MCP-1) and of cyclooxygenase-2 (COX-2). Parola and colleagues have found this aldehyde able to up-regulate the expression of the key C-C chemokine MCP-1 in primary cultures of human liver stellate cells, at very low steady-state concentration (1 μM).[38] Increased synthesis and activated secretion of MCP-1 by murine macrophages (J774A1) challenged with 1-10 μM HNE was then reported by Domenicotti et al.[39] Certainly, the demonstration that 4-hydroxynonenal is able to markedly up-regulate MCP-1 levels in tissues like liver and arterial wall where phagocytes and extracellular matrix cells are accumulating, provides important proofs in favour of this aldehyde as mediator molecule primarily involved in inflammatory processes, especially with chronic trend. As regards COX-2, treatment of rat liver epithelial RL34 cells with 5-25 μM HNE very strongly induced the expression of this inducible enzyme of the eicosanoid pathway. Such HNE-exerted induction of COX-2 was shown to be dependent upon the cell GSH status. In fact, it was inhibited by GSH depletion obtained with L-buthionine-sulfoximine (BSO).[40] Uchida *et al.*, besides demonstrating the possible induction of COX-2 by 4-hydroxynonenal, showed that in the same cells, expression of glutathione-S- transferase (GST) resulted to be stimulated as well. Moreover, the different temporal occurrence of the two events was suggestive of a correlation between COX-2 and the detoxification enzyme inductions. The authors indicated the cyclopentenone-type prostaglandins as the likely mediators of GST induction, at least in the conditions covered by their experimental model.[41]

The ability of HNE to induce the expression of genes coding for detoxification enzymes and substances is any way definitely supported by a number of recent findings from different laboratories. The treatment of human smooth muscle cells with HNE concentrations within the low micromolar range (1-10 μM) has been elegantly reported as able to strongly activate both expression and synthesis of aldose reductase. This paper interestingly pointed to this enzyme as potentially important in affording HNE disposition in various cell types.[42]

As regards glutathione, certainly a key detoxifying substance in cellular response against oxidative stress, the group of Forman clearly demonstrated that the aldehyde is also able to significantly induce the rate-limiting enzyme for its synthesis, namely the ligase commonly called γ-glutamyl-cysteine synthetase (GCS). The treatment of rat alveolar epithelial L2 cell suspensions with 5-20 μM HNE (x 3-6 hours) was sufficient to induce a strong increase of mRNA expression and stability of both enzyme subunits; i.e., catalytic heavy subunit and regulatory light subunit. Such overexpression was indeed followed by increased GCS synthesis and activity.[43] All these findings were later on confirmed by the same group on primary cultured rat alveolar epithelial type II (AT2) cells. Further, experiments with MAPK inhibitors were carried to analyse the potential pathway by which HNE was inducing the expression of both GCS subunits, that are by the way encoded by two separate genes. PD98059 (50 μM), which selectively blocks ERK pathway, was indeed able to prevent 20 μM HNE-induced expression of the catalytic subunit but not that of the regulatory one. SB202198, a selective p38 inhibitor was also employed up to the concentration of about 40 μM without showing any significant effect on the induction of both subunits.[44] A very recent report still by Forman's group further dealt with the investigation of HNE signaling in the induced overexpression of both GCS subunits this time using immortalized bronchial epithelial cells deriving from

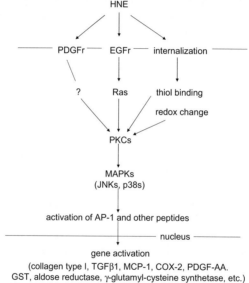

Figure 3. Hypothetical pathways of cell signaling and gene activation as induced by the lipid peroxidation product 4-hydroxynonenal.

normal individuals. By means of different MAPK inhibitors, they demonstrated a major role played by JNK1 but not ERKs and p38 in HNE-induced up-regulation of this enzyme as well,[45] in agreement with the above reported findings by others obtained in the analysis of genes related to fibrosis (see paragraph 4).

10.7 Conclusion

Figure 3 reports a tentative summary of the present knowledge about cell signaling reactions initiated by HNE as well as a list of genes whose expression has been demonstrated as significantly modulated by the aldehyde. Two biochemical events consistently appear as primarily involved in all cell types so far challenged with HNE concentrations of pathophysiological relevance: the selective activation of defined PKC isoforms and MAPK enzymes. MAPK up-regulation mainly concerns JNK1,2 with consequent activation of the transcription factor AP-1 and overexpression of AP-1 sensitive genes. Endogenously generated HNE may likelyinteract with PKCs and MAPKs through Michael reaction with –SH and –NH$_2$ groups but also forming Shiff bases with – NH$_2$ groups. As far as exogenous HNE is concerned, this could either be somehow taken up by the cell or signal from the cell surface by activation of EGF and/or EGF-like receptors. The high interest more recently earned by HNE signaling will probably lead to rapid clarification also of the upstream biochemical pathway/s triggered by the aldehyde.

Acknowledgements

Part of the reported research has been supported by grants from the Italian Ministry of the University, PRIN 1999, 2000, 2001, the National Research Center (CNR), Targeted Project on Biotechnology, the Regione Piemonte, the University of Torino).

References

1. Schauenstein, E., H. Esterbauer, G. Jaag and Taufer. 1964. The effect of aldehydes on normal and malignant cells. 1st report: Hydroxyoctenal, a new fat aldehyde. *Chemical Monthly* **95**:180-183.

2. Comporti, M. 1998. Lipid peroxidation and biogenic aldehydes: From the identification of 4-hydroxynonenal to further achievements in biopathology. *Free Rad Res* **28**:623-635.

3. Onorato, J. M., S. R. Thorpe and J. W. Baynes. 1998. Immunohistochemical and ELISA assays for biomarkers of oxidative stress in aging and disease. *Ann NY Acad. Sci* **854**:277-290.

4. Poli, G. and R. J. Schaur. 2000. 4-Hydroxynonenal in the pathomechanisms of oxidative stress. *IUBMB Life* **50**: 315-321.

5. Parola, M., G. Bellomo, G. Robino, G. Barrera and M. U. Dianzani. 1999. 4-Hydroxynonenal as a biological signal: molecular basis and pathophysiological implications. *Antiox Redox Signal* **1**: 255-284.

6. Esterbauer, H., R. J. Schaur and H. Zollner. 1991. Chemistry and biochemistry of 4-hydroxynonenal, malonaldehyde and related aldehydes. *Free Radic Biol Med* **11**:81-128.

7. Leonarduzzi, G., A. Scavazza, F. Biasi, E. Chiarpotto, S. Camandola, S. Vogl, R. Dargel and G. Poli. 1997. The lipid peroxidation end-product 4-hydroxy-2,3-nonenal up-regulates transforming growth factor-β1 expression in the macrophage lineage: a link between oxidative injury and fibrosclerosis. *FASEB J* **11**: 851-857.

8. Parola, M., G. Robino, F. Marra, M. Pinzani, G. Bellomo, G. Leonarduzzi, G. Chiarugi, S. Camandola, G. Poli, G. Waeg, P. Gentilini and M. U. Dianzani. 1998. HNE interacts

directly with JNK isoforms in human hepatic stellate cells. *J. Clin Invest* **102**:1942-1950.

9. Leonarduzzi, G., M. C. Arkan, H. Basaga, E. Chiarpotto, A. Sevanian and G. Poli. 2000. Lipid oxidation products in cell signaling. *Free Radic Biol Med* **28**:1370-1378.

10. Yan, R. T. and P. C. Maloney. 1995. Residues in the pathway through a membrane transporter. *Proc Natl Acad Sci USA* **92**:5973-5976.

11. Chiarpotto, E., C. Allasia, F. Biasi, G. Leonarduzzi, F. Ghezzo, G. Berta, G. Bellomo, G. Waeg and G. Poli. 2002. Down-modulation of nuclear localisation and pro-fibrogenic effect of 4-hydroxy-2,3-nonenal by thiol- and carbonyl-reagents. *BBA-Mol Cell Biol* **In press**.

12. Suc, I., O. Meilhac, I. Lajoie-Mazenc, J. Vandaele, G. Jurgens, R. Salvayre and A. Nègre-Salvayre. 1998. Activation of EGF receptor by oxidized LDL. *FASEB J* **12**:665-671.

13. Liu, W., A. A. Akhand, M. Kato, I. Yokoyama, T. Miyata, K. Kurokawa, K. Uchida and I. Nakashima. 1999. 4-hydroxynonenal triggers an epidermal growth factor receptor-linked signal pathway for growth inhibition. *J Cell Sci* **112**:2409-2417.

14. Escargueil-Blanc, I., R. Salvayre, N. Vacaresse, G. Jurgens, B. Darblade, J. F. Arnal, S. Parthasarathy and A. Nègre-Salvayre. 2001. Mildly oxidized LDL induces activation of platelet-derived growth factor β receptor pathway. *Circulation* **104**:1814-1821.

15. Chiarpotto, E., C. Domenicotti, D. Paola, A. Vitali, M. Nitti, M. A. Pronzato, F. Biasi, D. Cottalasso, U. M. Marinari, A. Dragonetti, P. Cesaro, C. Isidoro and G. Poli. 1999. Regulation of rat hepatocyte protein kinase C beta isoenzymes by the lipid peroxidation product 4-hydroxy-2, 3-nonenal: A signaling pathway to modulate vesicular transport of glycoproteins. *Hepatology* **29**:1565-1572.

16. Paola, D., C. Domenicotti, M. Nitti, A. Vitali, R. Borghi, D. Cottalasso, D. Zaccheo, P. Odetti, P. Strocchi, U. M. Marinari, M. Tabaton and M. A. Pronzato. 2000. Oxidative stress induces increase in intracellular amyloid beta-protein production and selective activation of betaI and betaII PKCs in NT2 cells. *Biochem.Biophys Res Commun* **268**:642-646.

17. Uchida, K., M. Shiraishi, Y. Naito, Y. Torii, Y. Nakamura and T. Osawa. 1999. Activation of stress signaling pathways by the end product of lipid peroxidation 4-hydroxy-2-nonenal, a potential inducer of intracellular peroxide production. *J Biol Chem* **274**:2234-2242.

18. Soy, H., K. S. Jeong, I. J. Lee, M. A. Bae, Y. C. Kim and B. J. Song. 2000. Selective activation of the c-Jun N-terminal protein kinase pathway during 4-hydroxynonenal-induced apoptosis of PC12 cells. *Mol Pharmacol* **58**:535-541.

19. Song, B. J., Y. Soh, M. Bae, J. Pie, J. Wan and K. Jeong. 2001. Apoptosis of PC12 cells by 4-hydroxy-2-nonenal is mediated through selective activation of the c-Jun N-terminal protein kinase pathway. *Chem Biol Interact* **132**:943-954.

20. Camandola, S., G. Poli and M. P. Mattson. 2000. The lipid peroxidation product 4-hydroxy-2,3-nonenal increases AP-1 binding activity through caspase activation in neurones. *J Neurochem* **74**:159-178.

21. Carini, R., G. Bellomo, L. Paradisi, M. U. Dianzani and E. Albano. 1996. 4-Hydroxynonenal triggers Ca^{2+} influx in isolated rat hepatocytes. *Biochem Biophys Res Commun* **218**:772-776.

22. Cheng, J. Z., S. S. Singhal, A. Sharma, M. Saini, Y. Yang, S. Awasthi, P. Zimniak and Y. C. Awasthi. 2001. Transfection of mGSTA4 in HL-60 cells protects against 4-hydroxynonenal-induced apoptosis by inhibiting JNK-mediated signaling. *Arch Biochem Biophys* **392**:197-207.

23. Kumagai, T., Y. Nakamura, T. Osawa and K. Uchida. 2002. Role of p38 mitogen-activated protein kinase in the 4-hydroxy-2-nonenal-induced cyclooxygenase-2 expression. *Arch Biochem Biophis* **397**:240-245.

24. Ruef, J., G. N. Rao, F. Li, C. Bode, C. Patterson, A. Bhatnagar and M. S. Runge. 1998. Induction of rat aortic smooth muscle cell growth by the lipid peroxidation product 4-hydroxy-2-nonenal. *Circulation* **97**:1071-1078.

25. Tournier C., G. Thomas, J. Pierre, C. Jacquemin, M. Pierre and B. Saunier. 1997. Mediation by arachidonic acid metabolites of the H_2O_2-induced stimulation of mitogen-activated protein kinases (extracellular-signal-regulated kinase and c-Jun NH_2-terminal kinase). *Eur J Biochem* **244**:587-595.

26. Arrigo, A. P. 1999. Gene expression and the thiol redox state. *Free Radic Biol Med* **27**:936-944.

27. Shaulian, E. and M. Karin. 2002. AP-1 as a regulator of cell life and death. *Nat Cell Biol* **4**:E131-136.

28. Rupec, R. A. and P. A. Baeuerle. 1995. The genomic response of tumor cells to hypoxia and reoxygenation. Differential activation of transcription factors AP-1 and NF-κB. *Eur. J. Biochem.* **234**, 632-640.

29. Camandola, S., M. A. Aragno, J. C. Cutrin, E. Tamagno, O. Danni, E. Chiarpotto, M. Parola, G. Leonarduzzi, F. Biasi and G. Poli. 1999. Liver AP-1 activation due to carbon tetrachloride is potentiated by 1,2-dibromoethane but is inhibited by α-tocopherol or gadolinium chloride. *Free Radic Biol Med.* **26**:1108-1116.

30. Camandola, S., A. Scavazza, G. Leonarduzzi, F. Biasi, E. Chiarpotto, A. Azzi and G. Poli. 1997. Biogenic 4-hydroxy-2-nonenal activates transcription factor AP-1 but not NF-κB in cells of the macrophage lineage. *Biofactors,* **6**:173-179.

31. Schreck, R., P. Rieber and P. A. Baeuerle. 1991. Reactive oxygen intermediates as apparently widely used messengers in the activation of the NF-κ B transcription factor and HIV-1. *EMBO J* **10**:2247-2258.

32. Camandola, S., G. Poli and M. P. Mattson. 2000. The lipid peroxidation product 4-hydroxy-2,3-nonenal inhibits constitutive and inducible activity of nuclear factor κ B in neurons. *Brain Res Mol Brain Res* **85**:53-60.

33. Page, S., C. Fischer, B. Baumgartner, M. Haas, U. Kreusel, G. Loidl, M. Hayn, H. W. Ziegler-Heitbrock, D. Neumeier and K. Brand. 1999. 4-Hydroxynonenal prevents NF-κB activation and tumour necrosis factor expression by inhibiting IκB phosphorylation and subsequent proteolysis. *J Biol Chem* **274**:11611-11618.

34. Ruef, J., M. Moser, C. Bode, W. Kubler and M. S. Runge. 2001. 4-Hydroxynonenal induces apoptosis, NF-κB activation and formation of 8-isoprostane in vascular smooth muscle cells. *Basic Res Cardiol* **96**:143-150.

35. Parola, M., R. Muraca, I. Dianzani, G. Barrera, G. Leonarduzzi, G. Bendinelli, R. Piccoletti and G. Poli. 1992. Vitamin E dietary supplementation inhibits transforming growth factor β1 gene expression in the rat liver. *FEBS Lett* **308**:267-270.

36. Parola, M., M. Pinzani, A. Casini, G. Leonarduzzi, F. Marra, A. Caligiuri, E. Cenni, P. Biondi, G. Poli and M. U. Dianzani. 1996. Induction of procollagen type I gene expression and synthesis in human hepatic stellate cells by 4-hydroxy-2,3-nonenal and other 4-hydroxy-2,3-alkenals is related to their molecular *structure.Biochem Biophys Res Commun* **222**:261-264.

37. Poli, G. 2000. Pathogenesis of liver fibrosis: role of oxidative stress. *Mol Aspects Med* **22**:287-305.

38. Marra, F., R. DeFranco, C. Grappone, M. Parola, S. Milani, G. Leonarduzzi, S. Pastacaldi, U. O. Wenzel, M. Pinzani, M. U. Dianzani, G. Laffi and P. Gentilini. 1999. Expression of monocyte chemotactic protein-1 precedes monocyte recruitment in a rat model of acute liver injury, and is modulated by Vitamin E. *J Invest Med* **47**:66-75.

39. Nitti, M., C. Domenicotti, C. d'Abramo, S. Assereto, D. Cottalasso, E. Melloni, G. Poli, F. Biasi, U. M. Marinari and M. A. Pronzato. 2002. Activation of PKC-β isoforms mediates HNE-induced MCP-1 release by macrophages. *Biochim Biophys Res Commun* **294**:547-552.

40. Kumagai, T., Y. Kawamoto, Y. Nakamura, I. Katayama, K. Satoh, T. Osawa and K. Uchida. 2000. 4-Hydroxy-2-nonenal, the end product of lipid peroxidation, is a specific inducer of cyclooxygenase-2 gene expression. *Biochem Biophys Res Commun* **273**:437-441.

41. Uchida, K. 2000. Cellular response to bioactive lipid peroxidation products. *Free Rad Res* **33**:731-737.

42. Spycher, S. E., S. Tabataba-Vakili, V. B. O'Donnell, L. Palomba and A. Azzi. 1997. Aldose reductase induction:a novel response to oxidative stress of smooth muscle cells. *FASEB J* **11**:181-188.

43. Liu, R. M., L. Gao, J. Choi and H. J. Forman. 1998. γ-Glutamylcysteine synthetase: mRNA stabilisation and independent subunit transcription by 4-hydroxy-2-nonenal. *Am J Physiol* **275**:L-861-L869.

44. Liu, R. M., Z. Borok and H. J. Forman. 2001. 4-Hydroxynonenal increases γ-glutamylcysteine synthetase gene expression in alveolar epithelial cells. *Am J Respir Cell Mol Biol* **24**:499-505.

· 45. Dickinson, D. A., K. E. Iles, N. Watanabe, T. Iwamoto, H. Zhang, D. M. Krzywansky and H. J. Forman. 2002. 4-Hydroxynonenal induces glutamate cysteine ligase though JNK in HBE1 cells. *Free Radic Biol Med* **33**: 974-987, 2002.

Chapter 11

CERAMIDE SIGNALING UNDER OXIDATIVE STRESS

Tzipora Goldkorn, Tommer Ravid and Edward A Medina

11.1 Introduction

Aerobic cells are constantly exposed to reactive oxygen and nitrogen species (ROS/RNS) that are generated under both physiologic and pathologic conditions. ROS include species such as superoxide anion (O_2^-), hydrogen peroxide (H_2O_2), and hydroxyl radicals (OH). Under basal conditions, human cells produce about 2 billion O_2^- and H_2O_2 molecules per cell per day (1). ROS were once considered as only waste byproducts of aerobic metabolism or molecules of defense produced by host inflammatory cells against invading organisms. Now ROS are recognized as controlling key steps in cellular signal transduction cascades.[2,3] Therefore, ROS are involved in diverse biologic processes that include embryogenesis, normal tissue homeostasis, aging, and many human diseases. For example, ROS have been implicated as mediators of lung injury in Acute Respiratory Distress Syndrome (ARDS) (alveolar damage), asthma (airway epithelial damage), chronic obstructive pulmonary disease (COPD), and interstitial pulmonary fibrosis.[4-8] The role of ROS in illness was previously explained by the chemistry in which critical cell proteins and lipids were randomly oxidized and rendered metabolically inactive.[9] However, several groups, including ours, have recently demonstrated that ROS function as signaling molecules. As one example of the signaling properties of ROS in the airway, treatment of airway epithelial cells with exogenous H_2O_2, the agent commonly produced during lung inflammatory processes, has been shown to

H. J. Forman, J. Fukuto, and M. Torres (eds.), Signal Transduction by Reactive Oxygen and Nitrogen Species: Pathways and Chemical Principles, ©2003, *Kluwer Academic Publishers. Printed in the Netherlands*

activate EGF receptor tyrosine kinase but not the receptor's trafficking;[10,11] this presents a mechanism by which ROS directly mediate transduction of mitogenic signals to the nucleus. Our laboratory has also examined the role of ROS in another key physiologic process, apoptosis.[12-14] We found that H_2O_2, at physiological concentrations [50-250 nmoles/mg protein],[15,16] induces epithelial apoptosis via the ceramide signal transduction pathway. This work provides a direct link between two important aspects of mammalian stress responses: the generation of ROS and activation of the sphingomyelin/ ceramide cycle leading to apoptosis.[12-14] Indeed, a role for ceramide as a transducer of signaling by oxidative stress, and vice versa, has been demonstrated by other laboratories in different cell types.[17-19] Clearly, the interaction between oxidative stress and the ceramide pathway is taking a prominent role in biological processes and diseases where ROS are found. The objective of this chapter is to summarize the work that established the coupling of oxidative stress and ceramide production and illustrate where this coupling plays a role in disease. This chapter will emphasize important developments, elucidate basic principles and address key questions for future studies.

11.2 Reactive Oxidants Induce Apoptosis

Excessive accumulation of reactive oxidants is toxic, and the intracellular level of reactive oxidants is therefore tightly regulated by several antioxidants. Although antioxidant defenses are constitutively expressed in mammalian cells,[20] additional responses are mounted when the amount of environmental oxidants exceeds a threshold level, thereby becoming a threat to overall tissue integrity. Apoptosis may be one such cellular adaptive response.

Conflicting data related to oxidative stress and apoptosis may be the result of missing information regarding the various targets for reactive oxygen species as well as lack of data related to the different chemical modifications of these targets. Moreover, the targets may be differently expressed in various cells, thus leading to different end-results in terms of cell apoptosis. Recently, there has been a growing consensus that ROS as well as nitric oxide and its congeners play a key role in apoptosis. However, the precise nature of this role is unclear.

The generation of oxidative stress has been proposed as a critical event for death-inducing agents in the process of initiating their apoptotic activity. Specifically, depletion of glutathione (GSH), the most abundant intracellular thiol-containing small peptide, has been suggested to precede the onset of apoptosis induced by various agents. Moreover, it has been also suggested that depletion of GSH could be an early event in the commitment to apoptosis.[21-24] However, how depletion of GSH transmits apoptotic signals is yet unknown.

One of the most reproducible inducers of apoptosis is mild oxidative stress produced by H_2O_2, which is a ubiquitous molecule, freely miscible and able to cross cell membranes readily. It is present in several air pollutants, including the vapor phase of cigarette smoke. It is detected in the exhaled air of humans[25] and amounts of exhaled H_2O_2 are greater in subjects with lung inflammation[26-28] or who smoke cigarettes[29]. Importantly, several agonists increase H_2O_2 generation by epithelial cells, including cytokines (TNFα, IL1, and Fas ligand), cytotoxic agents, ionizing radiation and infections (e.g., HIV or bacteria). With the demonstration that apoptosis can be triggered by ceramide generation as a result of H_2O_2 interaction with the cell membrane of lung airway epithelial cells,[12] the hypothesis that ceramide is a "coordinator" of eukaryotic stress responses[30] is strengthened. However, the mechanism by which ceramide induces apoptosis is still unknown.

11.3 Ceramide Metabolism

Initially, sphingolipids were thought to play predominantly a structural role as components of lipid bilayers. It is now apparent that sphingolipid metabolites, especially ceramide, sphingosine (SP), and sphingosine-1-phosphate (S1P), play essential roles in cell growth, survival and death.[31,32] Ceramide is a central molecule in sphingolipid metabolism (Figure 1). Sphingolipid biosynthesis begins in the endoplasmic reticulum (ER) with the condensation of serine and palmitoyl CoA, followed by reduction to dihydrosphingosine.   Then, the N-acylation of dihydrosphingosine by ceramide synthase produces dihydroceramide that is then desaturated to generate ceramide.[33,34]

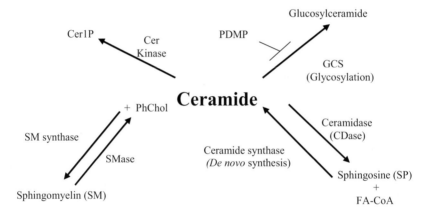

Figure 1: Metabolic pathways of ceramide.

Ceramide then serves as the precursor for all major sphingolipids in eukaryotes, such as sphingomyelin (SM), via SM-synthase, or glucosylceramide, via glucosyl ceramide synthase (GCS).[35]

The breakdown of complex sphingolipids results in the formation of ceramide through the action of sphingomyelinases (SMases). The catabolism of ceramide advances via the action of ceramidases (CDases), whereas the product sphingoid bases act as substrates for sphingosine kinases to form S1P or are recycled into ceramide and complex sphingolipids via the action of ceramide synthases[36,37] (see Figure1).

Ceramide and S1P appear to mediate different cellular functions. Ceramide has been implicated in differentiation, cell cycle arrest, cellular senescence and apoptosis.[32,38] In contrast, its metabolic products sphingosine and particularly S1P have been implicated in cell proliferation,[39,40] protection from apoptosis,[41] induction of mitogenesis[42] and angiogenesis[43] (Figure 2). Thus, a key part of sphingolipid metabolism is the balance between the relative and absolute cellular levels of the sphingolipid metabolites; the modulation of this balance by oxidants, cytokines and other factors has important life or death consequences for the cell.[44-46]

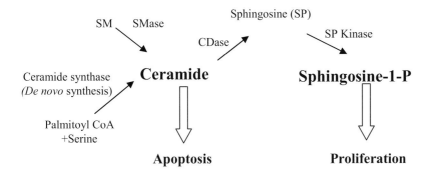

Figure 2: Biological effects of sphingolipid metabolites, ceramide and spingosine-1-phosphate (SIP).

11.4 Ceramide accumulation and apoptosis

Significant evidence suggests an important role for ceramide in mediating the apoptotic responses of several diverse agents, including oxidative stress. Importantly, the elevation in endogenous ceramide levels in response to these agents occurs prior to the onset of the execution phase of apoptosis when effector caspases are activated.[30,47,48] Furthermore, the addition of cell-permeable analogs of ceramide causes apoptosis in several cell lines.[49-52] The effects of ceramide are very specific such that dihydroceramide, which is the endogenous metabolic precursor to ceramide that lacks the 4-5 trans double bond, and has an uptake and metabolism similar to ceramide, does not cause apoptosis.[49,52-59] A role for ceramide in apoptosis is further supported by the demonstration that overexpression of glucosylceramide synthase (GCS) (Figure 1), which catabolizes ceramide by generating glucosylceramide, attenuated ceramide formation in response to TNF-α and several chemotherapeutic agents and protected the cells from apoptosis.[60,61]

Typically, ceramide accumulation results from the activation of one or more of the cellular SMases (Figure 3). Several of these sphingomyelin [SM]-specific phospholipase C (PLC) activities exist in mammalian tissues[52,55,62-65] and are distinguished according to their pH optimum and subcellular localization. At least eight SMases have been described. The best known includes the acidic SMase (aSMase), which displays a pH optimum of 4.5 and is confined to the lysosomes. Deficiency of this enzyme due to defects in the aSMase gene is responsible for the lysosomal disorder, Niemann-Pick disease,[66] which results in a massive accumulation of SM in the lysosomes and death in early childhood.[67] There are also various neutral SMases (nSMases) that act at neutral pH, are stimulated by Mg^{2+} or Mn^{2+} and are located in the plasma membrane, cytosol, endoplasmic reticulum or nuclear membranes.[68,69] Recently, an alkaline SMase in intestinal cells was also described.[70] Therefore, these isoenzymes differ not only in their catalytic properties and subcellular localization, but probably also in their modes of regulation. SM hydrolysis by these SMases, triggered by the binding of extracellular ligands to cell-surface receptors or by agents that induce cellular stress, was believed to be the major source of elevated cellular ceramide levels.[71] Indeed, the Mg^{2+}-dependent nSMase that resides in the plasma membrane was thought to generate most of the ceramide used as a second messenger for apoptosis.[18,72-74]

However, recent studies indicate that other mechanisms of ceramide accumulation can occur that lead to apoptosis, beside activation of the plasma membrane nSMase. In fact, all of the metabolic pathways involved in ceramide generation, breakdown and incorporation into more complex lipids (Figure 1) may be regulated and play distinct roles in apoptosis.[46,75-79] For example, stress response-induced ceramide accumulation can occur due to activation of the de-novo pathway, catalyzed by ceramide synthase,[80] or sometimes as a result of inhibition of ceramide clearance through SM-synthase or ceramidases (CDases).[81-83] Fumonisin B1, an inhibitor of ceramide synthase and consequently of the de novo pathway, was able to block ceramide generation and apoptosis in response to extracellular agents such as retinoic acid, etoposide, angiotensin II, or daunorubicin.[84-89] It was also shown that cells treated with B13, a ceramidase inhibitor, responded with elevated ceramide levels and activation of the apoptotic cascade.[82,90] However, it is still unknown whether these metabolic pathways, beside nSMase, are also involved in oxidant-induced ceramide accumulation.

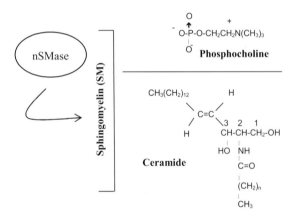

Figure 3: Hydrolysis of Sphingomyelin to ceramide.

11.5 Ceramide formation and reactive oxidants

Few reports have explored the role of ceramide in oxidant injury. Most studies that have used the apoptotic model of cell injury showed that increased ceramide generation due to diverse stimuli is mediated by activation of Smases.[30,91,92] Exposure of leukemic or endothelial cells to H_2O_2 resulted in increased ceramide generation with a concomitant decrease in SM content, which is suggestive of SMase activation.[93] Also, in an in vitro model of hypoxic injury to PC12 cells, reactive oxygen metabolites triggered ceramide generation via activation of nSMase.[91] In our laboratory, direct examination of nSMase activation during oxidative stress showed that H_2O_2 exposure concurrently induces both nSMase activity and elevated ceramide levels thereby leading to apoptosis in lung cells.[12,13]

However, Shah and co-workers[94,95] showed that hypoxia increases ceramide generation through ceramide synthase activation in LLC-PK1 cells. These authors further demonstrated that H_2O_2 stimulated ceramide synthase but not SMase activity, and demonstrated that inhibition of ceramide synthase prevented oxidant-induced ceramide production, DNA damage and cell death. This report provides evidence that ceramide synthase activation by oxidative stress leads to ceramide-mediated injury of renal epithelial tubule cells.

A role for ceramidases during oxidant exposure may also be critical as suggested recently.[96] These investigators demonstrated that exogenously added SMase significantly induced SM hydrolysis that resulted in only a modest ceramide increase in a human proximal tubule HK-2 cell line. This indicated that the vast majority of the generated ceramide was rapidly catabolized (presumably by ceramidases), and underscored the authors' previous conclusion that "the ceramide increments after acute tubular injury must stem, at least in part, from decreased ceramide catabolism, and not simply SM breakdown".[96,97] Therefore, activation of a SMase by it self may not be sufficient for cell death induction during oxidative stress; inhibition of ceramide catabolism (presumably by ceramidases) may also be required.

11.6 Ceramide and Nitric Oxide

Nitric oxide (NO) is a short-lived free radical gas described as a cytotoxic agent in several signaling pathways. Although NO can mediate apoptosis in various cell systems,[98] its mechanism of action is not completely understood. Some studies suggest that ceramide and NO interact to mediate cell death. Chronic exposure of renal mesangial or glomerular endothelial cells to NO donors resulted in a dose-dependent increase in ceramide levels and in the activation of both acidic and neutral Smases.[99,100] In contrast, acidic and neutral ceramidases, the ceramide-metabolizing enzymes, were shown to be inhibited by NO.

In addition to its role as a second messenger in apoptosis, NO, under different conditions, can also protect cells against apoptosis.[101] For example, in human monocytic cells, exogenous NO protects from cell death by inhibiting TNFα-induced TRADD recruitment, caspase-8 activity, and ceramide generation.[102] Likewise, NO inhibits p75[NTR]-induced apoptosis in neuroblastoma cells,[103] but the protective role of NO occurs downstream of ceramide accumulation. Finally, recent studies suggest that NO can act both upstream and downstream of ceramide generation.[104] Other reports describe NO production as a committed step in ceramide signaling. For example, cell-permeable ceramide potentiated the effects of TNFα on NO production and on the inducible-NO synthase expression in glioma cells, rat primary astrocytes and murine macrophages, and in smooth muscle cells, NO production was induced following treatments with exogenous Smase.[105-107] Therefore, although more work is needed to elucidate how the NO system and the ceramide pathway are connected in apoptosis signaling, the two are clearly coupled.

11.7 Ceramide Generation by nSMase is Modulated by GSH

Our laboratory demonstrated that ceramide accumulation following H_2O_2 treatment appears to occur via the activation of SMase pathway(s).[12,13] These pathways also respond to TNFα, Fas, and γ-irradiation-initiated apoptosis[58,59,108] wherein ceramide accumulation is concurrent with SM hydrolysis by activated SMases. However, whereas both the membrane-associated neutral and the acidic forms of SMases are activated by TNFα receptors through different domains, we demonstrated that only nSMase, and not aSMase, is affected by exposure to H_2O_2.[13]

The mechanism by which H_2O_2 stimulates SM hydrolysis to ceramide is unknown. We demonstrated that extracellular supplementation of GSH to A549 lung epithelial cells inhibited ceramide production, whereas depletion of intracellular GSH by H_2O_2 or DL-buthionine-[S,R]-sulfoximine (BSO) paralleled increases in ceramide levels and apoptosis induction.[14] When GSH was supplemented extracellularly, the H_2O_2-induced drop in cellular GSH was diminished and both ceramide elevation and apoptosis prevented. These were all specific effects of GSH and not of other thiol-containing molecules. Importantly, aminotriazole (inhibitor of Catalase) mimicked the effects of H_2O_2 treatment, and N-Acetyl Cysteine (NAC) inhibited the effects of intracellularly generated H_2O_2. These results suggested that in lung epithelial cells, H_2O_2 triggers the apoptotic pathway by inducing ceramide generation via depletion of GSH and that elevation of ceramide is sufficient and necessary for inducing apoptosis.

In several systems, ROS production has been shown to play a key and early role in ceramide-mediated apoptosis induced by serum starvation,[14] anthracyclins such as daunorubicin,[109] and cytokines such as TNF.[19] In these systems, generation of ROS precedes ceramide elevation, and GSH depletion is often evident;[19] supplementation with antioxidants such as GSH and NAC inhibits ceramide accumulation and apoptosis. Therefore, it appears that generation of ROS and increases in ceramide are coupled and mediate cell death due to several diverse apoptotic stimuli. However, how GSH regulates ceramide levels is not yet established.

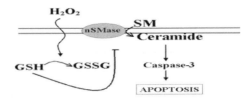

Figure 4: Schematic representation of the role of oxidants (H_2O_2) and antioxidants (GSH) in ceramide generation and apoptosis. GSH inhibits nSMase activity, thus maintaining low ceramide levels. Stress factors (i.e., extracellularly administered H_2O_2) result in ROS-mediated depletion of GSH. Therefore, the redox state of the cell determines the activity of nSMase and the levels of ceramide, thus modulating the apoptotic pathway in lung epithelial cells.

It was recently shown that GSH regulates the neutral Mg^{2+}-dependent nSMase. GSH elicits a direct inhibitory effect on nSMase from blood cells[21,68,110] and on purified nSMase from brain cells.[111] Moreover, decreases in cellular GSH levels induced by TNFα precede activation of nSMase.[21] These are among the observations that led us to propose (Figure 4) that in lung epithelial cells, the plasma membrane-bound nSMase may exist as an inactive form inhibited by high levels of both intra- and extracellular GSH present in epithelial lining fluid (ELF), thus maintaining low levels of ceramide. The inhibition of nSMase would render lung cells less sensitive and less susceptible to oxidants than they are ordinarily exposed to and the threshold for ceramide elevation required for the induction of apoptosis increases. However, when oxidant levels increase and GSH levels drop, the inhibitory effect of GSH on nSMase is overcome, therefore increasing ceramide elevation and the apoptotic pathway is initiated. This hypothesis is supported by our findings that the inhibitory effect of GSH on H_2O_2-induced ceramide production is specific for GSH and not for other thiol-containing molecules and most importantly not for GSSG. Therefore, oxidation of GSH by oxidants renders it incapable of inhibiting ceramide generation. It is interesting that even a short exposure of cells to H_2O_2 for 1 h, followed by incubation in regular media, is sufficient to induce apoptosis. This demonstrates that the events that control the fate of the cells occur within this hour, during which GSH is depleted and ceramide is generated.[14] We also observed that supplementation of GSH shortly before exposure to H_2O_2 was sufficient to inhibit the apoptotic effects of H_2O_2. It appears that providing GSH to replenish the decreased levels of GSH is sufficient to maintain ceramide below the threshold levels, thus preventing apoptosis.

11.8 Ceramide Signaling and Enzymatic Antioxidants

Consistent with the inhibitory effects of GSH on nSMase, [14,21] several studies indicate that antioxidant scavenging proteins can also interfere with ceramide signaling. For example, Mn(III) tetrakis (benzoic acid) porphyrin, a superoxide dismutases (SOD) mimic, inhibits ceramide-induced apoptosis in neuronal cells.[112] Gouaze et al[113] showed that overexpression of the cytosolic/mitochondrial selenium-dependent glutathione peroxidase (GPx) can prevent doxorubicin-induced ROS production, nSMase activation, SM hydrolysis and ceramide generation. In agreement with these observations, cell-permeant ceramide promoted apoptosis in T47D human breast cancer cells, bypassing the doxorubicin-induced SMase activation, but neither GPx overexpression nor treatment with exogenous NAC blocked this event, which indicates that GPx targets an event located upstream to ceramide generation.

11.9 Ceramide Generation is Ipstream of the Executioner Phase of Apoptosis

Few reports[73,113] have examined the temporal placement of ceramide accumulation with respect to caspases in oxidant-induced apoptosis. Caspases are described as the executioners of the program for cell death, but the literature conflicts as to whether ceramide generation triggered by apoptotic inducers precedes caspases or vice versa.[114,115] We demonstrated that ceramide accumulation due to SM hydrolysis (via exposures to either ROS or C6-ceramide), ceramide synthesis or inhibition of UDP-glucose-ceramide glucosyltransferase, induced caspase-3 activation.[48] Because the cleaved forms of both caspase-3 and PARP were detected subsequent to ceramide increases, this suggests that ceramide accumulation occurs earlier in the apoptotic cascade (Figure 5). These data agree with studies from non-lung cell systems,[116,117] and indicate that ceramide accumulation per se may serve as an initial trigger for apoptosis, though in some systems not sufficient to induce apoptosis without activation of the downstream caspase (s) signal.

When caspase-3 was identified as the mammalian analog of the C. elegans CED-3 gene product, it was suggested that this protease could be a common effector for all apoptotic pathways.[118] However, mice with a homozygous deletion of the caspase-3 gene still had a normal development of all of the organs except for the brain.[119] It was also found that a potent inhibitor of caspase-3 activation was ineffective in preventing ceramide-induced apoptosis in U937 (120). Furthermore, it was shown that caspase-3 deficient MCF7 cells were able to complete nuclear apoptosis in response to ceramide and no PARP cleavage was observed. Together, these studies suggest that caspase-3 is not always essential for apoptosis induction by the ceramide pathway. However, activation of caspase-3 by ceramide and induction of apoptosis were inhibited by overexpression of Bcl2,[91] which indeed has been proposed to act via its inhibition of caspase-3.[118]

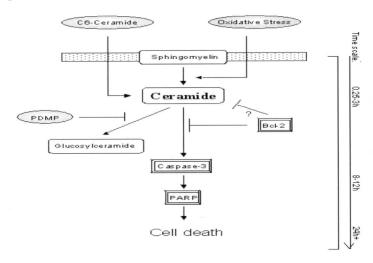

Figure 5: Model for the apoptotic events during ceramide-mediated apoptosis: Caspase-3 is activated following the induction of ceramide accumulation by diverse inducing agents. This results in the cleavage of PARP and other possible target molecules, followed by apoptotic cell death. Bcl2 functions to inhibit the activation of caspase-3 and may also prevent ceramide production.

To further investigate where endogenous ceramide generation is located in the apoptotic signaling cascade of lung epithelial cells exposed to oxidative stress, our laboratory[48] studied the role of Bcl2 overexpression on the ceramide-induced pathway. Some roles of Bcl2 modulating the ceramide pathway have been also suggested. For example, Tepper *et al*[117] demonstrated that Bcl2 overexpression in Jurkat cells reduced ceramide accumulation induced by CD95. On the other hand, Jaffrezou *et al*[121] showed that overexpression of Bcl2 in HL60 cells had no effect on ceramide generation induced by C6-ceramide. Only a few studies reported that Bcl2 could regulate ceramide generation,[91,122] whereas most studies suggested that Bcl2 blocks apoptosis but not ceramide generation.[47,123-125] We have demonstrated that Bcl2 overexpression protects airway epithelial cells against H_2O_2 and C6-ceramide-induced caspase-3 activation and cell death;[48] Bcl2 also inhibited ceramide generation in response to inducers of apoptosis and reduced basal cellular levels of ceramide. Thus, Bcl2 may exert its anti-apoptotic effects via targeting the ceramide pathway (Figure 5). Nevertheless, additional studies are needed to determine whether Bcl2 prevents caspase activation directly or by the inhibition of ceramide generation.

11.10 Ceramide Generates Reactive Oxidants: The Mitochondria Connection

Ceramide induces ROS production in intact mitochondria[16] and in cells,[126] and recent studies have begun to unravel the intimate connections between mitochondrial involvement in apoptosis and the ceramide pathway, including the determination of mitochondrial-specific actions of ceramide. These studies point to ROS generated in the mitochondrial respiratory chain as early mediators of ceramide-induced apoptosis, suggesting that coupling between oxidative stress and ceramide production is bi-directional: not only oxidants activate ceramide production, but ceramide may also induce generation of reactive oxidants.

Because mitochondria have a central role in the control of cell survival and death, the question arose as to whether the ceramide pathway and mitochondrial processes were coupled, and whether there was a mitochondrial pool of ceramide that is directly involved in the progression of the release of cytochrome C and other proteins, which lead to the activation of effector caspases and apoptosis execution.

These questions were addressed when it was reported that ceramide accumulation was detected in mitochondria,[127] and when the addition of exogenous ceramide to purified mitochondria was shown to inhibit the respiratory chain[128] generate ROS and release cytochrome C.[16,128-132] It is, therefore, possible that a mitochondrial pool of ceramide is involved in these processes.

Moreover, several activities of the ceramide-generating machinery appear to reside in the mitochondria, which supports the existence of a pool of SM in mitochondria.[82,83] For example, a ceramide synthase was partially purified from bovine liver mitochondria,[133] and several enzyme activities such as SMase and SM-synthase were identified in mitochondria-associated membranes. The expression of bacterial SMase in mitochondria, but not other subcellular compartments, resulted in induction of apoptosis[83] and points to a role for endogenous mitochondrial ceramide in regulating apoptosis. The recent identification and cloning of a mitochondrial ceramidase,[134] and its mitochondrial localization, has substantiated the existence of mitochondrial pathways of ceramide metabolism that may play a key role in mitochondrial functions and in the regulation of apoptosis.[82]

11.11 Diseases Related to Ceramide Signaling

Reactive oxidant-ceramide coupling appears to play a role in many different diseases such as Macular Degeneration,[135] diseases of the nervous system,[136-138] vascular diseases, insulin resistance and cancer. For the sake of space, we will focus only on some studies related to diseases involving ceramide signaling such as vascular disease, insulin resistance and multi-drug resistance in cancer.

11.11.1 Vasodilatation, Oxidative Stress and Ceramide Generation.

An understanding of how oxidant and ceramide coupling participates in vascular disease is emerging.[139] For example, recent studies showed that ceramide stimulates the production of O_2^- in vascular cells.[140-142] Since O_2^- can interact with NO and thereby decrease NO within endothelial cells,[143,144] it is possible that ceramide-stimulated O_2^- production in the vascular endothelium depletes NO resulting in impairment of endothelium-dependent vasodilatation in the coronary circulation.[145]

It was also shown that NO can induce ceramide production in glomerular mesangial and endothelial cells and that ratio of NO to O_2^- may determine whether cells live or die. Exposure of glomerular endothelial and mesangial cells to either NO donors or superoxide-generating substances led to a sustained ceramide formation that paralleled the induction of apoptosis in both cell types. Co-incubation of endothelial cells with NO and superoxide led to the generation of peroxynitrite and caused a synergistic enhancement of ceramide generation and apoptosis, but co-stimulation with superoxide neutralized not only NO-induced apoptosis but also NO-induced ceramide formation in mesangial cells, although O2- alone triggered ceramide formation and cell death. Furthermore, exposure of endothelial cells to glucose oxidase, which generates H_2O_2, or to exogenous H_2O_2, also showed a dose-dependently increased ceramide formation and apoptosis, although to a lesser extent than exposure to superoxide. Taken together this suggests that ceramide represents an important mediator of reactive oxygen and nitrogen species-triggered cell responses. There also seems to be cell type-specific protective mechanisms that critically depend on a fine-tuned redox balance between reactive nitrogen and oxygen species that determines whether a cell undergoes apoptosis or survives when exposed to oxidative and/or nitrosative stress conditions.[146]

11.11.2 Restenosis and Ceramide Generation

Strategic elevation of cellular ceramide is often used for therapies that aim to control growth or advance apoptosis. On the other hand, agents that reduce ceramide or elevate S1P tend to inhibit apoptosis and enhance proliferation. One remarkable demonstration of the first approach is presented in the work of Kester *et al* coworkers: they found that ceramide analogs applied directly to damaged arteries can be strongly antiproliferative.[147] Indeed, neointimal hyperplasia of vascular smooth muscle cells and secondary occlusion of coronary arteries, the cause of restenosis after balloon angioplasty or stenting, affects nearly 20% of the 1.5 million patients who undergo coronary angioplasty yearly. Proliferation of cultured vascular smooth muscle cells appears to involve the extracellular signal–regulated kinase

(ERK) and Akt kinase cascades, which can be inhibited by ceramide.[148] In rabbits, C_6-ceramide–coated balloon catheters prevent stretch-induced neointimal hyperplasia in carotid arteries[147] by inactivating ERK and Akt signaling and thereby inducing cell cycle arrest in stretch-injured vascular smooth muscle cells.[148]

11.11.3 Insulin Resistance, Oxidative Stress and Ceramide Generation

Despite enormous effort, the mechanism of insulin resistance that accompanies Type II diabetes remains elusive. Oxidative stress and ceramide have gained considerable attention for their potential roles in contributing to impaired insulin responsiveness.[149] Oxidant stress is directly correlated with metabolic control in patients with Type II diabetes; it is thought that hyperglycemia compromises endogenous antioxidants, which results in the enhanced generation of free radicals.[150,151] It has also been proposed that it is the insulin resistance and compensatory hyperinsulinemia that leads to increased oxidative stress.[152] Ceriello et al observed that insulin stimulates the production of H_2O_2 and proposed that a deleterious cycle exists where hyperglycemia and hyperinsulinemia leads to free radical production, which further impairs insulin action.[150] Studies with diabetic patients and animals indicate that disrupting such a cycle with anti-oxidative therapy improves metabolic parameters[149,153] but these studies have not shed light on the mechanism(s) by which oxidative stress leads to insulin resistance. A number of reports have demonstrated that ceramide inhibits insulin signaling in cultured muscle and fat.[154-157] In addition, ceramide levels have been reported to be elevated in the muscles tissues of genetically obese insulin-resistant rats.[158] Thus, it is possible that oxidative stress and ceramide coupling plays a role in insulin resistance.

Most of the work investigating the effects of oxidative stress and ceramide on insulin signaling was conducted with 3T3-L1 adipocytes. Rudich et al demonstrated that extended exposure to oxidative stress inhibited glucose transporter-4 (GLUT-4) translocation due to insulin in 3T3-L1 adipocytes. Further work by the same group provided evidence that the effects of oxidative stress on insulin responses were due to compromised phosphatidylinositol (PI)-3 kinase activation, insulin receptor substrate (IRS)-1 redistribution and Akt activation.[156,157] Anti-oxidants such as lipoic acid can reverse oxidant-impaired insulin responsive.[159,160] On the other hand, Long et al demonstrated that exposure to membrane-permeable ceramide decreases GLUT-4 mRNA levels[161] in 3T3-L1 adipocytes, and Peraldi et al showed that treatment with ceramide or with bacterial SMase inhibited insulin-stimulated IR and IRS-1 tyrosine phosphorylation.[162] Subsequent work by other groups confirmed that exogenous ceramide inhibited insulin-induced glucose uptake. Recent work by Summers et al showed that membrane-permeable ceramide suppressed Akt without affecting IRS-1 tyrosine phosphorylation or IRS-1 associated PI3 kinase activity.[163] Muscle cells exposed to oxidative stress or ceramide exhibit similar impairments in insulin responsiveness as adipocytes.[154,155] Unfortunately, there are no reports in the literature that have explored whether oxidative stress stimulates ceramide production in muscle and fat cells.

A deleterious cycle likely exists between proinflammatory cytokines, oxidative stress and ceramide in insulin resistance. TNFα is reportedly elevated in the muscle and adipose tissue of insulin resistant humans and animals and it has been demonstrated to increase endogenous ceramide and impair insulin signaling in cultured muscle and fat cells.[164,165] On the other hand, TNFα can also modulate the cellular redox status of muscle and adipocytes.[166,167] Such a loop makes determining the inciting factor in insulin resistance difficult to determine. Nevertheless, because TNFα and oxidative stress stimulate endogenous ceramide accumulation in a variety of different cell types, ceramide accumulation may very well act as the common pathway for those factors that leads to impaired insulin signaling. Thus, preventing ceramide accumulation may be a useful therapeutic strategy for improving insulin sensitivity in Type II diabetes.

11.11.4 Drug Resistance, Oxidative Stress and Ceramide Generation

Cancer cells develop multiple mechanisms to become resistant to chemotherapeutic agents.[168,169] Multi drug resistance may be caused by diverse mechanisms such as activation of efflux pumps, which lower drug levels within the cells[170,171] or modifications of glutathione metabolism, which may protect against oxidant stress.[172] A number of studies suggest that dysfunction of ceramide metabolism may contribute to multi drug resistance.[61,173,174] For example, excessive ceramide glycosylation by the enzyme glucosylceramide synthase (GCS) (see Figure 1) has been observed in some multi drug resistant cell lines.[175,176] Since glucosylceramide is a noncytotoxic metabolite of ceramide, this enzymatic reaction may be an important pathway for bypassing apoptosis induced by ceramide.

It was recently suggested that the resistance of cancer cells to drugs or other apoptosis-inducing agents could be reversed by targeting ceramide metabolism.[173] Notably, a number of clinically important chemotherapeutic agents inhibit tumor growth because of their ability to enhance ceramide formation in cancer cells[173] via the activation of various pathways of ceramide metabolism. This in turn can overcome an inhibition of ceramide accumulation, which may be conferred, for example by elevating GCS activity.[174] For example, anthracyclins such as doxorubicin and daunorubicin elevate ceramide levels by activating ceramide synthase as well as Smases.[115,177] Tamoxifen (an estrogen analog used to block estrogen receptor), on the other hand inhibits the conversion of ceramide to glucosylceramide by GCS,[178] thereby increasing cellular ceramide levels. Another chemotherapeutic drug that modulates ceramide metabolism is the synthetic retinoid N-(4-hydroxyphenyl)retinamide (4-HPR), which increases intracellular ceramide levels in diverse cell types.[179] The increase in ceramide by 4-HPR was suggested to involve ceramide synthase since ceramide increase was abrogated by inhibitors of de novo ceramide synthesis.[180] This drug also mediates p53-independent cytotoxiciy and can increase reactive oxygen species and ceramide in solid tumor cell lines.[180]

Modulation of ceramide levels may also augment the efficacy of some cancer treatments.[177] For example, Mehta et al showed that the addition of ceramide boosts taxol-mediated apoptotic death of Tu138 head and neck tumor cells.[181] Selzner *et al* showed that the cellular content of ceramide in human colon cancer is reduced by more than 50% relative to that of healthy colon mucosa.[90] The effective ceramidase inhibitor B13 increases the ceramide content of tumor cells and induces tumor cell apoptosis, without affecting the ceramide level or survival of normal liver cells. B13 also prevents growth of two aggressive human colon cancer cell lines metastatic to the liver, suggesting that ceramidase inhibition may provide a promising therapeutic strategy for selective toxicity toward malignant but not normal cells.

Interestingly, some of the chemotherapeutic agents that modulate ceramide levels also produce ROS. The anthracyclins, for example, elicit ROS formation[109] that may contribute to anthracyclin cytotoxiciy.[182] Gouaze *et al.*[113] have shown that apoptosis induced by doxorubicin is preceded by ROS production. Treatment of cancer cells with sodium nitroprusside, a nitric oxide donor, was also associated with elevation of ceramide via the activation of nSMase[183] that was followed by apoptosis, establishing a relationship between the ceramide pathway and NO-mediated apoptosis and pointing to a new strategy for chemotherapeutic intervention.

The overall importance of oxidants in cancer treatment was demonstrated by pretreatment of cancer cells with NAC, which resulted in inhibition of both ceramide production and cell death.[109] This suggests that ROS production may be a major constituent in chemotherapeutic agents-induced apoptosis via ceramide generating pathways, and illustrates the concept that cellular antioxidant defense can influence the clinical efficiency of such agents (i.e., anthracyclins) by modulating the coupling of ROS and ceramide.

In summary, recent studies have implied that dysfunction of ceramide metabolism, observed in multi drug resistant cancers, may be overcome by chemotherapeutic agents that modulate the ceramide pathway. The ability of some of these drugs to produce ROS may contribute to their ability to generate ceramide and induce apoptosis. However, for rational advances in chemotherapy to proceed, the exact mechanism and specific targets for ROS in the modulation of ceramide levels in cancer cells that turn drug resistant need to be defined.

11.12 Conclusion

Several lines of evidence indicate that oxidative stress and ceramide generation are intimately coupled in cell signaling: 1. Inducers of apoptosis trigger the generation of both ROS and ceramide; 2. Oxidative stress-induced apoptosis involves the SM-ceramide pathway.[3] Ceramide stimulates ROS production and may act directly on the mitochondrial respiratory chain; 4. Cellular antioxidants regulate ceramide buildup; 5. Growing evidence suggests that the pathologic states of several diseases that are affected by reactive oxidants involve the regulation of the ceramide pathway.

The kinetics of ROS and ceramide generation in cell death show that both occur early in the commitment phase of the apoptotic cascade; ceramide accumulation occurs prior to the execution phase initiated by caspases and may also be upstream of the site where Bcl2 acts. However, more work is needed to determine the exact cause-effect relationships between oxidative stress, ceramide, caspase activation and Bcl2 inhibition.

While it is unequivocal that oxidative stress modulates ceramide production, ROS generation and mitochondrial alterations may also characterize a committed phase in ceramide signaling, indicating that positive feedback may occur between ROS and ceramide generation in cell death. However, many of the links between oxidative stress and ceramide signaling are still correlative and require rigorous molecular and mechanistic studies. The exact molecular mechanisms and the

Chapter 11

subcellular localization of oxidative stress and ceramide pathway(s) interactions need to be defined. The molecular identification of SMases, SM-synthase, ceramidases that are modulated by oxidative stress is not complete. Recent studies reported on the cloning of two candidate nSMases,[184] but evidence was provided that one nSMase localizes to the endoplasmic reticulum and functions as a lyso-PAF phospholipase C,[185] whereas the other nSMase localizes to the Golgi and its physiologic substrates were not established. Because enzymatic and sub fractionation studies indicate the presence of a plasma membrane nSMase that is modulated by oxidative stress, effort is still being directed toward isolating and characterizing this SMase from the lung [Goldkorn *et al*, unpublished data]. Similarly, very little is known about the SM-synthase, which also has been difficult to purify and clone.[81] Because most of the key enzymes regulating ceramide metabolism have not been characterized yet there is still a lack of molecular and pharmacological tools to study these pathways and their functions.

On one hand, it is desired to prevent the killing of cells and tissue injury that oxidant-ceramide coupling leads to in diseases such as Asthma or ARDS. In cancer, the opposite effect is sought.

Decoding the exact molecular interactions between oxidative stress and ceramide pathways should lead to new strategies for pharmacological intervention in such diverse diseases.

References

1. Hoidal, J. R. 2001. Reactive oxygen species and cell signaling. *Am J Respir Cell Mol Biol* **25:**661-3.

2. Schreck, R., P. Rieber and P. A. Baeuerle. 1991. Reactive oxygen intermediates as apparently widely used messengers in the activation of the NF-κ B transcription factor and HIV-1. *Embo J* **10:**2247-58.

3. Sen, C. K. and L. Packer. 1996. Antioxidant and redox regulation of gene transcription. *Faseb J* **10:**709-20.

4. Adler, K. B., B. M. Fischer, D. T. Wright, L. A. Cohn and S. Becker. 1994. Interactions between respiratory epithelial cells and cytokines: relationships to lung inflammation. Ann N Y Acad Sci **725:**128-45.

5. Cross, C. E., A. van der Vliet, C. A. O'Neill and J. P. Eiserich. 1994. Reactive oxygen species and the lung. *Lancet.* **344:**930-3.

6. Yamaya, M., K. Sekizawa, T. Masuda, M. Morikawa, T. Sawai and H. Sasaki. 1995. Oxidants affect permeability and repair of the cultured human tracheal epithelium. *Am J Physiol* **268:**L284-93.

7. Jobsis, Q., H. C. Raatgeep, P. W. Hermans and J. C. de Jongste. 1997. Hydrogen peroxide in exhaled air is increased in stable asthmatic children. *Eur Respir J* **10:**519-21.

8. Worlitzsch, D., G. Herberth, M. Ulrich and G. Doring. 1998. Catalase, myeloperoxidase and hydrogen peroxide in cystic fibrosis. *Eur Respir J* **11:**377-83.

9. Halliwell, B. and J. M. Gutteridge. 1990. Role of free radicals and catalytic metal ions in human disease: an overview. *Methods Enzymol* **186**:1-85.

10. Goldkorn, T., N. Balaban, K. Matsukuma, V. Chea, R. Gould, J. Last, C. Chan and C. Chavez. 1998. EGF-Receptor phosphorylation and signaling are targeted by H2O2 redox stress. *Am J Respir Cell Mol Biol* **19**:786-98.

11. Ravid, T., C. Sweeney, P. Gee, K. R. K. Carraway and T. Goldkorn. 2002. EGF receptor activation under oxidative stress fails to promote c-Cbl mediated down regulation. *J Biol Chem* **12**:12.

12. Goldkorn, T., N. Balaban, M. Shannon, V. Chea, K. Matsukuma, D. Gilchrist, H. Wang and C. Chan. 1998. H2O2 acts on cellular membranes to generate ceramide signaling and initiate apoptosis in tracheobronchial epithelial cells. *J Cell Sci* **111**:3209-20.

13. Chan, C. and T. Goldkorn. 2000. Ceramide path in human lung cell death. *Am J Respir Cell Mol Biol* **22**:460-8.

14. Lavrentiadou, S. N., C. Chan, T. N. Kawcak, T. Ravid, A. Tsaba, A. van der Vliet, R. Rasooly and T. Goldkorn. 2001. Ceramide-Mediated Apoptosis in Lung Epithelial Cells Is Regulated by Glutathione. *Am. J. Respir. Cell Mol. Biol.* **25**:676-684.

15. Cadenas, E. and H. Sies. 2000. Formation of electronically excited states during the oxidation of arachidonic acid by prostaglandin endoperoxide synthase. *Methods Enzymol* **319**:67-77.

16. Garcia-Ruiz, C., A. Colell, M. Mari, A. Morales and J. C. Fernandez-Checa. 1997. Direct effect of ceramide on the mitochondrial electron transport chain leads to generation of reactive oxygen species. Role of mitochondrial glutathione. *J Biol Chem* **272**:11369-77.

17. Thannickal, V. J. and B. L. Fanburg. 2000. Reactive oxygen species in cell signaling. *Am J Physiol Lung Cell Mol Physiol* **279**:L1005-28.

18. Levade, T. and J. P. Jaffrezou. 1999. Signalling sphingomyelinases: which, where, how and why? *Biochim Biophys Acta* **1438**:1-17.

19. Singh, I., K. Pahan, M. Khan and A. K. Singh. 1998. Cytokine-mediated induction of ceramide production is redox-sensitive. Implications to proinflammatory cytokine-mediated apoptosis in demyelinating diseases. *J Biol Chem* **273**:20354-62.

20. Cross, C. E., A. van der Vliet, S. Louie, J. J. Thiele and B. Halliwell. 1998. Oxidative stress and antioxidants at biosurfaces: plants, skin, and respiratory tract surfaces. *Environ Health Perspect 106 Suppl* **5**:1241-51.

21. Liu, B., N. Andrieu-Abadie, T. Levade, P. Zhang, L. M. Obeid and Y. A. Hannun. 1998. Glutathione regulation of neutral sphingomyelinase in tumor necrosis factor-α -induced cell death. *J Biol Chem* **273**:11313-20.

22. Macho, A., T. Hirsch, I. Marzo, P. Marchetti, B. Dallaporta, S. A. Susin, N. Zamzami and G. Kroemer. 1997. Glutathione depletion is an early and calcium elevation is a late event of thymocyte apoptosis. *J Immunol* **158**:4612-9.

23. Ghibelli, L., S. Coppola, C. Fanelli, G. Rotilio, P. Civitareale, A. I. Scovassi and M. R. Ciriolo. 1999. Glutathione depletion causes cytochrome c release even in the absence of cell commitment to apoptosis. *Faseb J* **13:**2031-6.

24. Baker, A., B. D. Santos and G. Powis. 2000. Redox Control of Caspase-3 Activity by Thioredoxin and Other Reduced Proteins. *Biochem Biophys Res Commun* **268:**78-81.

25. Williams, K. R., E. K. Spicer, M. B. LoPresti, R. A. Guggenheimer and J. W. Chase. 1983. Limited proteolysis studies on the Escherichia coli single-stranded DNA binding protein. Evidence for a functionally homologous domain in both the Escherichia coli and T4 DNA binding proteins. *J Biol Chem* **258:**3346-55.

26. Sznajder, J. I., A. Fraiman, J. B. Hall, W. Sanders, G. Schmidt, G. Crawford, A. Nahum, P. Factor and L. D. Wood. 1989. Increased hydrogen peroxide in the expired breath of patients with acute hypoxemic respiratory failure. *Chest* **96:**606-12.

27. Loukides, S., I. Horvath, T. Wodehouse, P. J. Cole and P. J. Barnes. 1998. Elevated levels of expired breath hydrogen peroxide in bronchiectasis. *Am J Respir Crit Care Med* **158:**991-4.

28. Dekhuijzen, P. N., K. K. Aben, I. Dekker, L. P. Aarts, P. L. Wielders, C. L. van Herwaarden and A. Bast. 1996. Increased exhalation of hydrogen peroxide in patients with stable and unstable chronic obstructive pulmonary disease. *Am J Respir Crit Care Med* **154:**813-6.

29. Nowak, D., J. Heinrich, R. Jorres, G. Wassmer, J. Berger, E. Beck, S. Boczor, M. Claussen, H. E. Wichmann and H. Magnussen. 1996. Prevalence of respiratory symptoms, bronchial hyperresponsiveness and atopy among adults: west and east Germany. *Eur Respir J* **9:**2541-52.

30. Hannun, Y. A. 1996. Functions of ceramide in coordinating cellular responses to stress. *Science* **274:**1855-9.

31. Spiegel, S. and A. H. Merrill, Jr. 1996. Sphingolipid metabolism and cell growth regulation. *Faseb J* **10:**1388-97.

32. Hannun, Y. A. and C. Luberto. 2000. Ceramide in the eukaryotic stress response. Trends Cell *Biol* **10:**73-80.

33. Merrill, A. H., Jr., E. Wang, R. LaRocque, R. E. Mullins, E. T. Morgan, J. L. Hargrove, H. L. Bonkovsky and I. A. Popova. 1992. Differences in glycogen, lipids, and enzymes in livers from rats flown on COSMOS 2044. *J Appl Physiol* **73:**142S-147S.

34. Michel, C. and G. van Echten-Deckert. 1997. Conversion of dihydroceramide to ceramide occurs at the cytosolic face of the endoplasmic reticulum. *FEBS Lett* **416:**153-5.

35. Kolter, T., R. L. Proia and K. Sandhoff. 2002. Combinatorial Ganglioside Biosynthesis. *J. Biol. Chem.* **277:**25859-25862.

36. Hannun, Y. A., C. Luberto and K. M. Argraves. 2001. Enzymes of sphingolipid metabolism: from modular to integrative signaling. *Biochemistry* **40:**4893-903.

37. Riley, R. T., E. Enongene, K. A. Voss, W. P. Norred, F. I. Meredith, R. P. Sharma, J. Spitsbergen, D. E. Williams, D. B. Carlson and A. H. Merrill, Jr. 2001. Sphingolipid perturbations as mechanisms for fumonisin carcinogenesis. *Environ Health Perspect* **109** :301-8.

38. Kolesnick, R. and Y. A. Hannun. 1999. Ceramide and apoptosis. Trends *Biochem Sci* **24:**224-5; discussion 227.

39. Zhang, H., N. E. Buckley, K. Gibson and S. Spiegel. 1990. Sphingosine stimulates cellular proliferation via a protein kinase C- independent pathway. *J Biol Chem* **265:**76-81.

40. Zhang, H., N. N. Desai, A. Olivera, T. Seki, G. Brooker and S. Spiegel. 1991. Sphingosine-1-phosphate, a novel lipid, involved in cellular proliferation. *J Cell Biol* **114:**155-67.

41. Cuvillier, O., D. S. Rosenthal, M. E. Smulson and S. Spiegel. 1998. Sphingosine 1-phosphate inhibits activation of caspases that cleave poly(ADP-ribose) polymerase and lamins during Fas- and ceramide- mediated apoptosis in Jurkat T lymphocytes. *J Biol Chem* **273:**2910-6.

42. Su, Y., D. Rosenthal, M. Smulson and S. Spiegel. 1994. Sphingosine 1-phosphate, a novel signaling molecule, stimulates DNA binding activity of AP-1 in quiescent Swiss 3T3 fibroblasts. *J Biol Chem* **269:**16512-7.

43. Lee, M. J., S. Thangada, K. P. Claffey, N. Ancellin, C. H. Liu, M. Kluk, M. Volpi, R. I. Sha'afi and T. Hla. 1999. Vascular endothelial cell adherens junction assembly and morphogenesis induced by sphingosine-1-phosphate. *Cell* **99:**301-12.

44. Pyne, S. and N. Pyne. 2000. Sphingosine 1-phosphate signalling via the endothelial differentiation gene family of G-protein-coupled receptors. *Pharmacol Ther* **88:**115-31.

45. Spiegel, S. and S. Milstien. 2000. Sphingosine-1-phosphate: signaling inside and out. *FEBS Lett* **476:**55-7.

46. Olivera, A. and S. Spiegel. 2001. Sphingosine kinase: a mediator of vital cellular functions. *Prostaglandins Other Lipid Mediat* **64:**123-34.

47. Dbaibo, G. S., D. K. Perry, C. J. Gamard, R. Platt, G. G. Poirier, L. M. Obeid and Y. A. Hannun. 1997. Cytokine response modifier A (CrmA) inhibits ceramide formation in response to tumor necrosis factor (TNF)-α : CrmA and Bcl-2 target distinct components in the apoptotic pathway. *J Exp Med* **185:**481-90.

48. Ravid, T., A. Tsaba, R. Rasooly, E. Medina and T. Goldkorn. 2002. Ceramide accumulation precedes caspase-3 activation during apoptosis of A549 human lung adenocarcinoma cells. *Am. J. Physiol. Lung. Cell. Mol. Physiol* (submitted).

49. Obeid, L. M., C. M. Linardic, L. A. Karolak and Y. A. Hannun. 1993. Programmed cell death induced by ceramide. *Science* **259:**1769-71.

50. Jarvis, W. D., A. J. Turner, L. F. Povirk, R. S. Traylor and S. Grant. 1994. Induction of apoptotic DNA fragmentation and cell death in HL-60 human promyelocytic leukemia cells by pharmacological inhibitors of protein kinase C. *Cancer Res* **54**:1707-14.

51. Quintans, J., J. Kilkus, C. L. McShan, A. R. Gottschalk and G. Dawson. 1994. Ceramide mediates the apoptotic response of WEHI 231 cells to anti- immunoglobulin, corticosteroids and irradiation. *Biochem Biophys Res Commun* **202**:710-4.

52. Cifone, M. G., R. De Maria, P. Roncaioli, M. R. Rippo, M. Azuma, L. L. Lanier, A. Santoni and R. Testi. 1994. Apoptotic signaling through CD95 (Fas/Apo-1) activates an acidic sphingomyelinase. *J Exp Med* **180**:1547-52.

53. Bielawska, A., C. M. Linardic and Y. A. Hannun. 1992. Modulation of cell growth and differentiation by ceramide. *FEBS Lett* **307**:211-4.

54. Hannun, Y. A. 1994. The sphingomyelin cycle and the second messenger function of ceramide. *J Biol Chem* **269**:3125-8.

55. Jayadev, S., B. Liu, A. E. Bielawska, J. Y. Lee, F. Nazaire, M. Pushkareva, L. M. Obeid and Y. A. Hannun. 1995. Role for ceramide in cell cycle arrest. *J Biol Chem* **270**:2047-52.

56. Jarvis, W. D., R. N. Kolesnick, F. A. Fornari, R. S. Traylor, D. A. Gewirtz and S. Grant. 1994. Induction of apoptotic DNA damage and cell death by activation of the sphingomyelin pathway. *Proc Natl Acad Sci U S A* **91**:73-7.

57. Strum, J. C., G. W. Small, S. B. Pauig and L. W. Daniel. 1994. 1-beta-D-Arabinofuranosylcytosine stimulates ceramide and diglyceride formation in HL-60 cells. *J Biol Chem* **269**:15493-7.

58. Haimovitz-Friedman, A., C. C. Kan, D. Ehleiter, R. S. Persaud, M. McLoughlin, Z. Fuks and R. N. Kolesnick. 1994. Ionizing radiation acts on cellular membranes to generate ceramide and initiate apoptosis. *J Exp Med* **180**:525-35.

59. Tepper, C. G., S. Jayadev, B. Liu, A. Bielawska, R. Wolff, S. Yonehara, Y. A. Hannun and M. F. Seldin. 1995. Role for ceramide as an endogenous mediator of Fas-induced cytotoxicity. *Proc Natl Acad Sci U S A* **92**:8443-7.

60. Lavie, Y., H. Cao, A. Volner, A. Lucci, T. Y. Han, V. Geffen, A. E. Giuliano and M. C. Cabot. 1997. Agents that reverse multidrug resistance, tamoxifen, verapamil, and cyclosporin A, block glycosphingolipid metabolism by inhibiting ceramide glycosylation in human cancer cells. *J Biol Chem* **272**:1682-7.

61. Liu, Y. Y., T. Y. Han, A. E. Giuliano and M. C. Cabot. 1999. Expression of glucosylceramide synthase, converting ceramide to glucosylceramide, confers adriamycin resistance in human breast cancer cells. *J Biol Chem* **274**:1140-6.

62. Schutze, S., K. Potthoff, T. Machleidt, D. Berkovic, K. Wiegmann and M. Kronke. 1992. TNF activates NF-κ B by phosphatidylcholine-specific phospholipase C-induced "acidic" sphingomyelin breakdown. *Cell* **71**:765-76.

63. Lawler, J. F., Jr., M. Yin, A. M. Diehl, E. Roberts and S. Chatterjee. 1998. Tumor necrosis factor-α stimulates the maturation of sterol regulatory element binding protein-1 in human hepatocytes through the action of neutral sphingomyelinase. *J Biol Chem* **273**:5053-9.

64. Wiegmann, K., S. Schutze, T. Machleidt, D. Witte and M. Kronke. 1994. Functional dichotomy of neutral and acidic sphingomyelinases in tumor necrosis factor signaling. *Cell* **78**:1005-15.

65. Okazaki, T., A. Bielawska, N. Domae, R. M. Bell and Y. A. Hannun. 1994. Characteristics and partial purification of a novel cytosolic, magnesium-independent, neutral sphingomyelinase activated in the early signal transduction of 1 α ,25-dihydroxyvitamin D3-induced HL-60 cell differentiation [published erratum appears in J Biol Chem 1994 Jun 10;269(23):16518]. *J Biol Chem* **269**:4070-7.

66. Otterbach, B. and W. Stoffel. 1995. Acid sphingomyelinase-deficient mice mimic the neurovisceral form of human lysosomal storage disease (Niemann-Pick disease). *Cell* **81**:1053-61.

67. Quintern, L. E., E. H. Schuchman, O. Levran, M. Suchi, K. Ferlinz, H. Reinke, K. Sandhoff and R. J. Desnick. 1989. Isolation of cDNA clones encoding human acid sphingomyelinase: occurrence of alternatively processed transcripts. *Embo J* **8**:2469-73.

68. Liu, B. and Y. A. Hannun. 1997. Inhibition of the neutral magnesium-dependent sphingomyelinase by glutathione. *J Biol Chem* **272**:16281-7.

69. Samet, D. and Y. Barenholz. 1999. Characterization of acidic and neutral sphingomyelinase activities in crude extracts of HL-60 cells. *Chem Phys Lipids* **102** :65-77.

70. Cheng, Y., A. Nilsson, E. Tomquist and R. D. Duan. 2002. Purification, characterization, and expression of rat intestinal alkaline sphingomyelinase. *J Lipid Res* **43**:316-24.

71. Kolesnick, R. N. and M. Kronke. 1998. Regulation of ceramide production and apoptosis. *Annu Rev Physiol* **60**:643-65.

72. Segui, B., C. Bezombes, E. Uro-Coste, J. A. Medin, N. Andrieu-Abadie, N. Auge, A. Brouchet, G. Laurent, R. Salvayre, J. P. Jaffrezou and T. Levade. 2000. Stress-induced apoptosis is not mediated by endolysosomal ceramide. *Faseb J* **14**:36-47.

73. Andrieu-Abadie, N., V. Gouaze, R. Salvayre and T. Levade. 2001. Ceramide in apoptosis signaling: relationship with oxidative stress. *Free Radic Biol Med* **31**:717-28.

74. Bezombes, C., B. Segui, O. Cuvillier, A. P. Bruno, E. Uro-Coste, V. Gouaze, N. Andrieu-Abadie, S. Carpentier, G. Laurent, R. Salvayre, J. P. Jaffrezou and T. Levade. 2001. Lysosomal sphingomyelinase is not solicited for apoptosis signaling. *Faseb J* **15**:297-9.

75. El Bawab, S., J. Usta, P. Roddy, Z. M. Szulc, A. Bielawska and Y. A. Hannun. 2002. Substrate specificity of rat brain ceramidase. *J Lipid Res* **43**:141-8.

76. Mao, C., R. Xu, Z. M. Szulc, A. Bielawska, S. H. Galadari and L. M. Obeid. 2001. Cloning and characterization of a novel human alkaline ceramidase. A mammalian enzyme that hydrolyzes phytoceramide. *J Biol Chem* **276**:26577-88.

77. Gatt, S. 1966. Enzymatic hydrolysis of sphingolipids. I. Hydrolysis and synthesis of ceramides by an enzyme from rat brain. *J Biol Chem* **241**:3724-30.

78. Tani, M., N. Okino, S. Mitsutake, T. Tanigawa, H. Izu and M. Ito. 2000. Purification and Characterization of a Neutral Ceramidase from Mouse Liver. A Single Protein Catalyzes The Reversible Reaction In Which Ceramide Is Both Hydrolyzed And Synthesized. *J. Biol. Chem.* **275**:3462-3468.

79. El Bawab, S., H. Birbes, P. Roddy, Z. M. Szulc, A. Bielawska and Y. A. Hannun. 2001. Biochemical characterization of the reverse activity of rat brain ceramidase. A CoA-independent and fumonisin B1-insensitive ceramide synthase. *J Biol Chem* **276**:16758-66.

80. Perry, D. K. 2000. The role of de novo ceramide synthesis in chemotherapy-induced apoptosis. *Ann N Y Acad Sci* **905**:91-6.

81. Hannun, Y. A. and L. M. Obeid. 2002. The Ceramide-centric Universe of Lipid-mediated Cell Regulation: Stress Encounters of the Lipid Kind. *J Biol Chem* **277**:25847-50.

82. Birbes, H., S. E. Bawab, L. M. Obeid and Y. A. Hannun. 2002. Mitochondria and ceramide: intertwined roles in regulation of apoptosis. *Adv Enzyme Regul* **42**:113-29.

83. Birbes, H., S. El Bawab, Y. A. Hannun and L. M. Obeid. 2001. Selective hydrolysis of a mitochondrial pool of sphingomyelin induces apoptosis. *Faseb J* **15**:2669-79.

84. Bose, R., M. Verheij, A. Haimovitz-Friedman, K. Scotto, Z. Fuks and R. Kolesnick. 1995. Ceramide synthase mediates daunorubicin-induced apoptosis: an alternative mechanism for generating death signals. *Cell* **82**:405-14.

85. Kalen, A., R. A. Borchardt and R. M. Bell. 1992. Elevated ceramide levels in GH4C1 cells treated with retinoic acid. *Biochim Biophys Acta* **1125**:90-6.

86. Plo, I., S. Ghandour, A. C. Feutz, M. Clanet, G. Laurent and A. Bettaieb. 1999. Involvement of de novo ceramide biosynthesis in lymphotoxin-induced oligodendrocyte death. *Neuroreport* **10**:2373-6.

87. Liao, W. C., A. Haimovitz-Friedman, R. S. Persaud, M. McLoughlin, D. Ehleiter, N. Zhang, M. Gatei, M. Lavin, R. Kolesnick and Z. Fuks. 1999. Ataxia telangiectasia-mutated gene product inhibits DNA damage-induced apoptosis via ceramide synthase. *J Biol Chem* **274**:17908-17.

88. Lehtonen, J. Y., M. Horiuchi, L. Daviet, M. Akishita, and V. J. Dzau. 1999. Activation of the de novo biosynthesis of sphingolipids mediates angiotensin II type 2 receptor-induced apoptosis. *J Biol Chem* **274**:16901-6.

89. Xu, J., C. H. Yeh, S. Chen, L. He, S. L. Sensi, L. M. Canzoniero, D. W. Choi and C. Y. Hsu. 1998. Involvement of de novo ceramide biosynthesis in tumor necrosis factor-α /cycloheximide-induced cerebral endothelial cell death. *J Biol Chem* **273**:16521-6.

90. Selzner, M., A. Bielawska, M. A. Morse, H. A. Rudiger, D. Sindram, Y. A. Hannun and P. A. Clavien. 2001. Induction of apoptotic cell death and prevention of tumor growth by ceramide analogues in metastatic human colon cancer. *Cancer Res* **61**:1233-40.

91. Yoshimura, S., Y. Banno, S. Nakashima, K. Takenaka, H. Sakai, Y. Nishimura, N. Sakai, S. Shimizu, Y. Eguchi, Y. Tsujimoto and Y. Nozawa. 1998. Ceramide formation leads to caspase-3 activation during hypoxic PC12 cell death. Inhibitory effects of Bcl-2 on ceramide formation and caspase-3 activation. *J Biol Chem* **273**:6921-7.

92. De Maria, R., M. R. Rippo, E. H. Schuchman and R. Testi. 1998. Acidic Sphingomyelinase (ASM) Is Necessary for Fas-induced GD3 Ganglioside Accumulation and Efficient Apoptosis of Lymphoid Cells. *J. Exp. Med.* **187**:897-902.

93. Verheij, M., R. Bose, X. H. Lin, B. Yao, W. D. Jarvis, S. Grant, M. J. Birrer, E. Szabo, L. I. Zon, J. M. Kyriakis, A. Haimovitz-Friedman, Z. Fuks and R. N. Kolesnick. 1996. Requirement for ceramide-initiated SAPK/JNK signalling in stress- induced apoptosis. *Nature* **380**:75-9.

94. Ueda, N., S. M. Camargo, X. Hong, A. G. Basnakian, P. D. Walker and S. V. Shah. 2001. Role of ceramide synthase in oxidant injury to renal tubular epithelial cells. *J Am Soc Nephrol* **12**:2384-91.

95. Ueda, N., G. P. Kaushal, X. Hong and S. V. Shah. 1998. Role of enhanced ceramide generation in DNA damage and cell death in chemical hypoxic injury to LLC-PK1 cells. *Kidney Int* **54**:399-406.

96. Zager, R. A., K. M. Burkhart and A. Johnson. 2000. Sphingomyelinase and membrane sphingomyelin content: determinants ofProximal tubule cell susceptibility to injury. *J Am Soc Nephrol* **11**:894-902.

97. Zager, R. A., D. S. Conrad and K. Burkhart. 1998. Ceramide accumulation during oxidant renal tubular injury: mechanisms and potential consequences. *J Am Soc Nephrol* **9**:1670-80.

98. Jacobson, M. D. 1996. Reactive oxygen species and programmed cell death. *Trends Biochem Sci* **21**:83-6.

99. Huwiler, A., S. Dorsch, V. A. Briner, H. van den Bosch and J. Pfeilschifter. 1999. Nitric oxide stimulates chronic ceramide formation in glomerular endothelial cells. *Biochem Biophys Res Commun* **258**:60-5.

100. Huwiler, A., J. Pfeilschifter and H. van den Bosch. 1999. Nitric oxide donors induce stress signaling via ceramide formation in rat renal mesangial cells. *J Biol Chem* **274**:7190-5.

101. Wink, D. A. and J. B. Mitchell. 1998. Chemical biology of nitric oxide: Insights into regulatory, cytotoxic, and cytoprotective mechanisms of nitric oxide. *Free Radic Biol Med* **25**:434-56.

102. De Nadai, C., P. Sestili, O. Cantoni, J. P. Lievremont, C. Sciorati, R. Barsacchi, S. Moncada, J. Meldolesi and E. Clementi. 2000. Nitric oxide inhibits tumor necrosis factor-α -induced apoptosis by reducing the generation of ceramide. *Proc Natl Acad Sci U S A* **97**:5480-5.

103. Lievremont, J. P., C. Sciorati, E. Morandi, C. Paolucci, G. Bunone, G. Della Valle, J. Meldolesi and E. Clementi. 1999. The p75(NTR)-induced apoptotic program develops through a ceramide- caspase pathway negatively regulated by nitric oxide. *J Biol Chem* **274**:15466-72.

104. Paolucci, C., P. Rovere, C. De Nadai, A. A. Manfredi and E. Clementi. 2000. Nitric oxide inhibits the tumor necrosis factor α -regulated endocytosis of human dendritic cells in a cyclic GMP-dependent Way. *J. Biol. Chem.* **275**:19638-19644.

105. Vann, L. R., S. Twitty, S. Spiegel and S. Milstien. 2000. Divergence in regulation of nitric-oxide synthase and its cofactor tetrahydrobiopterin by tumor necrosis factor-alpha. Ceramide potentiates nitric oxide synthesis without affecting GTP cyclohydrolase I activity. *J Biol Chem* **275**:13275-81.

106. Knapp, K. M. and B. K. English. 2000. Ceramide-mediated stimulation of inducible nitric oxide synthase (iNOS) and tumor necrosis factor (TNF) accumulation in murine macrophages requires tyrosine kinase activity. *J Leukoc Biol* **67**:735-41.

107. Katsuyama, K., M. Shichiri, F. Marumo and Y. Hirata. 1998. Role of nuclear factor-κB activation in cytokine- and sphingomyelinase-stimulated inducible nitric oxide synthase gene expression in vascular smooth muscle cells. *Endocrinology* **139**:4506-12.

108. Dbaibo, G. S., L. M. Obeid and Y. A. Hannun. 1993. Tumor necrosis factor-alpha (TNF-alpha) signal transduction through ceramide. Dissociation of growth inhibitory effects of TNF-alpha from activation of nuclear factor-κ B. *J Biol Chem* **268**:17762-6.

109. Mansat-de Mas, V., C. Bezombes, A. Quillet-Mary, A. Bettaieb, D. D'Orgeix A, G. Laurent and J. P. Jaffrezou. 1999. Implication of radical oxygen species in ceramide generation, c-Jun N- terminal kinase activation and apoptosis induced by daunorubicin. *Mol Pharmacol* **56**:867-74.

110. Yoshimura, S., Y. Banno, S. Nakashima, K. Hayashi, H. Yamakawa, M. Sawada, N. Sakai and Y. Nozawa. 1999. Inhibition of neutral sphingomyelinase activation and ceramide formation by glutathione in hypoxic PC12 cell death. *J Neurochem* **73**:675-83.

111. Chatterjee, S., H. Han, S. Rollins and T. Cleveland. 1999. Molecular cloning, characterization, and expression of a novel human neutral sphingomyelinase. *J Biol Chem* **274**:37407-12.

112. Patel, M. 1998. Inhibition of neuronal apoptosis by a metalloporphyrin superoxide dismutase mimic. *J Neurochem* **71**:1068-74.

113. Gouaze, V., M. E. Mirault, S. Carpentier, R. Salvayre, T. Levade and N. Andrieu-Abadie. 2001. Glutathione peroxidase-1 overexpression prevents ceramide production and partially inhibits apoptosis in doxorubicin-treated human breast carcinoma cells. *Mol Pharmacol* **60**:488-96.

114. Janicke, R. U., M. L. Sprengart, M. R. Wati and A. G. Porter. 1998. Caspase-3 is required for DNA fragmentation and morphological changes associated with apoptosis. *J Biol Chem* **273**:9357-60.

115. Jaffrezou, J. P., T. Levade, A. Bettaieb, N. Andrieu, C. Bezombes, N. Maestre, S. Vermeersch, A. Rousse and G. Laurent. 1996. Daunorubicin-induced apoptosis: triggering of ceramide generation through sphingomyelin hydrolysis. *Embo. J.* **15:**2417-24.

116. Metkar, S. S., M. Anand, P. P. Manna, K. N. Naresh and J. J. Nadkarni. 2000. Ceramide-induced apoptosis in fas-resistant Hodgkin's disease cell lines is caspase independent. *Exp Cell Res* **255:**18-29.

117. Tepper, A. D., E. de Vries, W. J. van Blitterswijk and J. Borst. 1999. Ordering of ceramide formation, caspase activation, and mitochondrial changes during CD95- and DNA damage-induced apoptosis [published erratum appears in J Clin Invest 1999 May; 103(9):1363]. *J Clin Invest* **103:**971-8.

118. Reed, J. C. 1994. Bcl-2 and the regulation of programmed cell death. *J Cell Biol* **124:**1-6.

119. Kuida, K., T. S. Zheng, S. Na, C. Kuan, D. Yang, H. Karasuyama, P. Rakic and R. A. Flavell. 1996. Decreased apoptosis in the brain and premature lethality in CPP32-deficient mice. *Nature* **384:**368-72.

120. Belaud-Rotureau, M. A., F. Lacombe, F. Durrieu, J. P. Vial, L. Lacoste, P. Bernard and F. Belloc. 1999. Ceramide-induced apoptosis occurs independently of caspases and is decreased by leupeptin. *Cell Death Differ* **6:**788-95.

121. Jaffrezou, J. P., N. Maestre, V. de Mas-Mansat, C. Bezombes, T. Levade and G. Laurent. 1998. Positive feedback control of neutral sphingomyelinase activity by ceramide. *Faseb J* **12:**999-1006.

122. Sawada, M., S. Nakashima, Y. Banno, H. Yamakawa, K. Takenaka, J. Shinoda, Y. Nishimura, N. Sakai and Y. Nozawa. 2000. Influence of Bax or Bcl-2 overexpression on the ceramide-dependent apoptotic pathway in glioma cells. *Oncogene* **19:**3508-20.

123. Allouche, M., A. Bettaieb, C. Vindis, A. Rousse, C. Grignon and G. Laurent. 1997. Influence of Bcl-2 overexpression on the ceramide pathway in daunorubicin-induced apoptosis of leukemic cells. *Oncogene* **14:**1837-45.

124. El-Assaad, W., M. El-Sabban, C. Awaraji, N. Abboushi and G. S. Dbaibo. 1998. Distinct sites of action of Bcl-2 and Bcl-xL in the ceramide pathway of apoptosis. *Biochem J* **336:**735-41.

125. Zhang, J., N. Alter, J. C. Reed, C. Borner, L. M. Obeid and Y. A. Hannun. 1996. Bcl-2 interrupts the ceramide-mediated pathway of cell death. *Proc Natl Acad Sci U S A* **93:**5325-8.

126. Quillet-Mary, A., J. P. Jaffrezou, V. Mansat, C. Bordier, J. Naval and G. Laurent. 1997. Implication of mitochondrial hydrogen peroxide generation in ceramide- induced apoptosis. *J Biol Chem* **272:**21388-95.

127. Ardail, D., I. Popa, K. Alcantara, A. Pons, J. P. Zanetta, P. Louisot, L. Thomas and J. Portoukalian. 2001. Occurrence of ceramides and neutral glycolipids with unusual long-chain base composition in purified rat liver mitochondria. *FEBS Lett* **488:**160-4.

128. Gudz, T. I., K. Y. Tserng and C. L. Hoppel. 1997. Direct inhibition of mitochondrial respiratory chain complex III by cell-permeable ceramide. *J Biol Chem* **272**:24154-8.

129. Ghafourifar, P., S. D. Klein, O. Schucht, U. Schenk, M. Pruschy, S. Rocha and C. Richter. 1999. Ceramide induces cytochrome c release from isolated mitochondria. Importance of mitochondrial redox state. *J Biol Chem* **274**:6080-4.

130. Guidarelli, A., E. Clementi, C. De Nadai, R. Bersacchi and O. Cantoni. 2001. TNFalpha enhances the DNA single-strand breakage induced by the short- chain lipid hydroperoxide analogue tert-butylhydroperoxide via ceramide- dependent inhibition of complex III followed by enforced superoxide and hydrogen peroxide formation. *Exp Cell Res* **270**:56-65.

131. Richter, C. and P. Ghafourifar. 1999. Ceramide induces cytochrome c release from isolated mitochondria. *Biochem Soc Symp* **66**:27-31.

132. Zhou, G. and B. Roizman. 2000. Wild-Type Herpes Simplex Virus 1 Blocks Programmed Cell Death and Release of Cytochrome c but Not the Translocation of Mitochondrial Apoptosis-Inducing Factor to the Nuclei of Human Embryonic Lung Fibroblasts. *J. Virol.* **74**:9048-9053.

133. Shimeno, H., S. Soeda, M. Sakamoto, T. Kouchi, T. Kowakame and T. Kihara. 1998. Partial purification and characterization of sphingosine N- acyltransferase (ceramide synthase) from bovine liver mitochondrion- rich fraction. *Lipids* **33**:601-5.

134. El Bawab, S., P. Roddy, T. Qian, A. Bielawska, J. J. Lemasters and Y. A. Hannun. 2000. Molecular cloning and characterization of a human mitochondrial ceramidase. *J Biol Chem* **275**:21508-13.

135. Barak, A., L. S. Morse and T. Goldkorn. 2001. Ceramide: A potential mediator of apoptosis in human retinal pigment epithelial cells [In Process Citation]. *Invest Ophthalmol Vis Sci* **42**:247-54.

136. Ariga, T., W. D. Jarvis and R. K. Yu. 1998. Role of sphingolipid-mediated cell death in neurodegenerative diseases. *J Lipid Res* **39**:1-16.

137. Denisova, N. A., D. Fisher, M. Provost and J. A. Joseph. 1999. The role of glutathione, membrane sphingomyelin, and its metabolites in oxidative stress-induced calcium "dysregulation" in PC12 cells. *Free Radic Biol Med* **27**:1292-301.

138. Song, J. H., S. H. Shin, and G. M. Ross. 2001. Oxidative stress induced by ascorbate causes neuronal damage in an in vitro system. *Brain Res* **895**:66-72.

139. Levade, T., N. Auge, R. J. Veldman, O. Cuvillier, A. Negre-Salvayre and R. Salvayre. 2001. Sphingolipid mediators in cardiovascular cell biology and pathology. *Circ Res* **89**:957-68.

140. Bhunia, A. K., H. Han, A. Snowden and S. Chatterjee. 1997. Redox-regulated signaling by lactosylceramide in the proliferation of human aortic smooth muscle cells. *J Biol Chem* **272**:15642-9.

141. Harada-Shiba, M., M. Kinoshita, H. Kamido and K. Shimokado. 1998. Oxidized low density lipoprotein induces apoptosis in cultured human umbilical vein endothelial cells by common and unique mechanisms. *J Biol Chem* **273**:9681-7.

142. Bhunia, A. K., T. Arai, G. Bulkley and S. Chatterjee. 1998. Lactosylceramide mediates tumor necrosis factor-alpha-induced intercellular adhesion molecule-1 (ICAM-1) expression and the adhesion of neutrophil in human umbilical vein endothelial cells. *J Biol Chem* **273**:34349-57.

143. Omar, H. A., P. D. Cherry, M. P. Mortelliti, T. Burke-Wolin and M. S. Wolin. 1991. Inhibition of coronary artery superoxide dismutase attenuates endothelium-dependent and -independent nitrovasodilator relaxation. *Circ Res* **69**:601-8.

144. MacKenzie, A. and W. Martin. 1998. Loss of endothelium-derived nitric oxide in rabbit aorta by oxidant stress: restoration by superoxide dismutase mimetics. *Br J Pharmacol* **124**:719-28.

145. Zhang, D. X., A. P. Zou and P. L. Li. 2001. Ceramide reduces endothelium-dependent vasodilation by increasing superoxide production in small bovine coronary arteries. *Circ Res* **88**:824-31.

146. Pautz, A., R. Franzen, S. Dorsch, B. Boddinghaus, V. A. Briner, J. Pfeilschifter and A. Huwiler. 2002. Cross-talk between nitric oxide and superoxide determines ceramide formation and apoptosis in glomerular cells. *Kidney Int* **61**:790-6.

147. Charles, R., L. Sandirasegarane, J. Yun, N. Bourbon, R. Wilson, R. P. Rothstein, S. W. Levison and M. Kester. 2000. Ceramide-coated balloon catheters limit neointimal hyperplasia after stretch injury in carotid arteries. *Circ Res* **87**:282-8.

148. Bourbon, N. A., J. Yun, D. Berkey, Y. Wang and M. Kester. 2001. Inhibitory actions of ceramide upon PKC-epsilon/ERK interactions. *Am J Physiol Cell Physiol* **280**:C1403-11.

149. Shinomiya, K., M. Fukunaga, H. Kiyomoto, K. Mizushige, T. Tsuji, T. Noma, K. Ohmori, M. Kohno and S. Senda. 2002. A role of oxidative stress-generated eicosanoid in the progression of arteriosclerosis in type 2 diabetes mellitus model rats. *Hypertens Res* **25**:91-8.

150. Ceriello, A. 2000. Oxidative stress and glycemic regulation. *Metabolism* **49**:27-9.

151. Penckofer, S., D. Schwertz and K. Florczak. 2002. Oxidative stress and cardiovascular disease in type 2 diabetes: the role of antioxidants and pro-oxidants. *J Cardiovasc Nurs* **16**:68-85.

152. Facchini, F. S., N. W. Hua, G. M. Reaven and R. A. Stoohs. 2000. Hyperinsulinemia: the missing link among oxidative stress and age- related diseases? *Free Radic Biol Med* **29**:1302-6.

153. Midaoui, A. E. and J. de Champlain. 2002. Prevention of Hypertension, Insulin Resistance, and Oxidative Stress by {alpha}-Lipoic Acid. *Hypertension* **39**:303-307.

154. Hajduch, E., A. Balendran, I. H. Batty, G. J. Litherland, A. S. Blair, C. P. Downes and H. S. Hundal. 2001. Ceramide impairs the insulin-dependent membrane recruitment of protein kinase B leading to a loss in downstream signalling in L6 skeletal muscle cells. *Diabetologia* **44:**173-83.

155. Schmitz-Peiffer, C., D. L. Craig and T. J. Biden. 1999. Ceramide generation is sufficient to account for the inhibition of the insulin-stimulated PKB pathway in C2C12 skeletal muscle cells pretreated with palmitate. *J Biol Chem* **274:**24202-10.

156. Tirosh, A., R. Potashnik, N. Bashan and A. Rudich. 1999. Oxidative stress disrupts insulin-induced cellular redistribution of insulin receptor substrate-1 and phosphatidylinositol 3-kinase in 3T3- L1 adipocytes. A putative cellular mechanism for impaired protein kinase B activation and GLUT4 translocation. *J Biol Chem* **274:**10595-602.

157. Rudich, A., A. Tirosh, R. Potashnik, R. Hemi, H. Kanety and N. Bashan. 1998. Prolonged oxidative stress impairs insulin-induced GLUT4 translocation in 3T3-L1 adipocytes. *Diabetes* **47:**1562-9.

158. Turinsky, J., D. M. O'Sullivan and B. P. Bayly. 1990. 1,2-Diacylglycerol and ceramide levels in insulin-resistant tissues of the rat in vivo. *J Biol Chem* **265:**16880-5.

159. Greene, E. L., B. A. Nelson, K. A. Robinson and M. G. Buse. 2001. alpha-Lipoic acid prevents the development of glucose-induced insulin resistance in 3T3-L1 adipocytes and accelerates the decline in immunoreactive insulin during cell incubation. *Metabolism* **50:**1063-9.

160. Rudich, A., A. Tirosh, R. Potashnik, M. Khamaisi and N. Bashan. 1999. Lipoic acid protects against oxidative stress induced impairment in insulin stimulation of protein kinase B and glucose transport in 3T3-L1 adipocytes. *Diabetologia* **42:**949-57.

161. Long, S. D. and P. H. Pekala. 1996. Lipid mediators of insulin resistance: ceramide signalling down- regulates GLUT4 gene transcription in 3T3-L1 adipocytes. *Biochem J* **319:**179-84.

162. Peraldi, P., G. S. Hotamisligil, W. A. Buurman, M. F. White and B. M. Spiegelman. 1996. Tumor necrosis factor (TNF)-alpha inhibits insulin signaling through stimulation of the p55 TNF receptor and activation of sphingomyelinase. *J Biol Chem* **271:**13018- 22.

163. Summers, S. A., L. A. Garza, H. Zhou and M. J. Birnbaum. 1998. Regulation of insulin-stimulated glucose transporter GLUT4 translocation and Akt kinase activity by ceramide. *Mol Cell Biol* **18:**5457-64.

164. Hotamisligil, G. S. 1999. Mechanisms of TNF-alpha-induced insulin resistance. *Exp Clin Endocrinol Diabetes* **107:**119-25.

165. Begum, N. and L. Ragolia. 1996. Effect of tumor necrosis factor-alpha on insulin action in cultured rat skeletal muscle cells. Endocrinology 137:2441-6.

166. Reid, M. B. and Y. P. Li. 2001. Cytokines and oxidative signalling in skeletal muscle. *Acta Physiol Scand* **171:**225-32.

167. Jain, R. G., M. J. Meredith and P. H. Pekala. 1998. Tumor necrosis factor-alpha mediated activation of signal transduction cascades and transcription factors in 3T3-L1 adipocytes. *Adv Enzyme Regul* **38**:333-47.

168. Gottesman, M. M. 1993. How cancer cells evade chemotherapy: sixteenth Richard and Hinda Rosenthal Foundation Award Lecture. *Cancer Res* **53**:747-54.

169. Radin, N. S. 2001. Killing cancer cells by poly-drug elevation of ceramide levels: a hypothesis whose time has come? *Eur J Biochem* **268**:193-204.

170. Ueda, T. and D. G. Plagens. 1987. 3-Phosphoglycerate-dependent protein phosphorylation. *Proc Natl Acad Sci U S A* **84**:1229-33.

171. Deffie, A. M., J. K. Batra and G. J. Goldenberg. 1989. Direct correlation between DNA topoisomerase II activity and cytotoxicity in adriamycin-sensitive and -resistant P388 leukemia cell lines. *Cancer Res* **49**:58-62.

172. Morrow, C. S. and K. H. Cowan. 1990. Glutathione S-transferases and drug resistance. *Cancer Cells* **2**:15-22.

173. Senchenkov, A., D. A. Litvak and M. C. Cabot. 2001. Targeting ceramide metabolism--a strategy for overcoming drug resistance. *J Natl Cancer Inst* **93**:347-57.

174. Liu, Y. Y., T. Y. Han, A. E. Giuliano and M. C. Cabot. 2001. Ceramide glycosylation potentiates cellular multidrug resistance. *Faseb J* **15**:719-30.

175. Lavie, Y., H. Cao, S. L. Bursten, A. E. Giuliano and M. C. Cabot. 1996. Accumulation of glucosylceramides in multidrug-resistant cancer cells. *J Biol Chem* **271**:19530-6.

176. Lucci, A., W. I. Cho, T. Y. Han, A. E. Giuliano, D. L. Morton and M. C. Cabot. 1998. Glucosylceramide: a marker for multiple-drug resistant cancers. *Anticancer Res* **18**:475-80.

177. Ogretmen, B. and Y. A. Hannun. 2001. Updates on functions of ceramide in chemotherapy-induced cell death and in multidrug resistance. *Drug Resist Updat* **4**:368-77.

178. Cabot, M. C., A. E. Giuliano, A. Volner and T. Y. Han. 1996. Tamoxifen retards glycosphingolipid metabolism in human cancer cells. *FEBS Lett* **394**:129-31.

179. Maurer, B. J., L. S. Metelitsa, R. C. Seeger, M. C. Cabot and C. P. Reynolds. 1999. Increase of ceramide and induction of mixed apoptosis/necrosis by N-(4-hydroxyphenyl)- retinamide in neuroblastoma cell lines. J Natl Cancer Inst 91:1138- 46.

180. O'Donnell, P. H., W. X. Guo, C. P. Reynolds and B. J. Maurer. 2002. N-(4-hydroxyphenyl)retinamide increases ceramide and is cytotoxic to acute lymphoblastic leukemia cell lines, but not to non-malignant lymphocytes. Leukemia 16:902-10.

181. Mehta, S., D. Blackinton, I. Omar, N. Kouttab, D. Myrick, J. Klostergaard and H. Wanebo. 2000. Combined cytotoxic action of paclitaxel and ceramide against the human Tu138 head and neck squamous carcinoma cell line. Cancer Chemother Pharmacol 46:85-92.

182. Muller, I., D. Niethammer and G. Bruchelt. 1998. Anthracycline-derived chemotherapeutics in apoptosis and free radical cytotoxicity (Review). Int J Mol Med 1:491-4.

183. Takeda, Y., M. Tashima, A. Takahashi, T. Uchiyama and T. Okazaki. 1999. Ceramide generation in nitric oxide-induced apoptosis. Activation of magnesium-dependent neutral sphingomyelinase via caspase-3. J Biol Chem 274:10654-60.

184. Hofmann, K., S. Tomiuk, G. Wolff and W. Stoffel. 2000. Cloning and characterization of the mammalian brain-specific, Mg2+- dependent neutral sphingomyelinase. Proc Natl Acad Sci U S A 97:5895-900.

185. Sawai, H., N. Domae, N. Nagan and Y. A. Hannun. 1999. Function of the cloned putative neutral sphingomyelinase as lyso- platelet activating factor-phospholipase C. J Biol Chem 274:38131-9.

Chapter 12

MAP KINASES IN REDOX SIGNALING

Atsushi Matsuzawa, Hideki Nishitoh, Kohsuke Takeda and Hidenori Ichijo

12.1 Introduction

Both extra- and intracellular stimuli elicit a wide variety of responses, such as cell survival, proliferation, differentiation and apoptosis, through the intracellular signal transduction system. Recent work has revealed that various signal transduction pathways are regulated by the intracellular redox state.

Mitogen-activated protein (MAP) kinase cascades are the most common and important intracellular signal transduction system, which are typically composed of three kinases that establish a sequential activation pathway, comprising a MAP kinase kinase kinase (MAPKKK), MAP kinase kinase (MAPKK), and MAP kinase (MAPK). It has been demonstrated that the extra- and intracellular stimuli-triggered generation of reactive oxygen species (ROS) and reactive nitrogen species (RNS) leads to the modulation of various components in MAP kinase cascade, which is followed by the activation or inactivation of various downstream signal molecules, resulting in diverse biological responses.

This chapter focuses on the regulation of MAPK signal transduction pathways by the intracellular redox state and oxidative stress, and taken together, provides recent findings on the molecular mechanisms that the cell's fate is determined by cross talk between the MAPK signaling and the intracellular redox state.

12.2 Map Kinase Cascade and Redox Signaling

Mitogen-activated protein (MAP) kinase cascade is evolutionary well conserved in all eukaryotic cells and typically includes central three-tiered core signaling

223

H. J. Forman, J. Fukuto, and M. Torres (eds.), Signal Transduction by Reactive Oxygen and Nitrogen Species: Pathways and Chemical Principles, ©2003, Kluwer Academic Publishers. Printed in the Netherlands

modules comprising a MAP kinase kinase kinase (MAPKKK), MAP kinase kinase (MAPKK), and MAP kinase (MAPK).[1] All eukaryotic cells possess multiple MAP kinase pathways. JNK (c-Jun *N*-terminal kinase), p38 MAP kinase and ERK (extracellular signal-regulated kinase) are well-characterized subgroups of a large MAP kinase family. These kinase pathways are structurally similar, but functionally distinct. While ERK is rapidly activated by a variety of cell growth and differentiation stimuli and plays a central role in mitogenic signaling, JNK and p38 are primarily activated by various environmental stresses: osmotic shock, UV radiation, heat shock, oxidative stress, protein synthesis inhibitors, stimulation of Fas, and proinflammatory cytokines such as tumor necrosis factor α (TNFα) and interleukin-1 (IL-1).[2] Therefore, both JNK and p38 are stress-activated MAPK family members. ERK5/big MAP kinase 1 (BMK1) was recently identified as the novel class of stress-activated MAPK, which is also activated by oxidative stress and osmotic shock[3] (*Figure 1*).

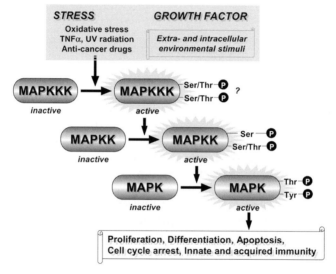

Figure 1. The MAP kinase signaling module. MAP kinase cascade is typically composed of three kinases that establish a sequential activation pathway comprising a MAP kinase kinase kinase (MAPKKK), MAP kinase kinase (MAPKK), and MAP kinase (MAPK). Various extra- and intracellular stimuli activate MAP kinase cascade, which is followed by the activations or inactivations of various downstream signal molecules, resulting in diverse and appropriate biological responses.

Recent studies have demonstrated that the activities of various signal transduction enzymes can be affected by oxidative stress caused by elevated concentration of reactive oxygen species (ROS) and reactive nitrogen species (RNS), such as superoxide and hydroperoxyl radicals, H_2O_2, organic peroxides, hydroxyl radical, ozone, singlet oxygen, hypochlorous acid, nitric oxide, nitrogen dioxide, and, peroxynitrite.[4] Moreover, when cells are activated by physiological stimuli, the cells often generate ROS or RNS, which in turn stimulate other signaling pathways, indicating that ROS and RNS act as second messengers, rather than as directly cytotoxic molecules. MAP kinase cascade is both directly and indirectly targeted by ROS and RNS, and the modification of MAPK signal transduction pathway by oxygen radicals generates a great variety of biological responses. From the next paragraph, we will review regulatory mechanisms of MAPK signal transduction pathways by the intracellular redox state and oxidative stress, and discuss how the redox modulation of MAPK, which is actually implicated in various diseases, controls a variety of cellular events from cell growth to apoptosis.

12.2.1 ERK1/2

MAP kinases are serine/threonine protein kinases, and various subfamilies of MAP kinases have been identified, including the ERK1/2, JNK, p38 MAP kinase, and BMK1/ERK5, which are activated/phosphorylated by specific upstream kinases (as depicted in *Figure 2*). The best studied MAP kinases are ERK1 and ERK2. The ERK1/2 pathway is generally triggered by growth factors and cytokines and its activation is related to the stimulation of tyrosine kinase receptors that elicits a signaling cascade involving Ras activation, recruitment of Raf-1 MAPKKK to the plasma membrane, and sequential activation/phosphorylation of MEK1/2, and ERK1/2.[1,5] ERK1 and ERK2 phosphorylate and activate various transcription factors and other protein kinases, thereby influencing a large variety of cellular processes, such as cell survival, differentiation and cell cycle regulation. ROS and RNS activate ERK1/2 pathway, especially through a Ras-dependent mechanism.[6-8]

Ras is a small G protein that transduces a signal from tyrosine kinase receptors to the ERK1/2 MAP kinase cascade. Ras is activated by oxidative stress, and it has been shown that nitric oxide (NO) binds to a cysteine residue (Cys118) exposed on the surface of the Ras molecule, leading to its activation.[9] NO regulates a broad functional spectrum of proteins by *S*-nitrosylation. Evidence has been accumulating that *S*-nitrosylation directly modulates the biological functions of many intracellular signaling molecules (see below for *S*-nitrosylation). Furthermore, it has been reported that Ras is not only *S*-nitrosylated but also *S*-thiolated by thiol oxidants such as H_2O_2, *S*-nitrosoglutathione, diamide and glutathione disulphide (GSSG) at reactive cysteine residues, and that such modification of Ras may directly affect the bound GTP/GDP ratio and palmitate lipid turnover,[10] resulting in the alteration of various biological activities in the Ras-ERK1/2 pathway.

12.3 ERK5/BMK1

The novel MAPK ERK5, a putative MEK5 target, was cloned as part of a two-hybrid screening that employed MEK5 as bait.[3] All MAPKs are activated by concomitant Tyr and Thr dual phosphorylation within a conserved Thr-X-Tyr motif in the activation loop. ERK5 is an approximately 90-kDa MAPK and has the same specific sequence Thr-Glu-Tyr in its phosphorylation loop as ERK1/2, whereas JNK and p38 have Thr-Pro-Tyr and Thr-Gly-Tyr motifs, respectively.

ERK5/BMK1 can be activated by stress stimuli such as oxidative stress and hyperosmolarity, but not by mitogens, phorbol esters, vasoactive peptides, or inflammatory cytokines such as TNFα. It was reported that ERK5/BMK1 is activated by fluid shear stress.[11] Although the upstream MAPKKK in ERK5/BMK1 pathway has not been yet identified, several studies have shown that a non-receptor tyrosine kinase, c-Src, is redox-sensitive and required for activation of ERK5/BMK1.[12,13] It has been demonstrated that MEKK3 directly regulates MEK5

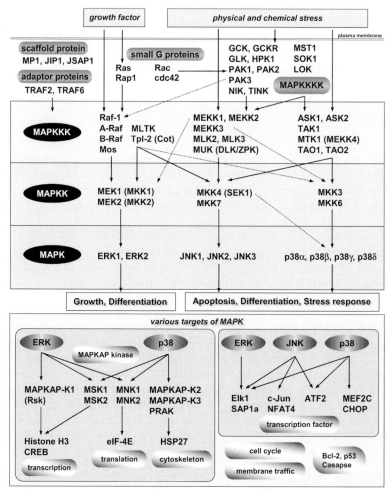

activity as part of the ERK5/BMK1 signaling pathway.[14]

Figure 2. Overview of MAPK cascade. In mammals, three major subgroups of MAP kinase, ERK, JNK, and p38, have been identified, which are structurally similar, but functionally distinct. Whereas ERK is activated by cell growth and differentiation stimuli, JNK and p38 are primarily activated by various types of physical and chemical environmental stress, such as UV radiation, heat shock, and oxidative stress. A novel class of stress-activated MAPK, ERK5/BMK1, is also activated by oxidative stress and osmotic shock. In response to environmental changes, upstream kinases (MAPKKKs and MAPKKKKs) and their various modulators (small G proteins and adaptor proteins) activate MAPK cascade, and then MAP kinases regulate and determine cell fate from survival to apoptosis through phosphorylation of downstream targets, such as transcription or translation factors, cytoskeletal proteins, and regulators of the cell cycle and apoptosis.

12.4 JNK and p38 MAPK

Both JNK and p38, which belong to the stress-activated MAPK family, are regulated by environmental stress and pro-inflammatory cytokines such as IL-1 and TNFα. JNK is activated by various pro-oxidants, such as H_2O_2, arsenite, cadmium chloride and ultraviolet-B radiation.[15-17] Recently, it has been reported that mice lacking expression of the p66 isoform of the ShcA adaptor protein (p66ShcA) are less susceptible to oxidative stress and have an extended life span.[18] Especially, phosphorylation of p66ShcA at serine 36 is critical for the cell death response elicited by light-induced oxidative stress such as UV radiation. JNK was found to be the kinase that phosphorylates p66ShcA at serine 36.[19] The p38 MAPKs also are activated by a wide variety of oxidative stress such as JNK-activating pro-oxidants, as well as other cellular stresses such as osmotic shock, heat shock and lipopolysaccharide.[15,20,21] Singlet oxygen and NO contribute to activate p38 MAPK pathway.[22,23] Activation of JNK and p38 MAPK pathways by inflammatory cytokines, such as IL-1, TNFα and IFNγ, are mediated in part through the generation of oxygen radicals.[24]

Specific inhibitors of JNK and p38 pathways, or expression of JNK and p38 dominant-negative mutants suppressed various stress-induced apoptosis.[2] In the case of the JNK3 knockout mouse or the JNK1/JNK2 double knockout mouse, glutamate-induced hippocampal cell death or UV radiation-induced apoptosis are remarkably prevented, respectively.[25,26] Thus, it has been suggested that JNK and p38 have a critical role in signal transduction of stress-induced apoptotic cell death. Recent studies have demonstrated that ROS, RNS and the resulting oxidative stress play a pivotal role in apoptosis. Antioxidants and thiol reductants, such as *N*-acetylcysteine and dithiothreitol, and overexpression of antioxidative enzymes, such as manganese superoxide dismutase (MnSOD), can block or delay apoptosis. But the mechanisms by which MAPK signaling molecules regulate oxidative stress-induced apoptosis had not yet been defined until quite recently.

Apoptosis signal-regulating kinase 1 (ASK1) is a member of the MAPKKK family, which activates both the JNK and p38 MAP kinase pathways and constitutes a pivotal signaling pathway in oxidative stress-induced apoptosis.[27] In the next section, we will discuss the redox-sensitive ASK1 pathway and the regulatory mechanisms of ASK1-mediated apoptosis.

12.5 The MAPKKK Ask1 in Redox Signaling

12.5.1 *ASK1 is a redox-sensor MAPKKK in oxidative stress-induced apoptosis signaling*

The reduction-oxidation (redox) state of the cell is a consequence of the precise balance between the levels of oxidizing and reducing equivalents, that is, ROS and RNS, such as superoxide anion, H_2O_2 and NO, and endogenous thiol buffers present in the cell, such as glutathione (GSH) and thioredoxin (Trx), which protect cells from oxidative stress-induced cell damage. Under oxidative conditions, increased highly reactive radicals can attack DNA, RNA, proteins, and lipid bilayer, which may compromise various cellular functions and homeostasis. Elevation of ROS and RNS in excess of the buffering capacity results in potentially cytotoxic oxidative stress, which leads to apoptosis as a final event.[28]

Apoptosis signal-regulating kinase 1 (ASK1) is structurally a MAPKKK family member that activates both the SEK1(MKK4)/MKK7-JNK and MKK3/MKK6-p38 MAPK signaling cascades.[27,29-32] ASK1 is activated in cells treated with inflammatory cytokines and various stresses. Overexpression of wild-type ASK1 or the constitutively active mutant strongly induced apoptosis in various cell types. Furthermore, cell death induced by various stresses, such as TNFα and oxidative stress, is remarkably reduced by a dominant-negative mutant form of ASK1, suggesting that ASK1 is a key element in cytokine- and stress-induced apoptosis.

Oxidative stress-induced activation of ASK1 leads to apoptosis. Trx was identified as a negative regulator of the ASK1-JNK/p38 pathway through yeast two-hybrid screening for ASK1-binding proteins.[29] In resting cells, ASK1 constantly forms an inactive complex with Trx, whereas upon treatment of cells with TNFα or H_2O_2 as a ROS donor, ASK1 dissociates from Trx and is fully activated by conformational changes and covalent modifications, such as oligomerization and auto- and/or cross-phosphorylation.[27,29,33] Trx is a redox -regulatory protein that has two redox-sensitive cysteine residues within the active center. Only the reduced form of Trx is associated with the *N*-terminal regulatory domain of ASK1 and silences the activity of ASK1, while the dissociation of ASK1 from oxidized Trx switches an inactive form of ASK1 to an active kinase; the ASK1-Trx complex is thought to be a redox-sensor, which functions as a molecular switch of external and internal redox status to the kinase signaling module (*Figure 3*).

Recently, protein serine/threonine phosphatase 5 (PP5) was identified as another negative regulator of ASK1.[34] PP5 binds to and dephosphorylates the activated form of ASK1 in response to oxidative stress, enabling inactivation of ASK1 by negative feedback. Both negative regulators, Trx and PP5, almost completely eliminate oxidative stress-induced apoptosis in an ASK1-dependent manner, suggesting that ASK1 plays a critical role in signal transduction of oxidative stress-induced apoptosis. In the last section, we will discuss the regulation of phosphatases by oxidative stress.

ROS and RNS are implicated in the regulation of diverse cellular functions including defense against pathogen, signal mediation, proliferation, and apoptosis. Although the precise mechanisms for the diverse responses remain to be fully understood, different cell types and different patterns of MAP kinase activation may result in distinct phenomena, such as survival, differentiation and apoptosis.

12.5.2   *ASK1-Deficient Mice*

To confirm that ASK1 is required for oxidative stress-induced apoptosis, we disrupted the ASK1 gene in mice, which were then analyzed in vivo and in vitro,[35] [36]. Mouse embryonic fibroblasts (MEFs) derived from ASK1-deficient mice were significantly resistant to oxidative stress-induced apoptosis. Moreover, it was observed that the oxidative stress-induced transient activation of JNK and p38 were indistinguishable between ASK1+/+ and ASK-/- MEFs, but that the sustained activation of JNK and p38 were dramatically diminished in ASK1-deficient cells. No significant change in activity was apparent for ERK1/2. Similar results were observed in cells treated not only with oxidative reagents, such as H_2O_2, diamide and tert-butylhydroperoxide, but also with TNFα. We found that TNFα-induced

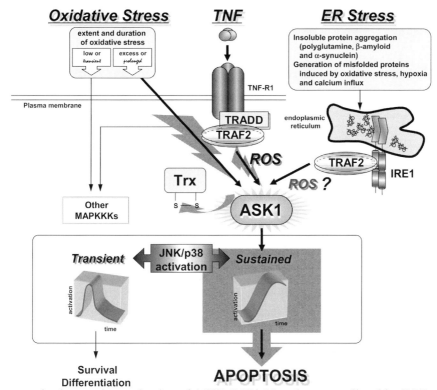

apoptosis also requires activation of ASK1-JNK/p38 pathways mediated by ROS as second messengers.

Figure 3. ASK1 is a redox-sensor MAPKKK and regulates apoptosis signaling pathway. ASK1 is a member of the MAPKKK family, which activates both JNK and p38 MAP kinase pathways and constitutes a pivotal signaling pathway in oxidative stress-induced apoptosis. In response to oxidative stress and TNFα, ASK1 is dissociated from Trx (thioredoxin), a negative regulator of the ASK1-JNK/p38 pathway, and is activated fully. The ASK1-Trx complex is a redox-sensor. TNFα- and oxidative stress-induced sustained activations of JNK and p38 may be responsible for apoptosis. Furthermore, the TRAF2-ASK1 module mediates not only TNFα-induced but also endoplasmic reticulum stress-induced apoptosis, which is implicated in various diseases, such as neurodegeneration and atherosclerosis.

The extent and/or duration of activation of MAP kinases may contribute to determination of cell fate, such as survival, differentiation and apoptosis.[36,37] It has been reported that early/transient and late/sustained activation of JNK and/or p38 induced by oxidative stress and TNFα correlate with various types of cellular processes from survival/differentiation to apoptosis, respectively.[38-40] For example, delayed and persistent JNK activation caused by ultraviolet C or γ-radiation is important to induce apoptosis.[38] Analysis of ASK1-deficient mice suggests that TNFα- and oxidative stress-induced sustained but not transient activation of JNK/p38 may be responsible for apoptosis, and that the ASK1-JNK/p38 pathway mainly mediates apoptosis. Transient activation of JNK/p38 (as well as ERK1/2 activation) may be mediated by other MAPKKKs, such as MEKKs and MLKs, and induce cell growth and differentiation but not apoptosis. As a redox sensor, ASK1 may sense the degree of oxidative stress and drive apoptosis signaling only when cells are damaged lethally by excess and prolonged exposure to oxidative stress. Thus, ASK1 is a determinant of cell fate, such as survival, differentiation or apoptosis, in redox signaling (*Figure 3*).

12.5.3   The TRAF2-ASK1 Pathway Mediates Endoplasmic Reticulum Stress

TRAF2 (TNF receptor associated factor 2) is an adaptor protein that connects TNF receptor with downstream signaling molecules. Overexpression of TRAF2 fosters the production of ROS in transfected cells, and the interaction between TRAF2 and ASK1 is redox-sensitive and can be prevented by free-radical scavengers. Although the mechanism by which oxygen radicals are generated downstream of TRAF2 is unknown, as described above, TNFα-induced apoptosis also is mediated by ROS-triggered activation of ASK1-JNK/p38 pathway. TRAF2 is a strong activator of ASK1 in both TNF receptor signaling and endoplasmic reticulum (ER) stress signaling.[30,41]

ASK1-/- MEFs were resistant to the ER stress-induced apoptosis, accompanied by the drastic suppression of the activation of JNK and p38.[41] Accumulation of unfolded and misfolded proteins in the ER induces cellular stress and triggers the expression of a number of molecular chaperones, such as Bip/GRP78 and GRP94, which assist protein folding and promote cell survival. However, the excessive extent and the long duration of ER stress lead to apoptosis, eventually.[42,43] It has been reported that the mutation or deletion of presenilin-1 involved in Alzheimer's disease influences profoundly the ER stress signaling pathway and facilitates neuronal apoptosis in the patients.[44-46] IRE1, a transmembrane sensor protein in ER stress signaling, is one of the downstream substrates of presenilin-1 and activates the JNK pathway through recruiting of TRAF2.[47] Recently, we found that ER stress, such as polyglutamine and β-amyloid aggregation, induces the formation of IRE1-TRAF2-ASK1 complex, and that the ASK1-JNK pathway regulates the ER stress-induced apoptosis and plays an important role in the progression of neurodegenerative disorders such as polyglutamine disease.[41] Although it is unknown whether or not ER stress-induced activation of the IRE1-TRAF2-ASK1-JNK signaling pathway is mediated by ROS generation, ASK1 is actually implicated in various diseases and the modulation of this pathway controls a variety of cellular events from cell survival to apoptosis (*Figure 3*).

12.6 Regulation of MAPK Cascade by Upstream Kinases and Protein Phosphatases

12.6.1 Regulation of MAPKKKs by Oxidative Stress

Similarly to ASK1, some of MAPKKKs are directly or indirectly activated by oxidative stress. It has been demonstrated that MEKK1 (MEK-kinase 1) is activated by a variety of oxidative stress. Recent genetic studies suggest that MEKK1-/- ES cell-derived cardiac myocytes are more prone to undergo apoptosis in response to H_2O_2, and that MEKK1 plays a protective rather than pro-apoptotic role in oxidative stress signaling.[48] MEKK4 has a binding site for GADD (growth arrest and DNA damage)-45 family proteins and activates both JNK and p38 pathways in response to UV radiation and other genotoxic stress.[49] Activation of TAK1 (TGFβ-activated kinase-1) by UV radiation has also been reported. MAPKKKs (and MAPKKK kinases, such as NIK) are important molecules to sense the first signals upstream of oxidative stress signaling pathway.

12.6.2 Regulation of Protein Phosphatases by Oxidative Stress

Recently, it has been indicated that serine/threonine and tyrosine phosphatases can be regulated by altering the oxidation state of active-site functional groups, such as an active-site Fe ion in serine/threonine phosphatases or cysteine residues in tyrosine phosphatases. Redox regulation of not only MAP kinases but also protein phosphatases might provide at least one mechanism by which the cell senses a cellular redox state altered by generation of ROS and RNS. The expression of a dual protein phosphatase CL100/MKP-1, which is capable of dephosphorylating MAP kinase, is potently induced by oxidative stress.[50] Wip1, a serine/threonine protein phosphatase, is expressed by UV radiation and H_2O_2 through the activation of p53, leading to the specific inactivation of p38 MAPK.[51] Thus, protein phosphatases are key mediators for regulation of signal transduction by oxidative stress and cytoplasmic redox.[4]

12.7 Modulation of MAPK Components by *S*-Nitrosylation

NO is a diffusible and short-lived free radical that can be endogenously produced by various types of cells by the family of NO synthases. As described above, NO exerts many of its functions through *S*-nitrosylation of proteins, and signal transduction molecules often are modulated by *S*-nitrosylation, including Ras, JNK, protein kinase C, caspase-3, NF-κB, ryanodine receptor and *N*-methyl-D-aspartate (NMDA) receptor-coupled channel.[52,53] In MAPK cascade, Ras is activated by *S*-nitrosylation of a cysteine residue (Cys118).

JNK is also modified by NO, and *S*-nitrosylation might occur at Cys116, which has been shown to be a cysteine residue that is sensitive to thiol-modifying agents.[54] Whereas JNK signal pathway is positively regulated by NO-induced activation of upstream kinases such as ASK1 and MEKK1, *S*-nitrosylation of JNK by endogenously produced NO can directly suppress its activity. Thus, NO may positively or negatively regulate the JNK pathway as a function of the extent of NO production, the cell type or the presence of other stimuli. In the situation where conflicting signals are integrated into cells, the diverse biological responses, such as survival, differentiation and apoptosis, must be controlled by balancing of different modification of signaling molecules, such as modification of JNK pathway by NO.

Although it has been reported that there are various modifications of signal proteins by RNS such as tyrosine nitration, it remains uncertain whether such modification plays a role in cell signaling. Peroxynitrite, a highly reactive nitrogen radical, modifies tyrosines in proteins, such as α-synuclein, which are particularly prone to aggregate in neurons of Parkinson's disease patients.[55]

12.8 Conclusions

Redox regulation of cellular signaling has multiple functions in cell physiology. The dosage of the redox signal is an important factor in the modulation of cell function. The mode of cell reaction is shifted from proliferation to apoptosis, corresponding to increases in the level of oxidative stress. Moreover, different cell types respond to the same stress in different ways, and different pro-oxidants in the same cell elicit different signaling pathways. Cells determine their fate, such as survival, differentiation and apoptosis, by cross talk between the redox state and the intracellular signaling systems, especially the MAPK cascades. Further studies are necessary to understand the mechanism by which the redox state regulates cell signaling at the molecular level.

References

1. Kyriakis, J. M. and J. Avruch J. 2001. Mammalian mitogen-activated protein kinase signal transduction pathways activated by stress and inflammation. *Physiol Rev* **81**:807-869.

2. Tibbles, L. A. and J. R. Woodgett. 1999. The stress-activated protein kinase pathways. *Cell Mol Life Sci* **55**:1230-1254.

3. Zhou, G., Z. Q. Bao and J. E. Dixon. 1995. Components of a new human protein kinase signal transduction pathway. *J Biol Chem* **270**:12665-12669.

4. Rusnak, F. and T. Reiter. 2000. Sensing electrons: protein phosphatase redox regulation. *Trends Biochem Sci* **25**:527-529.

5. Cobb, M. H. 1999. MAP kinase pathways. *Prog Biophys Mol Biol* **71**:479-500.

6. Rao, G. N. 1996. Hydrogen peroxide induces complex formation of SHC-Grb2-SOS with receptor tyrosine kinase and activates Ras and extracellular signal-regulated protein kinases group of mitogen-activated protein kinases. *Oncogene* **13**:713-719.

7. Aikawa, R., I. Komuro, T. Yamazaki, Y. Zou, S. Kudoh, M. Tanaka, I. Shiojima, Y. Hiroi and Y. Yazaki. 1997. Oxidative stress activates extracellular signal-regulated kinases through Src and Ras in cultured cardiac myocytes of neonatal rats. *J Clin Invest* **100**:1813-1821.

8. Guyton, K. Z., Y. Liu, M. Gorospe, Q. Xu and N. J. Holbrook. 1996. Activation of mitogen-activated protein kinase by H_2O_2. Role in cell survival following oxidant injury. *J Biol Chem* **271**:4138-4142.

9. Lander, H. M., A. J. Milbank, J. M. Tauras, D. P. Hajjar, B. L. Hempstead, G. D. Schwartz, R. T. Kraemer, U. A. Mirza, B. T. Chait, S. C. Burk and L. A. Quilliam. 1996. Redox regulation of cell signalling. *Nature* **381**:380-381.

10. Mallis, R. J., J. E. Buss and J. A. Thomas. 2001. Oxidative modification of H-ras: *S*-thiolation and *S*-nitrosylation of reactive cysteines. *Biochem J* **355**:145-153.

11. Yan, C., M. Takahashi, M. Okuda, J. D. Lee and B. C. Berk. 1999. Fluid shear stress stimulates big mitogen-activated protein kinase 1 (BMK1) activity in endothelial cells. Dependence on tyrosine kinases and intracellular calcium. *J Biol Chem* **274**:143-150.

12. Abe, J., M. Takahashi, M. Ishida, J. D. Lee and B. C. Berk. 1997. c-Src is required for oxidative stress-mediated activation of big mitogen-activated protein kinase 1. *J Biol Chem* **272**:20389-20394.

13. Suzaki, Y., M. Yoshizumi, S. Kagami, A. H. Koyama, Y. Taketani, H. Houchi, K. Tsuchiya, E. Takeda and T. Tamaki. 2002. Hydrogen peroxide stimulates c-Src-mediated big mitogen-activated protein kinase 1 (BMK1) and the MEF2C signaling pathway in PC12 cells: potential role in cell survival following oxidative insults. *J Biol Chem* **277**:9614-9621.

14. Chao, T. H., M. Hayashi, R. I. Tapping, Y. Kato and J. D. Lee. 1999. MEKK3 directly regulates MEK5 activity as part of the big mitogen-activated protein kinase 1 (BMK1) signaling pathway. *J Biol Chem* **274**:36035-36038.

15. Guyton, K. Z., Y. Liu, M. Gorospe, Q. Xu and N. J. Holbrook. 1996. Activation of mitogen-activated protein kinase by H_2O_2. Role in cell survival following oxidant injury. *J Biol Chem* **271**:4138-4142.

16. Verheij, M., R. Bose, X. H. Lin, B. Yao, W. D. Jarvis, S. Grant, M. J. Birrer, E. Szabo, L. I. Zon, J. M. Kyriakis, A. Haimovitz-Friedman, Z. Fuks and R. N. Kolesnick. 1996. Requirement for ceramide-initiated SAPK/JNK signalling in stress-induced apoptosis. *Nature* **380**:75-79.

17. Iordanov, M. S. and B. E. Magun. 1999. Different mechanisms of c-Jun NH_2-terminal kinase-1 (JNK1) activation by ultraviolet-B radiation and by oxidative stressors. *J Biol Chem* **274**:25801-25806.

18. Migliaccio, E., M. Giorgio, S. Mele, G. Pelicci, P. Reboldi, P. P. Pandolfi, L. Lanfrancone and P. G. Pelicci. 1999. The p66shc adaptor protein controls oxidative stress response and life span in mammals. *Nature* **402**:309-313.

19. Le, S., T. J. Connors and A. C. Maroney. 2001. c-Jun *N*-terminal kinase specifically phosphorylates p66ShcA at serine 36 in response to ultraviolet irradiation. *J Biol Chem* **276**:48332-48336.

20. Han, S. J., K. Y. Choi, P. T. Brey and W. J. Lee. 1998. Molecular cloning and characterization of a Drosophila p38 mitogen-activated protein kinase. *J Biol Chem* **273**:369-374.

21. Galan, A., M. L. Garcia-Bermejo, A. Troyano, N. E. Vilaboa, E. de Blas, M. G. Kazanietz and P. Aller. 2000. Stimulation of p38 mitogen-activated protein kinase is an early

regulatory event for the cadmium-induced apoptosis in human promonocytic cells. *J Biol Chem* **275**:11418-11424.

22. Zhuang, S., J. T. Demirs and I. E. Kochevar. 2000. p38 mitogen-activated protein kinase mediates bid cleavage, mitochondrial dysfunction, and caspase-3 activation during apoptosis induced by singlet oxygen but not by hydrogen peroxide *J Biol Chem* **275**:25939-25948.

23. Cheng, A., S. L. Chan, O. Milhavet, S. Wang and M. P. Mattson. 2001. p38 MAP kinase mediates nitric oxide-induced apoptosis of neural progenitor cells. *J Biol Chem* **276**:43320-43327.

24. Lo, Y. Y., J. M. Wong and T. F. Cruz. 1996. Reactive oxygen species mediate cytokine activation of c-Jun NH_2-terminal kinases. *J Biol Chem* **271**:15703-15707.

25. Yang, D. D., C. Y. Kuan, A. J. Whitmarsh, M. Rincon, T. S. Zheng, R. J. Davis, P. Rakic and R. A. Flavell. 1997. Absence of excitotoxicity-induced apoptosis in the hippocampus of mice lacking the Jnk3 gene. *Nature* **389**:865-870.

26. Tournier, C., P. Hess, D. D. Yang, J. Xu, T. K. Turner, A. Nimnual, D. Bar-Sagi, S. N. Jones, R. A. Flavell and R. J. Davis. 2000. Requirement of JNK for stress-induced activation of the cytochrome *c*-mediated death pathway. *Science* **288**:870-874.

27. Ichijo, H., E. Nishida, K. Irie, P. ten Dijke, M. Saitoh, T. Moriguchi, M. Takagi, K. Matsumoto, K. Miyazono and Y. Gotoh. 1997. Induction of apoptosis by ASK1, a mammalian MAPKKK that activates SAPK/JNK and p38 signaling pathways. *Science* **275**:90-94

28. Davis, W. Jr, Z. Ronai and K. D. Tew. 2001. Cellular thiols and reactive oxygen species in drug-induced apoptosis. *J Pharmacol Exp Ther* **296**:1-6.

29. Saitoh, M., H. Nishitoh, M. Fujii, K. Takeda, K. Tobiume, Y. Sawada, M. Kawabata, K. Miyazono and H. Ichijo. 1998. Mammalian thioredoxin is a direct inhibitor of apoptosis signal-regulating kinase (ASK) 1. *EMBO J* **17**: 2596-2606.

30. Nishitoh, H., M. Saitoh, Y. Mochida, K. Takeda, H. Nakano, M. Rothe, K. Miyazono and H. Ichijo. 1998. ASK1 is essential for JNK/SAPK activation by TRAF2. *Mol Cell* **2**:389-395.

31. Chang, H. Y., H. Nishitoh, X. Yang, H. Ichijo and D. Baltimore. 1998. Activation of apoptosis signal-regulating kinase 1 (ASK1) by the adapter protein Daxx. *Science* **281**:1860-1863.

32. Ichijo, H. 1999. From receptors to stress-activated MAP kinases. *Oncogene* **18**:6087-6093.

33. Gotoh, Y. and J. A. Cooper. 1998. Reactive oxygen species- and dimerization-induced activation of apoptosis signal-regulating kinase 1 in tumor necrosis factor-α signal transduction. *J Biol Chem* **273**:17477-17482.

34. Morita, K., M. Saitoh, K. Tobiume, H. Matsuura, S. Enomoto, H. Nishitoh and H. Ichijo. 2001. Negative feedback regulation of ASK1 by protein phosphatase 5 (PP5) in response to oxidative stress. *EMBO J* **20**:6028-6036.

35. Tobiume, K., A. Matsuzawa, T. Takahashi, H. Nishitoh, K. Morita, K. Takeda, O. Minowa, K. Miyazono, T. Noda and H. Ichijo. 2001. ASK1 is required for sustained activations of JNK/p38 MAP kinases and apoptosis. *EMBO Rep* **2**:222-228.

36. Matsuzawa, A., H. Nishitoh, K. Tobiume, K. Takeda and H. Ichijo. 2002. Physiological roles of ASK1-mediated signal transduction in oxidative stress- and endoplasmic reticulum stress-induced apoptosis: advanced findings from ASK1 knockout mice. *Antioxid Redox Signal* **4**:415-425.

37. Matsuzawa, A. and H. Ichijo. 2001. Molecular mechanisms of the decision between life and death: regulation of apoptosis by apoptosis signal-regulating kinase 1. *J Biochem* **130**:1-8.

38. Chen, Y. R., X. Wang, D. Templeton, R. J. Davis and T. H. Tan. 1996. The role of c-Jun *N*-terminal kinase (JNK) in apoptosis induced by ultraviolet C and γ radiation. Duration of JNK activation may determine cell death and proliferation. *J Biol Chem* **271**:31929-31936.

39. Guo, Y. L., K. Baysal, B. Kang, L. J. Yang and J. R. Williamson. 1998. Correlation between sustained c-Jun *N*-terminal protein kinase activation and apoptosis induced by tumor necrosis factor-α in rat mesangial cells. *J Biol Chem* **273**:4027-4034.

40. Roulston, A., C. Reinhard, P. Amiri and L. T. Williams. 1998. Early activation of c-Jun *N*-terminal kinase and p38 kinase regulate cell survival in response to tumor necrosis factor α. *J Biol Chem* **273**:10232-10239.

41. Nishitoh, H., A. Matsuzawa, K. Tobiume, K. Saegusa, K. Takeda, K. Inoue, S. Hori, A. Kakizuka and H. Ichijo. 2002. ASK1 is essential for endoplasmic reticulum stress-induced neuronal cell death triggered by expanded polyglutamine repeats. *Genes Dev* **16**:1345-1355.

42. Mori, K. 2000. Tripartite management of unfolded proteins in the endoplasmic reticulum. *Cell* **101**:451-454.

43. Travers, K. J., C. K. Patil, L. Wodicka, D. J. Lockhart, J. S. Weissman and P. Walter. 2000. Functional and genomic analyses reveal an essential coordination between the unfolded protein response and ER-associated degradation. *Cell* **101**:249-258.

44. Sherman, M. Y. and A. L. Goldberg. 2001. Cellular defenses against unfolded proteins: a cell biologist thinks about neurodegenerative diseases. *Neuron* **29**:15-32.

45. Niwa, M., C. Sidrauski, R. J. Kaufman and P. Walter. 1999. A role for presenilin-1 in nuclear accumulation of Ire1 fragments and induction of the mammalian unfolded protein response. *Cell* **99**:691-702.

46. Katayama, T., K. Imaizumi, N. Sato, K. Miyoshi, T. Kudo, J. Hitomi, T. Morihara, T. Yoneda, F. Gomi, Y. Mori, Y. Nakano, J. Takeda, T. Tsuda, Y. Itoyama, O. Murayama, A. Takashima, P. St George-Hyslop, M. Takeda and M. Tohyama. 1999. Presenilin-1 mutations downregulate the signalling pathway of the unfolded-protein response. *Nat Cell Biol* **1**:479-485.

47. Urano, F., X. Wang, A. Bertolotti, Y. Zhang, P. Chung, H. P. Harding, and D. Ron. 2000. Coupling of stress in the ER to activation of JNK protein kinases by transmembrane protein kinase IRE1. *Science* **287**:664-666.

48. Minamino, T., T. Yujiri, P. J. Papst, E. D. Chan, G. L. Johnson, and N. Terada. 1999. MEKK1 suppresses oxidative stress-induced apoptosis of embryonic stem cell-derived cardiac myocytes. *Proc Natl Acad Sci U S A* **96**:15127-15132.

49. Takekawa, M. and H. Saito. 1998. A family of stress-inducible GADD45-like proteins mediate activation of the stress-responsive MTK1/MEKK4 MAPKKK. *Cell* **95**:521-530.

50. Keyse, S. M. and E. A. Emslie. 1992. Oxidative stress and heat shock induce a human gene encoding a protein-tyrosine phosphatase. *Nature* **359**:644-647.

51. Takekawa, M., M. Adachi, A. Nakahata, I. Nakayama, F. Itoh, H. Tsukuda, Y. Taya and K. Imai. 2000. p53-inducible wip1 phosphatase mediates a negative feedback regulation of p38 MAPK-p53 signaling in response to UV radiation. *EMBO J* **19**:6517-6526.

52. Stamler, J. S. Redox signaling: nitrosylation and related target interactions of nitric oxide. 1994. *Cell* **78**:931-936.

53. Hess, D. T., A. Matsumoto, R. Nudelman and J. S. Stamler. *S*-nitrosylation: spectrum and specificity. 2001. *Nat Cell Biol* **3**:E46-E49.

54. Park, H. S., S. H. Huh, M. S. Kim, S. H. Lee and E. J. Choi. 2000. Nitric oxide negatively regulates c-Jun *N*-terminal kinase/stress-activated protein kinase by means of *S*-nitrosylation. 2000. *Proc Natl Acad Sci U S A* **97**:14382-14387.

55. Spillantini, M. G., R. A. Crowther, R. Jakes, M. Hasegawa and M. Goedert. 1998. α-Synuclein in filamentous inclusions of Lewy bodies from Parkinson's disease and dementia with Lewy bodies. *Proc Natl Acad Sci U S A* **95**:6469-6473.

Chapter 13

ACTIVATION OF PROSTAGLANDIN BIOSYNTHESIS: PEROXYNITRITE VS HYDROPEROXIDES

Lisa Landino

13.1 Introduction

13.1.1 Overview of PGHS catalysis

Prostaglandins are an important family of lipid-derived molecules that play critical roles in physiological processes including platelet aggregation, vasoconstriction and inflammation. The committed step in prostaglandin biosynthesis is the oxygenation of the polyunsaturated fatty acid, arachidonic acid, by the enzyme prostaglandin endoperoxide synthase (PGHS), a membrane-bound hemeprotein. PGHS is a bifunctional enzyme with both peroxidase and cyclooxygenase activities.[45,46] The reactions catalyzed occur at two structurally distinct sites on the PGHS protein called the cyclooxygenase and peroxidase active sites. The requisite heme prosthetic group is positioned between the two separate active sites.[47,48,49] The cyclooxygenase activity of PGHS catalyzes the oxygenation of arachidonic acid to form the initial product, the hydroperoxy endoperoxide, PGG_2 (Scheme I). The peroxidase activity of PGHS reduces PGG_2 to the hydroxy endoperoxide, PGH_2.

PGH_2 serves as the precursor for all prostaglandins and thromboxanes. The specific prostaglandin product formed in a certain cell type depends on which PGH_2-metabolizing enzyme is expressed therein. For example, platelets convert PGH_2 almost exclusively to thromboxane A_2, a potent activator of platelet

237

H. J. Forman, J. Fukuto, and M. Torres (eds.), Signal Transduction by Reactive Oxygen and Nitrogen Species: Pathways and Chemical Principles, ©2003, Kluwer Academic Publishers. Printed in the Netherlands

aggregation.[50,51,52] Inhibition of the cyclooxygenase activity of PGHS is the basis for the pharmacological action of non-steriodal antiinflammatory drugs including aspirin, ibuprofen and indomethacin. Given its importance as a drug target, the cyclooxygenase activity of PGHS has received considerably more attention than the enzyme's peroxidase activity; thus, the PGHS is often referred to simply as "cyclooxygenase" or COX.[45,46]

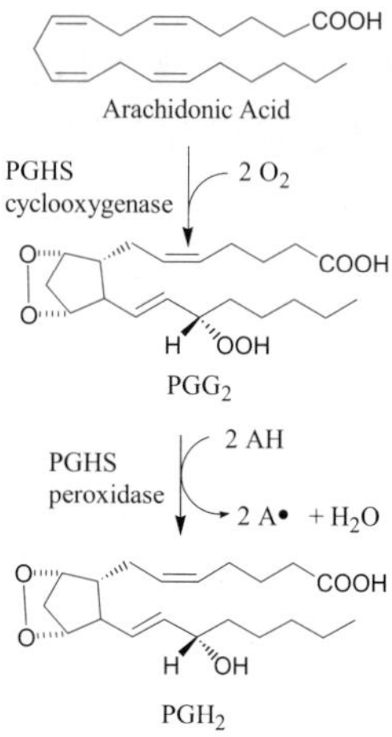

Scheme I. Synthesis of PGH_2 from arachidonic acid by PGHS

13.1.2   Peroxidase Activation of the Cyclooxygenase Reaction

The peroxidase activity of PGHS is essential for activation of the cyclooxygenase reaction. The peroxidase activity of PGHS is typical of heme-containing peroxidases and spectroscopically distinct Compound I and Compound II intermediates have been detected (Scheme II).[53] Hydroperoxides react with the resting Fe^{3+} form of the heme prosthetic group of PGHS to generate the corresponding alcohol and the higher oxidation state of the heme called Compound I that is consistent with the $Fe^{4+}=O$ porphyrin cation radical. One electron reduction of Compound I by an electron donor, respresented by AH in Schemes I and II, leads to formation of Compound II, a $Fe^{4+}=O$ species with the porphyrin fully covalent. One electron reduction of Compound II regenerates the resting Fe^{3+} state of the enzyme. As depicted, reduction of the oxidized enzyme requires the input of two electrons in a sequential fashion (i.e. two one-electron transfers). For PGHS, two-electron reduction of a hydroperoxide to form Compound I, followed by intramolecular electron transfer, is required to generate a tyrosyl radical, $Y\bullet$, that serves as the cyclooxygenase oxidant (Scheme II). The tyrosyl radical abstracts a hydrogen atom from arachidonic acid (AA) to initiate PGG_2 synthesis. Thus, an oxidized intermediate of peroxidase catalysis is required to activate cyclooxygenase catalysis.[46]

Site-directed mutants of PGHS that have decreased peroxidase activity, and thus react much more slowly with hydroperoxides, typically have pronounced cyclooxygenase lag phases that can be overcome by adding high concentrations of hydroperoxides.[54,55] Notably, once the lag phase is overcome, the rate and extent of prostaglandin formation is nearly identical to the wild-type enzyme; thus, the peroxidase activity of PGHS is only required for activation of cyclooxygenase, not for continued product formation.[56] In fact, only nanomolar concentrations of fatty acid hydroperoxide are required for full activation. PGG_2, the initial fatty acid hydroperoxide product of the PGHS reaction, can serve as an excellent hydroperoxide activator of additional PGHS proteins to propagate PGH_2 synthesis (Scheme II).

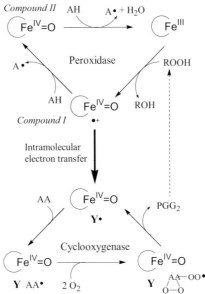

Scheme II. Mechanism of PGHS peroxidase and cyclooxygenase catalysis

13.1.3   Glutathione Peroxidase Regulation of PGHS Activity

The requirement for peroxidase activation of the cyclooxygenase activity of PGHS is also supported by the observation that glutathione peroxidase (GPx) in conjunction with reduced glutathione (GSH) can completely inhibit the cyclooxygenase activity by reducing hydroperoxide activators to the corresponding alcohols.[56,57,58] Potential hydroperoxide activators of PGHS including hydrogen peroxide and fatty acid hydroperoxides such as PGG_2 and 15-HPETE, 15-hydroperoxyeicosatetraenoic acid, are excellent substrates for GPx/GSH. The reaction catalyzed by GPx is shown:

$$ROOH + 2GSH \rightarrow ROH + GSSG + H_2O \qquad (1)$$

Recently, it has been confirmed that two isoforms of PGHS exist, each of which catalyzes the cyclooxygenase and peroxidase reactions.[59,60] PGHS-1 is constitutively expressed in a variety of cells such as gastrointestinal epithelial cells and platelets. In these cells, PGH_2 is processed to prostaglandins essential for gastric cytoprotection and to thromboxane A_2, a powerful activator of platelet aggregation.[50,51,52]   Conversely, PGHS-2 is an inducible enzyme expressed transiently in a variety of cells including fibroblasts, macrophages, endothelial cells, vascular smooth muscle cells, neurons, and astrocytes in response to pathophysiological challenge.[61] For example, PGHS-2 plays an essential role in the propagation of the inflammatory response. Isoform-selective drugs that inhibit PGHS-2, but not PGHS-1, hold great promise because they are both anti-inflammatory and have fewer side effects than traditional NSAIDs.[45]

13.2 Nitric Oxide and PGHS

13.2.1   Overview of Nitric Oxide Synthesis and Function in vivo

Like prostaglandins, nitric oxide (NO) has been implicated in a wide variety of physiological processes including vasodilation, neurotransmission, and inflammation.[62,63,64]   NO has received considerable attention since it was first identified as an enzymatically produced signalling molecule in the late 1980's. A family of NO synthase (NOS) enzymes derived from distinct genes have been identified that produce NO from arginine in numerous cell types including endothelial cells (eNOS), neurons (nNOS) and macrophages (inducible NOS or iNOS). iNOS, like PGHS-2, is regulated at the transcriptional level whereas eNOS and nNOS are constitutively expressed and their enzymatic activity is regulated by the availability of protein cofactors and by Ca^{2+} ions.[65]   Like PGHS-2, iNOS is associated primarily with inflammation and other immunological processes.[63,64]

Stimulation of a wide variety of cell types with cytokines and other messengers induces rapid expression of the iNOS and PGHS-2 genes.[66,67,68,69,70]   Several investigators have also verified that in cells expressing both inducible PGHS and NOS, NO activates prostaglandin biosynthesis.[71,72] For example, Salvemini and co-workers have shown that addition of the NOS inhibitors N^G-monomethyl-L-arginine (L-NMMA) or aminoguanidine to lipopolysaccharide (LPS) -stimulated RAW 264.7 cells, a mouse macrophage cell line, results not only in a decrease in nitrite formation, the stable end product of NO, but also a four- to fivefold decrease in prostaglandin E_2 (PGE_2) biosynthesis.[67] Conversely, the addition of the PGHS inhibitor indomethacin only inhibits PGE_2 synthesis, and not nitrite formation.

These workers concluded that the connection between prostaglandin biosynthesis and NO results from a direct interaction between NO and the PGHS-2 protein rather than an interaction between NO and an intermediate target such as soluble guanylate cyclase, a well-known protein receptor for NO.[67] Also in support of a direct interaction between NO and PGHS-2 protein, Eling and coworkers showed that NO from several different sources does not increase either PGHS-2 mRNA or protein levels in RAW264.7 cells, eliminating transcriptional control as a mechanism for NO-stimulated prostaglandin biosynthesis in these cells.[73]

13.2.2   Interactions between PGHS Enzymes and NO in vitro

Numerous studies in cells and in vivo models support the hypothesis that NO stimulates prostaglandin biosynthesis. Thus, biochemists have sought to examine the chemical nature of the interaction between NO and PGHS. Several in vitro studies have been undertaken to address this issue; however, the results of these investigations have varied widely. Some investigators note that NO stimulates prostaglandin synthesis by PGHS enzymes.[74,75] Others observe either no stimulation or slight inhibition of the enzyme by NO.[73,76,77] These conflicting observations have spawned considerable discussion of the possible mechanisms by which NO could interact with PGHS and enhance prostaglandin biosynthesis by PGHS in vivo. The possible interactions between NO and PGHS that have been examined are: 1) NO as an electron donor during peroxidase turnover; 2) NO binding to the heme prosthetic group of PGHS; 3) nitrosylation of free cysteines of PGHS; and 4) reactions of NO-derivatives, rather than NO itself, with PGHS.

NO served as a peroxidase electron donor (see peroxidase catalytic cycle in Scheme II) in a study by Eling and coworkers and thus could support peroxidase catalysis but NO did not affect cyclooxygenase catalysis.[73] Kulmacz and others established that NO does bind to the resting ferric state of the PGHS heme.[76,78,79,80] However, the K_d value for the NO-ferric heme interaction was nearly 1 mM - a low affinity interaction that would not occur in vivo where only picomolar to low micromolar NO concentrations are reached.

With respect to S-nitrosylation of PGHS cysteines, Hajjar et al. observed nitrosylation of the three free cysteines of PGHS-1 when the enzyme was exposed to NO.[74] S-nitrosylation of PGHS-1 did correlate with an increase in catalytic activity and no dramatic shifts in the heme absorption spectrum were detected following NO treatment. However, mutagenesis of the three cysteines of PGHS-1 to serine as well as modification of the cysteines by maleimides resulted in substantial decreases in PGHS activity.[81] Given these conflicting in vitro observations, it is not clear at this time if S-nitrosylation of PGHS could serve to activate prostaglandin biosynthesis in vivo.

13.3 Peroxynitrite and PGHS Activation

Our in vitro and cell work examining the link between NO and PGHS has focused on peroxynitrite (ONOO⁻), the product of the reaction of NO with superoxide (O_2^-). We believe that ONOO⁻ rather than NO itself, serves as a potential physiologic activator of prostaglandin biosynthesis in vivo.[82] Much of the cell and in vivo work demonstrating a link between NO and prostaglandins has been undertaken using inflammation models where neutrophils or macrophages are involved. In addition to producing NO, these cells produce copious amounts of O_2^- using an NADPH oxidase[83,84] or NOS itself.[85] NO reacts rapidly ($k = 1.9 \times 10^{10}$ M⁻¹ s⁻¹) with O_2^- to form ONOO⁻, an inorganic peroxide.[86] Like other peroxides, ONOO⁻ reacts rapidly with Fe^{3+} (5×10^7 M⁻¹ s⁻¹) and Mn^{3+} (1.8×10^6 M⁻¹ s⁻¹) porphyrins to form the corresponding $M^{4+}=O$ complexes.[87,88,89] Likewise, Fe^{3+}-containing myelo-, lacto-, and horseradish peroxidases react with ONOO⁻ at comparable rate constants to form ferryl-oxo complexes, most probably compound I (see Scheme II).[90] Once formed, these reactive porphyrins (free and protein bound) rapidly oxidize reductants such as guaiacol, glutathione, trolox, and ascorbate to regenerate the Fe^{3+} porphyrin at rates at least three orders of magnitude higher than the direct reaction between ONOO⁻ and these electron donors.[87,89,91,92] This constitutes typical peroxidase activity where ONOO⁻ acts as the peroxide source of oxidizing equivalents. As discussed, the peroxidase activity of PGHS is requisite for activation of the PGHS cyclooxygenase reaction (Scheme II). Therefore, it is reasonable to suggest that ONOO⁻ may activate cyclooxygenase catalysis by serving as a typical peroxide substrate.

13.3.1   ONOO⁻ is a PGHS Peroxidase Substrate

ONOO⁻ supports PGHS-catalyzed guaiacol oxidation, and therefore is a substrate for PGHS peroxidase activity.[82] The PGHS peroxidase can use a wide variety of peroxides to support the oxidation of small organic reducing substrates[93,94] and the kinetic parameters for ONOO⁻ reaction with PGHS are consistent with those observed for other hydroperoxides (Table 1). Although the K_m of PGHS for peroxynitrite is considerably larger than that of the fatty acid hydroperoxide, 15-HPETE, it is lower than that of H_2O_2. Moreover, ONOO⁻ supports a greater extent of guaiacol oxidation than either 15-HPETE or H_2O_2. Given the bolus method of addition and the rapid decomposition of ONOO⁻ under the assay conditions used, these kinetic values should be taken as estimates.

Table 1. Substrate specificity of PGHS peroxidases. The units of kcat* are mol guaiacol oxidized per mol PGHS. From Landino et al.[82]

Substrate	PGHS-1			PGHS-2		
	K_m (μM)	k_{cat}*ᵇ	k_{cat}/K_m	K_m (μM)	k_{cat}*	k_{cat}/K_m
Peroxynitrite	140	759	5.4	100	287	2.9
15-HPETE	24	237	9.7	9	59	6
H_2O_2	287	692	2.4	109	167	1.5

13.3.2   Peroxynitrite Activation of GPx/GSH Inhibited PGHS

Peroxides are required to activate PGHS, and commercial preparations of arachidonic acid contain sufficient quantities of fatty acid hydroperoxide impurities to accomplish full activation of cyclooxygenase. To block activation by these peroxides, PGHS must be placed in a reaction with high concentrations of glutathione peroxidase/glutathione (GPx/GSH).[56,95,96] Any trace hydroperoxides present in the arachidonic acid are consumed readily by GPx/GSH and the cyclooxygenase remains inactive. Under these conditions, the ability of various exogenously added hydroperoxides to accomplish activation can be evaluated.

In our experiments, ONOO⁻, but not H_2O_2 or 15-HPETE was able to initiate cyclooxygenase turnover even in the presence of GPx/GSH.[82] Because only ONOO⁻ was capable of activating the cyclooxygenase activity of PGHS, these data suggests that ONOO⁻ may be a more effective cyclooxygenase activator in vivo than fatty acid hydroperoxides (e.g., PGG₂, HPETEs). Of note, ONOO⁻ reacts rapidly with GSH and thus bolus addition of ONOO⁻ may have overwhelmed the GPx/GSH system. However, because arachidonic acid was already present and all trace hydroperoxides would have been reduced, ONOO⁻ still served as a peroxidase substrate and thus the cyclooxygenase activator.

Bolus addition of ONOO⁻ does not accurately reflect the continuous production of this oxidant in the cellular environment. The continuous production of ONOO⁻ by the combination of a NO donor and xanthine/xanthine oxidase (X/XO) reverses GPx/GSH inhibition of PGHS (Figure 1), but withholding either component prevents PGHS activation. Similarly, decomposition of SIN-1 (to produce both NO and O_2^-) in the presence of PGHS, GPx, up to 2.5 mM GSH, and arachidonic acid results in the synthesis of prostaglandins (Figure 2). The decrease in prostaglandin synthesis in the presence of higher GSH concentrations was expected because ONOO⁻ reacts directly with GSH and this reaction competes with the reaction of ONOO⁻ with PGHS.

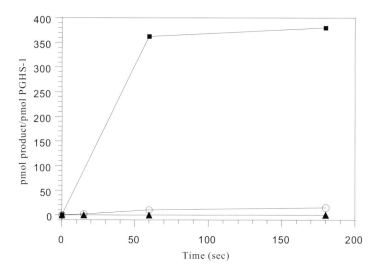

Figure 1. Activation of PGHS cyclooxygenase by X/XO and SNAP. PGHS was reacted with arachidonic acid in the presence of GPx/GSH. Reactions also contained X/XO alone (open circles), SNAP alone (filled triangles) or both X/XO and SNAP. From Landino et al.[82]

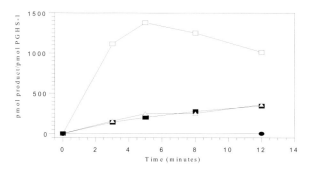

Figure 2. Activation of GPx/GSH-inhibited PGHS by SIN-1. Reactions contained PGHS-1, SIN-1, catalase, GPx, arachidonic acid, and 0.25 mM GSH (open squares), 1.0 mM GSH (filled squares), 2.5 mM GSH (open triangles) or 0.25 mM GSH and SOD (filled circles). From Landino et al.[82]

The presence of superoxide dismutase (SOD) inhibits ONOO⁻ production from SIN-1 and thus prevents cyclooxygenase activation. These data confirm that ONOO⁻ produced even at a low steady state can evade GPx/GSH to initiate arachidonic acid oxygenation by PGHS. Neither NO nor O_2^{-} alone is sufficient to reverse inhibition by GPx/GSH. Similarly, nitrate, nitrite, and S-nitrosoglutathione are unable to activate PGHS.[82]

13.3.3   *Glutathione Peroxidase, ONOO⁻ and PGHS Regulation*

By controlling the cellular levels of peroxide activators, the ubiquitous enzyme, GPx, may play a pivotal role in regulating prostaglandin biosynthesis. We propose that ONOO⁻ is able to bypass GPx and react with PGHS. This implies that ONOO⁻ reacts faster with PGHS than with GPx under conditions where the concentrations of the two enzymes are similar. Recent investigations by Sies and co-workers show that, like its heme-containing counterparts, GPx also reacts with ONOO⁻ to form nitrite.[97] Concomitantly, 2 mol of GSH are oxidized, indicating that ONOO⁻ behaves as a typical peroxide in the GPx catalytic cycle.[97] It also has been demonstrated by stopped-flow kinetic methods that the reduced form of GPx reacts with ONOO⁻ at rates comparable to other peroxidases (k = 2×10^6 M⁻¹ s⁻¹ per monomer).[98] Given the magnitude of this rate constant, it may be somewhat surprising that ONOO⁻ (either by bolus addition or in situ generation) is capable of activating PGHS in the presence of GPx/GSH.

13.3.4   *Kinetic Studies of the Reaction of ONOO⁻ with PGHS*

In support of our hypothesis that ONOO⁻ reacts with PGHS rather than GPx, we performed some preliminary stopped-flow studies to determine the rate constant for the reaction of ONOO⁻ with PGHS. From our preliminary stopped-flow work, we measured a rate constant of 1.5 x 10^7 M⁻¹ s⁻¹ at pH 7 and at 8 °C for the reaction of ONOO⁻ with PGHS-1.[99] The rate constant for PGHS-1 and ONOO⁻ is comparable to the value determined for the reaction of myeloperoxidase with ONOO⁻. Myeloperoxidase reduces ONOO⁻ with a pH-independent rate constant of 2.0 x10^7 M⁻¹ s⁻¹ at 12 °C.[90] The rate constants for these two heme peroxidases is nearly one order of magnitude faster than the reaction of ONOO⁻ with GPx. Also of note, the rate constants for ONOO⁻ reduction by these peroxidases were calculated under different temperature conditions (GPx at 25°C, myeloperoxidase at 12°C and PGHS-1 at 8 °C). Moreover, the rate constant for ONOO⁻ reduction by the Fe(III) porphyrin 5,10,15,20-tetrakis(N-methyl-4'-pyridyl)porphinatoiron(III) is around 5 x10^7 M⁻¹ s⁻¹,[87] thus, it is reasonable that the reaction of ONOO⁻ with PGHS may be considerably faster than with GPx.

In assessing the potential for ONOO⁻ to serve as a PGHS activator, it is important to recognize that for full activation of PGHS, only one equivalent of ONOO⁻ or any other peroxide is necessary, because once activated, the peroxide activator is not required to sustain the cyclooxygenase catalytic cycle. It is entirely plausible that GPx/GSH is unable to scavenge ONOO⁻ to this level before PGHS can be activated, especially considering the possibility that PGHS reacts faster with ONOO⁻ than GPx/GSH.

Recently Kulmacz and coworkers compared the reaction rates of the two PGHS isoforms with organic hydroperoxides by stopped-flow spectroscopy.[100] They measured not only the rate of Compound I formation but also the rate of conversion of Compound I to Compound II. Although the rates of Compound I formation with 15-HPETE were comparable (2.3×10^7 M^{-1} s^{-1} for PGHS-1 and 2.5×10^7 M^{-1} s^{-1} for PGHS-2 at 4 °C), the rates of conversion of Compound I to Compound II were significantly different with PGHS-2 reacting faster. In the case of PGHS, it is likely that Compound II is the one electron oxidized heme and the tyrosyl radical that serves as the cyclooxygenase oxidant (Scheme II). They concluded that PGHS-2 required about a 10-fold lower concentration of hydroperoxide activator than PGHS-1 to form the tyrosyl radical and thus to initiate the cyclooxygenase reaction. Their data suggests that the activity of the two isoforms may be regulated in vivo as a function of the cellular hydroperoxide activator concentration.[101] Consistent with these observations, they noted that a higher concentration of GPx was required to suppress PGHS-2 activity relative to PGHS-1.[100]

13.3.4   Cellular Model of PGHS Activation by ONOO⁻

Upon activation, RAW267.4 cells produce NO and O_2^{-}. As many investigators have demonstrated, prostaglandin biosynthesis is also dramatically enhanced in these activated mouse macrophage-like cells, and inhibition of NO production at the level of expression or by selective inhibition of iNOS dramatically reduces prostaglandin biosynthesis. If ONOO⁻ is the direct activator of PGHS rather than NO, one would expect that the removal of O_2^{-} would also inhibit prostaglandin biosynthesis. We observed the dose-dependent inhibition of prostaglandin biosynthesis in activated RAW267.4 cells by two SOD mimetic agents, CuDIPS (Figure 3) and MnTMPyP.[82] The extent of inhibition by these agents (about a 4–5 fold decrease in PGD_2 and PGE_2) is similar to the effect observed upon removing NO.[67] Furthermore, the parallel decrease in PGE_2 and PGD_2 suggests that the effect of the SOD mimics is not on the enzymes that convert PGH_2 to PGD_2 or PGE_2. Several control experiments support the conclusion that O_2^{-} removal is the only effect of these compounds: 1) neither CuDIPS nor MnTMPyP inhibits expression of NOS or PGHS-2; 2) they do not directly inhibit PGHS-1, PGHS-2, or NOS; and 3) neither appears to inhibit phospholipase A_2 because both CuDIPS and MnTMPyP inhibit prostaglandin synthesis whether arachidonic acid is derived from endogenous or exogenous sources. These results with SOD mimetic agents are consistent with the effect of SOD on cyclooxygenase activation by peroxynitrite in vitro with purified PGHS (see Figure 2).

Based on these results, we propose that NO enhances PGHS activity under inflammatory conditions by acting as a precursor to ONOO-, a potent peroxide activator of PGHS (Scheme III). In response to inflammatory signals (e.g., LPS, cytokines, and other messengers), cells express iNOS and PGHS-2. Furthermore, inflammatory cells such as neutrophils and macrophages produce large quantities of O2.- using NADPH oxidase. The coupling of NO and O2.- from these sources produces ONOO-, which can then activate PGHS. ONOO- activates PGHS as a typical peroxide substrate, oxidizing the Fe3+ heme of PGHS to the Fe4+=O porphyrin cation radical intermediate. Intramolecular electron transfer from tyrosine 385 to the heme produces the tyrosyl radical necessary for arachidonic acid oxygenation. The net effect is the enhancement of prostaglandin biosynthesis. Thus, blocking ONOO⁻ formation by removal of either NO or O_2^- inhibits prostaglandin biosynthesis, and replacement of NO or O_2^- with donor compounds restores PGHS activity.

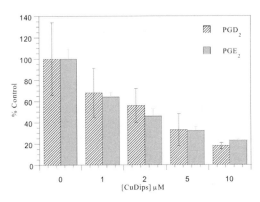

Figure 3. Inhibition of prostaglandin biosynthesis in activated RAW264.7 cells by the SOD mimetic CuDIPS. From Landino *et al.*[82]

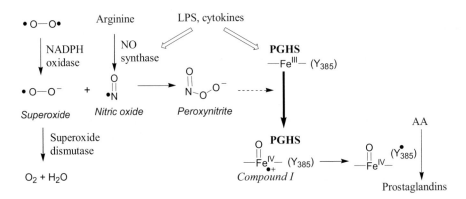

Scheme III. Proposed role of NO and O_2^- in prostaglandin biosynthesis.[82,102]

13.4 Effects of NO and Reactive Oxygen Species on PGHS Activation

A number of recent studies provide additional support for a role of NO and NO-derived species in the activation of prostaglandin biosynthesis in cell culture models and in vivo. Among the cell types examined were: 1) macrophages and microglia; 2) endothelial cells; 3) epithelial cells; and 4) chondrocytes. These studies have the potential to provide clues concerning the role of prostaglandins in several diseases including Alzheimer's disease, Parkinson's disease, arthritis and atherosclerosis. By understanding the mechanisms by which NO and NO-derived species activate prostaglandin biosynthesis, potential therapies to reduce inflammation and tissue damage may be considered.

13.4.1 NO Regulation of Prostaglandin Biosynthesis in Macrophages

Several groups have observed that macrophages isolated from old mice (24 months) produce more prostaglandins than young mice (6 months) following LPS stimulation and the increased production of PGE_2 in old mice is attributed to increased expression of PGHS-2.[103] Meydani and coworkers demonstrated that vitamin E was effective in reducing PGE_2 production in macrophages isolated from old mice and they concluded that vitamin E affects levels of ONOO⁻ by scavenging both NO and O_2^-.[104] NO donors or O_2^- donors alone could not enhance PGE_2 production in their experiments.

Marnett and coworkers recently measured PGE_2 levels in peritoneal macrophages isolated from iNOS knockout mice.[105] PGE_2 levels produced by iNOS-deficient mouse macrophages were reduced by 80% relative to control mouse macrophages following induction with γ-interferon and LPS. Of note, PGHS-2 expression was comparable in iNOS-deficient and control cells. This study clearly strengthens the link between iNOS and PGHS activation in macrophages.

However, not all macrophage-specific studies aimed at understanding the effects of NO on prostaglandin production are as clear-cut. Patel et al. studied the effect of the NO synthase inhibitor, L-NMMA on PGE_2 levels in LPS-activated RAW264.7 cells.[106] They observed no change in PGE_2 levels when cells were treated with up to 200 μM L-NMMA. Treatment of cells with 500 μM L-NMMA decreased nitrite production by 85% but PGE_2 levels increased. They also noted a change in the subcellular localization of PGHS-2 following L-NMMA addition.

Guastadisegni et al compared the effects of reactive nitrogen species on PGE_2 synthesis in cultured rat microglial cells and RAW264.7 cells.[107] Microglial cells, expressed in brain, did not respond to NO donors in the same manner as RAW264.7 cells. Whereas SIN-1 treatment led to an increase in PGE_2 production in RAW264.7 cells, SIN-1 inhibited microglial PGHS-2 expression. Thus, even using similar cells and identical cell lines, researchers have observed very different effects of NO on prostaglandin production.

In experiments using RAW264.7 cells and a rat paw model of inflammation, Salvemini and coworkers examined the ability of ONOO⁻ decomposition catalysts to block cell death induced by ONOO⁻.[108] They demonstrated that two redox active Fe(III) porphyrins could inhibit cell death in RAW264.7 cells and reduce inflammation in the rat paw model. Although these researchers did not measure the effects of the metalloporphyrins on prostaglandin levels, we have shown that MnTMPyP, a similar ONOO⁻ decomposition catalyst and superoxide scavenger, inhibits prostaglandin biosynthesis in RAW264.7 cells.[82]

13.4.2 NO Regulation of Prostaglandin Biosynthesis in other Cell Types

Several examples of cell-based studies examining NO/PGHS interactions are highlighted below to illustrate the variable effects reported recently. The effects of two iNOS-selective inhibitors on PGE_2 production in murine colonic epithelial cells were evaluated in a study by Phang and coworkers.[109] They observed that the iNOS inhibitors affected PGHS-2 expression in response to induction by LPS and interferon-γ. Consequently the levels of PGE$_2$ produced by the treated cells were decreased.

Chondrocytes, expressed in mature cartilage, have also been examined for a connection between NO and prostaglandin biosynthesis. In cases of osteoarthritis, there are alterations in the ability of these cells to produce the cartilage matrix and there is often a decrease in the number of viable chondrocytes. Treatment of human chondrocytes with NO donors induced cell death by activating PGHS-2.[110] NO donors enhance the expression of PGHS-2 and thus increase the production of prostaglandins. Although NSAIDs blocked cell death presumably by inhibiting PGHS cyclooxygenase activity, direct addition of PGE$_2$ alone did not induce cell death. In another study using rat chondrocytes, NOS inhibitors and CuDIPS, an SOD mimetic, inhibited prostaglandin production following interleukin-1β induction.[111] Notably, no changes in PGHS-2 protein expression were observed in rat chondrocytes.

The roles of the PGHS isoforms and of reactive oxygen species in vascular system function has recently been reviewed.[112,113] As observed in other cell types, effects of NO on PGHS function in vascular endothelial cells are complex with varied responses depending on the quantity and source of NO.[68,114,115,116] NO donors enhanced PGHS-1 activity in unstimulated bovine aortic endothelium but PGHS-2 activity was inhibited in activated endothelial cells.[116]

It is clear that a large group of investigators have observed NO-stimulated prostaglandin biosynthesis, apparently by direct interaction with PGHS; however, the field is by no means unified regarding the relationship between NO and prostaglandin biosynthesis. Matters are not helped by the fact that NO does not consistently enhance or inhibit prostaglandin biosynthesis across cellular or in vivo models. This is compounded by complex effects of NO on transcription, post-transcriptional processing, translation, post-translational processing, and PGHS catalysis. This wide variability in observations may result from the use of differing cell lines and tissues, as well as differences in methods of cell activation and NO delivery.

13.5 PGHS Activation and Inflammation in Disease

The plentiful data gathered regarding the effects of NO and NO-derived species on prostaglandin biosynthesis in vivo provides insight into the etiology of several debilitating diseases. Inflammation, oxidative damage to proteins, overexpression of PGHS-2, and increased prostaglandin production are hallmarks of numerous diseases including Alzheimer's (AD), Parkinson's, arthritis and athlerosclerosis.[117,118,119,120] ONOO⁻ is formed in neurons that contain neurofibrillary tangles, a lesion characteristic of AD, because nitrotyrosine, a marker of oxidative damage predominantly induced by ONOO⁻ has been detected within these neurons in AD brains.[121,122] Likewise, protein tyrosine nitration has been detected in Parkinson's disease.[123] Thus, ONOO⁻ may activate PGHS-2 to initiate prostaglandin biosynthesis in these disease states. Elevated levels of PGE$_2$ were detected in the cerebrospinal fluid of probable AD patients vs. age-matched controls.[124] These findings suggest NO synthase inhibitors, O$_2$⁻ scavengers, and ONOO⁻ decomposition catalysts may reduce inflammation by limiting the ability of ONOO⁻ to activate PGHS-2.

13.6 Concluding remarks

Considerable evidence points to ONOO⁻ and fatty acid hydroperoxides as the most likely activators of PGHS enzymes in vivo. Oxidative stress in challenged cells contributes to the formation of lipid-derived hydroperoxides that could serve as PGHS activators;[120] however, lipid-derived hydroperoxides do not leave an obvious cellular "footprint" like ONOO⁻ (in the form of nitrotyrosine). The best scientists can do to understand PGHS activation in vivo is to: 1) use kinetic methods to measure the rates of reaction of the PGHS isoforms with potential activators; 2) estimate the cellular concentrations of the possible activators and the PGHS proteins; 3) continue to perform cell-based studies using NO synthase inhibitors, $O_2^{.-}$ scavengers and antioxidants.

The discovery of a second PGHS isoform in the last decade has reinvigorated research efforts to understand the mechanisms of PGHS activation and catalysis. The development of isoform-selective inhibitors has given scientists the tools to dissect the function of the two isoforms in different cell types. The connections between NO and prostaglandin biosynthesis have sparked controversy, but also have the potential to uncover novel drug therapies designed to reduce inflammation in a number of debilitating diseases. Twenty years ago, no one would have imagined that a free radical like NO would be an important signalling molecule but it is. Why not ONOO⁻?

References

1. Smith, W. L., D. L. DeWitt and R. M. Garavito. 2000. Cyclooxygenases: structural, cellular, and molecular biology. *Ann. Rev. Biochem.* **69**:145-182.

2. Marnett, L. J., S. W. Rowlinson, D. C. Goodwin, A. S. Kalgutkar and C. A. Lanzo. 1999. Arachidonic acid oxygenation by COX-1 and COX-2: mechanisms of catalysis and inhibition. *J. Biol. Chem.* **274**:22903-22906

3. Ogino, N., S. Ohki, S. Yamamoto and O. Hayaishi. 1978. Prostaglandin endoperoxide synthetase from bovine vesicular gland microsomes. Inactivation and activation by heme and other metalloporphyrins. *J. Biol. Chem.* **253**:5061-5068.

4. Picot, D., P. J. Loll and R .M. Garavito. 1994. The X-ray crystal structure of the membrane protein prostaglandin H_2 synthase-1. *Nature (London)* **367**:243-249.

5. Kurumbail, R. G., A. M. Stevens, J. K. Gierse, J. J. McDonald, R. A. Stegeman, J. Y. Pak, D. Gildehaus, J. M. Miyashiro, T. D. Penning, K. Seibert, P. C. Isakson and W. C. Stallings. 1996. Structural basis for selective inhibition of cyclooxygenase-2 by anti-inflammatory agents. *Nature (London)* **384**:644-648.

6. Vane, J. R., Y. S. Bakhle and R. M. Botting. 1998. Cyclooxygenases 1 and 2. *Annu. Rev. Pharmacol. Toxicol.* **38**:97-120.

7.   Hamberg, M., J. Svensson and B. Samuelsson. 1975. Thromboxanes: a new group of biologically active compounds derived from prostaglandin endoperoxides. *Proc. Natl. Acad. Sci. USA* **72**:2994-2998.

8.   Lambeir, A. M., C. M. Markey, H. B. Dunford and L. J. Marnett. 1985. Spectral properties of the higher oxidation states of prostaglandin H synthase. *J. Biol. Chem.* **260**:14894-14896.

9.   Landino, L. M., B. C. Crews, J. K. Gierce, S. D. Hauser and L. J. Marnett. 1997. Mutational analysis of the role of the distal histidine and glutamine residues of prostaglandin endoperoxide synthase-2 in peroxidase catalysis, hydroperoxide reduction, and cyclooxygenase activation. *J. Biol. Chem.* **272**:21565-21574

10. Goodwin, D. C., S. W. Rowlinson and L. J. Marnett. 2000. Substitution of tyrosine for the proximal histidine ligand to the heme of prostaglandin endoperoxide synthase-2: implications for the mechanism of cyclooxygenase activation and catalysis. *Biochemistry* **39**:5422-5432.

11. Kulmacz, R. J. and W. E. M. Lands. 1983. Requirements for hydroperoxides by the cyclooxygenase and peroxidase activities of prostaglandin H synthase. *Prostaglandins* **25**:531-540.

12. Hemler, M. E., G. Graff and W. E. M. Lands. 1978. Accelerative autoactivation of prostaglandin biosynthesis by PGG_2. *Biochem. Biophys. Res. Commun.* **85**:1325-1331.

13. Hemler, M. E., H. W. Cook and W. E. M. Lands. 1979. Prostaglandin biosynthesis can be triggered by lipid peroxides. *Arch. Biochem. Biophys.* **193**:340-345.

14. Xie, W., J. G. Chipman, D. L. Robertson, R. L. Erikson and D. L. Simmons. 1991. Expression of a mitogen-responsive gene encoding prostaglandin synthase is regulated by mRNA splicing. *Proc. Natl. Acad. Sci. USA* **88**:2692-2696.

15. Kujubu, D. A., B. S. Fletcher, B. C. Varnum, R. W. Lim and H. R. Herschman. 1991. TIS10, a phorbol ester tumor promoter-inducible mRNA from Swiss 3T3 cells, encodes a novel prostaglandin synthase/cyclooxygenase homologue. *J. Biol. Chem.* **266**:12866-12872.

16. Herschman, H. R. 199) Prostaglandin synthase 2. Biochim. Biophys. Acta Lipids & Lipid Metab. 1299,125-140.

17. Bredt, D. S. and S. H. Snyder. 1994. Nitric oxide: a physiologic messenger molecule. *Annu. Rev. Biochem.* **63**:175-195.

18. Nathan, C. 1992. Nitric oxide as a secretory product of mammalian cells. *FASEB J* **6**:3051-3064.

19. Moncada, S., R. M. J. Palmer and E. A. Higgs. 1991. Nitric oxide: physiology, pathophysiology, and pharmacology. *Pharmacol. Rev.* **43**:109-142.

20. Stuehr, D., S. Pou and G. M. Rosen. 2001. Oxygen reduction by nitric oxide synthases. *J. Biol. Chem.* **276**:14533-14536.

21. Corbett, J. A., G. Kwon, J. Turk and M. L. McDaniel. 1993. IL-1ß induces the coexpression of both nitric oxide synthase and cyclooxygenase by islets of Langerhans: activation of cyclooxygenase by nitric oxide. *Biochemistry* **32**:13767-13770.

22. Salvemini, D., T. P. Misko, J. L. Masferrer, K. Seibert, M. G. Currie and P. Needleman. 1993. Nitric oxide activates cyclooxygenase enzymes. *Proc. Natl. Acad. Sci. USA* **90**:7240-7244.

23. Davidge, S. T., P. N. Baker, M. K. McLaughlin and J. M. Roberts. 1995. Nitric oxide produced by endothelial cells increases production of eicosanoids through activation of prostaglandin H synthase. *Circ. Res.* **77**:274-283.

24. Swierkosz, T. A., J. A. Mitchell, T. D. Warner, R. M. Botting and J. R. Vane. 1995. Co-induction of nitric oxide synthase and cyclooxygenase: interactions between nitric oxide and prostanoids. *Br. J. Pharmacol.* **114**:1335-1342.

25. Okamoto, H., O. Ito, R. J. Roman and A. G. Hudetz. 1998. Role of inducible nitric oxide synthase and cyclooxygenase-2 in endotoxin-induced cerebral hyperemia. *Stroke* **29**:1209-1218.

26. Salvemini, D., S. L. Settle, J. L. Masferrer, K. Seibert, M. G. Currie and P. Needleman. 1995. Regulation of prostaglandin production by nitric oxide: an in vivo analysis. *Br. J. Pharmacol.* **114**:1171-1178.

27. Hughes, F. J., L. D. K. Buttery, M. V. J. Hukkanen, A. O'Donnell, J. Maclouf and J. M. Polak. 1999. Cytokine-induced prostaglandin E_2 synthesis and cyclooxygenase-2 activity are regulated both by a nitric oxide-dependent and -independent mechanism in rat osteoblasts in vitro. *J. Biol. Chem.* **274**:1776-1782.

28. Curtis, J. F., N. G. Reddy, R. P. Mason, B. Kalyanaraman and T. E. Eling. 1996. Nitric oxide: a prostaglandin H synthase 1 and 2 reducing cosubstrate that does not stimulate cyclooxygenase activity or prostaglandin H synthase expression in murine macrophages. *Arch. Biochem. Biophys.* **335**:369-376.

29. Hajjar, D. P., H. M. Lander, S. F. A. Pearce, R. K. Upmacis and K. B. Pomerantz. 1995. Nitric oxide enhances prostaglandin-H synthase-1 activity by a heme-independent mechanism: evidence implicating nitrosothiols. *J. Am. Chem. Soc.* **117**:3340-3346.

30. Maccarrone, M., S. Putti and A. F. Agro. 1997. Nitric oxide donors activate the cyclooxygenase and peroxidase activities of prostaglandin H synthase. *FEBS Lett* **410**:470-476.

31. Tsai, A. L., C. Wei and R. J. Kulmacz. 1994. Interaction between nitric oxide and prostaglandin H synthase. *Arch. Biochem. Biophys.* **313**:367-372.

32. Kanner, J., S. Harel and R. Granit. 1992. Nitric oxide, an inhibitor of lipid oxidation by lipoxygenase, cyclooxygenase and hemoglobin. *Lipids* **27**:46-49.

33. Holzhütter, H. G., R. Wiesner, J. Rathmann, R. Stösser and H. Kühn. 1997. A kinetic model for the interaction of nitric oxide with a mammalian lipoxygenase. *Eur. J. Biochem.* **245**:608-616.

34. Tsai, A. L. 1994. How does NO activate hemeproteins?. *FEBS Lett* **341**:141-145.

35. Stone, J. R. and M. A. Marletta. 1994. Soluble guanylate cyclase from bovine lung: activation with nitric oxide and carbon monoxide and spectral characterization of the ferrous and ferric states. *Biochemistry* **33**:5636-5640.

36. Kennedy, T. A., C. J. Smith and L. J. Marnett. 1994. Investigation of the role of cysteines in catalysis by prostaglandin endoperoxide synthase. *J. Biol. Chem.* **269**:27357-27364.

37. Landino, L. M., B. C. Crews, M. D. Timmons, J. D. Morrow and L. J. Marnett. 1996. Peroxynitrite, the coupling product of nitric oxide and superoxide, activates prostaglandin biosynthesis. *Proc. Natl. Acad. Sci. USA.* **93**:15069-15074.

38. Cassatella, M. A., F. Bazzoni, R. M. Flynn, S. Dusi, G. Trinchieri and F. Rossi. 1990. Molecular basis of interferon-γ and lipopolysaccharide enhancement of phagocyte respiratory burst capability. *J. Biol. Chem.* **265**:20241-20246.

39. Rossi, F. 1986. The O_2^--forming NADPH oxidase of the phagocytes: nature, mechanisms of activation and function. *Biochem. Biophys. Acta* **853**:65-89.

40. Xia, Y., L. J. Roman, B. S. S. Masters and J. L. Zweier. 1998. Inducible nitric-oxide synthase generates superoxide from the reductase domain. *J. Biol. Chem.* **273**:22635-22639.

41. Koppenol. W. H. 1998. The basic chemistry of nitrogen monoxide and peroxynitrite. *Free Rad. Biol. Med.* **25**:385-391.

42. Lee, J., J. A. Hunt and J. T. Groves. 1998. Mechanisms of iron porphyrin reactions with peroxynitrite. *J. Am. Chem. Soc.* **120**:7493-7501.

43. Stern, M. K., M. P. Jensen and K. Kramer. 1996. Peroxynitrite decomposition catalysts. *J. Am. Chem. Soc.* **118**:8735-8736.

44. Lee, J., J. A. Hunt and J. T. Groves. 1998. Manganese porphyrins as redox-coupled peroxynitrite reductases. *J. Am. Chem. Soc.* **120**:6053-6061.

45. Floris, R., S. R. Piersma, G. Yang, P. Jones and R. Wever. 1993. Interaction of myeloperoxidase with peroxynitrite: a comparison with lactoperoxidase, horseradish peroxidase and catalase. *Eur. J. Biochem.* **215**:767-775.

46. Dunford, H. B. and J.S. Stillman. 1976. On the function and mechanism of action of peroxidases. *Coord. Chem. Rev.* **19**:187-251.

47. Job, D. and H. B. Dunford. 1976. Substituent effect on the oxidation of phenols and aromatic amines by horseradish peroxidase compound I. *Eur. J. Biochem.* **66**:607-614.

48. Ohki, S., N. Ogino, S. Yamamoto and O. Hayaishi, O. 1979. Prostaglandin hydroperoxidase, an integral part of prostaglandin endoperoxide synthetase from bovine vesicular gland microsomes. *J. Biol. Chem.* **254**:829-836.

49. Marnett, L. J. and K. R. Maddipati. 1991. Prostaglandin H synthase. Everse, J. Everse, K. E. Grisham, M. B. eds. *Peroxidases in Chemistry and Biology Vol. 1* pp. 293-334 CRC Press Boca Raton, Florida.

50. Hemler, M. E., G. Graff and W. E. M. Lands. 1978. Accelerative autoactivation of prostaglandin biosynthesis by PGG_2. *Biochem. Biophys. Res. Commun.* **85**:1325-1331.

51. Hemler, M. E., H. W. Cook and W. E. M. Lands. 1979. Prostaglandin biosynthesis can be triggered by lipid peroxides. *Arch. Biochem. Biophys.* **193**:340-345.

52. Sies, H., V. S. Sharov, L. Klotz and K. Briviba. 1997. Glutathione peroxidase protects against peroxynitrite-mediated oxidations. *J. Biol. Chem.* **272**:27812-27817.

53. Briviba, K, R. Kissner, W. H. Koppenol and H. Sies. 1998. Kinetic study of the reaction of glutathione peroxidase with peroxynitrite. *Chem Res Toxicol.* **11**:1398-1401.

54. Marnett, L. J., D. C. Goodwin, S. W. Rowlinson, A. S. Kalgutkar and L. M. Landino. 1999. "Structure, Function, and Inhibition of Prostaglandin Endoperoxide Synthases," *In Comprehensive Natural Products Chemistry, Vol. 5*, (Poulter, C.D. Ed.) pp. 225-261, Elsevier Science, Ltd., Oxford, UK.

55. Lu, G., A. H. Tsai, H. E. Van Wart and R. J. Kulmacz. 1999. Comparison of the peroxidase reaction kinetics of prostaglandin H synthase-1 and -2. *J. Biol. Chem.* **274**:16162-16167.

56. Kulmacz, R. J. 1998. Cellular regulation of prostaglandin H synthase catalysis. *FEBS Lett.* **430**:154-157.

57. Goodwin, D. C., L. M. Landino and L. J. Marnett. 1999. Effects of nitric oxide and nitric oxide-derived species on prostaglandin endoperoxide synthase and prostaglandin biosynthesis. *FASEB J.* **13**:1121-1136.

58. Wu D., M. G. Hayek and S. Meydani. 2001. Vitamin E and macrophage cyclooxygenase regulation in the aged. *J. Nutrition.* **131**:382S-388S.

59. Beharka, A. A., D. Wu, M. Serafini and S. N. Meydani. 2002. Mechanism of vitamin E inhibition of cyclooxygenase activity in macrophages from old mice: role of peroxynitrite. *Free Rad. Biol. Med.* **32**:503-511.

60. Marnett, L. J., T. L. Wright, B. C. Crews, S. R. Tannenbaum and J. D. Morrow. 2000. Regulation of prostaglandin biosynthesis by nitric oxide is revealed by targeted deletion of inducible nitric oxide synthase. *J. Biol. Chem.* **275**:13427-13430.

61. Patel, R., M. G. Attur, M. Dave, S. B. Abramson and A. R. Amin. 1999. Regulation of cytosolic COX-2 and prostaglandin E_2 production by nitric oxide in activated murine macrophages. *J. Immunol.* **162**:4191-4197.

62. Guastadisegni, C., L. Minghetti, A. Nicolini, E. Polazzi, P. Ade, M. Balduzzi and G. Levi. 1997. Prostaglandin E2 synthesis is differentially affected by reactive nitrogen intermediates in cultured rat microglia and RAW264.7 cells. *FEBS Lett.* **413**:314-318.

63. Salvemini, D., Z. Wang, M. K. Stern, M. G. Currie and T. P. Misko. 1998. Peroxynitrite decomposition catalysts: therapeutics for peroxynitrite-mediated pathology. *Proc. Natl. Acad. Sci. USA.* **95**:2659-2663.

64. Mei, J. M., N. G. Hord, D. F. Winterstein, S. P. Donald and J. M. Phang. 2000. Expression of prostaglandin endoperoxide H synthase-2 induced by nitric oxide in conditionally immortalized murine colonic epthelial cells. *FASEB J.* **14**:1188-1201.

65. Notoya, K., D. V. Jovanovic, P. Reboul, J. Martel-Pelletier, F. Mineau and J. P. Pelletier. 2000. The induction of cell death in human osteoarthritis chondrocytes by nitric oxide is related to the production of prostaglandin E_2 via the induction of cyclooxygenase-2. *J. Immunol.* **165**:3402-3410.

66. Nedelec, E., A. Abid, C. Cipolletta, N. Presle, B. Terlain, P. Netter and J. Y. Jouzeau. 2001. Stimulation of cyclooxygenase-2 activity by nitric oxide-derived species in rat chondrocyte: lack of contribution to loss of cartilage anabolism. *Biochem. Pharmacol.* **61**:965-978.

67. Davidge, S. T. 2001. Prostaglandin H synthase and vascular function. *Circ. Res.* **89**:650-660.

68. Wolin, M. S. 2000. Interactions of oxidants with vascular signaling systems. *Arterioscler. Thromb Vasc. Biol.* **20**:1430-1442.

69. Pueyo, M. E., J. F. Arnal, J. Rami and J. B. Michel. 1998. Angiotensin II stimulates the production of NO and peroxynitrite in endothelial cells. *Am. J. Physiol.* **274**: C214-C220.

70. Eligini, S., A. Habib, M. Lebret, C. Creminon, S. Levy-Toledano and J. Maclouf. 2001. Induction of cyclooxygenase-2 in human endothelial cells by SIN-1 in the absence of prostaglandin production. *Brit. J. Pharmacol.* **133**:1163-1171.

71. Onodera, M., I. Morita, Y. Mano and S. Murota. 2000. Differential effects of nitric oxide on the activity of prostaglandin endoperoxide H synthase-1 and -2 in vascular endothelial cells. *Prostag. Leukotr Ess.* **62**:161-167.

72. Pasinetti, G. M. 1998. Cyclooxygenase and inflammation in Alzheimer's disease: experimental approaches and clinical interventions. *J. Neurosci. Res.* **54**:1-6.

73. Knott, C., G. Stern and G. P. Wilkin. 2000. Inflammatory regulators in Parkinson's disease: iNOS, lipocortin-1, and cyclooxygenases-1 and -2. *Mol. Cell. Neurosci.* **16**:724-739.

74. Jang, D. and G. A. C. Murrell. 1998. Nitric oxide in arthritis. *Free Rad. Biol. Med.* **24**:1511-1519.

75. O'Donnell, V. B. and B. A. Freeman. 2001. Interactions between nitric oxide and lipid oxidation pathways. *Circ. Res.* **88**:12-21.

76. Good, P. F., P. Werner, A. Hsu, C. W. Olanow and D. P. Perl. 1996. Evidence of neuronal oxidative damage in Alzheimer's disease. *Am J Pathol.* **49**:21-28.

77. Smith, M. A., P. L. Richey Harris, L. M. Sayre, J. S. Beckman and G. Perry. 1997. Widespread peroxynitrite-mediated damage in Alzheimer's disease. *J Neurosci* **17**:2653-7

78. Good, P. F., A. Hsu, P. Werner, D. P. Perl and C. W. Olanow. 1998. Protein nitration in Parkinson's disease. *J Neuropathol Exp Neurol* **57**:338-342.

79. Montine, T. J., K. R. Sidell, B. C. Crews, W. R. Markesbery, L. J. Marnett, L. J. Roberts and J. D. Morrow. 1999. Elevated CSF prostaglandin E_2 levels in patients with probable AD. *Neurology* **53**:1495-1498.

Chapter 14

PHYLOGENETIC CONSERVATION OF THE NRF2-KEAP1 SIGNALING SYSTEM

Xue Zhang, Mark Garfinkel, and Douglas Ruden

14.1 Introduction

Aerobic life necessarily entails a continuous production of metabolites of molecular oxygen known as reactive oxygen species (ROS), a major cause of malignancy and cellular damage. In normal situations, organisms are capable of eliminating or reducing the negative effects of ROS and electrophilic insults by antioxidants such as glutathione (GSH) and phase 2 detoxification enzymes such as glutathione S-transferase (Gst) and NAD(P)H: quinone oxidoreductase (Nqo-1). Once the balance between the oxidant burden and the antioxidant defense is overcome, the deleterious accumulation of ROS can contribute to a wide variety of diseases, including diabetes, cancer, and Alzheimer's disease.[1] Therefore, many previous studies were focused on the pathological aspects of ROS.[2,3] However, ROS are not always harmful depending on the severity of the oxidant burden. Increasing evidence shows that ROS, such as superoxide and hydrogen peroxide (H_2O_2), act as second messengers in signal transduction. With the increase in the oxidant burden, ROS plays roles in hypoxic signaling, physiological redox signaling, adaptive signaling, apoptotic signaling, and ultimately necrosis.[4] Signals triggered by low levels of oxidative stress, which is referred to as physiological redox signaling, can activate the defense system so that the cells will be more resistant to the subsequent challenges of even greater oxidative stress. The goals of this review are to discuss the state-of-the art knowledge of the regulation of the antioxidant/electrophile response enhancer element (ARE/EpRE) upstream of many

H. J. Forman, J. Fukuto, and M. Torres (eds.), Signal Transduction by Reactive Oxygen and Nitrogen Species: Pathways and Chemical Principles, ©2003, Kluwer Academic Publishers. Printed in the Netherlands

phase 2 genes and to discuss the phylogenic conservation of the ARE/EpRE-mediated signaling system in mammals, zebrafish, and *Drosophila*.

14.2 The Antioxidant Response Signaling Pathway

Previous studies showed that AP-1 and Cnc-family bZIP proteins are potential components in the transcriptional complex that induce the *glutamate-cysteine ligase catalytic subunit* (*gclc*) gene that is required for GSH biosynthesis and several phase 2 genes during oxidative/electrophilic stress (Fig. 1).[5] Although there is still a controversy regarding the identification or involvement of components of the complex, it is generally agreed that nuclear factor-erythroid 2-related factor 2 (Nrf2), a cap-'n-collar (Cnc)-family basic leucine zipper (bZIP) protein, is the most likely candidate.[5-9] Nrf2 was first cloned in 1994 when the transcriptional factors regulating β-globin genes were sought.[10] However, Nrf2 is not restricted to the erythroid cells. Nrf2 is also widely expressed in various tissues, although with different levels of expression. Later, Chan *et al.* found that Nrf2 is not essential for murine erythropoiesis, growth and development.[11] Instead, Nrf2 is found to be associated with the transcriptional regulation of genes of many phase 2 and antioxidant enzymes.[12]

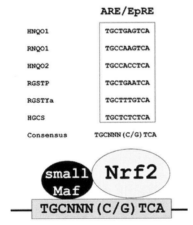

Figure 14-1. The ARE/EpRE sequences are shown for six genes.

A common regulatory motif in the enhancers of many of the antioxidant response genes, such as gclc, NAD(P)H: quinone oxidoreductase (nqo-1 and nqo-2), glutathione S-transferase-Ya (gst-Ya), cystine membrane transporters (x_c^-), heme oxygenase (ho-1 and ho-2), and peroxyredoxin (prx), is a DNA sequence called the ARE/EpRE (Fig. 1).[13,14] The ARE/EpRE was first discovered in the rat gst-Ya gene.[15,16] Genes with the ARE/EpRE sequence are induced by xenobiotics like the phorbol ester 12-O-tetradecanoylphorbol-13-acetate (TPA),[17] oxidants like H_2O_2,[18] and phenolic "antioxidants" like tert-butyl hydroxyquinoline (tBHQ).[16,19,20] The ARE/EpRE contains two or more imperfect or perfect AP-1 binding sequences. Fine mutational analysis of the rat gst-Ya ARE showed that the DNA sequence 5'-TGACnnnGC-3' is essential for ARE activity,[21] which contains one AP-1 like element (TGAC) and one GC box (GCA) that is a potential small Maf binding site (Fig. 1).

The mechanism controlling Nrf2 activity is not known. It was first found that, upon exposure to electrophilic agents and tBHQ, the DNA binding activity of Nrf2 is markedly induced, whereas the steady-state mRNA levels remain constant.[22,23] Also, the induction of Nrf2 activity is rapid (~2 hours) and occurs in the presence of protein synthesis inhibitors.[22,23] Therefore, efforts were made to look for inhibitory factors that are present in uninduced cells. In 1999 and 2001, a cytosolic inhibitory molecule of Nrf2 was cloned in mouse and rat, respectively, that were named Keap1 and INrf2.[22,23] Their highly conserved amino acid sequences suggest that they are interspecies orthologs.[23] To avoid confusion, this inhibitory factor will be referred to as Keap1 in this review. By co-immunoprecipitation, Jaiswal's laboratory demonstrated the interaction of Nrf2 and Keap1. Furthermore, by transiently overexpressing both a portion of Nrf2 fused to green fluorescent protein (GFP) and Keap1 proteins in tissue culture cells, Yamamoto's and Jaiswal's laboratories observed that, under non-stressful conditions, Nrf2 binds to this inhibitory factor and stays in the cytosol. Upon addition of reagents that cause oxidative stress, Nrf2-GFP dissociates from Keap1 and translocates into the nucleus.

The translocation of cytosolic Nrf2 to the nucleus is not apparent in Western blot analyses with antibodies generated against native Nrf2 because cytosolic or total cellular Nrf2 levels also increase when cells are exposed to reagents that cause oxidative stress.[24,25] Thus, Kwak et al. proposed that the nuclear accumulation of Nrf2 might initially be the result of translocation from the cytoplasm, followed by new synthesis of Nrf2 protein.[25] It has been shown that the transcription of Nrf2 is inducible in some tissues by some agents,[24-26] and that nrf2 transcription is auto-regulated by Nrf2 itself through an ARE-like sequence in the nfr2 promoter.[25]

Yamamoto's laboratory recently generated Nrf2 knockout mice and found that nrf2 -/- cells from these mice, and the mice themselves, are hypersensitive to oxidative stress.[27] Interestingly, some genes that are coordinately regulated by AREs/EpREs, such as the x_c^- gene, are barely induced by the oxidative stress agents diethylmaleate (DEM), paraquat, glucose oxidase, and $CdCl_2$ in nrf2 -/- cell lines, but are still activated to 60-70% of wild-type activity by lipopolysaccharide (LPS) in these cells.[27] This suggests that LPS induction of the x_c^- gene is regulated through an alternative regulatory pathway in addition to the Nrf2-mediated pathway. This work is important and relevant to this review because it indicates that two toxicants that both lead to oxidative stress, such as $CdCl_2$ and LPS, cause the transcriptional activation of different, but overlapping, groups of genes.

Many questions are just beginning to be answered about the regulation of Nrf2 activity. What signal caused by the oxidative stress triggers the dissociation of Nrf2 from Keap1? Is loss of Keap1, the cytosolic partner, enough for Nrf2 to change compartments from the cytoplasm to the nucleus? Is Nrf2 released from Keap1 functional or are post-translational modification(s) also necessary? Is Keap1 sensitive to upstream signals or oxidants, or is Keap1 just a binding site for Nrf2? While none of these questions can yet be answered definitively, research that addresses these questions will be summarized in this review.

14.3 Separating Facts and Artifacts in Nrf2-Keap1 Signaling Studies

Probably because the Nrf2-Keap1 signaling field is still in its infancy, many laboratories have published confusing and often contradictory results, especially concerning the regulation of phase 2 genes by the ARE/EpRE regulatory element. The results that are generally believed by most researchers in the field to be the least amenable to making meaningful conclusions are those done with tissue culture cells transiently transfected with both a reporter gene, such as ARE/EpRE driving chloramphenical acetyl transferase (CAT), and the activating or repressing transcription factor, such as Nrf2 or small Mafs. The reason that this type of experiment is often misleading is that transcription factors often form heteromeric or multimeric complexes, such as AP-1 (Jun/Fos) or Nrf2/Maf. Also, overexpressing an activator can actually cause a decrease in transcription by altering its location in a cell or by "squelching"- the titration of components of the general transcription machinery [28,29]. Therefore, transient overexpression of transcription factors and reporter genes can often allow a researcher to obtain either high levels of activation or high levels of repression depending on the amount that was transfected. This effect, compounded with the fact that reporter genes have transcription factor binding sites in heterologous contexts, causes the conclusions that one makes with these experiments to be tentative at best.

As summarized in Fig. 2, it is generally believed by most investigators that the best experiments are those in which a transcription factor has been knocked out and endogenous gene expression is analyzed. These "highest confidence" experiments have recently been done for Nrf2, for example, as described in the previous section [27]. There are also generally agreed upon distinctions made between genetic knockouts being superior to anti-sense or other inhibitory RNA techniques, and whole animal experiments being superior to tissue culture research. Nevertheless, even the "highest confidence" experiments can provide misleading results if compensatory mechanisms are not taken into account, for instance.

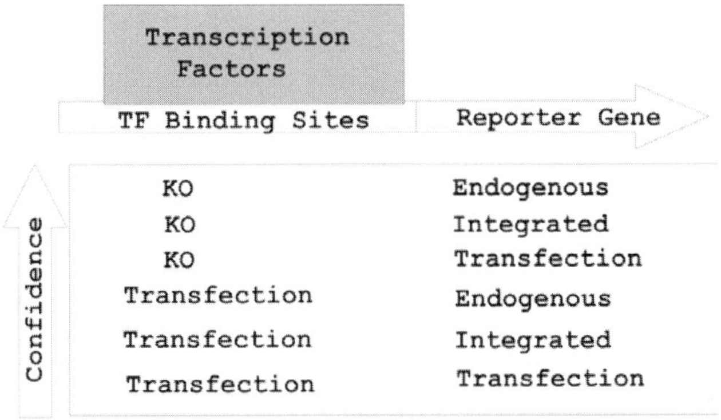

	TF Binding Sites	Reporter Gene
	KO	Endogenous
	KO	Integrated
	KO	Transfection
	Transfection	Endogenous
	Transfection	Integrated
	Transfection	Transfection

Figure 14.2. Confidence in the results of experiments related to understanding the transcriptional regulatory regions of a gene can be predicted by experimental design.

Another consideration that one must make in interpreting experiments is whether the inducer used, such as H_2O_2, is at a physiological level. Generally speaking, H_2O_2

levels in the 10 to 50 μM range are considered by most researchers to be in the "physiological" range. One should probably be wary of conclusions presented by a researcher who uses 1 mM or higher concentrations of H_2O_2 or some other inducer. Such high concentrations of inducers of oxidative stress are probably unrealistic with respect to what happens in an organism and likely cause apoptosis or even necrosis. This being said, one should also consider that what might be a high level of H_2O_2 for one cell line might not be sufficient to induce the phase 2 genes in another cell line. The sensitivities to H_2O_2 differ not only among distantly related taxa as one might expect, but even among cells lines derived from the same species.

14.4 The MAPK Cascade, PKC, and the Activation of Nrf2

The genes of the oxidative stress defense system enzymes, consisting of the phase 2 and the antioxidant enzymes, are inducible by sub-lethal oxidative stress. Studies showed that Nrf2, a bZIP protein, is one of the critical transcription factors in the up-regulation of these genes by the ARE/EpRE. But the signaling mechanisms upstream of Nrf2 remain for the most part a mystery.

Papers published so far suggest that multiple mitogen-activated protein kinase (MAPK) cascades (Fig. 3),[30] protein kinase C (PKC)[31,32] and phosphoinositol-3-kinase (PI3K) are involved in activating Nrf2. Kong *et al.*, for example, showed that xenobiotics such as tBHQ activate Nrf2 through c-Jun N-terminal kinase 1 (Jnk1), extracellular signal regulated kinase 2 (Erk2), and p38.[30] These authors also showed that the map kinase kinase kinases (MAPKKK) involved are probably the extracellular signal-regulated kinase kinase kinase 1 (MEKK1), the transforming growth factor-beta-activated kinase (TAK1), and the apoptosis signal-regulating kinase (ASK1).[30] Using the same cell line, HepG2, Zipper and Mulcahy showed that xenobiotics can activate Erk1/2 and p38, and that this causes an increase in the binding of Nrf2 to the ARE/EpRE.[34] Jnk1, however, was activated only weakly and transiently in this cell line by all of the xenobiotics tested.[34] Using a PKC activator and a PKC inhibitor in HepG2 cells, Huang *et al.*[31,32] reported that PKC is an upstream that is responsible for the phosphorylation of Nrf2 and that tBHQ causes a persistent activation of Nrf2. In IMR-32 human neuroblastoma cells treated with tBHQ, Lee et al. found that PI3 kinase probably mediates the up-regulation of the Nqo-1 via its ARE/EpRE regulatory sequence.[33]

There is evidence that phospolipase C γ (PLCγ) is activated by H_2O_2 through the activation of receptor tyrosine kinases (RTKs) like the Egfr (Fig. 3).[35-38] Interestingly, the activation of PLCγ and PKC is associated with an increased protection of cells against oxidative injury,[39] and an increased resistance to cancer chemotherapeutic drugs.[40] It has been reported that PKC can phosphorylate Nrf2 and increase its transcriptional activity.[31,32] Mutating the phosphorylated residue on the PKC site from a serine to an alanine in Nrf2 is not sufficient to cause its release from Keap1, but it does decrease its ability to activate transcription.[31,32] However, the phosphorylated serine residue on Nrf2 is not conserved in *Drosophila*, thus throwing some doubt on whether PKC phosphorylation at this site is an evolutionarily conserved mechanism (see below). Therefore, it remains to be determined whether H_2O_2 activates Nrf2 through the Egfr-PLC-PKC pathway and, if it does occur, the significance of this event. Another question is whether PKC activates Nrf2 directly or through its ability to activate the Erk pathway since the involvement of Erk pathway in the regulation of Nrf2 activity has been suggested,[30] and the relationship of PKC and the Erk pathway is well established. Further studies on these and other problems will help to identify the signaling pathways through which H_2O_2 activates Nrf2.

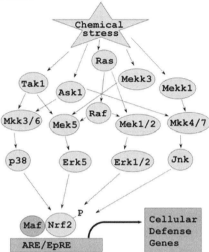

Figure 14.3 A model for how oxidative stress leads to the activation of the MAPK signaling pathways and the Nrf2/Maf-dependent anti-oxidant response genes.

Taken together, there is still no agreement about the signaling pathways regulating Nrf2. One common aspect in all the studies is the application of H_2O_2 producing quinones or xenobiotics that cause H_2O_2 accumulation. This implies that H_2O_2 is one of the inducers of Nrf2 activity. Although the cell type specificity, the chemical specificity, and the difference in concentrations of H_2O_2 caused by different chemicals make the pathway complicated, some potential pathways can still be proposed. Possible mechanisms of H_2O_2 and electrophiles inducing signaling pathways upstream of Nrf2 activation is discussed in other chapters of this book.

14.5 Oxidation of Keap1 and the Translocation of Nrf2 to the Nucleus

Both xenobiotics and oxidants can trigger the translocation of Nrf2 from the cytosol to the nucleus. Using Western blot analyses, Lee et al. showed that the protein level of nuclear Nrf2 increased in IMR-32 cells after treated by 10 μM tBHQ.[41] However, they did not mention any changes in the protein level of cytosolic Nrf2. By transiently expressing a green fluorescent protein-Nrf2 (GFP-Nrf2) fusion protein in quail fibroblast cells, Jaiswal *et al.* observed the compartmental change of Nrf2 directly.[23] These results, while preliminary because they are done with a transient transfection assay, suggest that the translocation of Nrf2 from the cytoplasm to the nucleus could be a mechanism for its activation. Talalay's laboratory has recently shown that dexamethasone 21-mesylate (Dex-mes) and other inducers of phase 2 genes react primarily with four cysteines in purified mouse Keap1 at rates that are closely related to their potencies (Fig. 4).[42] Eventually, all of the cysteines in Keap1 react with Dex-mex, but these four "reactive cysteines" are the most sensitive to chemical crosslinking. These four cysteines cluster in a less conserved region between the two highly conserved domains of Keap1 (see below). The authors propose that this region is the sensor of oxidative stress.[42]

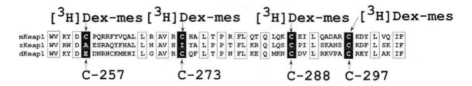

Figure 14.4. The four most reactive sulfhydryl groups of mouse Keap1 (C-257, 273, 288, and 297) are indicated. The homologous regions of zebrafish and *Drosophila* Keap1 are aligned for comparison.

A phylogenetic comparison of Keap1 from several species casts doubt that at least two of these four "reactive cysteines" are sensors of oxidant stress. Examination of 34 aligned amino acid sequences from Keap1-related proteins from diverse species reveals that the residue labeled C-288 in Fig. 4 is a cysteine in only six Keap1 family members. In none of the other 28 proteins is this position a cysteine (12 have glutamate, 6 aspartate, 2 histidine, 2 serine, 3 asparagine and three are gapped or truncated). However, the proteins that have C-288, rat Keap1, zebrafish Keap1, human Keap1, *Drosophila* Keap1, and mosquito Keap1, are the most closely related to mouse Keap1. Therefore, it is possible that C-288 is indeed a "reactive cysteine", because this sub-family of Keap1 is, presumably, the only one that senses oxidative stress. Likewise, the C-273 site is cysteine in only six Keap1 family members, five of which are in the sub-family that is closest in homology to mouse Keap1. In the remaining 28 sequences, 21 are leucine at this position, 4 are methionine and 3 are gapped or truncated. The remaining cysteine site, C-257, is found only in human, mouse and rat Keap1 sequences. Nine of the other 31 sequences have a valine in this position, which is the plurality residue at this position. The cysteine residue labeled C-297 in Fig. 4 is present in 33 of the 34 aligned sequences, thus suggesting a structural role rather an a role as an "oxidant sensor".

To summarize these results, two of the four putative "reactive cysteines" are conserved in most of the members of the "oxidant sensing" subfamily of Keap1, C-273 and C-288. These two cysteines are not likely to be conserved structural motifs because they are not present in the other Keap1 homologous proteins. For these reasons, if Keap1 is a sensor of oxidative stress, it is likely that C-273 and C-288 are the "reactive cysteines" that sense oxidative stress (Fig. 5).[42] However, for the above phylogenetic reasons, it is unlikely that C-257 and C-273 have this role.

Mulcahy's laboratory used a mutagenesis approach to identify the hypothesized "oxidant sensor" amino acids on Keap1 and proposed a model that differs significantly from the above model.[43] They show that mutation of a conserved serine (S104A) in the conserved N-terminal region of Keap1 disrupts Keap1 dimerization and eliminates the ability of Keap1 to sequester Nrf2 in the cytoplasm.[43] The authors propose a model whereby oxidative stress disrupts the dimerization function of Keap1, presumably by post-translational modification of S-104, and that this causes the release of Nrf2 and the activation of the genes regulated by the ARE/EpRE.[43] Phylogenic analysis of S-104 reveals that 32 out of 34 Keap1-related proteins have serine at this position, thus suggesting that this serine has a structural roll in the dimerization of all members of this family. Since it is unlikely that all Keap1 homologous proteins are sensors of oxidative stress, we think that it is also unlikely that a presumed structural amino acid such as S-104 is used as an "oxidant sensor". Needless to say, the two "oxidant sensor" models presented in this section are preliminary and require further experiments to either verify or refute

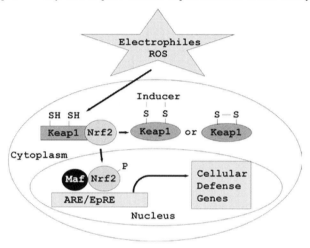

them.

Figure 14.5. Model for phase 2 response regulation. Nrf2 binds to Keap1 in the cytoplasm and Nrf2 is released when the inducer binds to the reactive sulfhydryl groups on Keap1. In the nucleus, small Maf and Nrf2 activate transcription of the cellular defense genes.

14.6 Nrf2 is Phosphorylated after Oxidative Stress

In transient transfection experiments, Huang *et al.*[44] found that ARE/EpRE-directed transcription is activated by phorbol 12-myristate (PMA) which activates PKC, but is completely suppressed by selective inhibitors of PKC, such as staurosporine and Ro-32-0432. PMA was also found to cause the phosphorylation of Nrf2 and trigger its translocation to the nucleus in HepG2 cells, while staurosporine and Ro-32-0432 inhibited these processes. Therefore, they conclude that PKC is a key kinase regulating Nrf2 signaling. Alternatively, Yu *et al.*[45-47] propose that MAP kinase pathways that are activated by MEKK1, TAK1 and ASK1 may link chemical signals to Nrf2 post-translational regulation. Downstream of these kinases, MKK4, MKK6 and JNK1 are involved, but not MKK3 and p38. However, Zipper and Mulcahy report that inhibition of ERK and p38 inhibits binding of Nrf2 and induction of the GCS genes.[48] In spite of the conflicting conclusions, all these results suggest that phosphorylation is likely important for Nrf2 function.

14.7 Small Maf Proteins Bind with Nrf2 at the ARE/EpRE

Jaiswal's laboratory recently attempted to determine whether small Maf (MafG and MafK) proteins positively or negatively regulate ARE/EpRE-mediated tBHQ induction of the Nqo-1 and Gst-Ya genes in transfected Hep-G2 cells. They showed that over-expression of small Maf proteins repressed Nrf2-mediated up-regulation of ARE/EpREs from these genes fused to a reporter gene in a transient-transfection experiment. Using bandshift and supershift assays with the Nqo1 ARE/EpRE, they showed that small MafG and MafK proteins bind to the ARE/EpRE as Maf-Maf homodimers and Maf-Nrf2 heterodimers.[49] They propose that Maf-Maf homodimers, and possibly Maf-Nrf2 heterodimers, play a role in negative regulation of transcription and anti-oxidant induction of NQO1 and other phase 2 genes.[49]

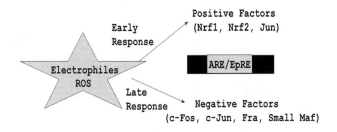

Figure 14.6. Model showing the possible role of small Maf, c-Fos, c-Jun, and Fra in down-regulation of ARE/EpRE-mediated expression of anti-oxidant genes.

While these results are intriguing, one must consider the fact that transient-transfection experiments provide conclusions whose validity under physiological conditions are probably the least reliable. Other laboratories have shown that Maf-Maf homodimers repress, whereas Maf-Nrf2 heterodimers activate transcription of the β-globin gene.[50,51] Jaiswal's laboratory also argue in this paper that Jun-Nrf2 heterodimers bind to the ARE/EpRE and are active complexes, while Nrf2-c-Fos, Nrf2-c-Jun, and Nrf2-Fra form negative complexes (Fig. 6). Obviously, it will be important to make mouse and zebrafish knockouts of small Maf and other proteins, as was done with Nrf2,[27] to determine whether they have a positive or negative role in regulating phase 2 gene induction by oxidants and electrophiles. If there are more than one isoform of small Maf proteins that bind to ARE/EpREs, as is likely the case, then multiple knockout strains will be required for this experiment.

14.8 The Nrf2-Keap1 System is Present in Zebrafish

To understand how cells receive oxidative and electrophilic signals and transduce them to Nrf2, Yamamoto's laboratory recently developed a zebrafish model system for Nrf2-Keap1 signaling.[52] They cloned several zebrafish cytoprotective cDNAs using an extensive expressed sequence tag (EST) database and found their expression to be efficiently induced by tBHQ.[52] They also cloned Nrf2 and Keap1 from zebrafish and used loss- and gain-of-function analyses to demonstrate that Nrf2 is a primary regulator of a subset of cytoprotective enzyme genes, while Keap1 represses Nrf2 activity by binding to it in the cytoplasm.[52] Also in this paper, the authors screened for missense mutations that prevent the Nrf2-Keap1 interaction. They identified an ETGE motif in the Nrf2-ECH homology 2 (Neh2) domain of Nrf2 that is required for binding to Keap1 (Fig. 7). Taken together, these results suggest that the Nrf2-Keap1 signaling system is highly conserved among vertebrates and that the Nrf2-Keap1 interaction is an important means of regulating the antioxidant/electrophilic signaling system.[52]

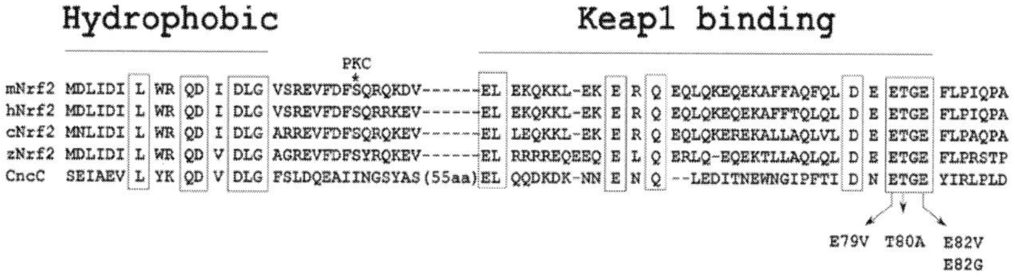

Hydrophobic   **Keap1 binding**

```
                                           PKC
                                            *
mNrf2  MDLIDI L WR QD I DLG VSREVFDFSQRQKDV------EL EKQKKL-EK E R Q EQLQKEQEKAFFAQFQL D E ETGE FLPIQPA
hNrf2  MDLIDI L WR QD I DLG VSREVFDFSQRRKEV------EL EKQKKL-EK E R Q EQLQKEQEKAFFTQLQL D E ETGE FLPIQPA
cNrf2  MNLIDI L WR QD I DLG ARREVFDFSQRQKEV------EL LEQKKL-EK E R Q EQLQKEREKALLAQLVL D E ETGE FLPAQPA
zNrf2  MDLIDI L WR QD V DLG AGREVFDFSYRQKEV------EL RRRREQEEQ E L Q ERLQ-EQEKTLLAQLQL D E ETGE FLPRSTP
CncC   SEIAEV L YK QD V DLG FSLDQEAIINGSYAS(55aa)EL QQDKDK-NN E N Q --LEDITNEWNGIPFTI D N ETGE YIRLPLD
```

                                                                       E79V T80A E82V
                                                                                   E82G

Figure 14.7. Comparison of the Neh2 domains from mouse, human, C. elegans, zebrafish, and *Drosophila* Nrf2 (CncC). The absolutely conserved amino acids are boxed, the PKC phosphorylation site is indicated, and the mutations in the ETGE motif that disrupts binding to Keap1 are indicated.

Zebrafish is superior in many ways to mouse or tissue culture models. Recently, an *in vivo* gene targeting strategy using morpholino phosphorodiamidate oligonucleotides (MOs) was established that works well in zebrafish.[53] Yamamoto's laboratory used this technique to reduce the endogenous expression level of Nrf2 in zebrafish embryos and showed that GST-π expression is abolished even after treating the embryos with tBHQ.[52] They also overexpressed zKeap1 in embryos and showed that it reduces GST-π expression.[52] However, they did not mention using the MO technique to reduce zKeap1 levels, so it remains to be determined whether loss of Keap1 is sufficient for GST-π expression or whether post-translational modification of Nrf2 is required before it is able to activate transcription of this gene.

14.9 *Drosophila* and Nrf2-Keap1 Signaling

Drosophila is an excellent model to study the mechanisms of oxidative signaling because most of the phase 2 enzymes and antioxidant molecules such as GSH are present in *Drosophila*, and mutant and overexpressing strains exist for many of these genes. Furthermore, in contrast to vertebrates, many of the phase 2 enzymes are present in only one or two copies in *Drosophila*, thus greatly simplifying the genetic and proteomics analyses. *Drosophila* is a powerful experimental model because the genome is completely sequenced, there is a mature database, handling the animal is simple, and techniques for making mutations are convenient.[54] Its highly conserved signaling pathways compared with higher animals provide a basis and reasonable starting point for similar studies in higher animals. After searching the *Drosophila* genomic sequence database, one can find likely *Drosophila* orthologs of mammalian

Nrf2 and Keap1, CncC and dKeap1.[52] This allows one the possibility of using *Drosophila* to study the Nrf2-Keap1 signaling mechanism.

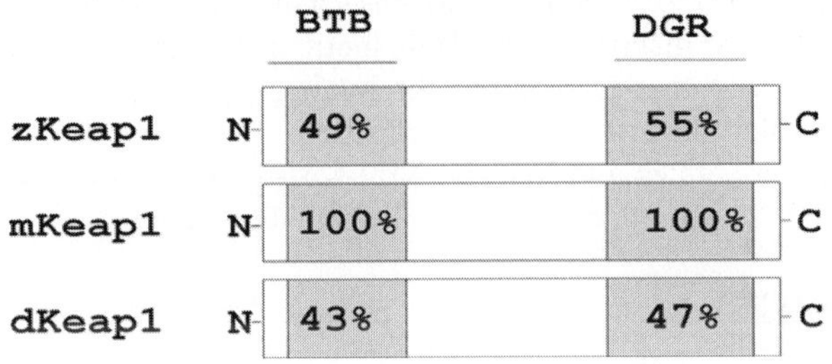

Figure 14.8. Comparison of the BTB and DGR motifs from zebrafish, mouse, and *Drosophila* Keap1 proteins, with respect to the mouse Keap1 protein.

Both mouse Keap1 and rat Keap1 have BTB/POZ and double-glycine repeat/Kelch (DGR) domains that Cooley proposed is the domain for protein-protein interaction (Fig. 8).[55] Itoh *et al.* demonstrated that the DGR domain is critical in the interaction of Keap1 with Nrf2. In addition to the DGR domain, Jaiswal showed that the 27 amino acids C-terminal to the DGR domain are sufficient for the interaction.[23] The alignment in Fig. 8 reveals high similarity between mouse, zebrafish, and *Drosophila* Keap1 in both critical domains.

Nrf2 is also highly conserved in species varying from *C. elegans* to humans. Based on the comparison of mouse and chicken Nrf2, Itoh *et al.* proposed six domains that comprise three functional parts.[22] The three parts are the Keap1 interacting part (KB), the transcriptional activation part (TA), and the Cnc-bZIP part (Fig. 9).[22] BLAST comparison of human Nrf2 to the *Drosophila* genome suggests that Cap'n'collar isoform C (CncC) is the *Drosophila* homologue to Nrf2. The *Drosophila cnc* gene encodes three isoforms due to different transcription start sites and RNA splicing. They are isoforms A, B and C, among which A is the shortest, and C is the longest. The alignment of Cnc proteins to human Nrf2 reveals that all Cnc isoforms share the bZIP domain with human Nrf2, but the Keap1-interacting and transactivation domains are present only in CncC, albeit in a different arrangement (Fig. 9). In CncC, the putative transcription activation domain is amino-terminal to the putative Keap1 interacting domain, whereas in all other sequenced genomes except mosquito, these domains are arranged in the opposite manner (Fig. 9).

14.10 ARE/EpRE Elements in *Drosophila* Genes

Drosophila has a single Nrf2 gene (*cnc*) and a single Keap1 gene (*dKeap1*), but a critical question is whether these proteins function in a conserved antioxidant manner as in mammalian cells. *Drosophila* Nrf2 has three alternative transcript isoforms Cnc-A, -B, and –C, but only the isoform Cnc-C has all of the domains required for Nrf2 binding and transcriptional activation (Fig. 9). If the antioxidant response is conserved between mammals and *Drosophila*, then there should be

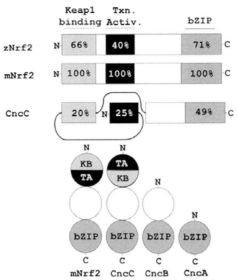

Nrf2-Maf binding sites in the enhancer regions of *Drosophila* genes that encode anti-oxidant enzymes. As described below, searching the *Drosophila* genomic database reveals that this is indeed the case.

Figure 14.9. Comparison of the Transcriptional Activation, Keap1 binding, and bZIP domains from zebrafish, mouse, and *Drosophila* Nrf2 proteins, with respect to the mouse Nrf2 protein.

In a random nucleotide selection experiment, McGinnis' laboratory determined the CncA-Maf-S (Nrf2-small Maf) heterodimer binding site in *Drosophila* is 5'-TGCnnnnTCAT-3'.[57] We searched 3-kb-long 5'-flanking regions of all 13,254 annotated genes in the Release 2.0 (October, 2000) version of the *Drosophila* genome sequence for matches to the Nrf2-Maf consensus DNA recognition site. One of the *Drosophila* thioredoxin genes, *deadhead* (*dhd*), has two matches to the 5'-TGCnnnnTCAT-3' sequence, the gene encoding the extracellular isoform of Cu/Zn-Sod, *ecSod*, also has two matches, and several other anti-oxidant genes have one match to the consensus, including *Catalase 2* (*Cat2*), *glutathione peroxidase 2* (*gp2*), *gst2, gstD1, thioredoxin-2 and –3* (*trx-2* and *trx-3*), *gclc*, and *ggt*. Consistent with these genes being coordinately regulated by antioxidant response-like elements, Jans' laboratory found using microarrays that represent less than 10% of the *Drosophila* genome that *GstD1* and several other *Gst* genes are upregulated when flies are fed paraquat, a superoxide-generating molecule [58]. While these whole genome analyses provide suggestive evidence that *Drosophila* has an anti-oxidant response system utilizing Nrf2, Keap1, and small Maf proteins, this has not yet been demonstrated experimentally.

14.11 CncB is required During *Drosophila* Embryonic Head Development

As described above, the *Drosophila* Cnc gene encodes three isoforms that have different N-terminal amino acids (Fig. 9). McGinnis' laboratory found that only CncB is required for proper development of certain embryonic head structures.[57] They also show that CncB transcriptional repression activity requires a *Drosophila* homolog of the mammalian small Maf protein, Maf-S.[57] McGinnis' laboratory characterized four alleles of Cnc and found that only one of them, cnc^{K6}, does not cause embryonic head defects. As diagrammed in Fig. 10, DNA sequencing experiments reveal that the three alleles that cause head defects, cnc^{C29}, cnc^{C5}, and cnc^{26}, disrupt the protein in the CncA region that is common to all three Cnc isoforms. However, the allele cnc^{K6} that does not cause head defects introduces a nonsense mutation in the coding region that is unique for the CncC isoform.[57] The cnc^{K6} allele encodes a truncated CncC protein that still encodes the N-terminal Neh2 domain that interacts with Keap1, but it does not affect expression or function of CncB or CncA.[57] The absence of head defects in cnc^{K6} (or more precisely, $cncC^{K6}$) mutant embryos strongly suggests that the CncC isoform is not required for the development of the embryonic head skeleton.[57] Since cnc^{K6} is lethal over a null allele of cnc, cnc^{VL110}, they propose that CncC is apparently required for other

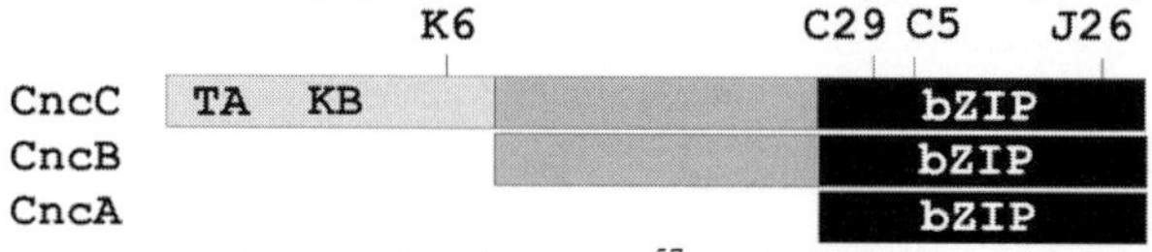

essential functions later during development.[57] It is attractive to speculate that CncC might function as a transcriptional regulator of phase 2 detoxifying genes, but, as described in the next section, it might have other developmental functions as well.

Figure 14.10. A representation of the CncC, CncB, and CncA proteins and the locations of four mutations and the transcriptional activation (TA) and Keap1-binding (KB) domains.

14.12 Cnc is required During *Drosophila* Oogenesis

Recently, Roth's laboratory showed that Cnc also has a role in anchoring the germinal vesicle (the oocyte nucleus) to the cortical actin network at the dorsal-anterior region of the oocyte during *Drosophila* oogenesis.[59] They found that egg chambers mutant for two alleles of cnc, cnc^{03921} and cnc^{5561}, have normal cytoskeletal-mediated migration of the germinal vesicle from the posterior of the oocyte to the anterior of the oocyte during the middle (stages 8 to 9) of oogenesis. However, at the end of stage 9 in the mutant oocytes, the nucleus drifts away from its dorsal-anterior position towards the posterior pole.[59] Interestingly, the mislocalization of the nucleus is accompanied by changes in the microtubule and actin cytoskeletal networks.[59] Unfortunately, both of the alleles of cnc that Roth's laboratory used in their paper carry a P-element transposon inserted in the exon common to all three isoforms.[59] Therefore, these experiments do not provide information about which isoform of Cnc (A, B, or C) is required for anchoring the oocyte to the cortical actin cytoskeleton during oogenesis.

Again, it is attractive to speculate that the CncC isoform is the only one that is required for anchoring the germinal vesicle to the cortical actin cytoskeleton because CncC is the only one that has the Neh2 domain that is required for binding to Keap1. The reason for this speculation is that Keap1 has multiple actin binding motifs in the DGR domain. Recently, Keap1 has been found to bind to Myosin-VIIa in specialized adhesion junctions,[60] thus supporting the idea that Keap1 might have a role in actin structural dynamics. However, immunoprecipitation of transiently-transfected GFP-Keap1 with anti-GFP antibodies reportedly does not pull down actin,[61] thus providing some controversy for Keap1 having a role in actin dynamics.

If CncC is the only isoform that is required for oogenesis, as we speculate, then developmental genetic studies might be useful in identifying new components of the Nrf2-Keap1 anti-oxidant signaling pathway. Because of its sophisticated genetics, especially developmental genetics, *Drosophila* promises to be an extremely powerful system for teasing apart the evolutionarily conserved aspects of Nrf2-Keap1 signaling.

14.13 Future Prospects and Outstanding Questions on Keap1-Nrf2 Signaling

In this review, the major proposition is that a phylogenetic comparison of the components of the Nrf2-Keap1 signaling system might allow a better understanding of the essential components of this system. Research investigating the mechanisms of Nrf2-Keap1 signaling is still in its infancy and many questions are still outstanding. What is upstream of Nrf2-Keap1? Do other transcription factors bind to the ARE? Are Small Mafs activators or repressors? What are the post-translational modifications on Nrf2 and Keap1? Can genetic analyses of the role of CncC during oogenesis provide a better understanding of the Nrf2-Keap1 signaling system? At the rate that this field is progressing, all of these questions will probably be answered in the next few years.

References

1. Finkel, T. and N. J. Holbrook. 2000. Oxidants, oxidative stress and the biology of ageing. *Nature* **408**:239-247.

2. Forman, H. J. and M. Torres. 2001. Redox signaling in macrophages. *Mol Aspects Med* **22**:189-216.

3. Forman, H. J. and M. Torres. 2001. Signaling by the respiratory burst in macrophages. *IUBMB Life* **51**:365-371.

4. Forman, H. J., M. Torres and J. Fukuto. 2002. Redox signaling. *Mol Cell Biochem* **234-235**:49-62.

5. Chan, J. Y. and M. Kwong. 2000. Impaired expression of glutathione synthetic enzyme genes in mice with targeted deletion of the Nrf2 basic-leucine zipper protein. *Biochim Biophys Acta* **1517**:19-26.

6. McMahon, M., K. Itoh, M. Yamamoto, S. A. Chanas, C. J. Henderson, L. I. McLellan, C. R. Wolf, C. Cavin and J. D. Hayes. 2001. The Cap'n'Collar basic leucine zipper transcription factor Nrf2 (NF-E2 p45-related factor 2) controls both constitutive and inducible expression of intestinal detoxification and glutathione biosynthetic enzymes. *Cancer research* **61**:3299-3307.

7. Jeyapaul, J. and A. K. Jaiswal. 2000. Nrf2 and c-Jun regulation of antioxidant response element (ARE)-mediated expression and induction of gamma-glutamylcysteine synthetase heavy subunit gene. *Biochem Pharmacol* **59**:1433-1439.

8. Wild, A. C., H. R. Moinova and R. T. Mulcahy. 1999. Regulation of gamma-glutamylcysteine synthetase subunit gene expression by the transcription factor Nrf2. *J Biol Chem* **274**:33627-33636.

9. Moinova, H. R. and R. T. Mulcahy. 1999. Up-regulation of the human gamma-glutamylcysteine synthetase regulatory subunit gene involves binding of Nrf-2 to an electrophile responsive element. *Biochem Biophys Res Commun* **261**:661-668.

10. Moi, P., K. Chan, I. Asunis, A. Cao and Y. W. Kan. 1994. Isolation of NF-E2-related factor 2 (Nrf2), a NF-E2-like basic leucine zipper transcriptional activator that binds to the tandem NF-E2/AP1 repeat of the beta-globin locus control region. *Proc Natl Acad Sci U S A* **91**:9926-9930.

11. Chan, K., R. Lu, J. C. Chang and Y. W. Kan. 1996. NRF2, a member of the NFE2 family of transcription factors, is not essential for murine erythropoiesis, growth, and development. *Proc Natl Acad Sci U S A* **93**:13943-13948.

12. Itoh, K., T. Chiba, S. Takahashi, T. Ishii, K. Igarashi, Y. Katoh, T. Oyake, N. Hayashi, K. Satoh, I. Hatayama, M. Yamamoto and Y. Nabeshima. 1997. An Nrf2/small Maf heterodimer mediates the induction of phase II detoxifying enzyme genes through antioxidant response elements. *Biochem Biophys Res Commun* **236**:313-322.

13. Mulcahy, R. T. and J. J. Gipp. 1995. Identification of a putative antioxidant response element in the 5'-flanking region of the human gamma-glutamylcysteine synthetase heavy subunit gene. *Biochem Biophys Res Commun* **209**:227-233.

14. Jaiswal, A. K. 1994. Antioxidant response element. *Biochem Pharmacol* **48**:439-444.

15. Rushmore, T. H., R. G. King, K. E. Paulson and C. B. Pickett. 1990. Regulation of glutathione S-transferase Ya subunit gene expression: identification of a unique xenobiotic-responsive element controlling inducible expression by planar aromatic compounds. *Proc Natl Acad Sci U S A* **87**:3826-3830.

16. Rushmore, T. H. and C. B. Pickett. 1990. Transcriptional regulation of the rat glutathione S-transferase Ya subunit gene. Characterization of a xenobiotic-responsive element controlling inducible expression by phenolic antioxidants. *J Biol Chem* **265**:14648-14653.

17. Nguyen, T., T. H. Rushmore and C. B. Pickett. 1994. Transcriptional regulation of a rat liver glutathione S-transferase Ya subunit gene. Analysis of the antioxidant response element and its activation by the phorbol ester 12-O-tetradecanoylphorbol-13-acetate. *J Biol Chem* **269(18):13656-62**:13656-13662.

18. Rahman, I., A. Bel, B. Mulier, M. F. Lawson, D. J. Harrison, W. Macnee and C. A. Smith. 1996. Transcriptional regulation of gamma-glutamylcysteine synthetase-heavy subunit by oxidants in human alveolar epithelial cells. *Biochem Biophys Res Commun* **229**:832-837.

19. Li, J., J. M. Lee and J. A. Johnson. 2002. Microarray analysis reveals an antioxidant responsive element-driven gene set involved in conferring protection from an oxidative stress-induced apoptosis in IMR-32 cells. *J Biol Chem* **277**:388-394.

20. Kong, A. N., E. Owuor, R. Yu, V. Hebbar, C. Chen, R. Hu and S. Mandlekar. 2001. Induction of xenobiotic enzymes by the map kinase pathway and the antioxidant or electrophile response element (ARE/EpRE). *Drug Metabolism Review* **33**:255-271.

21. Rushmore, T. H., M. R. Morton and C. B. Pickett. 1991. The antioxidant responsive element. Activation by oxidative stress and identification of the DNA consensus sequence required for functional activity. *J Biol Chem* **266**:11632-11639.

22. Itoh, K., N. Wakabayashi, Y. Katoh, T. Ishii, K. Igarashi, J. D. Engel and M. Yamamoto. 1999. Keap1 represses nuclear activation of antioxidant responsive elements by Nrf2 through binding to the amino-terminal Neh2 domain. *Genes Dev* **13**:76-86.

23. Dhakshinamoorthy, S. and A. K. Jaiswal. 2001. Functional characterization and role of INrf2 in antioxidant response element-mediated expression and antioxidant induction of NAD(P)H:quinone oxidoreductase1 gene. *Oncogene* **20**:3906-3917.

24. Braun, S., C. Hanselmann, M. Gassmann, U. auf dem Keller, C. Born-Berclaz, K. Chan, Y. Kan and S. Werner. 2002. Nrf2 transcription factor, a novel target of keratinocyte growth factor action which regulates gene expression and inflammation in the healing skin wound. *Mol Cell Biol* **22**:5492-5505.

25. Kwak, M., K. Itoh, M. Yamamoto and T. Kensler. 2002. Enhanced expression of the transcription factor Nrf2 by cancer chemopreventive agents: role of antioxidant response element-like sequences in the nrf2 promoter. *Mol Cell Biol* **22**:2883-2892.

26. Kwak, M., K. Itoh, M. Yamamoto, T. Sutter and T. Kensler. 2001. Role of transcription factor Nrf2 in the induction of hepatic phase 2 and antioxidative enzymes in vivo by the cancer chemoprotective agent, 3H-1, 2-dimethiole-3-thione. *Mol Med* **7**:135-145.

27. Ishii, T., K. Itoh, S. Takahashi, H. Sato, T. Yanagawa, Y. Katoh, S. Bannai and M. Yamamoto. 2000. Transcription factor Nrf2 coordinately regulates a group of oxidative stress-inducible genes in macrophages. *J Biol Chem* **275**:16023-16029.

28. Yu, Y., Y. Wang, M. Li and P. Kannan. 2002. Tumorigenic effect of transcription factor hAP-2alpha and the intricate link between hAP-2alpha activation and squelching. *Mol Carcinog* **34**:172-179.

29. Gill, G. and M. Ptashne. 1988. Negative effect of the transcriptional activator GAL4. *Nature* **334**:721-724.

30. Kong, A. N., E. Owuor, R. Yu, V. Hebbar, C. Chen, R. Hu and S. Mandlekar. 2001. Induction of xenobiotic enzymes by the MAP kinase pathway and the antioxidant or electrophile response element (ARE/EpRE). *Drug Metab Rev* **33**:255-271.

31. Huang, H. C., T. Nguyen and C. B. Pickett. 2000. Regulation of the antioxidant response element by protein kinase C- mediated phosphorylation of NF-E2-related factor 2. *Proc Natl Acad Sci U S A* **97**:12475-12480.

32. Huang, H. C., T. Nguyen and C. B. Pickett. 2002. Phosphorylation of Nrf2 at Ser40 by protein kinase C regulates antioxidant response element-mediated transcription. *J Biol Chem* **26**:26.

33. Lee, J. M., J. M. Hanson, W. A. Chu and J. A. Johnson. 2001. Phosphatidylinositol 3-kinase, not extracellular signal-regulated kinase, regulates activation of the antioxidant-responsive element in IMR-32 human neuroblastoma cells. *J Biol Chem* **276**:20011-20016.

34. Zipper, L. M. and R. T. Mulcahy. 2000. Inhibition of ERK and p38 MAP kinases inhibits binding of Nrf2 and induction of GCS genes. *Biochem Biophys Res Commun* **278**:484-492.

35. Goldkorn, T., N. Balaban, K. Matsukuma, V. Chea, R. Gould, J. Last, C. Chan and C. Chavez. 1998. EGF-Receptor phosphorylation and signaling are targeted by H2O2 redox stress. *Am J Respir Cell Mol Biol* **19**:786-798.

36. Rezaul, K., K. Sada and H. Yamamura. 1998. Involvement of reactive oxygen intermediates in lectin-induced protein- tyrosine phosphorylation of Syk in THP-1 cells. *Biochem Biophys Res Commun* **246**:863-867.

37. Huang, R. P., J. X. Wu, Y. Fan and E. D. Adamson. 1996. UV activates growth factor receptors via reactive oxygen intermediates. *J Cell Biol* **133**:211-220.

38. Banan, A., J. Z. Fields, Y. Zhang and A. Keshavarzian. 2001. Phospholipase C-gamma inhibition prevents EGF protection of intestinal cytoskeleton and barrier against oxidants. *Am J Physiol Gastrointest Liver Physiol* **281**:G412-423.

39. Wang, X. T., K. D. McCullough, X. J. Wang, G. Carpenter and N. J. Holbrook. 2001. Oxidative stress-induced phospholipase C-gamma 1 activation enhances cell survival. *J Biol Chem* **276**:28364-28371.

40. Yang, J. M., A. D. Vassil and W. N. Hait. 2001. Activation of phospholipase C induces the expression of the multidrug resistance (MDR1) gene through the Raf-MAPK pathway. *Mol·Pharmacol* **60**:674-680.

41. Lee, J. M., J. D. Moehlenkamp, J. M. Hanson and J. A. Johnson. 2001. Nrf2-dependent activation of the antioxidant responsive element by tert-butylhydroquinone is independent of oxidative stress in IMR-32 human neuroblastoma cells. *Biochem Biophys Res Commun* **280**:286-292.

42. Dinkova-Kostova, A. T., W. D. Holtzclaw, R. N. Cole, K. Itoh, N. Wakabayashi, Y. Katoh, M. Yamamoto and P. Talalay. 2002. Direct evidence that sulfhydryl groups of Keap1 are the sensors regulating induction of phase 2 enzymes that protect against carcinogens and oxidants. *Proc Natl Acad Sci U S A* **99**:11908-11913.

43. Zipper, L. M. and R. T. Mulcahy. 2002. The Keap1 BTB/POZ dimerization function is required to sequester Nrf2 in cytoplasm. *J Biol Chem* **26**:26.

44. Huang, H. C., T. Nguyen and C. B. Pickett. 2000. Regulation of the antioxidant response element by protein kinase C-mediated phosphorylation of NF-E2-related factor 2. *Proc Natl Acad Sci U S A* **97**:12475-12480.

45. Yu, R., W. Lei, S. Mandlekar, M. J. Weber, C. J. Der, J. Wu and A. T. Kong. 1999. Role of a mitogen-activated protein kinase pathway in the induction of phase II detoxifying enzymes by chemicals. *J Biol Chem* **274**:27545-27552.

46. Yu, R., C. Chen, Y. Y. Mo, V. Hebbar, E. D. Owuor, T. H. Tan and A. N. Kong. 2000. Activation of mitogen-activated protein kinase pathways induces antioxidant response element-mediated gene expression via a Nrf2-dependent mechanism. *Journal of Biological chemistry* **275**:39907-39913.

47. Yu, R., S. Mandlekar, W. Lei, W. E. Fahl, T. H. Tan and A. T. Kong. 2000. p38 mitogen-activated protein kinase negatively regulates the induction of phase II drug-metabolizing enzymes that detoxify carcinogens. *J Biol Chem* **275**:2322-2327.

48. Zipper, L. M. and R. T. Mulcahy. 2000. Inhibition of ERK and p38 MAP kinases inhibits binding of Nrf2 and induction of GCS genes. *Biochem Biophys Res Commun* **278**:484-492.

49. Dhakshinamoorthy, S. and A. K. Jaiswal. 2000. Small maf (MafG and MafK) proteins negatively regulate antioxidant response element-mediated expression and antioxidant induction of the NAD(P)H:Quinone oxidoreductase1 gene. *J Biol Chem* **275**:40134-40141.

50. Marini, M. G., K. Chan, L. Casula, Y. W. Kan, A. Cao and P. Moi. 1997. hMAF, a small human transcription factor that heterodimerizes specifically with Nrf1 and Nrf2. *J Biol Chem* **272**:16490-16497.

51. Kataoka, K., M. Nishizawa and S. Kawai. 1993. Structure-function analysis of the maf oncogene product, a member of the b-Zip protein family. *J Virol* **67**:2133-2141.

52. Kobayashi, M., K. Itoh, T. Suzuki, H. Osanai, K. Nishikawa, Y. Katoh, Y. Takagi and M. Yamamoto. 2002. Identification of the interactive interface and phylogenic conservation of the Nrf2-Keap1 system. *Genes Cells* **7**:807-820.

53. Ekker, S. C. and J. D. Larson. 2001. Morphant technology in model developmental systems. *Genesis* **30**:89-93.

54. Beal, M. F. 2001. Experimental models of Parkinson's disease. *Nat Rev Neurosci* **2**:325-334.

55. Adams, J., R. Kelso and L. Cooley. 2000. The kelch repeat superfamily of proteins: propellers of cell function. *Trends Cell Biol* **10**:17-24.

56. McGinnis, N., E. Ragnhildstveit, A. Veraksa and W. McGinnis. 1998. A cap 'n' collar protein isoform contains a selective Hox repressor function. *Development* **125**:4553-4564.

57. Veraksa, A., N. McGinnis, X. Li, J. Mohler and W. McGinnis. 2000. Cap 'n' collar B cooperates with a small maf subunit to specify pharyngeal development and suppress deformed homeotic function in the drosophila head. *Development* **127**:4023-4037.

58. Zou, S., S. Meadows, L. Sharp, L. Y. Jan and Y. N. Jan. 2000. Genome-wide study of aging and oxidative stress response in Drosophila melanogaster. *Proc Natl Acad Sci U S A* **97**:13726-13731.

59. Guichet, A., F. Peri and S. Roth. 2001. Stable anterior anchoring of the oocyte nucleus is required to establish dorsoventral polarity of the Drosophila egg. *Dev Biol* **237**:93-106.

60. Velichkova, M., J. Guttman, C. Warren, L. Eng, K. Kline, A. W. Vogl and T. Hasson. 2002. A human homologue of Drosophila kelch associates with myosin-VIIa in specialized adhesion junctions. *Cell Motil Cytoskeleton* **51**:147-164.

61. T'Jampens, D., L. Devriendt, V. De Corte, J. Vandekerckhove and J. Gettemans. 2002. Selected   BTB/POZ-kelch   proteins   bind   ATP.   *FEBS   Lett*   **516**:20-26.

Chapter 15

THE NO-CYTOCHROME C OXIDASE SIGNALING PATHWAY; MECHANISMS AND BIOLOGICAL IMPLICATIONS

Sruti Shiva, Anna-Liisa Levonen, Maria Cecilia Barone, and Victor M. Darley-Usmar

15.1 Introduction

The first unequivocal evidence that reactive oxygen or nitrogen species (ROS/RNS) could act as primary or second messengers in signal transduction was the finding that endothelium derived relaxing factor was the free radical nitric oxide (NO).[1] In comparison to the protein, peptide, or lipid macromolecules that were known to initiate signaling, NO at first glance seemed completely unsuitable for this purpose. How such a small, chemically reactive molecule could exhibit specificity, induce amplification of a signal or show controlled production and reversibility of activation was at first far from clear. However, although the chemical properties of NO are quite different to many signaling molecules it is now evident that these characteristics are entirely consistent with the ability to initiate signal transduction. Indeed, the properties of NO such as it being a free radical, low molecular weight with few structural features and high lipid solubility make it particularly suitable for signaling between cells in the vasculature and central nervous system.[2,3,4,3,5] Since the discovery that NO can activate the cGMP-soluble guanylate cyclase pathway[6] it has become clear that NO can also initiate other signal transduction pathways. Identifying the molecular targets in these cases has turned out to be more difficult but also resulted in the emergence of new aspects of what is now called cellular redox signaling.[2,3]

H. J. Forman, J. Fukuto, and M. Torres (eds.), Signal Transduction by Reactive Oxygen and Nitrogen Species: Pathways and Chemical Principles, ©2003, Kluwer Academic Publishers. Printed in the Netherlands

Cellular redox signaling is mediated by the post-translational modification of proteins by ROS/RNS.[7,8] Two important modifications of proteins by NO have emerged which are able to induce conformational changes leading modulation of function and hence have the potential to contribute to cell signaling. The first of these is the nitrosylation of metalloproteins by NO at the metal active site of enzymes. Nitrosylation, not to be confused with S-nitros(yl)ation, is the addition of NO to heme or a metal site which results in ligand induced changes in the co-ordination of the protein. The two examples for which this mode of action has been defined are soluble guanylate cyclase and cytochrome c oxidase. The soluble guanylate cyclase-cGMP pathway is well understood and here the focus will be cytochrome c oxidase and more general aspects of the interaction of NO with mitochondria. In addition to nitrosylation of heme the modification of thiols by NO to form an S-nitrosated species is also emerging as a potentially important NO-dependent post translational modification but will not be discussed in detail here.

Interestingly, the mitochondrion is one of the best examples of a cellular target within which several forms of NO signaling seemingly coexist. For example, NO directly reacts with the transition metals in aconitase[9] and cytochrome c oxidase,[10,11] causing inhibition of these enzymes. However, NO derivatives, such as peroxynitrite, inhibit other components of the respiratory chains including complexes I and V,[12,13] III[14,15] and cytochrome c.[16] While the mechanisms responsible for the nitration of cytochrome c and the supposed S-nitrosation of complex I are still emerging, much more has been uncovered about the NO dependent inhibition of cytochrome c oxidase since its initial discovery by several groups in 1994.[10,11,17]

The main characteristics which quickly emerged in these studies were that inhibition of cytochrome c oxidase by NO was rapid, reversible, and increased as the oxygen concentration was decreased in the isolated enzyme,[18] rat mitochondria,[11,17] and astrocytes.[19] Although this inhibition was known to be reversible, it was viewed primarily as a pathway leading to uncontrolled cytotoxicity. This is due to the fact that prolonged inhibition of respiration by NO can deplete ATP levels to cause cell death.[20,21] This view was strengthened by the finding that NO-dependent inhibition of the enzyme increases formation of reactive oxygen species (ROS) from the mitochondrion.[22,23] Clearly, in these early studies these findings were entirely consistent with the prevailing notion that ROS led solely to damage to biomolecules and cell death. This view is now tempered by the knowledge that ROS can participate in cell signaling and have diverse effects on biological responses at low concentrations and are not necessarily cytotoxic.[24,25] Similarly, it is now clear that the interaction of NO with cytochrome c oxidase does not always result in a depletion of ATP and cell death.

These findings have led to the idea that the interaction between NO and cytochrome c oxidase may be an important component of a signaling pathway, initiated by the production of NO and resulting in the regulation of functions such as ATP production, redox cell signaling, and apoptosis. In this chapter, we will examine the molecular mechanism responsible for the binding of NO to cytochrome c oxidase, the direct consequences of this interaction, and finally its implications for cell signaling.

15.2 Molecular Mechanisms Of NO-Cytochrome C Oxidase Interaction Leading To Specificity And Reversibility

15.2.1 NO binding to cytochrome c oxidase.

Cytochrome c oxidase (complex IV) contributes to the maintenance of the electrochemical gradient across the inner mitochondrial membrane necessary for the synthesis of ATP. The enzyme does this by coupling proton transfer across the membrane to the reduction of molecular oxygen to water by reduced cytochrome c.[26,27] The stoichiometry of the overall process is described by the following equation. The subscripts indicate the matrix (in) and the intermembrane side (out) of the inner membrane:

$$4 \; cyt \; c^{2+} + O_2 + 8 \; H^{+}_{(in)} \rightarrow 4 \; cyt \; c^{3+} + 2H_2O + 4H^{+}_{(out)}$$

Cytochrome c and oxygen each bind to two distinct sites of the complex, located on either side of the inner membrane. Electrons enter oxidase via cytochrome c through a diatomic copper center (Cu_A,), and are in rapid equilibrium with a hexacoordinated $heme_a$. The electrons are then transferred to the O_2 active site, which is formed by a pentacoordinated $heme_a$ ($heme_{aa3}$) and a copper atom (Cu_B) and is known as the binuclear center. Once this site is fully reduced, O_2 binds and is reduced to water without the formation of radical intermediates from the enzyme active site.[27] From these data alone it is most likely that NO would then only have the potential to bind to partially or fully reduced intermediates in the bi-nuclear center which also binds oxygen. Indeed, early studies where NO was used simply as a probe of the structure of the redox centers in the enzyme and before the biological importance of this molecule was recognized support this idea.[28]

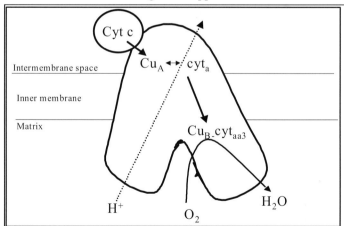

Figure 1: Reduction of O_2 by cytochrome c oxidase. Solid arrows show the movement of electrons from cytochrome c through Cu_A/cyt_a to the binuclear center to reduce O_2 to H_2O. Dotted arrow shows the pumping of H^+ from the matrix to the intermembrane space.

Indeed, more detailed recent studies have shown that despite the potential ability of NO to react with several different targets on cytochrome c oxidase (including an exposed cysteine residue on the protein surface and a tyrosine residue in the oxygen active site) only its reaction with the binuclear center has been observed. Using the purified enzyme it was shown that the NO binds most likely to the reduced Cu_B center before oxygen binds[29] (Figure 2).

Interestingly, the rate of dissociation of NO from the reduced heme of oxidase is surprisingly fast ($k_{off} = 0.0039s^{-1}$ at 20°C, 0.01at $37C^{18}$ and can account for a high affinity but *reversible* binding, leading to the recovery of enzymatic activity when the concentration of NO is decreased. The two processes that appear to contribute to reversibility are the partition of NO into the mitochondrial inner membrane and the possible reduction of NO by the enzyme.[30,31] Both these mechanisms will be discussed next since they both will influence key elements of the NO-cytochrome c oxidase pathway in signaling. Interestingly, they cannot explain why the degree of inhibition of oxidase by NO is higher at lower oxygen tensions.

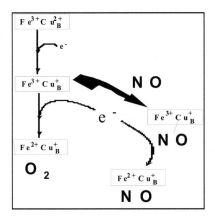

Figure 2: Inhibition of respiration by NO. NO binds the partially reduced binuclear center ($Fe^{3+}Cu_B^+$) to inhibit respiration, while O_2 binds the fully reduced active site ($Fe^{2+}Cu_B^+$).

15.2.2 The importance of the lipid phase of the phospholipid membrane in the reversibility of NO binding to cytochrome c oxidase.

The scheme shown in Figure 2 in which NO binds to a form of the enzyme which oxygen does not implies that the increased potency of NO as an inhibitor of cytochrome c oxidase at low oxygen must be indirect. An example of this behavior is shown in Figure 3. The localization of cytochrome c oxidase within the inner mitochondrial membrane coupled with the capability of NO to partition into biological membranes provides a mechanism by which the concentration of NO binding to the enzyme can be modulated.

The partitioning of NO into biological membranes results in an eightfold increase in the concentration of the molecule in the lipid phase[4] and a corresponding increase in the ability to inhibit a membrane bound target like the oxygen binding site of cytochrome c oxidase. However, O_2 also partitions into membrane thus accelerating the reaction between the two molecules to form nitrite.[4] The resulting competition between NO and oxygen with the binding site of cytochrome c oxidase is then the most likely explanation for both the reversibility of the reaction and the increased potency of NO as an inhibitor of oxidase at lower oxygen tension (Figure 2).[31] An important aspect of this mechanism is the implied reciprocal relationship between the O_2 gradient and NO gradients within a tissue. Paradoxically, although NO inhibits the enzyme responsible for oxygen-dependent energy formation within the cell this process will result in prevention of hypoxia within tissues by regulating oxygen consumption as a function of the NO gradient.[5,31] This theoretical aspect of NO diffusibility and its impact on oxygen gradients was first appreciated by Dr. Lancaster in a series of landmark publications.[4,32,5]

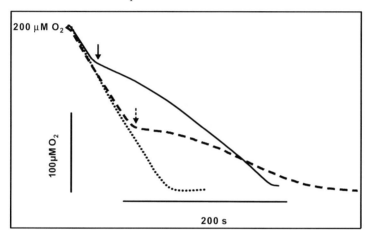

Figure 3: Inhibition of respiration by NO is O_2 dependent. Typical traces of O_2 consumption by isolated rat liver mitochondria. Arrows denote the addition of 1µM NO. Thin dashed line is in the absence of NO. Solid line is with the addition of NO at 80% O_2, and thick dashed line is the addition of NO at 20% O_2.

15.2.3 Metabolism of NO by cytochrome c oxidase.

While NO is known to react with both O_2 and O_2^-, other mechanisms of NO metabolism within the mitochondrion could involve cytochrome c oxidase under conditions of low electron flux.[33,34]. The low affinity binding of NO to the oxidized enzyme can result in the reduction of Cu_B^{2+}, to form NO^+ that may then be hydrated to form nitrite bound to the active site.[35] The nitrite is then released from the binuclear center by a process that is accelerated in the presence of electrons flowing through the enzyme.[30] Whether this process occurs in vivo has yet to be established.

15.3   Direct effects of NO Binding to Cytochrome C Oxidase

15.3.1   NO Inhibits Mitochondrial Respiration

The most well characterized effect of NO binding to cytochrome c oxidase is the inhibition of mitochondrial respiration. Once this phenomenon was shown to occur in isolated mitochondria[36,11,17] and several cell lines,[36,37,19,38,39] one question that quickly surfaced was whether a sufficient concentration of NO was able to gain access to the cytochrome c oxidase binding site to inhibit respiration in vivo.

Several groups have found evidence that cytochrome c oxidase is indeed a target for NO in vivo. Brown *et al.* added cytokines to astrocytes to activate iNOS, which subsequently caused an inhibition of respiration that could be reversed by NOS inhibitors.[19] Dai *et al* found that respiration in myocytes from hypertrophied hearts is significantly inhibited in comparison to control myocytes, an effect that is also reversed by NOS inhibitors.[40] In endothelial cells it was found that stimulation of NOS by bradykinin resulted in respiratory inhibition.[38] Perhaps the most conclusive evidence is the finding that administration of L-NAME, a NOS inhibitor, to conscious dogs increases respiration rate.[41] These data not only provided evidence that NO inhibits respiration in vivo, but also implies that even under basal conditions respiration is regulated by low concentrations of NO.

Initially, it was unclear how inhibition of a respiratory complex can be anything but detrimental to the cell, since it ultimately results in the depletion of ATP and cell death. However, when considered in the context of thresholds, it becomes easier to understand how inhibition of respiration could be a physiological rather than pathological phenomenon and how this inhibition may serve as part of a signaling pathway. While each respiratory complex possesses some degree of control over the process of respiration, this control is distributed unequally among all four complexes and the re-entry of protons into the respiratory chain.[42,43] This type of distribution allows an individual complex to be inhibited to some degree, a threshold, without having a significant effect on respiration.[43] For example Brookes et al showed that in heart mitochondria, cytochrome c oxidase is inhibited by 20% by NO before having a significant effect on respiration.[36] This demonstrates that NO-dependent inhibition of respiration is not an "all or none" response, but rather a graded modulator of respiration.[36] Only when high concentrations of NO are produced does respiration become completely inhibited and cell death ensues.

An important point to consider in the context of thresholds is that the degree of control each complex possesses may change in different situations, producing a new threshold level for a specific inhibitor. Such is the case for mitochondria in differing activity states.[43] It has been found that as mitochondrial respiratory activity increases, so does the potency of NO as an inhibitor of this process.[44] This differential response between respiring and quiescent mitochondria may serve as a regulatory mechanism for the modulation of respiration by NO to protect against cytotoxicity. When the respiration rate is high, cytochrome oxidase is more readily inhibited by NO, but as the respiration rate drops, higher concentrations of NO are needed to affect O_2 consumption, thus avoiding complete inhibition of respiration and depletion of ATP.

15.3.2 NO Regulates Permeability Transition and Cytochrome C Release

NO is a well-known regulator of apoptosis, with high levels inducing apoptosis, and low concentrations protecting against cell death induced by several different stimuli.[45,46,47,48] For example, cells lacking eNOS have been found to become more susceptible to apoptosis.[49] Many mechanisms of NO-mediated regulation of apoptosis have been found. Most recently, the inhibition of the downstream effectors, the caspases, by S-nitrosation has gained attention.[50,51] Several findings support a mitochondrial involvement in the cytoprotective effect of NO via the up-regulation of the anti-apoptotic protein Bcl-2.[52] The finding that cells lacking actively respiring mitochondria lose the protective effect of NO against H_2O_2 induced apoptosis is also consistent with mechanisms targeted at the respiratory chain.[39]

The role of the mitochondrion in apoptotic signaling also extends to components of the respiratory chain. Most notably the discovery that cytochrome c release is a pro-apoptotic signal has focused attention on other functions of the mitochondrial respiratory chain beyond ATP synthesis.[53] Complex I has been implicated in apoptosis since rotenone, a specific inhibitor of the enzyme, prevented TNF-induced apoptosis of endothelial cells.[15] More recently, a new subunit of complex I was found to be a homolog of GRIM-19, the product of a cell death regulatory gene.[54] The role of cytochrome c oxidase in cell death signaling has come under scrutiny since recent findings have demonstrated that susceptibility to apoptosis is dependent on the interaction of cytochrome c with its binding sites in the respiratory chain.[55] Recent studies[56,55] suggest that a competition may exist between the binding sites for cytochrome c on complex III and IV and the pool available to be released from the organelle (Figure 4).

A phenomenon that has been closely associated with cytochrome c release is the opening of the mitochondrial permeability transition pore.[57] In isolated mitochondria, NO has been shown to inhibit permeability transition (PT) and cytochrome c release in a concentration dependent manner.[58,59] This is a controversial area but it is now clear that the release of cytochrome c from the mitochondrion is a very early event not associated with mitochondrial swelling.[59,60,55] The inhibition of PT is rapid and is reversed by the addition of hemoglobin, implying that NO must be rapidly and reversibly affecting a site within the mitochondrion.[59] One possible target for this action may be cytochrome c oxidase. Furthermore, nanomolar levels of NO are sufficient to inhibit PT and cytochrome c release.[59] These concentrations are lower than the concentrations needed to affect ATP production, suggesting that apoptotic signaling can be regulated without affecting other cellular metabolic processes.

It is also important to note that high concentrations of NO are cytotoxic presumably via mechanisms that cause complete, rather than partial, inhibition of ATP synthesis.[20,21] NO has also been found to mediate apoptosis by activating p53, causing cytochrome c release.[61,62] It is worth noting that NO-dependent cytotoxicity is exacerbated by low glucose or oxygen tension; these results are perhaps explained by the inability of anaerobic glycolysis to provide ATP in low glucose conditions and greater inhibition of cytochrome c oxidase by NO in low oxygen tension lowering the threshold between control of cytochrome oxidase activity resulting in diminished ATP synthesis.[63,64,56]

15.3.3   Transduction of Nitrosative to Oxidative Signaling in the Mitochondrion.

It is widely accepted that complex III of the mitochondrion is a significant generator of superoxide (O_2^-) within the cell and that changing the level of respiratory chain reduction can alter the level of production of this free radical.[65] NO, by binding to cytochrome c oxidase inhibits electron flux through the complex, which increases the level of reduction throughout the respiratory chain and hence increases O_2^- generation.[65] Generation of O_2^- by mitochondria has traditionally been viewed as uncontrolled, and the O_2^- generated has been considered to be a toxic by-product of respiration.[66] However, closer examination of this process reveals that this may be a unique mechanism of signal transduction.[67]

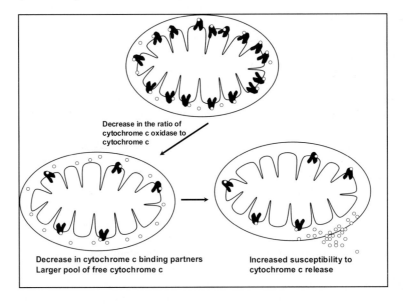

Figure 4: Changing the sensitivity of mitochondria to cytochrome c release. A change in the ratio of cytochrome c to cytochrome c oxidase leads to a larger pool of free cytochrome c, which is more susceptible to being released.

In intact mitochondria, Mn Superoxide Dismutase (MnSOD) is present and converts the O_2^- produced into H_2O_2.[68] This has previously been considered to be the first step in the detoxification of the free radical. However, this conversion may be viewed more as a mechanism of converting O_2^- into a functional signaling molecule. H_2O_2 is a versatile signaling molecule due to its ability to diffuse throughout the cell, and has been implicated in several signaling pathways.[69,70,2471] Considering the inhibition of cytochrome c oxidase as a mechanism to generate H_2O_2 implies that the production of this molecule is coupled to ATP production. However, this is not necessarily the case. Poderoso *et al* have shown that inhibition of isolated mitochondria by NO at concentrations that do not affect ATP synthesis are enough to generate H_2O_2 levels that are high enough to initiate signal transduction.[65] This transduction of a nitrosative to oxidative signal by NO may be a crucial step in many signaling pathways, which ultimately rely on H_2O_2 as discussed in the next section.

15.4 Implications of the Pathway: Mitochondria, NO, and Cellular Redox Signaling

An emerging area of interest is the role of mitochondria-derived reactive oxygen species (ROS) in cell signaling processes. Various extracellular stimuli can cause an increase in the production of ROS by the mitochondrial respiratory chain causing a shift in redox balance and subsequent initiation of downstream signaling cascades. One of the earliest findings was the cytotoxic and gene regulating effects of TNFα, shown to be mediated by mitochondrial ROS formation.[72,73] Increased mitochondrial production of ROS is also crucial in the execution of apoptosis caused by Fas-receptor or p53,[74] the latter being inhibited by the antioxidants vitamin E and ubiquinol conjugated to methyltriphenylphosphonium cation (TPMP) to target them to the mitochondrion.[75] Interestingly, TPMP-ubiquinol has been shown to inhibit apoptosis caused by H_2O_2, suggesting a role of secondary production of ROS by mitochondria also in H_2O_2 induced apoptosis.[76] More recently, a number of other cell signaling pathways have been suggested to be relayed through mitochondria. These include the activation of protein kinase C, nuclear factor κB (NF-κB), and p38 mitogen-activated protein kinase (MAPK) by high glucose,[76,77,78] c-Jun NH2 terminal kinase (JNK)-dependent activation of glycogen synthase,[79] and activation of NF-κB by angiotensin II.[80] Thus mitochondria are not only involved in the pro-apoptotic pathways but also more subtle adaptive responses can be mediated through mitochondrial production of ROS.

What is the role of NO in mediating mitochondrial signaling events? Nitric oxide has been observed to increase ROS production by several investigators in isolated mitochondria,[81] cultured cells[82,65] and isolated rat hearts.[83] While the studies examining the specific sites in the respiratory chain that account for increased ROS production rely upon the use of inhibitors and are therefore difficult to interpret, two putative sites of ROS production, cytochrome bc_1 segment of complex III[65] and cytochrome c oxidase[81] have been identified. An interesting new concept is the possibility of mitochondria producing NO themselves.[84,85] NO produced within mitochondria would then regulate mitochondrial ATP synthesis, oxygen consumption and the production of ROS.[86] Indeed, it has been suggested that mtNOS-derived NO can modulate mitochondrial ROS generation.[81]

Apart from increased production of ROS participating in cell signaling events, inhibition of respiration by NO and thus ATP synthesis may impact on cellular processes, especially in tissues that rely heavily on mitochondrial energy production. Many signaling cascades include protein kinases, which require ATP as a substrate. The notion that decreased cellular ATP level has an impact on kinase activity is supported by recent studies, in which intact mitochondrial respiration and ATP synthesis is needed for the activation of extracellular signal regulated kinases (ERK1/2) by H_2O_2 or ischemia-reperfusion in cardiac myocytes.[87,69]

In conclusion, the mitochondrion is gaining increasing interest as a putative mediator of cell signaling events. Nitric oxide, through its ability to increase the production of ROS within the mitochondria and on the other hand by inhibiting ATP synthesis and thus ATP-dependent processes, has an important role in modifying cell signaling processes.

Acknowledgments:

We are grateful for funding from the National Institutes of Health (s AA12613, HL58031, ES10167). SS is in receipt of an NIH postgraduate training fellowship. A-LL receives support from the Academy of Finland, the Finnish Foundation for Cardiovascular Research, the Emil Aaltonen Foundation, and the Finnish Medical Society Duodecim. .

References

1. Ignarro, L. J., R. E. Byrns, G. M. Buga and K. S. Wood. 1987. Endothelium-derived relaxing factor from pulmonary artery and vein possesses pharmacologic and chemical properties identical to those of nitric oxide radical. *Circ Res* **61**:866-879.879.

2. Patel, R. P., J. McAndrew, H. Sellak, C. R. White, H. Jo, B. A. Freeman and V. M. Darley-Usmar. 1999. Biological aspects of reactive nitrogen species. *Biochim Biophys Acta* **1411** :385-400.400.

3. Stamler, J. S. 1994. Redox signaling: nitrosylation and related target interactions of nitric oxide. *Cell* **78**:931-936.936.

4. Liu, X., M. J. S. Miller, M. S. Joshi, D. D. Thomas and J. R. Lancaster Jr. 1998. Accelerated reaction of nitric oxide with O2 within the hydrophobic interior of biological membranes. *Proc Natl Acad Sci U S A* **95**:2175-2179.2179.

5. Thomas, D. D., X. Liu, S. P. Kantrow and J. R. Lancaster Jr. 2001. The biological lifetime of nitric oxide: implications for the perivascular dynamics of NO and O2. *Proc Natl Acad Sci U S A* **98**:355-360.360.

6. Knowles, R. G., M. Palacios, R. M. Palmer and S. Moncada. 1989. Formation of nitric oxide from L-arginine in the central nervous system: a transduction mechanism for stimulation of the soluble guanylate cyclase. *Proc Natl Acad Sci U S A* **86**:5159-5162.5162.

7. Levonen, A. L., R. P. Patel, P. Brookes, Y. M. Go, H. Jo, S. Parthasarathy, P. G. Anderson and V. Darley-Usmar. 2001. Mechanisms of cell signaling by nitric oxide and peroxynitrite: from mitochondria to MAP kinases. *Antioxid Redox Signal* **3**:215-229.229.

8. Stamler, J. S., S. Lamas and F. Fang. 2001. Nitrosylation. the prototypic redox-based signaling mechanism. *Cell* **106**:675-683.683.

9. Hibbs Jr., J. B., R. R. Taintor, Z. Vavrin and E. M. Rachlin. 1988. Nitric oxide: a cytotoxic activated macrophage effector molecule. *Biochem Biophys Res Commun* **157**:87-94.94.

10. Brown, G. C. and C. E. Cooper. 1994. Nanomolar concentrations of nitric oxide reversibly inhibit synaptosomal respiration by competing with oxygen at cytochrome oxidase. *FEBS Lett* **356**:295-298.298.

11. Cleeter, M. W., J. M. Cooper, V. M. Darley-Usmar, S. Moncada and A. H. Schapira. 1994. Reversible inhibition of cytochrome c oxidase, the terminal enzyme of the mitochondrial respiratory chain, by nitric oxide. Implications for neurodegenerative diseases. *FEBS Lett* **345**:50-54.54.

12. Borutaite, V., A. Budriunaite and G. C. Brown. 2000. Reversal of nitric oxide-, peroxynitrite- and S-nitrosothiol-induced inhibition of mitochondrial respiration or complex I activity by light and thiols. *Biochim Biophys Acta* **1459**:405-412.412.

13. Clementi, E.,G. C. Brown, M. Feelisch and S. Moncada. 1998. Persistent inhibition of cell respiration by nitric oxide: crucial role of S-nitrosylation of mitochondrial complex I and protective action of glutathione. *Proc Natl Acad Sci U S A* **95**:7631-7636.7636.

14. Cassina, A. and R. Radi. 1996. Differential inhibitory action of nitric oxide and peroxynitrite on mitochondrial electron transport. *Arch Biochem Biophys* **328**:309-316.316.

15. Geng, Y., G. K. Hansson and E. Holme. 1992. Interferon-gamma and tumor necrosis factor synergize to induce nitric oxide production and inhibit mitochondrial respiration in vascular smooth muscle cells. *Circ Res* **71**:1268-1276.1276.

16. Cassina, A. M., R. Hodara, J. M. Souza, L. Thomson, L. Castro, H. Ischiropoulos, B. A. Freeman and R. Radi. 2000. Cytochrome c nitration by peroxynitrite. *J Biol Chem* **275**:21409-21415.21415.

17. Schweizer, M. and C. Richter. 1994. Nitric oxide potently and reversibly deenergizes mitochondria at low oxygen tension. *Biochem Biophys Res Commun* **204**:169-175.175.

18. Sarti, P., A. Giuffre, E. Forte, D. Mastronicola, M. C. Barone and M. Brunori. 2000. Nitric oxide and cytochrome c oxidase: mechanisms of inhibition and NO degradation. *Biochem Biophys Res Commun* **274**:183-187.187.

19. Brown, G. C., J. P. Bolanos, S. J. Heales and J. B. Clark. 1995. Nitric oxide produced by activated astrocytes rapidly and reversibly inhibits cellular respiration. *Neurosci Lett* **193**:201-204.204.

20. Brookes, P. S., J. P. Bolanos and S. J. Heales. 1999. The assumption that nitric oxide inhibits mitochondrial ATP synthesis is correct. *FEBS Lett* **446**:261-263.263.

21. Lelli Jr., J. L., L. L. Becks, M. I. Dabrowska and D. B. Hinshaw. 1998. ATP converts necrosis to apoptosis in oxidant-injured endothelial cells. *Free Radic Biol Med* **25**:694-702.702.

22. Cadenas, E. and K. J. Davies. 2000. Mitochondrial free radical generation, oxidative stress, and aging. *Free Radic Biol Med* **29**:222-230.230.

23. Turrens, J. F. 1997. Superoxide production by the mitochondrial respiratory chain. *Biosci Rep* **17**:3-8.8.

24. Suzuki, Y. J., H. J. Forman and A. Sevanian. 1997. Oxidants as stimulators of signal transduction. *Free Radic Biol Med* **22**:269-285.285.

25. Whisler, R. L., M. A. Goyette, I. S. Grants and Y. G. Newhouse. 1995. Sublethal levels of oxidant stress stimulate multiple serine/threonine kinases and suppress protein phosphatases in Jurkat T cells. *Arch Biochem Biophys* **319**:23-35.35.

26. Capaldi, R. A., F. Malatesta and V. M. Darley-Usmar. 1983. Structure of cytochrome c oxidase. *Biochim Biophys Acta* **726**:135-148.148.

27. Cooper, C. E. 2002. Nitric oxide and cytochrome oxidase: substrate, inhibitor or effector? *Trends Biochem Sci* **27**:33-39.39.

28. Stevens, T. H., G. W. Brudvig, D. F. Bocian and S. I. Chan. 1979. Structure of cytochrome a3-Cua3 couple in cytochrome c oxidase as revealed by nitric oxide binding studies. *Proc Natl Acad Sci U S A* **76**:3320-3324.3324.

29. Torres, J., V. Darley-Usmar and M. T. Wilson. 1995. Inhibition of cytochrome c oxidase in turnover by nitric oxide: mechanism and implications for control of respiration. *Biochem J* **312 (Pt 1)**:169-173.173.

30. Giuffre, A., M. C. Barone, D. Mastronicola, E. D'Itri, P. Sarti and M. Brunori. 2000. Reaction of nitric oxide with the turnover intermediates of cytochrome c oxidase: reaction pathway and functional effects. *Biochemistry* **39**:15446-15453.15453.

31. Shiva, S., P. S. Brookes, R. P. Patel, P. G. Anderson and V. M. Darley-Usmar. 2001. Nitric oxide partitioning into mitochondrial membranes and the control of respiration at cytochrome c oxidase. *Proc Natl Acad Sci U S A* **98**:7212-7217.7217.

32. Liu, X., A. Samouilov, J. R. Lancaster Jr. and J. L. Zweier. 2002. Nitric oxide uptake by erythrocytes is primarily limited by extracellular diffusion not membrane resistance. *J Biol Chem* **277**:26194-26199.26199.

33. Beckman, J. S., T. W. Beckman, J. Chen, P. A. Marshall and B. A. Freeman. 1990. Apparent hydroxyl radical production by peroxynitrite: implications for endothelial injury from nitric oxide and superoxide. *Proc Natl Acad Sci U S A* **87**:1620-1624.1624.

34. Wink, D. A., J. F. Darbyshire, R. W. Nims, J. E. Saavedra and P. C. Ford. 1993. Reactions of the bioregulatory agent nitric oxide in oxygenated aqueous media: determination of the kinetics for oxidation and nitrosation by intermediates generated in the NO/O2 reaction. *Chem Res Toxicol* **6**:23-27.27.

35. Cooper, C. E., J. Torres, M. A. Sharpe and M. T. Wilson. 1997. Nitric oxide ejects electrons from the binuclear centre of cytochrome c oxidase by reacting with oxidised copper: a general mechanism for the interaction of copper proteins with nitric oxide? *FEBS Lett* **414**:281-284.284.

36. Brookes, P. S., J. Zhang, L. Dai, F. Zhou, D. A. Parks, V. M. Darley-Usmar and P. G. Anderson. 2001. Increased sensitivity of mitochondrial respiration to inhibition by nitric oxide in cardiac hypertrophy. *J Mol Cell Cardiol* **33**:69-82.82.

37. Brown, G. C., P. L. Lakin-Thomas and M. D. Brand. 1990. Control of respiration and oxidative phosphorylation in isolated rat liver cells. *Eur J Biochem* **192**:355-362.362.

38. Clementi, E., G. C. Brown, N. Foxwell and S. Moncada. 1999. On the mechanism by which vascular endothelial cells regulate their oxygen consumption. *Proc Natl Acad Sci U S A* **96**:1559-1562.1562.

39. Paxinou, E., M. Weisse, Q. Chen, J. M. Souza, C. Hertkorn, M. Selak, E. Daikhin, M. Yudkoff, G. Sowa, W. C. Sessa and H. Ischiropoulos. 2001. Dynamic regulation of metabolism and respiration by endogenously produced nitric oxide protects against oxidative stress. *Proc Natl Acad Sci U S A* **98**:11575-11580.11580.

40. Dai, L., P. S. Brookes, V. M. Darley-Usmar and P. G. Anderson. 2001. Bioenergetics in cardiac hypertrophy: mitochondrial respiration as a pathological target of NO*. *Am J Physiol Heart Circ Physiol* **281**:H2261-H2269.H2269.

41. Shen, W., T. H. Hintze and M. S. Wolin. 1995. Nitric oxide. An important signaling mechanism between vascular endothelium and parenchymal cells in the regulation of oxygen consumption. *Circulation* **92**:3505-3512.3512.

42. Brand, M. D., B. P. Vallis and A. Kesseler. 1994. The sum of flux control coefficients in the electron-transport chain of mitochondria. *Eur J Biochem* **226**:819-829.829.

43. Groen, A. K., R. J. Wanders, H. V. Westerhoff, Van der, R. Meer and J. M. Tager. 1982. Quantification of the contribution of various steps to the control of mitochondrial respiration. *J Biol Chem* **257**:2754-2757.2757.

44. Borutaite, V. and G. C. Brown. 1996. Rapid reduction of nitric oxide by mitochondria, and reversible inhibition of mitochondrial respiration by nitric oxide. *Biochem J* **315 (Pt 1)**:295-299.299.

45. Dimmeler, S., C. Hermann, J. Galle and A. M. Zeiher. 1999. Upregulation of superoxide dismutase and nitric oxide synthase mediates the apoptosis-suppressive effects of shear stress on endothelial cells. *Arterioscler Thromb Vasc Biol* **19**:656-664.664.

46. Mannick, J. B., X. Q. Miao and J. S. Stamler. 1997. Nitric oxide inhibits Fas-induced apoptosis. *J Biol Chem* **272**:24125-24128.24128.

47. Oyadomari, S., K. Takeda, M. Takiguchi, T. Gotoh, M. Matsumoto, I. Wada, S. Akira, E. Araki and M. Mori. 2001. Nitric oxide-induced apoptosis in pancreatic beta cells is mediated by the endoplasmic reticulum stress pathway. *Proc Natl Acad Sci U S A* **98**:10845-10850.10850.

48. Taimor, G., B. Hofstaetter and H. M. Piper. 2000. Apoptosis induction by nitric oxide in adult cardiomyocytes via cGMP- signaling and its impairment after simulated ischemia. *Cardiovasc Res* **45**:588-594.594.

49. Hoffmann, J., J. Haendeler, A. Aicher, L. Rossig, M. Vasa, A. M. Zeiher and S. Dimmeler. 2001. Aging enhances the sensitivity of endothelial cells toward apoptotic stimuli: important role of nitric oxide. *Circ Res* **89**:709-715.715.

50. Li, J., T. R. Billiar, R. V. Talanian and Y. M. Kim. 1997. Nitric oxide reversibly inhibits seven members of the caspase family via S-nitrosylation. *Biochem Biophys Res Commun* **240**:419-424.424.

51. Rossig, L., B. Fichtlscherer, K. Breitschopf, J. Haendeler, A. M. Zeiher, A. Mulsch and S. Dimmeler. 1999. Nitric oxide inhibits caspase-3 by S-nitrosation in vivo. *J Biol Chem* **274**:6823-6826.6826.

52. Rossig, L., J. Haendeler, C. Hermann, P. Malchow, C. Urbich, A. M. Zeiher and S. Dimmeler. 2000. Nitric oxide down-regulates MKP-3 mRNA levels: involvement in endothelial cell protection from apoptosis. *J Biol Chem* **275**:25502-25507.25507.

53. Hengartner, M. O. 2000. The biochemistry of apoptosis.*Nature* **407**:770-776.776.

54. Fearnley, I. M., J. Carroll, R. J. Shannon, M. J. Runswick, J. E. Walker and J. Hirst. 2001. GRIM-19, a cell death regulatory gene product, is a subunit of bovine mitochondrial NADH:ubiquinone oxidoreductase (complex I). *J Biol Chem* **276**:38345-38348.38348.

55. Ott, M., J. D. Robertson, V. Gogvadze, B. Zhivotovsky and S. Orrenius. 2002. Cytochrome c release from mitochondria proceeds by a two-step process. *Proc Natl Acad Sci U S A* **99**:1259-1263.1263.

56. Ramachandran, A., D. R. Moellering, E. Ceaser, S. Shiva, J. Xu and V. Darley-Usmar. 2002. Inhibition of mitochondrial protein synthesis results in increased endothelial cell susceptibility to nitric oxide-induced apoptosis. *Proc Natl Acad Sci U S A* **99**:6643-6648.6648.

57. Crompton, M. 1999. The mitochondrial permeability transition pore and its role in cell death. *Biochem J* **341 (Pt 2)**:233-249.249.

58. Balakirev, M. Y., V. V. Khramtsov and G. Zimmer. 1997. Modulation of the mitochondrial permeability transition by nitric oxide. *Eur J Biochem* **246**:710-718.718.

59. Brookes, P. S., E. P. Salinas, K. Darley-Usmar, J. P. Eiserich, B. A. Freeman, V. M. Darley-Usmar and P. G. Anderson. 2000. Concentration-dependent Effects of Nitric Oxide on Mitochondrial Permeability Transition and Cytochrome c Release. *J Biol Chem* **275** :20474-20479.20479.

60. Eskes, R., B. Antonsson, A. Osen-Sand, S. Montessuit, C. Richter, R. Sadoul, G. Mazzei, A. Nichols and J. C. Martinou. 1998. Bax-induced cytochrome C release from mitochondria is independent of the permeability transition pore but highly dependent on Mg2+ ions. *J Cell Biol* **143**:217-224.224.

61. Kim, Y. M., C. A. Bombeck and T. R. Billiar. 1999. Nitric oxide as a bifunctional regulator of apoptosis. *Circ Res* **84**:253-256.256.

62. Yabuki, M., K. Tsutsui, A. A. Horton, T. Yoshioka and K. Utsumi. 2000. Caspase activation and cytochrome c release during HL-60 cell apoptosis induced by a nitric oxide donor. *Free Radic Res* **32**:507-514.514.

63. Dijkmans, R. and A. Billiau. 1991. Interferon-gamma/lipopolysaccharide-treated mouse embryonic fibroblasts are killed by a glycolysis/L-arginine-dependent process accompanied by depression of mitochondrial respiration. *Eur J Biochem* **202**:151-159.159.

64. Lee, V. Y., D. S. McClintock, M. T. Santore, G. R. Budinger and N. S. Chandel. 2002. Hypoxia sensitizes cells to nitric oxide-induced apoptosis. *J Biol Chem* **277**:16067-16074.16074.

65. Poderoso, J. J., M. C. Carreras, C. Lisdero, N. Riobo, F. Schopfer and A. Boveris. 1996. Nitric oxide inhibits electron transfer and increases superoxide radical production in rat heart mitochondria and submitochondrial particles. *Arch Biochem Biophys* **328**:85-92.92.

66. Boveris, A. and B. Chance. 1973. The mitochondrial generation of hydrogen peroxide. General properties and effect of hyperbaric oxygen. *Biochem J* **134**:707-716.716.

67. Brookes, P. and V. M. Darley-Usmar. 2002. Hypothesis: the mitochondrial NO(?) signaling pathway, and the transduction of nitrosative to oxidative cell signals: an alternative function for cytochrome C oxidase. *Free Radic Biol Med* **32**:370-374.374.

68. Weisiger, R. A. and I. Fridovich. 1973. Superoxide dismutase. Organelle specificity. *J Biol Chem* **248** :3582-3592.3592.

69. Bogoyevitch, M. A., D. C. Ng, N. W. Court, K. A. Draper, A. Dhillon and L. Abas. 2000. Intact mitochondrial electron transport function is essential for signalling by hydrogen peroxide in cardiac myocytes. *J Mol Cell Cardiol* **32**:1469-1480.1480.

70. Iles, K. E., D. A. Dickinson, N. Watanabe, T. Iwamoto and H. J. Forman. 2002. AP-1 activation through endogenous H(2)O(2) generation by alveolar macrophages. *Free Radic Biol Med* **32**:1304-1313.1313.

71. Ramachandran, A., D. Moellering, Y. M. Go, S. Shiva, A. L. Levonen, H. Jo, R. P. Patel, S. Parthasarathy and V. M. Darley-Usmar. 2002. Activation of c-Jun N-terminal kinase and apoptosis in endothelial cells mediated by endogenous generation of hydrogen peroxide. *Biol Chem* **383**:693-701.701.

72. Goossens, V., J. Grooten, K. De Vos and W. Fiers. 1995. Direct evidence for tumor necrosis factor-induced mitochondrial reactive oxygen intermediates and their involvement in cytotoxicity. *Proc Natl Acad Sci U S A* **92**:8115-8119.8119.

73. Schulze-Osthoff, K., A. C. Bakker, B. Vanhaesebroeck, R. Beyaert, W. A. Jacob and W. Fiers. 1992. Cytotoxic activity of tumor necrosis factor is mediated by early damage of mitochondrial functions. Evidence for the involvement of mitochondrial radical generation. *J Biol Chem* **267**:5317-5323.5323.

74. Li, P. F., R. Dietz and R. Von Harsdorf. 1999. p53 regulates mitochondrial membrane potential through reactive oxygen species and induces cytochrome c-independent apoptosis blocked by Bcl-2. *EMBO J* **18**:6027-6036.6036.

75. Hwang, P. M., F. Bunz, J. Yu, C. Rago, T. A. Chan, M. P. Murphy, G. F. Kelso, R. A. Smith, K. W. Kinzler and B. Vogelstein. 2001. Ferredoxin reductase affects p53-dependent, 5-fluorouracil-induced apoptosis in colorectal cancer cells. *Nat Med* **7**:1111-1117.1117.

76. Di Lisa, F. and M. Ziegler. 2001. Pathophysiological relevance of mitochondria in NAD(+) metabolism. *FEBS Lett* **492**:4-8.8.

77. Hsieh, T. J., S. L. Zhang, J. G. Filep, S. S. Tang, J. R. Ingelfinger and J. S. Chan. 2002. High glucose stimulates angiotensinogen gene expression via reactive oxygen species generation in rat kidney proximal tubular cells. *Endocrinology* **143**:2975-2985.2985.

78. Nishikawa, T., D. Edelstein, X. L. Du, S. Yamagishi, T. Matsumura, Y. Kaneda, M. A. Yorek, D. Beebe, P. J. Oates, H. P. Hammes, I. Giardino and M. Brownlee. 2000. Normalizing mitochondrial superoxide production blocks three pathways of hyperglycaemic damage. *Nature* **404**:787-790.790.

79. Nemoto, S., K. Takeda, Z. X. Yu, V. J. Ferrans and T. Finkel. 2000. Role for mitochondrial oxidants as regulators of cellular metabolism. *Mol Cell Biol* **20**:7311-7318.7318.

80. Pueyo, M. E., W. Gonzalez, A. Nicoletti, F. Savoie, J. F. Arnal and J. B. Michel. 2000. Angiotensin II stimulates endothelial vascular cell adhesion molecule-1 via nuclear factor-κB activation induced by intracellular oxidative stress. *Arterioscler Thromb Vasc Biol* **20**:645-651.651.

81. Sarkela, T. M., J. Berthiaume, S. Elfering, A. A. Gybina and C. Giulivi. 2001. The modulation of oxygen radical production by nitric oxide in mitochondria. *J Biol Chem* **276**:6945-6949.6949.

82. Beltran, B., A. Orsi, E. Clementi and S. Moncada. 2000. Oxidative stress and S-nitrosylation of proteins in cells. *Br J Pharmacol* **129**:953-960.960.

83. Poderoso, J. J., J. G. Peralta, C. L. Lisdero, M. C. Carreras, M. Radisic, F. Schopfer, E. Cadenas and A. Boveris. 1998. Nitric oxide regulates oxygen uptake and hydrogen peroxide release by the isolated beating rat heart. *Am J Physiol* **274**:C112-C119.C119.

84. Ghafourifar, P. and C. Richter. 1997. Nitric oxide synthase activity in mitochondria. *FEBS Lett* **418**:291-296.296.

85. Giulivi, C., J. J. Poderoso and A. Boveris. 1998. Production of nitric oxide by mitochondria. *J Biol Chem* **273**:11038-11043.11043.

86. Giulivi, C. 1998. Functional implications of nitric oxide produced by mitochondria in mitochondrial metabolism. *Biochem J* **332 (Pt 3)**:673-679.679.

87. Abas, L., M. A. Bogoyevitch and M. Guppy. 2000. Mitochondrial ATP production is necessary for activation of the extracellular-signal-regulated kinases during ischaemia/reperfusion in rat myocyte-derived H9c2 cells. *Biochem J* **349**:119-126.126.

Chapter 16

THE CONCEPT OF COMPARTMENTALIZATION IN SIGNALING BY REACTIVE OXYGEN SPECIES

Victor J. Thannickal and Barryl Fanburg

16.1 Introduction

A number of cellular metabolic reactions result in the formation of reactive oxygen species (ROS), such as superoxide anion (O_2^-) and hydrogen peroxide (H_2O_2), by both enzymatic and nonenzymatic pathways. To protect against the harmful effects of ROS, antioxidant defences effectively scavenge or rapidly neutralize these highly reactive biomolecules. Antioxidant defence systems are strategically localized in specific cellular compartments and organelles in close proximity to the site of ROS production. An example of this *compartmentalization* is the presence of three different mammalian isoforms of superoxide dismutase (SOD): Mn-SOD, Cu,Zn-SOD and extracellular (EC)-SOD localized in the mitochondria, cytoplasm/nucleus and extracellular matrix, respectively. There is substantial evidence now that, in addition to their capacity to cause cell damage, ROS generated by receptor-ligand interactions mediate diverse physiological responses such as proliferation, differentiation, and apoptosis.[1-3] However, the potential for these effects to be mediated by compartmentalized generation of ROS capable of targeting specific substrates is currently not well appreciated. Moreover, the precise mechanisms by which ROS signals are transmitted to regulate gene expression, a process commonly referred to as "redox signaling", is poorly understood.

H. J. Forman, J. Fukuto, and M. Torres (eds.), Signal Transduction by Reactive Oxygen and Nitrogen Species: Pathways and Chemical Principles, ©2003, Kluwer Academic Publishers. Printed in the Netherlands

Specificity in signal transduction is accomplished, in large part, by the multi-molecular organization of key signaling molecules, including enzymes and adapter proteins, into signaling complexes in specific cellular compartments.[4] Such spatial and temporal constraints on the high reactivity of ROS will have to be reconciled and better understood before "redox signaling" becomes incorporated into conventional concepts of signal transduction. In this brief review, we will examine the evidence to support the *principle of compartmentalization* in redox signaling.

16.2 Compartmentalization of Ros Production

A number of cellular organelles generate ROS in a compartmentalized manner, and in general, appear to exert their effects in the local microenvironment in which they are produced.

16.2.1 *Mitochondria*

The best recognized cellular source of ROS is the mitochondrion. Electron-transferring proteins in mitochondria can generate ROS as by-products of electron transfer reactions. This accounts for approximately 1-2% of total oxygen consumption required for aerobic metabolism under reducing conditions.[5] Due to the high concentrations of mitochondrial Mn-SOD, the intramitochondrial concentrations of O_2^- are maintained at very low steady state levels.[6] The potential role of mitochondrial ROS to mediate cell signaling has assumed greater importance in recent years, particularly with regard to the regulation of apoptosis.[7-12] TNF-α and IL-1-induced apoptosis appear to involve mitochondria-derived ROS.[13-15] The mitochondrion may also function as an "oxygen sensor" to regulate the transcription of hypoxia-inducible genes.[16,17] Mitochondrial ROS may be involved in hypoxia-induced phosphorylation of p38 MAP kinase in cardiomyocytes.[18] It is noteworthy that "signaling" by mitochondria-derived ROS mediates cellular responses that are essentially linked to mitochondrial function, namely apoptosis and oxygen sensing, providing support to the concept of compartmentalization.

16.2.2 *Peroxisomes*

Peroxisomes account for a large fraction of total cellular H_2O_2 production.[19] Peroxisomes contain H_2O_2-generating enzymes including glycollate oxidase, D-amino acid oxidase, urate oxidase, L-α-hydroxyacid oxidase, and fatty acyl-CoA oxidase. Peroxisomal catalase utilizes H_2O_2 produced by these oxidases to oxidize a variety of substrates in "peroxidative" reactions.[20] In liver and kidney cells, peroxisomes detoxify a variety of toxic molecules, including ethanol, that enter the circulation. Oxidative reactions in peroxisomes also mediate β-oxidation of fatty acids.[21] Due to high concentrations of peroxisomal catalase, minimal amounts of H_2O_2 are capable of escaping from these intracellular organelles.[19,22] Thus, although compartmentalized generation of ROS in peroxisomes serves to mediate important cellular functions, specific "signaling' roles for peroxisome-derived ROS have not been established.

16.2.3 Endoplasmic reticulum

The endoplasmic reticulum (ER) is another intracellular organelle that is primarily involved in lipid and protein biosynthesis. *Smooth* ER contains enzymes that catalyze a series of reactions to detoxify lipid-soluble drugs and other harmful metabolic products. Cytochrome p450 and *b5* enzymes oxidize unsaturated fatty acids and xenobiotics and, in the process, generate $O_2^{\cdot-}$ and/or H_2O_2.[5,23] As yet another example of a compartmentalized role for ROS, there is strong evidence of redox regulation of ER-related functions such as protein folding and secretion.[24-27] Moreover, the ER transmembrane receptor tyrosine kinase, Ltk, is activated by disulfide-linked multimerization involving redox-dependent, but ligand-independent mechanisms.[24]

16.2.4 Nucleus

Nuclear membranes contain cytochrome oxidases and electron transport systems that resemble those of ER but whose function is unknown.[5,28] Electron "leaks" from these enzymatic systems give rise to ROS that can induce DNA damage *in-vivo* given its proximity to the site of ROS production.[28] There is more information on the *antioxidant* enzymatic systems that maintain and regulate nuclear redox state. This primarily involves the glutathione (GSH)-related enzymes, thioredoxin (TRX) and redox factor-1 (Ref-1). There is significant overlap in their functions and they cooperatively regulate the activity of a number of transcription factors. Activator protein-1 (AP-1), a transcriptional complex formed by the dimerization of Fos-Jun or Jun-Jun proteins, is regulated by redox mechanisms.[29-31] In addition to pro-oxidant stimuli, AP-1 is also activated by some antioxidants.[32,33] One potential explanation for this apparent paradox is the different cellular compartments in which these redox effects are mediated. In the nucleus, a "reducing" environment appears to be critical for promoting DNA binding and transactivation of AP-1. *In-vitro* experiments suggest that a single cysteine residue in the highly conserved triamino acid sequence (Lys-Cys-Arg) of the DNA binding domain of Fos and Jun proteins confers redox-sensitivity to AP-1.[34] This study suggested that since this single cysteine residue was unlikely to account for the formation of inter- or intra-molecular disulfide bonds, the conversion of this cysteine to sulfenic acid, a reversible oxidation product, might be a more plausible mechanism for redox regulation. Redox regulation of AP-1 DNA binding is also facilitated by the reducing activity of Ref-1 protein that may act directly on this critical cysteine residue.[35] Furthermore, there is evidence that TRX, a small multifunctional protein with two redox-active cysteines within a conserved active site (Cys-Gly-Pro-Cys) (reviewed in ref. 36), can translocate from the cytosol to the nucleus in response to oxidant stress to regulate gene expression through Ref-1. Binding and activation of Ref-1 by TRX facilitates DNA binding of Jun-Fos complex to the AP-1 site to mediate transcription.[37] Similar effects of TRX on DNA binding of NF-κB,[38] p53[39,40] and PEBP2/CBF,[41] a transcription factor that contains two conserved redox-sensitive cysteines in its Runt domain, have been demonstrated. Other transcription factors whose DNA binding is regulated by redox mechanisms include Sp-1,[42,43] c-Myb,[44] and egr-1.[45]

16.2.5   Cytoplasm

The cytosol, similar to the nucleus, is normally maintained under strong reducing conditions by the redox buffering capacity of intracellular thiols, primarily GSH and TRX. GSH reductase and TRX reductase maintain the high ratio of reduced:oxidized GSH and TRX. Both of these thiol redox systems are capable of reducing H_2O_2 and lipid peroxides, reactions that are catalyzed by GSH and TRX peroxidases. Accumulating evidence suggests that, in addition to their antioxidant functions, GSH and TRX participate in redox signaling. GSH may be regulated by alterations in both the level of total GSH[46-48] and in the ratio of its oxidized (GSSG):reduced (GSH) forms.[42,49] Decreased levels of cellular GSH have been associated with inhibition of cell proliferation in vascular endothelial cells[50,51] and enhanced proliferation of fibroblasts.[52] PDGF receptor autophosphorylation appears to be inhibited under conditions of decreased cellular GSH levels,[53] suggesting a potential mechanism for the observed anti-proliferative effects in some cells.

TRX can also regulate the activity of certain proteins by directly binding to them. TRX inhibits apoptosis signal-regulating kinase-1 (ASK-1) by binding to its amino-terminal domain and its dissociation triggered by pro-oxidant stimuli, such as TNF-α and ROS, promotes apoptosis.[54]   Studies by Liu et al. suggest that dissociation of TRX is followed by the binding of TNF receptor associated factor-2 (TRAF-2) to ASK-1 and subsequent ASK-1 multimerization/activation.[55] Interestingly, regulation of such signaling events by TRX in the cytoplasm appear to involve "pro-oxidant" stimuli; whereas, binding and activation of transcription factors in the nucleus is promoted by the "reducing" activity of TRX. Thus, the thiol antioxidant, TRX, appears to have multifunctional effects in different cellular compartments, another example of the *compartmentalized* nature of redox signaling.

Other cytoplasmic sources of oxidants include soluble enzymes such as xanthine oxidase, aldehyde oxidase and flavoprotein dehydrogenase that can generate ROS during catalytic cycling.[5]   The most extensively studied of these is the $O_2^{\cdot-}$-generating xanthine oxidase, which is formed from xanthine dehydrogenase following tissue exposure to hypoxia.[56,57] Autooxidation of small molecules such as dopamine, epinephrine, and reduced flavins are a potentially important source of intracellular ROS production.[5]   Such reactions may alter cell redox state and contribute to cellular oxidative stress. It has been suggested that the pro-oxidant effects of dopamine autooxidation is involved in the dopamine-induced apoptosis and neurodegenerative changes in Parkinson's disease.[58,59]

16.2.6   Plasma membrane

The plasma membrane serves as a barrier separating the intracellular compartment from the relatively oxidizing state of the extracellular environment. It is also the site of a number of enzymatic systems implicated as sources of ligand-stimulated ROS production in nonphagocytic cells.[60-66] Several studies suggest that such enzymatic systems resemble the phagocytic NADPH oxidase and that functional components of this oxidase are present in nonphagocytic cells.[67-74] $p22^{phox}$, a membrane component of the phagocytic oxidase, has been shown to be present and functional in angiotensin-II-stimulated oxidase in vascular smooth muscle cells.[75,76] TNF-α also stimulates O_2^- production in vascular smooth muscle cells by a p22^{phox}-based NADH oxidase and appears to upregulate p22^{phox} gene expression in these cells.[77] Suh *et al.*[78] showed that Mox-1 (renamed Nox-1), a gene encoding a homologue of the catalytic subunit of the phagocytic gp91^{phox}, was expressed in a number of tissues including vascular smooth muscle. In this study, Nox-1 expression in NIH/3T3 cells was associated with increased O_2^- production, serum-stimulated cell growth, and a transformed phenotype. Since this earlier report, expression of other Nox homologues, with presumably diverse functions, have been demonstrated in a variety of mammalian tissues.[79-82]

There also appear to be some similarities between these enzymatic systems and the phagocytic NADPH oxidase with regard to the regulatory molecules and mechanisms involved in enzyme activation. In neutrophils, the membrane components, gp91^{phox} and p22^{phox}, along with the cytosolic components- p67^{phox}, p47^{phox}, and p40^{phox}, make up the core protein subunits of the phagocytic NADPH oxidase complex.[83-85] Activation of the NADPH oxidase, however, requires the additional participation of Rac2 (Rac1 in mouse macrophages) and Rap1A, members of the Ras superfamily of small guanine triphosphate (GTP) binding proteins. During activation, Rac2 binds GTP and migrates to the plasma membrane along with the cytosolic components to form the active complex. A similar requirement for Rac1 in the activation of the mitogenic oxidase in nonphagocytic cells has been demonstrated.[86-88] Another GTP-binding protein, p21^{Ras}, appears to function upstream of Rac1 in oxidant-dependent mitogenic signalling.[89,90] Dominant-negative expression of Rac1 inhibits not only the ligand-generated rise in intracellular ROS in NIH/3T3 cells, but also the ROS production in cells overexpressing a constitutively active isoform of Ras (H-Ras^{V12}).[89] Stable transfection of the same Ras plasmid (H-Ras^{V12}) in fibroblasts induces cellular transformation and constitutive production of large amounts of O_2^-.[90] In this study, mitogenic signaling in Ras-transformed fibroblasts was demonstrated to be redox-sensitive, but independent of the MAP kinase or c-Jun N-terminal kinase (JNK) pathways. Studies in primary human lung fibroblasts also suggest the requirement for p21^{Ras} in the generation of *intracellular* O_2^- by mitogenic growth factors including PDGF.[91]

Plasma membrane enzymes involved in phospholipid metabolism provide another potential source of second messengers that mediate intracellular, paracrine, and intercellular effects. The first step in these diverse pathways requires the action of the phopholipases: phospholipase A_2, phospholipase C, and phospholipase D. Although these enzymes have generally not been associated with redox signaling, a report by Touyz *et al.*[92] suggests that A-II-induced O_2^- production in smooth muscle cells is dependent on the phopholipase-D pathway. Phopholipase A_2 hydrolyzes phospholipids to generate arachidonic acid. Arachidonic acid then forms the substrate for cyclooxygenase (COX)- and lipoxygenase (LOX)-dependent synthesis of lipid mediators. These synthetic pathways involve a series of oxidation steps that involve a number of free radical intermediates.[5] LOX activity has also been implicated in redox-regulated signaling by angiotensin-II,[93] EGF,[94] and IL-1.[95]

16.2.7   Extracellular space

The extracellular space is composed of a complex and dynamic milieu of matrix molecules and various signaling components, including growth factors that are sequestered within the extracellular matrix (ECM). This microenvironment and its redox state are altered in many disease processes such as inflammation and cancer. Although a number of antioxidant enzymes that provide protection against oxidative stress are secreted into the extracellular space, such defences may be overwhelmed or become deficient in chronic disease states. Potential sources of extracellular ROS include activated phagocytic cells, soluble/secreted enzymes (e.g. xanthine oxidase), matrix-associated enzymes (e.g. lysyl oxidase) and certain nonphagocytic cells capable of releasing ROS from plasma membrane-associated NAD(P)H oxidases. Our results show that the pro-fibrotic cytokine, TGF-β1, is capable of inducing extracellular generation of H_2O_2 in cultured human lung fibroblasts by a cell surface-associated oxidase,[63] independently of the ability of this cytokine to upregulate lysyl oxidase activity.[91,96]

16.3 Intracellular Ros Generation by Mitogens

A number of mitogenic growth factors and vasoactive substances that bind receptor tyrosine kinases (RTKs) and G protein-coupled receptors have been shown to generate ROS in a variety of non-phagocytic cells (reviewed in ref. 2).

Many of these agents have been reported to stimulate a cellular response through ligand-mediated activation of an ROS-generating enzyme, such as that of an plasma membrane-associated NAD(P)H oxidase.   Subsequently, one or more protein kinases are activated including those belonging to the MAP kinase family or a specific phosphatase is inactivated, presumably by an oxidation-reduction reaction. As a result of these reactions, a transcription factor(s) is (are) activated to regulate gene expression.[97,98]

Studies in our laboratory on the mitogenic effects of serotonin (5-HT) are consistent with the concept of the requirement for *intracellular* ROS to mediate signaling of proliferative responses in vascular smooth muscle. 5-HT is a vasoactive peptide that has been implicated in the pathogenesis of pulmonary hypertension. Although cells contain a wide variety of receptors for this peptide, they can also be stimulated to proliferate via the action of the membrane-bound 5-HT transporter.[99] In this case, 5-HT is internalized through an energy-dependent transport process. Internalization of 5-HT is coupled to the generation of O_2^{-} probably as a result of activation of NAD(P)H oxidase.[100] The O_2^{-}, in turn, is dismutated to H_2O_2 and this serves to activate MAP kinase.[101] Extracellular H_2O_2 is not detectable in response to 5-HT in these cells. Unlike responses to angiotensin-II, where p38 MAP kinase is primarily involved,[97] 5-HT activates the p42/p44 MAP kinase.[100] Subsequently, transcription factors are activated and smooth muscle cells are stimulated to proliferate and become hypertrophic.[102]   The specific mechanisms for these responses are uncertain.   Nevertheless, redox signaling pathways of 5-HT are prototypic, in general, of a variety of other mitogenic growth factors and vasoactive substances. A consistent theme of this and other similar redox-regulated *mitogenic*

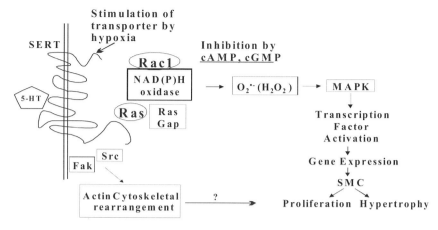

Figure 1: Proposed pathway for stimulation of SMC growth by 5-HT. Generation of intracellular ROS by 5-HT activates the mitogen-activated protein kinases (MAPKs) that mediate SMC proliferation and hypertrophy. See text for details.

signaling pathways is the compartmentalized generation of *intracellular* ROS (Figure 1).

16.4   Extracellular Ros Generation by Tgf-β1

TGF-β1 is the prototype of a large family of polypeptide growth factors that bind receptor serine-threonine kinases and include the TGF-βs, activins, inhibins, bone morphogenetic proteins and Mullerian-inhibiting substance.[103,104]   Each member of the TGF-β superfamily activates a heteromeric complex of type I/type II receptors in a combinatorial manner leading to the phosphorylation/activation of the Smad proteins that translocate to the nucleus to regulate gene transcription.[104] Unlike RTKs-linked growth factors, TGF-β1 typically inhibits growth of most target cells. Proliferative responses to TGF-β1 in some cells appear to be primarily related to indirect effects on the autocrine production of mitogenic growth factors[105,106] or their receptors.[107,108]   These multifunctional effects of TGF-β on cellular growth regulation, differentiation and extracellular matrix production are critical in complex biological processes such as embryogenesis, inflammation, fibrosis and carcinogenesis. TGF-β1, in contrast with most mitogenic growth factor ligands, is able to stimulate *extracellular* H_2O_2 production.[63,109,110]

Our studies on human lung fibroblasts show that TGF-β1 induces delayed, sustained generation of extracellular H_2O_2, while RTK-linked mitogenic growth factors rapidly induce O_2^- production that appears to be localized intracellularlly since neither extracellular O_2^- or H_2O_2 is detectable.[91]   TGF-β1-induced extracellular H_2O_2 production is not associated with cellular proliferation and exogenous addition of H_2O_2 over a wide range of concentrations does not mediate

mitogenic responses in these cells. Moreover, the mechanisms of TGF-β1- and
mitogen-induced ROS appear to involve different regulatory pathways.[91] TGF-β1
has been shown to stimulate ROS production in a variety of other cell types.[109,111-114]
In mouse osteoblastic cells, cell cycle-dependent extracellular release of H_2O_2
appears to be cell cycle-dependent and mediates growth-inhibitory effects of TGF-
β1.[111]   Ohba *et al.*[110] showed that the induction of the egr-1 gene by TGF-β1 is
inhibited by extracellular catalase, suggesting that H_2O_2 may be released
extracellularly before diffusing back into the cell. Some studies suggest that TGF-
β1, similar to mitogenic growth factors, is able to induce intracellular ROS in certain
other nonphagocytic cells to induce apoptosis,[113] IL-6 expression[115] or p38 MAP
kinase activation.[116] The source(s) of ROS production in response to TGF-β1 is
(are) currently unclear.   The TGF-β1-responsive oxidase in lung fibroblasts
generates extracellular H_2O_2 related to cell surface-associated NADH:flavin:O_2
oxidoreductase activity[63] and its activation is regulated by tyrosine phosphorylation,
likely involving nonreceptor tyrosine kinases[117] (Figure 2).

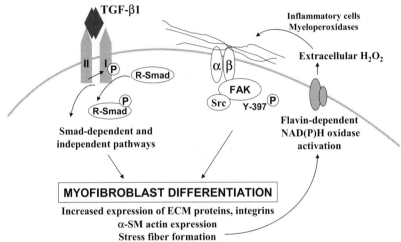

Figure 2: Proposed pathways for myofibroblast differentiation by TGF-β1. Stimulation of
extracellular H_2O_2 production by TGF-β1, in the presence of heme peroxidases (e.g.
myeloperoxidase), induces ECM protein cross-linking. This may activate integrin-ECM
signalling required for differentiation. See test for details.

Interestingly, TGF-β1 may also modulate cellular redox state by suppressing the
expression of antioxidant enzymes.[114,118] Kayanoki et al. showed that TGF-β1
inhibits the expression of manganese SOD, copper-zinc SOD, and catalase in rat
hepatocytes leading to increased cellular oxidative stress.[118] It lowers cellular
concentrations of intracellular glutathione (GSH) in endothelial and epithelial
cells.[50,51,119] This effect appears to be mediated by an inhibitory effect of TGF-β1 on
transcription of the rate-limiting enzyme involved in GSH synthesis, γ-
glutamylcysteine synthetase.[119,120] The reduction in cellular GSH levels by TGF-β1

in vascular endothelial cells is closely associated with its growth-inhibitory effect and appears to be modulated by thiol availability.[50] Another study suggests that intracellular levels of GSH may be important in discriminating an "oxidative stress" from a "signaling" response to TGF-β1.[121] In this study, endogenously generated H_2O_2 appears to mediate TGF-β1 autoinduction, but only under conditions in which intracellular GSH concentrations are high.

A number of other physiological effects of TGF-β1 have also been linked to ROS generation in various cells. These include apoptosis,[113,114,122-126] activation of latent TGF-β,[127] cellular transformation,[128] cytokine production,[115] collagen synthesis [129] and stellate cell activation.[130,131]

The mechanisms for these mutiple redox-regulated effects of TGF-β1 are currently unclear. It remains to be determined if the diversity in these responses is related to differences in compartmentalization and specificity of molecular targets. Our studies show that TGF-β1-induced extracellular H_2O_2, in the presence of heme peroxidases, has the potential to mediate dityrosine-dependent crosslinking of ECM proteins.[132] Thus, the nature and reactivity of cell surface/extracellular substrates and extracellular antioxidants may ultimately determine if H_2O_2 can diffuse from the extracellular compartment and target intracellular proteins. More likely, extracellular effects of ROS on the ECM may influence the response of the cell to other signals. Such combinatorial signaling events involving integrin-ECM signals and growth factors is a well recognized and evolving concept in cell signaling.[133]

16.5 Conclusion

Specificity of ROS and their mechanisms of action in cell signaling and regulation are just beginning to be understood. Several factors are likely to be important in determining their biological effects and functions. These include the nature, concentration and kinetics of the specific ROS produced, and, most importantly perhaps, the microenvironment in which they are released. Such factors may also determine if ROS induce cell/tissue injury or mediate physiologic signaling responses. In this review, we have discussed the differential regulation and compartmentalized generation of ROS in nonphagocytic cells in response to 5-HT and TGF-β1.

Mitogenic stimulation by 5-HT in vascular smooth muscle cells appears to be dependent on *intracellular* generation of ROS. This mitogenic effect is associated with tyrosine phosphorylation of GTPase-activating protein,[134] generation of intracellular O_2^- and activation of MAP kinases.[100,134,135] The mechanism(s) by which O_2^- or its dismutation product, H_2O_2, induces MAP kinase phopshorylation/activation is unclear. Reversible inactivation of protein tyrosine phosphatases (PTPs) by ROS is emerging as a generalized mechanism by which intracellular ROS mediate mitogenic signalling.[66,136] Meng *et al.*[136] recently showed that this effect may require association of SHP-2, an SH2 containing PTP, with the ligand-receptor complex. This *compartmentalized* response involving the formation of a receptor complex with the substrate PTP may be essential in conferring specificity to the action of intracellular ROS.

The *extracellular* generation of H_2O_2 induced by TGF-β1 in fibroblasts is not associated with cellular proliferation,[91] but rather a differentiation response.[137,138]

This cellular differentiation program leads to the phenotypic transformation of fibroblasts into myofibroblasts that are active in remodelling of the extracellular matrix.[139] In complex disease processes such as inflammation, fibrosis and cancer, oxidative changes in the ECM are likely to alter both its biochemical and biophysical properties.[132] Such alterations of the ECM are important in the determination of cell phenotype and cellular differentiation.[139] Thus, ROS have distinctive, and sometimes opposing, effects on cellular responses that are, in part, explained by their compartmentalization and targeting of redox-sensitive proteins.

Acknowledgments:

Supported by NIH research grants: HL 67967 (V.J.T.); HL 32723 and HL 42376 (B.L.F.)

References

1. Finkel, T. 1998. Oxygen radicals and signaling. *Curr Opin Cell Biol* **10**:248-253.

2. Thannickal, V. J. and B. L. Fanburg. 2000. Reactive oxygen species in cell signaling. *Am J Physiol Lung Cell Mol Physiol* **279**:L1005-1028.

3. Droge, W. 2002. Free radicals in the physiological control of cell function. *Physiol Rev* **82**:47-95.

4. Smith, F. D. and J. D. Scott. 2002. Signaling complexes: junctions on the intracellular information super highway. *Curr Biol* **12**:R32-40.

5. Freeman, B. A. and J. D. Crapo. 1982. Biology of disease: free radicals and tissue injury. *Lab Invest* **47**:412-426.

6. Tyler, D. D. 1975. Polarographic assay and intracellular distribution of superoxide dismutase in rat liver. *Biochem J* **147**:493-504.

7. Lee, H. C. and Y. H. Wei. 2000. Mitochondrial role in life and death of the cell. *J Biomed Sci* **7**:2-15.

8. Cai, J. and D. P. Jones. 1998. Superoxide in apoptosis. Mitochondrial generation triggered by cytochrome c loss. *J Biol Chem* **273**:11401-11404.

9. Li, A. E., H. Ito, Rovira, II, K. S. Kim, K. Takeda, Z. Y. Yu, V. J. Ferrans and T. Finkel. 1999. A role for reactive oxygen species in endothelial cell anoikis. *Circ Res* **85**:304-310.

10. Chandel, N. S. and P. T. Schumacker. 1999. Cells depleted of mitochondrial DNA (rho0) yield insight into physiological mechanisms. *FEBS Lett* **454**:173-176.

11. von Harsdorf, R., P. F. Li and R. Dietz. 1999. Signaling pathways in reactive oxygen species-induced cardiomyocyte apoptosis. *Circulation* **99**:2934-2941.

12. Banki, K., E. Hutter, N. J. Gonchoroff and A. Perl. 1999. Elevation of mitochondrial transmembrane potential and reactive oxygen intermediate levels are early events and

occur independently from activation of caspases in Fas signaling. *J Immunol* **162**:1466-1479.

13. Sidoti-de Fraisse, C., V. Rincheval, Y. Risler, B. Mignotte and J. L. Vayssiere. 1998. TNF-alpha activates at least two apoptotic signaling cascades. *Oncogene* **17**:1639-1651.

14. Singh, I., K. Pahan, M. Khan and A. K. Singh. 1998. Cytokine-mediated induction of ceramide production is redox-sensitive. Implications to proinflammatory cytokine-mediated apoptosis in demyelinating diseases. *J Biol Chem* **273**:20354-20362.

15. Wissing, D., H. Mouritzen and M. Jaattela. 1998. TNF-induced mitochondrial changes and activation of apoptotic proteases are inhibited by A20. *Free Radic Biol Med* **25**:57-65.

16. Chandel, N. S., E. Maltepe, E. Goldwasser, C. E. Mathieu, M. C. Simon and P. T. Schumacker. 1998. Mitochondrial reactive oxygen species trigger hypoxia-induced transcription. *Proc Natl Acad Sci U S A* **95**:11715-11720.

17. Duranteau, J., N. S. Chandel, A. Kulisz, Z. Shao and P. T. Schumacker. 1998. Intracellular signaling by reactive oxygen species during hypoxia in cardiomyocytes. *J Biol Chem* **273**:11619-11624.

18. Kulisz, A., N. Chen, N. S. Chandel, Z. Shao and P. T. Schumaker. 2002. Mitochondrial ROS initiate phosphorylation of p38 MAP kinase during hypoxia in cardiomyocytes. *Am J Physiol Lung Cell Mol Physiol* **282**:L1324-1329.

19. Boveris, A., N. Oshino and B. Chance. 1972. The cellular production of hydrogen peroxide. *Biochem J* **128**:617-630.

20. Tolbert, N. E. and E. Essner. 1981. Microbodies: peroxisomes and glyoxysomes. *J Cell Biol* **91**:271s-283s.

21. Alberts, B., A. Johnson, J. Lewis, M. Raff, K. Roberts and P. Walter, eds 2002. *Molecular Biology of the Cell*, Garland Science, New York

22. Poole, B. 1975. Diffusion effects in the metabolism of hydrogen peroxide by rat liver peroxisomes. *J Theor Biol* **51**:149-167.

23. Capdevila, J., L. Parkhill, N. Chacos, R. Okita, B. S. Masters and R. W. Estabrook. 1981. The oxidative metabolism of arachidonic acid by purified cytochromes P- 450. *Biochem Biophys Res Commun* **101**:1357-1363.

24. Bauskin, A. R., I. Alkalay and Y. Ben-Neriah. 1991. Redox regulation of a protein tyrosine kinase in the endoplasmic reticulum. *Cell* **66**:685-696.

25. Hwang, C., A. J. Sinskey and H. F. Lodish. 1992. Oxidized redox state of glutathione in the endoplasmic reticulum. *Science* **257**:1496-1502.

26. Pahl, H. L. and P. A. Baeuerle. 1997. The ER-overload response: activation of NF-κ B. *Trends Biochem Sci* **22**:63-67.

27. Bader, M., W. Muse, D. P. Ballou, C. Gassner and J. C. Bardwell. 1999. Oxidative protein folding is driven by the electron transport system. *Cell* **98**:217-227.

28. Halliwell, B. and J. M. C. Gutteridge. 1989. *Free Radicals in Biology and Medicine*, Oxford University Press, New York.

29. Nose, K., M. Shibanuma, K. Kikuchi, H. Kageyama, S. Sakiyama and T. Kuroki. 1991. Transcriptional activation of early-response genes by hydrogen peroxide in a mouse osteoblastic cell line. *Eur J Biochem* **201**:99-106.

30. Lo, Y. Y. and T. F. Cruz. 1995. Involvement of reactive oxygen species in cytokine and growth factor induction of c-fos expression in chondrocytes. *J Biol Chem* **270**:11727-11730.

31. Puri, P. L., M. L. Avantaggiati, V. L. Burgio, P. Chirillo, D. Collepardo, G. Natoli, C. Balsano and M. Levrero. 1995. Reactive oxygen intermediates (ROIs) are involved in the intracellular transduction of angiotensin II signal in C2C12 cells. *Ann N Y Acad Sci* **752**:394-405.

32. Choi, H. S. and D. D. Moore. 1993. Induction of c-fos and c-jun gene expression by phenolic antioxidants. *Mol Endocrinol* **7**:1596-1602.

33. Yoshioka, K., T. Deng, M. Cavigelli and M. Karin. 1995. Antitumor promotion by phenolic antioxidants: inhibition of AP-1 activity through induction of Fra expression. *Proc Natl Acad Sci U S A* **92**:4972-4976.

34. Abate, C., L. Patel, F. J. Rauscher, 3rd and T. Curran. 1990. Redox regulation of fos and jun DNA-binding activity in vitro. *Science* **249**:1157-1161.

35. Xanthoudakis, S., G. Miao, F. Wang, Y. C. Pan and T. Curran. 1992. Redox activation of Fos-Jun DNA binding activity is mediated by a DNA repair enzyme. *Embo J* **11**:3323-3335.

36. Nakamura, H., K. Nakamura and J. Yodoi. 1997. Redox regulation of cellular activation. *Annu Rev Immunol* **15**:351-369.

37. Hirota, K., M. Matsui, S. Iwata, A. Nishiyama, K. Mori and J. Yodoi. 1997. AP-1 transcriptional activity is regulated by a direct association between thioredoxin and Ref-1. *Proc Natl Acad Sci U S A* **94**:3633-3638.

38. Hirota, K., M. Murata, Y. Sachi, H. Nakamura, J. Takeuchi, K. Mori and J. Yodoi. 1999. Distinct roles of thioredoxin in the cytoplasm and in the nucleus. A two-step mechanism of redox regulation of transcription factor NF-κB. *J Biol Chem* **274**:27891-27897.

39. Rainwater, R., D. Parks, M. E. Anderson, P. Tegtmeyer and K. Mann. 1995. Role of cysteine residues in regulation of p53 function. *Mol Cell Biol* **15**:3892-3903.

40. Pearson, G. D. and G. F. Merrill. 1998. Deletion of the Saccharomyces cerevisiae TRR1 gene encoding thioredoxin reductase inhibits p53-dependent reporter gene expression. *J Biol Chem* **273**:5431-5434.

41. Akamatsu, Y., T. Ohno, K. Hirota, H. Kagoshima, J. Yodoi and K. Shigesada. 1997. Redox regulation of the DNA binding activity in transcription factor PEBP2. The roles of two conserved cysteine residues. *J Biol Chem* **272**:14497-14500.

42. Knoepfel, L., C. Steinkuhler, M. T. Carri and G. Rotilio. 1994. Role of zinc-coordination and of the glutathione redox couple in the redox susceptibility of human transcription factor Sp1. *Biochem Biophys Res Commun* **201**:871-877.

43. Wu, X., N. H. Bishopric, D. J. Discher, B. J. Murphy and K. A. Webster. 1996. Physical and functional sensitivity of zinc finger transcription factors to redox change. *Mol Cell Biol* **16**:1035-1046.

44. Myrset, A. H., A. Bostad, N. Jamin, P. N. Lirsac, F. Toma and O. S. Gabrielsen. 1993. DNA and redox state induced conformational changes in the DNA-binding domain of the Myb oncoprotein. *Embo J* **12**:4625-4633.

45. Huang, R. P. and E. D. Adamson. 1993. Characterization of the DNA-binding properties of the early growth response-1 (Egr-1) transcription factor: evidence for modulation by a redox mechanism. *DNA Cell Biol* **12**:265-273.

46. Droge, W., K. Schulze-Osthoff, S. Mihm, D. Galter, H. Schenk, H. P. Eck, S. Roth and H. Gmunder. 1994. Functions of glutathione and glutathione disulfide in immunology and immunopathology. *Faseb J* **8**:1131-1138.

47. Sen, C. K., S. Khanna, A. Z. Reznick, S. Roy and L. Packer. 1997. Glutathione regulation of tumor necrosis factor-alpha-induced NF-κ B activation in skeletal muscle-derived L6 cells. *Biochem Biophys Res Commun* **237**:645-649.

48. Esposito, F., V. Agosti, G. Morrone, F. Morra, C. Cuomo, T. Russo, S. Venuta and F. Cimino. 1994. Inhibition of the differentiation of human myeloid cell lines by redox changes induced through glutathione depletion. *Biochem J* **301**:649-653.

49. Henschke, P. N. and S. J. Elliott. 1995. Oxidized glutathione decreases luminal Ca^{2+} content of the endothelial cell ins(1,4,5)P3-sensitive Ca^{2+} store. *Biochem J* **312**:485-489.

50. Das, S. K., A. C. White and B. L. Fanburg. 1992. Modulation of transforming growth factor-beta 1 antiproliferative effects on endothelial cells by cysteine, cystine, and N-acetylcysteine. *J Clin Invest* **90**:1649-1656.

51. White, A. C., S. K. Das and B. L. Fanburg. 1992. Reduction of glutathione is associated with growth restriction and enlargement of bovine pulmonary artery endothelial cells produced by transforming growth factor-beta 1. *Am J Respir Cell Mol Biol* **6**:364-368.

52. Cantin, A. M., P. Larivee and R. O. Begin. 1990. Extracellular glutathione suppresses human lung fibroblast proliferation. *Am J Respir Cell Mol Biol* **3**:79-85.

53. Rigacci, S., T. Iantomasi, P. Marraccini, A. Berti, M. T. Vincenzini and G. Ramponi. 1997. Evidence for glutathione involvement in platelet-derived growth-factor- mediated signal transduction. *Biochem J* **324**:791-796.

54. Saitoh, M., H. Nishitoh, M. Fujii, K. Takeda, K. Tobiume, Y. Sawada, M. Kawabata, K. Miyazono and H. Ichijo. 1998. Mammalian thioredoxin is a direct inhibitor of apoptosis signal- regulating kinase (ASK) 1. *Embo J* **17**:2596-2606.

55. Liu, H., H. Nishitoh, H. Ichijo and J. M. Kyriakis. 2000. Activation of apoptosis signal-regulating kinase 1 (ASK1) by tumor necrosis factor receptor-associated factor 2 requires prior dissociation of the ASK1 inhibitor thioredoxin. *Mol Cell Biol* **20**:2198-2208.

56. McKelvey, T. G., M. E. Hollwarth, D. N. Granger, T. D. Engerson, U. Landler and H. P. Jones. 1988. Mechanisms of conversion of xanthine dehydrogenase to xanthine oxidase in ischemic rat liver and kidney. *Am J Physiol* **254**:G753-760.

57. Parks, D. A., T. K. Williams and J. S. Beckman. 1988. Conversion of xanthine dehydrogenase to oxidase in ischemic rat intestine: a reevaluation. *Am J Physiol* **254**:G768-774.

58. Yoshikawa, T., Y. Minamiyama, Y. Naito and M. Kondo. 1994. Antioxidant properties of bromocriptine, a dopamine agonist. *J Neurochem* **62**:1034-1038.

59. Offen, D., I. Ziv, H. Panet, L. Wasserman, R. Stein, E. Melamed and A. Barzilai. 1997. Dopamine-induced apoptosis is inhibited in PC12 cells expressing Bcl-2. *Cell Mol Neurobiol* **17**:289-304.

60. Meier, B., H. H. Radeke, S. Selle, M. Younes, H. Sies, K. Resch and G. G. Habermehl. 1989. Human fibroblasts release reactive oxygen species in response to interleukin-1 or tumour necrosis factor-alpha. *Biochem J* **263**:539-545.

61. Satriano, J. A., M. Shuldiner, K. Hora, Y. Xing, Z. Shan and D. Schlondorff. 1993. Oxygen radicals as second messengers for expression of the monocyte chemoattractant protein, JE/MCP-1, and the monocyte colony-stimulating factor, CSF-1, in response to tumor necrosis factor-alpha and immunoglobulin G. Evidence for involvement of reduced nicotinamide adenine dinucleotide phosphate (NADPH)-dependent oxidase. *J Clin Invest* **92**:1564-1571.

62. Griendling, K. K., C. A. Minieri, J. D. Ollerenshaw and R. W. Alexander. 1994. Angiotensin II stimulates NADH and NADPH oxidase activity in cultured vascular smooth muscle cells. *Circ Res* **74**:1141-1148.

63. Thannickal, V. J. and B. L. Fanburg. 1995. Activation of an H_2O_2-generating NADH oxidase in human lung fibroblasts by transforming growth factor beta 1. *J Biol Chem* **270**:30334-30338.

64. Krieger-Brauer, H. I. and H. Kather. 1995. Antagonistic effects of different members of the fibroblast and platelet-derived growth factor families on adipose conversion and NADPH- dependent H_2O_2 generation in 3T3 L1-cells. *Biochem J* **307**:549-556.

65. Sundaresan, M., Z. X. Yu, V. J. Ferrans, K. Irani and T. Finkel. 1995. Requirement for generation of H_2O_2 for platelet-derived growth factor signal transduction. *Science* **270**:296-299.

66. Bae, Y. S., S. W. Kang, M. S. Seo, I. C. Baines, E. Tekle, P. B. Chock and S. G. Rhee. 1997. Epidermal growth factor (EGF)-induced generation of hydrogen peroxide. Role in EGF receptor-mediated tyrosine phosphorylation. *J Biol Chem* **272**:217-221.

67. Jones, S. A., V. B. O'Donnell, J. D. Wood, J. P. Broughton, E. J. Hughes and O. T. Jones. 1996. Expression of phagocyte NADPH oxidase components in human endothelial cells. *Am J Physiol* **271**:H1626-1634.

68. Fukui, T., N. Ishizaka, S. Rajagopalan, J. B. Laursen, Q. t. Capers, W. R. Taylor, D. G. Harrison, H. de Leon, J. N. Wilcox and K. K. Griendling. 1997. p22^{phox} mRNA

expression and NADPH oxidase activity are increased in aortas from hypertensive rats. *Circ Res* **80**:45-51.

69. Hiran, T. S., P. J. Moulton and J. T. Hancock. 1997. Detection of superoxide and NADPH oxidase in porcine articular chondrocytes. *Free Radic Biol Med* **23**:736-743.

70. Bayraktutan, U., N. Draper, D. Lang and A. M. Shah. 1998. Expression of functional neutrophil-type NADPH oxidase in cultured rat coronary microvascular endothelial cells. *Cardiovasc Res* **38**:256-262.

71. Moulton, P. J., M. B. Goldring and J. T. Hancock. 1998. NADPH oxidase of chondrocytes contains an isoform of the gp91^{phox} subunit. *Biochem J* **329**:449-451.

72. Meyer, J. W., J. A. Holland, L. M. Ziegler, M. M. Chang, G. Beebe and M. E. Schmitt. 1999. Identification of a functional leukocyte-type NADPH oxidase in human endothelial cells :a potential atherogenic source of reactive oxygen species. *Endothelium* **7**:11-22.

73. Lavigne, M. C., H. L. Malech, S. M. Holland and T. L. Leto. 2001. Genetic requirement of p47^{phox} for superoxide production by murine microglia. *Faseb J* **15**:285-287.

74. Bayraktutan, U., L. Blayney and A. M. Shah. 2000. Molecular characterization and localization of the NAD(P)H oxidase components gp91-phox and p22-phox in endothelial cells. *Arterioscler Thromb Vasc Biol* **20**:1903-1911.

75. Zafari, A. M., M. Ushio-Fukai, M. Akers, Q. Yin, A. Shah, D. G. Harrison, W. R. Taylor and K. K. Griendling. 1998. Role of NADH/NADPH oxidase-derived H_2O_2 in angiotensin II-induced vascular hypertrophy. *Hypertension* **32**:488-495.

76. Marumo, T., V. B. Schini-Kerth, R. P. Brandes and R. Busse. 1998. Glucocorticoids inhibit superoxide anion production and p22 phox mRNA expression in human aortic smooth muscle cells. *Hypertension* **32**:1083-1088.

77. De Keulenaer, G. W., R. W. Alexander, M. Ushio-Fukai, N. Ishizaka and K. K. Griendling. 1998. Tumour necrosis factor alpha activates a p22^{phox}-based NADH oxidase in vascular smooth muscle. *Biochem J* **329**:653-657.

78. Suh, Y. A., R. S. Arnold, B. Lassegue, J. Shi, X. Xu, D. Sorescu, A. B. Chung, K. K. Griendling and J. D. Lambeth. 1999. Cell transformation by the superoxide-generating oxidase Mox1. *Nature* **401**:79-82.

79. Arbiser, J. L., J. Petros, R. Klafter, B. Govindajaran, E. R. McLaughlin, L. F. Brown, C. Cohen, M. Moses, S. Kilroy, R. S. Arnold and others. 2002. Reactive oxygen generated by Nox1 triggers the angiogenic switch. *Proc Natl Acad Sci U S A* **99**:715-720.

80. Lambeth, J. D., G. Cheng, R. S. Arnold and W. A. Edens. 2000. Novel homologs of gp91phox. *Trends Biochem Sci* **25**:459-461.

81. Geiszt, M., J. B. Kopp, P. Varnai and T. L. Leto. 2000. Identification of renox, an NAD(P)H oxidase in kidney. *Proc Natl Acad Sci U S A* **97**:8010-8014.

82. Cheng, G., Z. Cao, X. Xu, E. G. van Meir and J. D. Lambeth. 2001. Homologs of gp91^{phox}: cloning and tissue expression of Nox3, Nox4, and Nox5. *Gene* **269**:131-140.

83. Diekmann, D., A. Abo, C. Johnston, A. W. Segal and A. Hall. 1994. Interaction of Rac with p67^{phox} and regulation of phagocytic NADPH oxidase activity. *Science* **265**:531-533.

84. De Leo, F. R., K. V. Ulman, A. R. Davis, K. L. Jutila and M. T. Quinn. 1996. Assembly of the human neutrophil NADPH oxidase involves binding of p67^{phox} and flavocytochrome b to a common functional domain in p47^{phox}. *J Biol Chem* **271**:17013-17020.

85. Dang, P. M., A. R. Cross and B. M. Babior. 2001. Assembly of the neutrophil respiratory burst oxidase: a direct interaction between p67^{phox} and cytochrome *b558*. *Proc Natl Acad Sci U S A* **98**:3001-3005.

86. Kheradmand, F., E. Werner, P. Tremble, M. Symons and Z. Werb. 1998. Role of Rac1 and oxygen radicals in collagenase-1 expression induced by cell shape change. *Science* **280**:898-902.

87. Cool, R. H., E. Merten, C. Theiss and H. Acker. 1998. Rac1, and not Rac2, is involved in the regulation of the intracellular hydrogen peroxide level in HepG2 cells. *Biochem J* **332**:5-8.

88. Joneson, T. and D. Bar-Sagi. 1998. A Rac1 effector site controlling mitogenesis through superoxide production. *J Biol Chem* **273**:17991-17994.

89. Sundaresan, M., Z. X. Yu, V. J. Ferrans, D. J. Sulciner, J. S. Gutkind, K. Irani, P. J. Goldschmidt-Clermont and T. Finkel. 1996. Regulation of reactive-oxygen-species generation in fibroblasts by Rac1. *Biochem J* **318**:379-382.

90. Irani, K., Y. Xia, J. L. Zweier, S. J. Sollott, C. J. Der, E. R. Fearon, M. Sundaresan, T. Finkel and P. J. Goldschmidt-Clermont. 1997. Mitogenic signaling mediated by oxidants in Ras-transformed fibroblasts [see comments]. *Science* **275**:1649-1652.

91. Thannickal, V. J., R. M. Day, S. G. Klinz, M. C. Bastien, J. M. Larios and B. L. Fanburg. 2000. Ras-dependent and -independent regulation of reactive oxygen species by mitogenic growth factors and TGF-beta1. *Faseb J* **14**:1741-1748.

92. Touyz, R. M. and E. L. Schiffrin. 1999. Ang II-stimulated superoxide production is mediated via phospholipase D in human vascular smooth muscle cells. *Hypertension* **34**:976-982.

93. Wen, Y., S. Scott, Y. Liu, N. Gonzales and J. L. Nadler. 1997. Evidence that angiotensin II and lipoxygenase products activate c-Jun NH2-terminal kinase. *Circ Res* **81**:651-655.

94. Mills, E. M., K. Takeda, Z. X. Yu, V. Ferrans, Y. Katagiri, H. Jiang, M. C. Lavigne, T. L. Leto and G. Guroff. 1998. Nerve growth factor treatment prevents the increase in superoxide produced by epidermal growth factor in PC12 cells. *J Biol Chem* **273**:22165-22168.

95. Bonizzi, G., J. Piette, S. Schoonbroodt, R. Greimers, L. Havard, M. P. Merville and V. Bours. 1999. Reactive oxygen intermediate-dependent NF-κB activation by interleukin-1beta requires 5-lipoxygenase or NADPH oxidase activity. *Mol Cell Biol* **19**:1950-1960.

96. Kagan, H. M. and P. C. Trackman. 1991. Properties and function of lysyl oxidase [see comments]. *Am J Respir Cell Mol Biol* **5**:206-210.

97. Ushio-Fukai, M., R. W. Alexander, M. Akers and K. K. Griendling. 1998. p38 Mitogen-activated protein kinase is a critical component of the redox-sensitive signaling pathways activated by angiotensin II. Role in vascular smooth muscle cell hypertrophy. *J Biol Chem* **273**:15022-15029.

98. Viedt, C., U. Soto, H. I. Krieger-Brauer, J. Fei, C. Elsing, W. Kubler and J. Kreuzer. 2000. Differential activation of mitogen-activated protein kinases in smooth muscle cells by angiotensin II: involvement of p22^{phox} and reactive oxygen species. *Arterioscler Thromb Vasc Biol* **20**:940-948.

99. Fanburg, B. L. and S. L. Lee. 1997. A new role for an old molecule: serotonin as a mitogen. *Am J Physiol* **272**:L795-806.

100. Lee, S. L., W. W. Wang, G. A. Finlay and B. L. Fanburg. 1999. Serotonin stimulates mitogen-activated protein kinase activity through the formation of superoxide anion. *Am J Physiol* **277**:L282-291.

101. Lee, S. L., A. R. Simon, W. W. Wang and B. L. Fanburg. 2001. H$_2$O$_2$ signals 5-HT-induced ERK MAP kinase activation and mitogenesis of smooth muscle cells. *Am J Physiol Lung Cell Mol Physiol* **281**:L646-652.

102. Lee, S. L., W. W. Wang, J. J. Lanzillo and B. L. Fanburg. 1994. Serotonin produces both hyperplasia and hypertrophy of bovine pulmonary artery smooth muscle cells in culture. *Am J Physiol* **266**:L46-52.

103. Piek, E., C. H. Heldin and P. Ten Dijke. 1999. Specificity, diversity, and regulation in TGF-beta superfamily signaling. *Faseb J* **13**:2105-2124.

104. Massague, J. 1996. TGFbeta signaling: receptors, transducers, and Mad proteins. *Cell* **85**:947-950.

105. Leof, E. B., J. A. Proper, A. S. Goustin, G. D. Shipley, P. E. DiCorleto and H. L. Moses. 1986. Induction of c-sis mRNA and activity similar to platelet-derived growth factor by transforming growth factor beta: a proposed model for indirect mitogenesis involving autocrine activity. *Proc Natl Acad Sci U S A* **83**:2453-2457.

106. Battegay, E. J., E. W. Raines, R. A. Seifert, D. F. Bowen-Pope and R. Ross. 1990. TGF-beta induces bimodal proliferation of connective tissue cells via complex control of an autocrine PDGF loop. *Cell* **63**:515-524.

107. Rosenbaum, J., S. Blazejewski, A. M. Preaux, A. Mallat, D. Dhumeaux and P. Mavier. 1995. Fibroblast growth factor 2 and transforming growth factor beta 1 interactions in human liver myofibroblasts. *Gastroenterology* **109**:1986-1996.

108. Thannickal, V. J., K. D. Aldweib, T. Rajan and B. L. Fanburg. 1998. Upregulated expression of fibroblast growth factor (FGF) receptors by transforming growth factor-beta1 (TGF-beta1) mediates enhanced mitogenic responses to FGFs in cultured human lung fibroblasts. *Biochem Biophys Res Commun* **251**:437-441.

109. Thannickal, V. J., P. M. Hassoun, A. C. White and B. L. Fanburg. 1993. Enhanced rate of H$_2$O$_2$ release from bovine pulmonary artery endothelial cells induced by TGF-beta 1. *Am J Physiol* **265**:L622-626.

308   *Chapter 16*

110. Ohba, M., M. Shibanuma, T. Kuroki and K. Nose. 1994. Production of hydrogen peroxide by transforming growth factor-beta 1 and its involvement in induction of egr-1 in mouse osteoblastic cells. *J Cell Biol* **126**:1079-1088.

111. Shibanuma, M., T. Kuroki and K. Nose. 1991. Release of H_2O_2 and phosphorylation of 30 kilodalton proteins as early responses of cell cycle-dependent inhibition of DNA synthesis by transforming growth factor beta 1. *Cell Growth Differ* **2**:583-591.

112. Kayanoki, Y., S. Higashiyama, K. Suzuki, M. Asahi, S. Kawata, Y. Matsuzawa and N. Taniguchi. 1999. The requirement of both intracellular reactive oxygen species and intracellular calcium elevation for the induction of heparin-binding EGF-like growth factor in vascular endothelial cells and smooth muscle cells. *Biochem Biophys Res Commun* **259**:50-55.

113. Sanchez, A., A. M. Alvarez, M. Benito and I. Fabregat. 1996. Apoptosis induced by transforming growth factor-beta in fetal hepatocyte primary cultures: involvement of reactive oxygen intermediates. *J Biol Chem* **271**:7416-7422.

114. Islam, K. N., Y. Kayanoki, H. Kaneto, K. Suzuki, M. Asahi, J. Fujii and N. Taniguchi. 1997. TGF-beta1 triggers oxidative modifications and enhances apoptosis in HIT cells through accumulation of reactive oxygen species by suppression of catalase and glutathione peroxidase. *Free Radic Biol Med* **22**:1007-1017.

115. Junn, E., K. N. Lee, H. R. Ju, S. H. Han, J. Y. Im, H. S. Kang, T. H. Lee, Y. S. Bae, K. S. Ha, Z. W. Lee and others. 2000. Requirement of hydrogen peroxide generation in TGF-beta 1 signal transduction in human lung fibroblast cells: involvement of hydrogen peroxide and Ca2+ in TGF-beta 1-induced IL-6 expression. *J Immunol* **165**:2190-2197.

116. Chiu, C., D. A. Maddock, Q. Zhang, K. P. Souza, A. R. Townsend and Y. Wan. 2001. TGF-beta-induced p38 activation is mediated by Rac1-regulated generation of reactive oxygen species in cultured human keratinocytes. *Int J Mol Med* **8**:251-255.

117. Thannickal, V. J., K. D. Aldweib and B. L. Fanburg. 1998. Tyrosine phosphorylation regulates H2O2 production in lung fibroblasts stimulated by transforming growth factor beta1. *J Biol Chem* **273**:23611-23615.

118. Kayanoki, Y., J. Fujii, K. Suzuki, S. Kawata, Y. Matsuzawa and N. Taniguchi. 1994. Suppression of antioxidative enzyme expression by transforming growth factor-beta 1 in rat hepatocytes. *J Biol Chem* **269**:15488-15492.

119. Arsalane, K., C. M. Dubois, T. Muanza, R. Begin, F. Boudreau, C. Asselin and A. M. Cantin. 1997. Transforming growth factor-beta1 is a potent inhibitor of glutathione synthesis in the lung epithelial cell line A549: transcriptional effect on the GSH rate-limiting enzyme gamma-glutamylcysteine synthetase. *Am J Respir Cell Mol Biol* **17**:599-607.

120. Jardine, H., W. MacNee, K. Donaldson and I. Rahman. 2002. Molecular mechanism of transforming growth factor (TGF)-beta1-induced glutathione depletion in alveolar epithelial cells. Involvement of AP-1/ARE and Fra-1. *J Biol Chem* **277**;21158-21166

121. De Bleser, P. J., G. Xu, K. Rombouts, V. Rogiers and A. Geerts. 1999. Glutathione levels discriminate between oxidative stress and transforming growth factor-beta signaling in activated rat hepatic stellate cells. *J Biol Chem* **274**:33881-33887.

122. Lafon, C., C. Mathieu, M. Guerrin, O. Pierre, S. Vidal and A. Valette. 1996. Transforming growth factor beta 1-induced apoptosis in human ovarian carcinoma cells: protection by the antioxidant N-acetylcysteine and bcl- 2. *Cell Growth Differ* **7**:1095-1104.

123. Langer, C., J. M. Jurgensmeier and G. Bauer. 1996. Reactive oxygen species act at both TGF-beta-dependent and -independent steps during induction of apoptosis of transformed cells by normal cells. *Exp Cell Res* **222**:117-124.

124. Bauer, G. 1996. Elimination of transformed cells by normal cells: a novel concept for the control of carcinogenesis. *Histol Histopathol* **11**:237-255.

125. Sanchez, A., A. M. Alvarez, M. Benito and I. Fabregat. 1997. Cycloheximide prevents apoptosis, reactive oxygen species production, and glutathione depletion induced by transforming growth factor beta in fetal rat hepatocytes in primary culture. *Hepatology* **26**:935-943.

126. Haufel, T., S. Dormann, J. Hanusch, A. Schwieger and G. Bauer. 1999. Three distinct roles for TGF-beta during intercellular induction of apoptosis: a review. *Anticancer Res* **19**:105-111.

127. Barcellos-Hoff, M. H. and T. A. Dix. 1996. Redox-mediated activation of latent transforming growth factor-beta 1. *Mol Endocrinol* **10**:1077-1083.

128. Jurgensmeier, J. M., J. Panse, R. Schafer and G. Bauer. 1997. Reactive oxygen species as mediators of the transformed phenotype. *Int J Cancer* **70**:587-589.

129. Garcia-Trevijano, E. R., M. J. Iraburu, L. Fontana, J. A. Dominguez-Rosales, A. Auster, A. Covarrubias-Pinedo and M. Rojkind. 1999. Transforming growth factor beta1 induces the expression of alpha1(I) procollagen mRNA by a hydrogen peroxide-C/EBPbeta-dependent mechanism in rat hepatic stellate cells. *Hepatology* **29**:960-970.

130. Svegliati-Baroni, G., S. Saccomanno, H. van Goor, P. Jansen, A. Benedetti and H. Moshage. 2001. Involvement of reactive oxygen species and nitric oxide radicals in activation and proliferation of rat hepatic stellate cells. *Liver* **21**:1-12.

131. Poli, G. 2000. Pathogenesis of liver fibrosis: role of oxidative stress. *Mol Aspects Med* **21**:49-98.

132. Larios, J. M., R. Budhiraja, B. L. Fanburg and V. J. Thannickal. 2001. Oxidative protein cross-linking reactions involving L-tyrosine in transforming growth factor-beta1-stimulated fibroblasts. *J Biol Chem* **276**:17437-17441.

133. Assoian, R. K. and M. A. Schwartz. 2001. Coordinate signaling by integrins and receptor tyrosine kinases in the regulation of G1 phase cell-cycle progression. *Curr Opin Genet Dev* **11**:48-53.

134. Lee, S. L., W. W. Wang and B. L. Fanburg. 1997. Association of Tyr phosphorylation of GTPase-activating protein with mitogenic action of serotonin. *Am J Physiol* **272**:C223-230.

135. Lee, S. L., W. W. Wang and B. L. Fanburg. 1998. Superoxide as an intermediate signal for serotonin-induced mitogenesis. *Free Radic Biol Med* **24**:855-858.

136. Meng, T. C., T. Fukada and N. K. Tonks. 2002. Reversible oxidation and inactivation of protein tyrosine phosphatases in vivo. *Mol Cell* **9**:387-399.

137. Thannickal, V. J., J. M. Larios and B. L. Fanburg. 2001. H_2O_2 production by myofibroblasts is dependent on Src kinase(s) and actin cytoskeletal regulation. *Chest* **120**:32S-33S

138. Serini, G., M. L. Bochaton-Piallat, P. Ropraz, A. Geinoz, L. Borsi, L. Zardi and G. Gabbiani. 1998. The fibronectin domain ED-A is crucial for myofibroblastic phenotype induction by transforming growth factor-beta1. *J Cell Biol* **142**:873-881.

139. Tomasek, J. J., G. Gabbiani, B. Hinz, C. Chaponnier and R. A. Brown. 2002. Myofibroblasts and mechano-regulation of connective tissue remodelling. *Nat Rev Mol Cell Biol* **3**:349-363.

Chapter 17

ROLE OF MITOCHONDRIAL OXYGEN AND NITROGEN REACTIVE SPECIES IN SIGNALING

Cecilia Giulivi and Merry Jo Oursler

17.1 Introduction

The four-electron reduction of oxygen to water by cytochrome c oxidase is a reaction performed with high precision and rapidity; however, during the transfer of electrons throughout the electron transport chain, oxygen is partially reduced, yielding oxygen species that readily react with a variety of cellular components. The term "reactive oxygen species" (ROS) includes superoxide anion, hydrogen peroxide, and hydroxyl radical. Superoxide anion radical is the one-electron reduction product of oxygen and is a precursor of other reactive species. Hydrogen peroxide is formed from superoxide anion dismutation. Probably the most potent oxygen species in biological systems is the hydroxyl radical. This species forms from relatively harmless hydrogen peroxide. Although most free radicals are extremely short-lived (1 ns to 1 µs), they react readily with other molecules, transferring the radical character, converting them to free radicals and thereby initiating chain reactions.

On the basis that uni- or divalent reduction products of oxygen are produced during normal oxidative metabolism, it is possible to speculate that either an increase in the production of these species, or decrease in the antioxidant defenses, will lead to ROS injury. Considering the manifestations of ROS toxicity, it is likely that sequential processes would be involved; these include (a) the primary injury at the molecular level followed by (b) the defence mechanisms that normally respond to the initial injury. If the defence mechanisms are exhausted or after what may be

H. J. Forman, J. Fukuto, and M. Torres (eds.), Signal Transduction by Reactive Oxygen and Nitrogen Species: Pathways and Chemical Principles, ©2003, *Kluwer Academic Publishers. Printed in the Netherlands*

accounted for as the accumulation of errors in the handling of oxygen intermediates, the biological target (enzyme, cell, or tissue) cannot adequately perform their physiologic function. Considering the concentration of the normal biological antioxidant defences -usually 2-3 orders of magnitude higher than those of the corresponding substrates- it seems that the free-radical reactions arising during the course of normal oxidative metabolism are at least partially responsible for the aging process.[1,2] Because the mitochondrion is the site of the bulk of the cell's oxidative metabolism, mitochondrial biomolecules (e.g., DNA, lipids, and proteins) probably sustain most of free-radical damage. In this regard, several degenerative diseases, including Parkinson's, Alzheimer's, and Huntington's diseases, are associated with oxidative damage to mitochondria.

The high reactivity of free radicals makes it difficult to characterize their reaction products, for all classes of biological molecules are susceptible to oxidative damage by free radicals. As a consequence of their high reactivity, their short half-life limits the target radius. Thus, a role for ROS as initiators or participants in signal transduction pathways has been controversial. However, accumulating evidence indicates some important functions for reactive oxygen and nitrogen species (RONS) in signalling pathways.[3-5] Their actions seem to be attributed either to reduction-oxidation reactions or binding to target molecules. These signalling mechanisms are entirely different from those attributed to, for example, calcium-activated and phosphorylation-dependent pathways.

In this Chapter we will review the sources of oxygen- and nitrogen species in mitochondria, the modulation of their production, and describe some examples in which mitochondrial RONS have been implicated in signal transduction pathways.

17.2   Mitochondrial Oxygen Free Radicals

Almost 40 years ago, an increased formation of hydrogen peroxide was observed from mitochondria exposed to hyperbaric conditions [6]. In the following years, the generation of hydrogen peroxide was demonstrated in mitochondria isolated from various sources also under normoxic conditions.[7-10] Both NAD- and FAD-linked substrates support rates of hydrogen peroxide production (0.2-0.8 nmol hydrogen peroxide/min mg protein [9]) modulated by various metabolic states.[9,10] The production of hydrogen peroxide by submitochondrial particles (i.e., constituted only by inner membrane fraction) suggested that a member of the respiratory chain was responsible for hydrogen peroxide generation. Studies performed with inhibitors of the mitochondrial respiratory chain pointed to a carrier on the substrate side of succinate dehydrogenase-ubiquinone segment[9,11] as a potential generator of hydrogen peroxide. Among the different components of this segment, a main role for ubiquinone for hydrogen peroxide generation in mitochondria was supported by the linear correlation between ubiquinone supplementation and rate of for hydrogen peroxide production in ubiquinone-depleted mitochondria.[8,9] Later, a second site, albeit quantitatively less significant, was found at the NADH dehydrogenase segment.[12]

Besides mitochondria, other subcellular fractions have been identified as intracellular sources of for hydrogen peroxide: microsomes, peroxisomes, and soluble enzymes, among others.[8] However, the rate of ROS production by mitochondria at biologically relevant pO_2 constitutes the most important contribution to the cellular/organ rate of ROS production, therefore, these organelles should be considered as the main source of ROS under physiological conditions.[13] Supporting this view, experimental rates of for hydrogen peroxide production by perfused rat liver were found to be mostly of mitochondrial origin [14].

17.3 Modulation of Mitochondrial ROS Production

Mitochondria are dynamic organelles whose morphology, composition and function adapt to changes in physiological signals.[14] These physiological signals include nutritional variations, different workloads, oxygen availability, and development. Responses to physiological signals are fundamental at maintaining homeostasis and as such, are typically reversible and serve to optimise energy production relative to energy demand. In this regard, formation of hydrogen peroxide by mitochondria shows generally similar characteristics across species/tissues. All mitochondria exhibit lower (or negligible) rates of hydrogen peroxide production in State 3 (active ATP production), whereas is maximum in State 4 (resting, nonphosphorylating mitochondria). Hyperbaric oxygen and hyperoxia exposure results in a marked increase in for hydrogen peroxide production by isolated heart and liver mitochondria.[7,8,15] However, owed to mitochondrial plasticity and the consequent heterogeneity, assessment of ROS production by mitochondria necessitates consideration of specific aspects of mitochondrial characteristics. For example, presence of uncoupler and inhibitors are required to exert maximal rates of ROS production in heart but not in liver mitochondria. Phenotypic changes occur in cellular energetics during cell culture, thus results obtained with mitochondria isolated from organs might not be comparable to those obtained with mitochondria isolated from cultured cells. In addition, rates of ROS production obtained with isolated mitochondria are lacking the cell-specific and tissue-specific interaction, availability of chemically different substrates and the presence of regulatory pathways.

Focusing on isolated mitochondria, and considering the succinate dehydrogenase-ubiquinone segment as the most important source of ROS (60-80%), then the rate of ROS production is modulated by the steady-state concentrations of ubisemiquinone and oxygen (eq. 1; assuming that superoxide anion is the chemical precursor of for hydrogen peroxide) for the production of hydrogen peroxide is formulated as a non-enzymatic oxidation of ubisemiquinone (UQ•−) by oxygen:[16]

$$+d[O_2^{•-}]/dt = 2 \times +d[H_2O_2]/dt = k\,[UQ^{•-}]\,[O_2] \qquad [1]$$

As a consequence, several metabolic conditions that result in an increased level of ubiquinone (fasting, chronic treatment with dinitrophenol, cortisone treatment[17]) or higher availability of oxygen (e.g., hyperoxia) are expected to have increased rates of ROS.

17.3.1   Availability of Intracellular Oxygen and the role of nitric oxide

Hyperoxia and hyperbaric oxygen enhance hydrogen peroxide generation at the subcellular and cellular levels at different extents (from 60 to 200%). However, the hyperbaric response appears to be greatly diminished at the organ level *in vivo*. The level of oxygen in a tissue may be limited by the microvascular response, probably modulated by endogenous factors such as nitric oxide.

Nitric oxide, a reactive free radical molecule, is generated *in vivo* by nitric oxide synthase (NOS) during the conversion of L-Arg to citrulline. Several isoforms of NOS have been isolated,[18-21] namely the neuronal, endothelial, and macrophage forms. The constitutive forms, *i.e.*, neuronal and endothelial, account for the rapid, transient, Ca^{2+}-dependent production of nitric oxide;[19-21] the inducible form, *i.e.,* macrophage NOS, causes the slow onset of Ca^{2+}-independent nitric oxide synthesis in inflammatory cells (after stimulation with cytokines or lipopolysaccharides.[18,22]

Nitric oxide functions as an intercellular signal in regulating blood vessel dilation (among other important functions),[23-25] thus facilitating oxygen delivery and removal of metabolic end products from tissues. The stimulatory action of vasodilators on the phosphoinositide signalling system in endothelial cells produces an influx of calcium resulting in the synthesis of nitric oxide. Nitric oxide rapidly diffuses across membranes,[26] although its high reactivity prevents it from getting far away from its site of synthesis (e.g., it reacts efficiently with heme proteins and oxygen).[27] The physiological target of nitric oxide in smooth muscle cells is guanylate cyclase, which catalyzes the reaction of GTP to yield cGMP,[28] an intracellular second messenger similar to cAMP. Nitric oxide reacts with guanylate's cyclase heme prosthetic group to yield nitrosoheme, increasing the enzyme's activity by 50- to 200-fold,[29] presumably via a conformation change that allows the release of transaxial histidine ligand, which presumably, participates in catalysis. In turn, cGMP causes smooth muscle relaxation through its stimulation of protein phosphorylation by cGMP-dependent protein kinase. In other biological settings, nerve stimulation causes calcium increases at nerve terminals, thereby stimulating NOS activity. The resultant nitric oxide diffuses to nearby smooth muscle cells, where it binds to guanylate cyclase and activates it to synthesize cGMP as described above.[30,31]

The vasodilating effects mediated by nitric oxide are understood as part of intercellular signalling pathways, in which nitric oxide is produced by one cell type and acts on another one. However, the recent finding that mitochondria are endowed with a NOS has extended the role of this free radical as an intracellular signal molecule.

17.4 Mitochondrial Nitrogen Reactive Species

Our[32-37] and others'[38-41] studies provided evidence for the occurrence of a NOS (mtNOS) in mitochondria. Based on MALDI-ToF and Q-ToF analyses of tryptic digests, mtNOS has been identified as nNOS isoform alpha.[36] mtNOS is constitutively expressed, following a particulate distribution,[32,35] probably favoured by its acylation with myristic acid.[36] The rate of nitric oxide production by intact, coupled mitochondria is L-Arg-dependent; the apparent K_m for L-Arg and V_{max}

values were 5 µM and 1.4 nmol/min per mg protein, respectively, in agreement with the values published for the purified brain isoform.[32] Given that the experimental K_m is 30-40 times below the reported pool of L-Arg in mitochondrial matrix (about 200 µM), the modulation of mtNOS activity by L-Arg seems unlikely under physiological conditions. The production of nitric oxide is also sustained by endogenous NADPH, the specific electron donor for NOS. Under our experimental conditions, the energy-dependent transhydrogenase seems to represent the main source of NADPH because uncoupling conditions (overload of Ca^{2+} or FCCP) significantly decreased the levels of NADPH, halting nitric oxide production.[32]

Although nitric oxide is produced by mitochondria, the target molecule for this free radical was unknown for these organelles do not contain guanylate cyclase. Intact coupled mitochondria, when stimulated to produce nitric oxide, exhibited a decrease in their respiratory rates from 30 to 50% (at $pO_2 = 0.2$ atm), which was completely reversed upon the addition of an inhibitor of NOS or the removal of L-Arg.[33] Conversely, oxygen uptake increased in the presence of an inhibitor of NOS, N-monomethyl-L-arginine, indicating the involvement of nitric oxide in the modulation of oxygen consumption. Concomitantly to the decline in the respiratory rate, an inhibition of ATP synthesis was also observed (40-50%), a decrease not attributable to a direct effect of nitric oxide on Complex V.[33] The dependence of the respiratory rates of mitochondria in State 3 and cytochrome oxidase activities on oxygen concentrations indicated that both processes were linked and competitively inhibited by nitric oxide at the cytochrome oxidase level. Thus, the target molecule of nitric oxide in mitochondria is cytochrome oxidase: nitric oxide by binding to the oxidase, increases the apparent K_m for oxygen (decreasing the affinity for oxygen), thereby decreasing the oxygen consumption at a given pO_2. This inhibition of cytochrome oxidase by nitric oxide was explained primarily by the catabolism of nitric oxide to N_2O, and secondarily, by the direct binding of nitric oxide to the bimolecular center.

The main impact of this finding is that nitric oxide produced by mitochondria may constitute the main participant at regulating the oxygen uptake in organs. The intracellular level of oxygen may be estimated in 10-20 µM, thus, under normal physiological conditions, tissues are under hypoxic environments. A gradient of oxygen as high as 1000-fold[42] is expected from the capillaries toward mitochondria, and the steady-state concentration of oxygen in the latter compartment may be lower than those in either the peroxisomal and cytosolic spaces.[14] The critical oxygen concentration for bioenergetic function of mitochondria corresponds to approximately to 50% reduction of pyridine nucleotide being 60 and 80 nM in State 4 and 3, respectively.[42] Thus, when oxygen levels are adequate, the ratio [oxygen]/[nitric oxide] is high, and cytochrome oxidase activity is maximal. When oxygen levels are low, then the ratio [oxygen]/[nitric oxide] is lower, nitric oxide competitively inhibits cytochrome oxidase, thereby decreasing oxygen consumption. By this mechanism, the oxygen gradient is extended to other mitochondria/cells, thus allowing a more homogeneous distribution of oxygen throughout the cell/organ.

Therefore, the production of nitric oxide by mitochondria gains significance given the modulatory role that this molecule might have on energy metabolism, oxygen consumption, and consequently, on oxygen free radicals production (see below).

17.4.1   Mitochondrial Nitric Oxide and Cytochrome Oxidase Modulation

In the case of oxidative phosphorylation, the pathway from NADH to cytochrome *c* functions near equilibrium. In the cytochrome oxidase reaction, however, the terminal step of the electron transport chain is irreversible and is therefore a prime candidate as the control site of the pathway. It is believed that cytochrome oxidase, in contrast to most regulatory enzyme systems, appears to be controlled exclusively by the availability of one of its substrates, reduced cytochrome *c*. Since this substrate is in equilibrium with the rest of the coupled oxidation phosphorylation system, its concentration ultimately depends on the intramitochondrial [NADH]/[NAD$^+$] ratio and the ATP mass action ratio ([ATP]/[ADP][Pi]). Consequently, the higher the [NADH]/[NAD$^+$] ratio and the lower the ATP mass action ratio, the higher the concentration of ferrocytochrome *c* and thus the higher the cytochrome oxidase activity.

ATP production is controlled by mass action: an increased ATP utilization increases ATP synthesis, oxygen consumption, and substrate oxidation, simply by providing the substrates (ADP and Pi) for oxidative phosphorylation. By following this process, mass action stimulates the respiratory chain back to substrate oxidation. Several factors have been proposed to regulate respiration; among them Ca^{2+} mobilization, adenine nucleotide carriers, oxygen supply, and cytosolic ATP mass action ratio. However, cytochrome oxidase can serve as a potential control point of oxidative phosphorylation for it catalyzes an irreversible reaction (the reduction of oxygen to water. In this regard, several molecules have been proposed as regulators or inhibitors of cytochrome *c* oxidase, among them cyanide, azide, formate, sulfide, carbon monoxide and nitric oxide. From these, carbon monoxide has similar chemical characteristics to nitric oxide: it is synthesized by an enzyme (*i.e.*, heme oxygenase) and binds to guanylate cyclase. In contrast to nitric oxide, carbon monoxide has a higher affinity for myo- or hemo-globins than for cytochrome oxidase, thus the concentration required to exert a significant inhibition of cytochrome oxidase activity would be higher that that achieved under physiological conditions. Nitric oxide is the only one that satisfies all the requirements as an allosteric effector of cytochrome oxidase:[35] it is produced at a fair rate by a specific enzyme (mtNOS), located close to the target site (cytochrome oxidase), and it is formed at levels (10-220 nM) that effectively affect the respiratory rate (K_I = 6-10 nM).

As a consequence of the high affinity of nitric oxide for cytochrome oxidase, the cytotoxic role of nitric oxide observed in other systems could be explained as a sustained inhibition of cytochrome oxidase by high concentrations of nitric oxide - like those achieved by activated macrophages- or a formation of the powerful oxidant peroxynitrite (the product between superoxide anion and nitric oxide) which would favor mitochondrial dysfunction, leading probably to cell death.

17.4.2   Modulation of ROS Production by Mitochondrial Nitric Oxide

Changes in hydrogen peroxide production by L-Arg-supplemented mitochondria indicated that nitric oxide affected the rate of oxygen radical production by modulating the rate of oxygen consumption at the cytochrome oxidase level.[37] This mechanism was supported by two observations: (1) changes in hydrogen peroxide production correlated with the effect of nitric oxide on the respiratory rates, and (2) the pattern of oxidized/reduced carriers in the presence of nitric oxide indicated cytochrome oxidase as the crossover point according to the crossover theorem.[45] Although the rate of hydrogen peroxide production by mitochondria increased in the presence of L-Arg, the increase was not as high as that obtained with other inhibitors (e.g., antimycin A) used to stimulate ROS production. This could be explained by the catabolism of nitric oxide by cytochrome oxidase, which allowed electrons to go through the chain without resulting in the full reduction of respiratory carriers.

Elucidation of the effect of nitric oxide on the mitochondrial ROS production brought a more dynamic view at the way oxygen consumption and ROS formation occurs, considering not only availability of ADP but also other regulatory devices (Fig. 1). As it has been indicated in previous sections, historically the mitochondrial production of ROS has been considered as a side-process of the normal oxidative metabolism; its rate alternating between two levels determined by physiologically relevant mitochondrial metabolic states, namely States 4 (maximum) and 3 (minimum). However, our studies demonstrated that the mitochondrial production of ROS is not limited to these two values and may exhibit a degree of values modulated by endogenous nitric oxide.

The balance between producing slightly higher hydrogen peroxide when low levels of nitric oxide are available (which probably could be handle by endogenous peroxidases, minimizing organelle damage) to peroxynitrite when sustained production of nitric oxide are attained (which will produce damage to biomolecules) will be maintained by cellular conditions that would influence the availability of either Arg or other cofactors required for NOS activity as well as the concentration of oxygen.

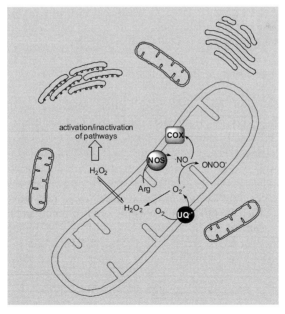

Figure 1: Mitochondrial Rons

17.5 Signal Transduction Elicited by RONS

A variety of extracellular and intracellular signals including the presence or absence of specific growth factors/cytokines and DNA damage result in elevation of mitochondria-derived ROS and subsequent apoptosis (reviewed in).[5,43,44] As discussed in detail elsewhere in this book, apoptosis is a controlled and regulated process involving caspase-mediated degradation of many cellular components that is often driven by elevated levels of mitochondria-derived ROS in the cytosol (reviewed in).[45] The mitochondria may serve to integrate multiple signalling events to focus the interactions that result in apoptosis and/or it may be the target of the apoptotic signal to drive caspase activation.[46] Many of these signalling pathways impact ROS levels by influencing Bcl-2 family member expression and/or localization. Pro-survival members of the Bcl2 family, including Bcl-2 and Bcl-xl, block apoptosis by blocking changes in the mitochondria membrane to prevent release of apoptosis-promoting factors and preserving mitochondria integrity.[47] In many instances, pro-apoptotic family members such as BAD, Bik, and Bid, alter the mitochondria membrane potential, resulting in elevations of cytosolic ROS. In addition, pathways that activate sphingomylinase to produce ceramide, disrupt mitochondria membrane integrity and influence ROS levels in the cytosol.[48-52] As discussed elsewhere in this book, activation of members of the MAPK kinase family of signalling molecules promotes apoptosis, to a large extent by influencing the Bcl-2 family members transcription or cellular localization. These members include MEK1/2, which activates ERK1/2, and ASK1, which activates both the JNK and p38 MAPK pathways.[53-56] The MEK/ERK pathway has been implicated in both survival and apoptosis induction, depending on the signal and the cell system under study.[57-61] ASK1 is an integral component of the apoptotic response following TNF-α and FAS treatment.[54,62-64] Recently, ASK1 has been implicated in the apoptotic response in sympathetic neurons following NGF withdrawal, supporting a role in stress-induced apoptosis as well.[65]

17.5.1 Molecular Basis for RONS Signalling

The starting point for signal transduction pathways is the ligand/receptor recognition step. This recognition is the result of a delicate balance among powerful countervailing forces of noncovalent interactions (electrostatic, hydrogen bonding, and hydrophobic forces). Conversely, the basis for free radical signalling does not entail protein-protein interaction but usually either a reduction-oxidation process or a binding reaction between the free radical and the target molecule. The specificity of free radical-mediated pathway then relies essentially on steric exposure, reduction potential (e.g., thiol group), rate constant, and binding capacity (dissociation constant of, for example, nitric oxide to heme) of the target molecule, in addition to spatial distribution of RONS production. This last condition is important for free radicals that diffuse to target molecules. Since the transport rate of diffusing substance varies inversely with the square of the distance it must diffuse, the diffusion rate of a free radical through a tissue thicker than 1 mm is too slow to support a reaction. Indeed, assuming a lifetime of a free radical of 1 ns, it can be calculated from Einstein's equation that the average distance a free radical can diffuse is about one-third of a membrane thickness or protein diameter. Thus, the target molecule must be in close proximity to the free radical source; otherwise pathway activation will be halted.

There are several mechanisms by which ROS can impact protein function. As discussed before, the nucleophilic cysteine side chains are reactive to RONS and modifications of these can alter protein functions (Reviewed in).[66] ROS can also modulate intracellular calcium ion concentrations or modify protein-protein interactions of redox-sensitive proteins.[67] There has been a massive influx of studies of the impact of ROS on cellular activity. Many of the studies that have been published on the impacts of ROS on cellular events have been done with relatively high treatment levels, so the physiological relevance of these effects remains unclear. Importantly, intracellular glutathione levels can determine whether a cell responds to ROS.[68-71] Thus discrepancies in published data may in part be the result of subtle shifts in the redox state of the cells under study.

17.5.2   *Molecular Targets of RONS*

The RONS-mediated signalling pathways can be divided on the basis of their molecular target. In this Chapter, we will focus on cytosolic and nuclear proteins of relevant signal transduction pathways as targets of RONS.

17.6 Cytosolic Proteins as Molecular Targets of RONS

17.6.1   *Tyrosine phosphatase*

Protein tyrosine phosphatases appear to be particularly sensitive to inactivation by RONS.[66,72,73] This inactivation is reversible, providing a basis for the speculation that regulation of phosphatase activity by ROS participates in regulation of cell signalling.[72] The impact of this is potentially enormous, as it could result in amplification of all activated signalling pathways since the removal of phosphates from phosphotyrosines in both receptors and downstream signalling molecules is integral to either activation or inactivation of several components of these pathways. This may be the case in many instances, particularly at high ROS levels used for cell response studies, but there is evidence of selective activation of specific pathways in response to ROS when studies are carried out using physiologically relevant ROS levels.

17.6.2   *Receptor Activation*

Although non-physiological hydrogen peroxide levels have been shown to activate several types of receptors, physiological levels do not have similar effects on receptor activation.[74] Physiological hydrogen peroxide levels can, however, amplify receptor responses under conditions where ligand levels are low.[74] Moreover, in the case of the insulin receptor, prolonged exposure to low hydrogen peroxide levels results in decreased signalling, suggesting that ROS impacts on receptor activity are complex [75].

17.6.3   MAPK pathways

The MAPK pathways are complex, intersecting pathways that impact a broad spectrum of cellular processes including proliferation, differentiation, and apoptosis (reviewed in).[53,76-78] Below are discussed components of these pathways that have been shown to be influenced by ROS. Many of the aspects of these pathways are covered in more detail elsewhere in this book.

17.6.4   c-Src family

The members of the c-Src family of tyrosine kinases act as second messengers in a broad spectrum of signalling pathways originating from both receptor kinases and receptors that do not have intrinsic kinase activity (reviewed in).[76,79,80] At millimolar levels, hydrogen peroxide activates several members of the Src family.[81-85] However, when examining lower levels of hydrogen peroxide, Lck is selectively activated.[83] The selective activation of Lck suggests that inactivation of tyrosine phosphatases is unlikely to be the Lck activation mechanism, although phosphorylation at the autophosphorylation site is required.[83] Nitric oxide selectively activates c-Src by causing aggregation of c-Src, promoting cross-phosphorylation and amplified activation.[86] Activation of members of the c-Src family can result in activation of several pathways. A key c-Src family member activation target is Ras (reviewed in[76]). In addition, *S*-nitrosylation of cysteine 118 in by nitric oxide can also lead to Ras activation independent of Src activation.[87-90] Hydrogen peroxide stimulates Ras and JAK2 activation, but only in mice with functional c-Fyn, another member of the c-Src family that can also activate Ras.[91] Activated Ras can target the MEK/ERK, JNK, and p38 MAPK pathways (reviewed in).[56,77,78,92-94] ROS have been documented to activate all of these pathways.[88,91,95-103] Treatment with hydrogen peroxide results in Src-mediated activation of JNK, ERK1/2, and p38 MAPK.[95,99,101,102] Others have shown that superoxide anion, but not hydrogen peroxide, activates ERK 1/2, so the responses are likely to be cell type or condition-specific.[98] Superoxide anion activates p38 MAPK, most likely by activation of c-Src.[102,104] Nitric oxide activates ERK1/2, JNK, and p38 MAPK pathways.[88,96,100] The mechanisms of hydrogen peroxide-mediated activation of JNK include targeted disruption of a complex of JNK and glutathione-*S*-transferase Pi.[103] While incorporated in this complex, JNK is inactive. Hydrogen peroxide caused oligomerization of the glutathione-*S*-transferase Pi, which dissociates from the complex with JNK, leading to JNK activation.

17.6.5   Protein Kinase C

The PKC pathway is mainly impacted by ROS influences on intracellular calcium levels. ROS have been shown to increase intracellular calcium ion levels by both mobilization of intracellular stores and increases influx of calcium from the extracellular environment.[105-112]

17.6.6   *Apoptosis Signalling-regulated Kinase 1 (ASK1)*

The amino terminus of ASK1 can bind with Trx, blocking ASK1 activation.[113] Trx is an oxidoreductase that in induced by oxidative stress and is redox sensitive.[114-116] ROS causes Trx dimerization, resulting in release of ASK1, which forms multimers and becomes active.[117,118]

17.7   Nuclear Proteins as Targets for RONS

17.7.1   *Transcription factor AP-1*

AP-1 is a dimer of members of the Fos and Jun families and regulates expression of a number of genes (reviewed in [119-124]). In yeast, hydrogen peroxide activates yAP-1, which activates genes involved in protection against oxidative stress.[125] In higher organisms, ROS also activate AP-1.[90,94,103,119,126-128] There are several levels of AP-1 activation that are potential ROS targets: expression of Fos and Jun family members, phosphorylation of Jun members by JNK, and oxidative state of Trx. ROS increases in c-Fos and c-Jun mRNA levels, perhaps through increasing intracellular calcium levels.[72,129-135] As noted above, oxidative-mediated JNK activation has been documented. JNK specifically phosphorylates the amino terminus of Jun, promoting dimerization of with Fos family members and AP-1 activation (reviewed in [136]). AP-1 is also sensitive to oxidized Trx levels and activity is thus impacted by this redox sensitive protein.[137] The role of AP-1 in promoting stress-induced apoptosis is not fully resolved, but has been shown to activate FAS-ligand, a well-known apoptosis inducing agent (reviewed in [120-124]). One possible mechanisms is that AP-1 has been implicated in repression of pro-survival Bcl-3[138] and activation of the pro-apoptotic Bcl-2 family member Bax.[139] Further, AP-1 has also been shown to promote survival of cells in some instances, but the mechanisms of this are also not fully determined.[120,122]

17.7.2   *Transcription factor STATS*

STAT transcription factors regulate a number of protooncogenes including c-fos and c-myc.[140-142] The JAK-STAT pathway is activated by hydrogen peroxide, raising the likelihood of regulation of transcriptional activity by ROS.[143]

17.7.3   *Transcription factor NF-κB*

ROS have been shown to activate NF-κB in many cell types.[89,94,103,126-128,144] The mechanisms by which activation takes place appears to vary with the cell type. Inactive NF-κB is sequestered in the cytosol in a complex with IκB and activation in response to external stimuli involves phosphorylation and targeted degradation of IκB (reviewed in[145]). A number of studies have supported ROS-mediated targeted degradation of IκB, either with or without increased IκB phosphorylation.[84,104,146-150] In several cell types, NF-κB responds directly to oxidative stress.[85,128,131,151] Additionally, physiological levels of glutathione disulfide decrease NF-κB binding to DNA, suggesting another mechanism by which NF-κB responds to the redox state of the cell.[137]   NF-κB has been most classically associated with pro-survival responses, but activation has also been associated with apoptosis induction in some instances (reviewed in[152-156]). Pro-survival Bcl-xL and Bcl-2 increase as a result of NF-κB activation in many of the systems in which NF-κB is involved in promoting survival [157]. In systems where NF-κB promotes apoptosis, this may be, at least in part, due to increase FAS transcription.[158]

17.8 Conclusions

Thus, a spectrum of signalling pathways both regulate mitochondria-derived RONS in the cytosol and are downstream targets of RONS. This creates the potential for a positive feedback amplification loop, significantly impacting the rate of apoptosis. In addition, the observations that several of these pathways may promote or suppress apoptosis and are regulated by RONS further suggests that the ultimate survival of a cell is dependent on the redox state of the intracellular environment.

Acknowledgements

This work was supported by grants ES011407-01 from the National Institutes of Health, 37470-B4 Petroleum Research Fund (American Chemical Society), and CC5675 Cottrell Research Corporation to C. G and the Department of the Army Grant DAMD17-00-1-0346, the Minnesota Medical Foundation, and the Lilly Center for Women's Health to M. J. O.

References

1. Menzel, D. B. 1970. Toxicity of ozone, oxygen, and radiation. *Annu Rev Pharmacol* **10**:379-94.

2. Gerschman, R. 1964. Biological Effects of Oxygen. in *Oxygen in the Animal Organism* (eds. Dickens, F. & Niel, E.), pp 475-494 Pergamon Press, London.

3. Lander, H. M. 1997. An essential role for free radicals and derived species in signal transduction. *Faseb J* **11**:118-24.

4. Finkel, T. 1998. Oxygen radicals and signaling. *Curr Opin Cell Biol* **10**:248-53.

5. Nose, K. 2000. Role of reactive oxygen species in the regulation of physiological functions. *Biol Pharm Bull* **23**:897-903.

6. Chance, B., D. Jamieson and H. Coles. 1965. Energy-linked pyridine nucleotide reduction: inhibitory effects of hyperbaric oxygen in vitro and in vivo. *Nature* **206**:257-63.

7. Loschen, G., L. Flohe and B. Chance. 1971. Respiratory Chain Linked H2O2 Production in Pigeon Heart Mitochondria. *FEBS Lett* **18**:261-264.

8. Boveris, A., N. Oshino and B. Chance. 1972. The Cellular Production of Hydorgen Peroxide. *Biochem J* **128**:617-630.

9. Boveris, A. and B. Chance. 1973. The Mitochondrial Generation of Hydrogen Peroxide: General Properties and Effect of Hyperbaric Oxygen. *Biochem J.* **134**:707-716.

10. Loschen, G., A. Azzim, C. Richter and L. Flohe. 1973. Mitochondial H2O2 Formation: Relationship with Energy Conservation. *FEBS Lett* **33**:84-88.

11. Loschen, G., A. Azzim and C. Richter. 1974. Superoxide Radicals as Precursors of Mitochondrial Hydrogen Peroxide. *FEBS Lett* **42**:68-72.

12. Turrens, J. F. and A. Boveris. 1980. Generation of superoxide anion by the NADH dehydrogenase of bovine heart mitochondria. *Biochem J* **191**:421-7.

13. Giulivi, C., A. Boveris and E. Cadenas. 1999. Oxygen Radicals in Mitochondria: Critical Evaluation of the Methodology Availabe for Estimating Steady-State Concentrations of Oxygen Radidals. in *Reactive Oxygen Species in Biological Systems: Selected Topics* (eds. Gilbert, D.L. & Colton, C.A.) pp. 77-102 Plenum Press, New York.

14. Oshino, N., D. Jamieson, T. Sugano and B. Chance. 1975. Optical measurement of the catalase-hydrogen peroxide intermediate (Compound I) in the liver of anaesthetized rats and its implication to hydrogen peroxide production in situ. *Biochem J* **146**:67-77.

15. Chance, B., A. Boveris, N. Oshino and G. Loschen. 1971. The Nature of the Catalase Intermediate in its Biological Function. in *Oxidases and Related Redox Systems*, Vol. I (eds. King, T.E., Mason, H.S. & Morrison, M.) pp. 350-353 University Park Press, Baltimore.

16. Boveris, A., E. Cadenas and A. O. Stoppani. 1976. Role of ubiquinone in the mitochondrial generation of hydrogen peroxide. *Biochem J* **156**:435-44.

17. Beyer, R. E., W. M. Noble and T. J. Hirschfeld. 1962. Alterations of Rat-Tissue Coenxymes Q (Ubiquinone) Levels by Various Treatments. *Biochim Biophys Acta* **57**:376-379.

18. Bredt, D. S. and S. H. Snyder. 1990. Isolation of nitric oxide synthetase, a calmodulin-requiring enzyme. *Proc Natl Acad Sci U S A* **87**:682-5.

19. Mayer, B., M. John and E. Bohme. Purification of a Ca2+/calmodulin-dependent nitric oxide synthase from porcine cerebellum. Cofactor-role of tetrahydrobiopterin. *FEBS Lett* **277**:215-9.

20. Schmidt, H. H. et al. 1991. Purification of a soluble isoform of guanylyl cyclase-activating-factor synthase. *Proc Natl Acad Sci U S A* **88**:365-9.

21. Stuehr, D. J. and M. A. Marletta. 1987. Induction of nitrite/nitrate synthesis in murine macrophages by BCG infection, lymphokines, or interferon-gamma. *J Immunol* **139**:518-25.

22. Forstermann, U., H. H. Schmidt, K. L. Kohlhaas and F. Murad. 1992. Induced RAW 264.7 macrophages express soluble and particulate nitric oxide synthase: inhibition by transforming growth factor-beta. *Eur J Pharmacol* **225**:161-5.

23. Feldman, P. L., O. W. Griffith and D. J. Stuehr. 1993. The surprising life of nitric oxide. *Chem. Eng. News* **20**:26-38.

24. Gally, J. A., P. R. Montague, G. N. Reeke, Jr. and G. M. Edelman. 1990. The NO hypothesis: possible effects of a short-lived, rapidly diffusible signal in the development and function of the nervous system. *Proc Natl Acad Sci U S A* **87**:3547-51.

25. Moncada, S., R. M. Palmer and E. A. Higgs. 1991. Nitric oxide: physiology, pathophysiology, and pharmacology. *Pharmacol Rev* **43**:109-42.

26. Wise, D. L. and G. Houghton. 1968. Diffusion coefficients of neon, krypton, xenon, carbon monoxide and nitric oxide in water at 10-60 degrees C. *Chem. Eng. Sci.* **23**:1211-1216.

27. Furchgott, R. F. and P. M. Vanhoutte. 1989. Endothelium-derived relaxing and contracting factors. *Faseb J* **3**:2007-18.

28. Craven, P. A. and F. R. DeRubertis. 1978. Restoration of the responsiveness of purified guanylate cyclase to nitrosoguanidine, nitric oxide, and related activators by heme and hemeproteins. Evidence for involvement of the paramagnetic nitrosyl-heme complex in enzyme activation. *J Biol Chem* **253**:8433-43.

29. Ignarro, L. J., J. B. Adams, P. M. Horwitz and K. S. Wood. 1986. Activation of soluble guanylate cyclase by NO-hemoproteins involves NO-heme exchange. Comparison of heme-containing and heme-deficient enzyme forms. *J Biol Chem* **261**:4997-5002.

30. Garthwaite, J., S. L. Charles and R. Chess-Williams. 1988. Endothelium-derived relaxing factor release on activation of NMDA receptors suggests role as intercellular messenger in the brain. *Nature* **336**:385-8.

31. Shibuki, K. and D. Okada. 1991. Endogenous nitric oxide release required for long-term synaptic depression in the cerebellum. *Nature* **349**:326-8.

32. Giulivi, C., J. J. Poderoso and A. Boveris. 1998. Production of nitric oxide by mitochondria. *J Biol Chem* **273**:11038-43.

33. Giulivi, C. 1998. Functional implications of nitric oxide produced by mitochondria in mitochondrial metabolism. *Biochem J* **332 (Pt 3)**:673-9.

34. Giulivi, C., T. M. Sarkela, J. Berthiaume and S. Elfering. 1999. Modulation of mitochondrial respiration by endogenous nitric oxide. *FASEb J* **13**:A1554.

35. Tatoyan, A. and C. Giulivi. 1998. Purification and characterization of a nitric-oxide synthase from rat liver mitochondria. *J Biol Chem* **273**:11044-8.

36. Elfering, S. L., T. M. Sarkela and C. Giulivi. 2002. Biochemistry of mitochondrial nitric-oxide synthase. *J Biol Chem*.

37. Sarkela, T. M., J. Berthiaume, S. Elfering, A. A. Gybina and C. Giulivi. 2001. The modulation of oxygen radical production by nitric oxide in mitochondria. *J Biol Chem* **276**:6945-9.

38. Bates, T. E., A. Loesch, G. Burnstock and J. B. Clark. 1995. Immunocytochemical evidence for a mitochondrially located nitric oxide synthase in brain and liver. *Biochem Biophys Res Commun* **213**:896-900.

39. Kobzik, L., B. Stringer, J. L. Balligand, M. B. Reid and J. S. Stamler. Endothelial type nitric oxide synthase in skeletal muscle fibers: mitochondrial relationships. *Biochem Biophys Res Commun* **211**:375-81.

40. Ghafourifar, P. and C. Richter. 1997. Nitric oxide synthase activity in mitochondria. *FEBS Lett* **418**:291-6.

41. Ghafourifar, P., U. Schenk, S. D. Klein and C. Richter. 1999. Mitochondrial nitric-oxide synthase stimulation causes cytochrome c release from isolated mitochondria. Evidence for intramitochondrial peroxynitrite formation. *J Biol Chem* **274**:31185-8.

42. Sugano, T., N. Oshino and B. Chance. 1974. Mitochondrial functions under hypoxic conditions. The steady states of cytochrome c reduction and of energy metabolism. *Biochim Biophys Acta* **347**:340-58.

43. Dormann, S. *et al.* 1999. Intercellular induction of apoptosis through modulation of endogenous survival factor concentration: a review. *Anticancer Res* **19**:87-103.

44. Carmody, R. J. and T. G. Cotter. 2001. Signalling apoptosis: a radical approach. *Redox Rep* **6**:77-90.

45. Leist, M. and M. Jaattela. 2001. Four deaths and a funeral: from caspases to alternative mechanisms. *Nat Rev Mol Cell Biol* **2**:589-98.

46. Ranger, A. M., B. A. Malynn and S. J. Korsmeyer. 2001. Mouse models of cell death. *Nat Genet* **28**:113-8.

47. Quaglino, D. and I. P. Ronchetti. 2001. Cell death in the rat thymus: A minireview. *Apoptosis* **6**:389-401.

48. Zha, J., H. Harada, E. Yang, J. Jockel and S. J. Korsmeyer. 1996. Serine phosphorylation of death agonist BAD in response to survival factor results in binding to 14-3-3 not BCL-X(L) [see comments]. *Cell* **87**:619-28.

49. Yang, E. et al. 1995. Bad, a heterodimeric partner for Bcl-XL and Bcl-2, displaces Bax and promotes cell death. *Cell* **80**:285-91.

50. Roberts, M. L., K. Virdee, C. P. Sampson, I. Gordon and A. M. Tolkovsky. 2000. The combination of bcl-2 expression and NGF-deprivation facilitates the selective destruction of BAD protein in living sympathetic neurons. *Mol Cell Neurosci* **16**:97-110.

51. Scheid, M. P., K. M. Schubert and V. Duronio. 1999 Regulation of bad phosphorylation and association with Bcl-x(L) by the MAPK/Erk kinase. *J Biol Chem* **274**:31108-13.

52. Zhou, X. M., Y. Liu, G. Payne, R. J. Lutz and T. Chittenden. 2000. Growth factors inactivate the cell death promoter BAD by phosphorylation of its BH3 domain on Ser155. *J Biol Chem* **275**:25046-51.

53. Cross, T. G. et al. 2000. Serine/threonine protein kinases and apoptosis. *Exp Cell Res* **256**:34-41.

54. Tobiume, K. et al. 2001. ASK1 is required for sustained activations of JNK/p38 MAP kinases and apoptosis. *EMBO Rep* **2**:222-8.

55. Davis, R. J. 1999. Signal transduction by the c-Jun N-terminal kinase. *Biochem Soc Symp* **64**:1-12.

56. Kolch, W. 2000. Meaningful relationships: the regulation of the Ras/Raf/MEK/ERK pathway by protein interactions. *Biochem J* **351(Pt 2)**:289-305.

57. Mitsui, H. et al. 2001. The MEK1-ERK map kinase pathway and the PI 3-kinase-Akt pathway independently mediate anti-apoptotic signals in HepG2 liver cancer cells. *Int J Cancer* **92**:55-62.

58. Wang, X., J. L. Martindale and N. J. Holbrook. 2000. Requirement for ERK activation in cisplatin-induced apoptosis. *J Biol Chem* **275**:39435-43.

59. Ishikawa, Y. and M. Kitamura. 1999. Dual potential of extracellular signal-regulated kinase for the control of cell survival. *Biochem Biophys Res Commun* **264**:696-701.

60. Moreno-Manzano, V., Y. Ishikawa, J. Lucio-Cazana and M. Kitamura. 1999. Suppression of apoptosis by all-trans-retinoic acid. Dual intervention in the c-Jun n-terminal kinase-AP-1 pathway. *J Biol Chem* **274**:20251-8.

61. Iryo, Y., M. Matsuoka, B. Wispriyono, T. Sugiura and H. Igisu. 2000. Involvement of the extracellular signal-regulated protein kinase (ERK) pathway in the induction of apoptosis by cadmium chloride in CCRF-CEM cells. *Biochem Pharmacol* **60**:1875-82.

62. Liu, Y., G. Yin, J. Surapisitchat, B. C. Berk and W. Min. 2001. Laminar flow inhibits TNF-induced ASK1 activation by preventing dissociation of ASK1 from its inhibitor 14-3-3. *J Clin Invest* **107**:917-23.

63. Liu, W. et al. 2000. Endothelial cell survival and apoptosis in the tumor vasculature. *Apoptosis* **5**:323-8.

64. Ichijo, H. et al. 1997. Induction of apoptosis by ASK1, a mammalian MAPKKK that activates SAPK/JNK and p38 signaling pathways. *Science* **275**:90-4.

65. Kanamoto, T. et al. 2000. Role of apoptosis signal-regulating kinase in regulation of the c-Jun N- terminal kinase pathway and apoptosis in sympathetic neurons. *Mol Cell Biol* **20**:196-204.

66. Barrett, W. C. et al. 1999. Regulation of PTP1B via glutathionylation of the active site cysteine 215. *Biochemistry* **38**:6699-705.

67. Droge, W. 2002. Free radicals in the physiological control of cell function. *Physiol Rev* **82**:47-95.

68. Roth, S. and W. Droge. 1987. Regulation of T-cell activation and T-cell growth factor (TCGF) production by hydrogen peroxide. *Cell Immunol* **108**:417-24.

69. Roth, S. and W. Droge. 1991. Regulation of interleukin 2 production, interleukin 2 mRNA expression and intracellular glutathione levels in ex vivo derived T lymphocytes by lactate. *Eur J Immunol* **21**:1933-7.

70. Roth, S., H. Gmunder and W. Droge. 1991. Regulation of intracellular glutathione levels and lymphocyte functions by lactate. *Cell Immunol* **136**:95-104.

71. Axline, S. G. 1970. Functional biochemistry of the macrophage. *Semin Hematol* **7**:142-60.

72. Beiqing, L., M. Chen and R. L. Whisler. 1996. Sublethal levels of oxidative stress stimulate transcriptional activation of c-jun and suppress IL-2 promoter activation in Jurkat T cells. *J Immunol* **157**:160-9.

73. Berlett, B. S., R. L. Levine and E. R. Stadtman. 1996. Comparison of the effects of ozone on the modification of amino acid residues in glutamine synthetase and bovine serum albumin. *J Biol Chem* **271**:4177-82.

74. Schmid, E., A. Holtz-Wagenblatt, V. Hack and W. Droge. 1999. Phosphorylation of the insulin receptor kinase by phosphocreatine in combination with hydrogen peroxide. The structural basis of redox priming. *FAseb J* **13**:1491-1500.

75. Tirosh, A., R. Potashnik, N. Bashan and A. Rudich. 1999. Oxidative stress disrupts insulin-induced cellular redistribution of insulin receptor substrate-1 and phosphatidylinositol 3-kinase in 3T3-L1 adipocytes. A putative cellular mechanism for impaired protein kinase B activation and GLUT4 translocation. *J Biol Chem* **274**:10595-602.

76. Porter, A. C. and R. R. Vaillancourt. 1998. Tyrosine kinase receptor-activated signal transduction pathways which lead to oncogenesis. *Oncogene* **17**:1343-52.

77. Avruch, J. et al. 2001. Ras activation of the Raf kinase: tyrosine kinase recruitment of the MAP kinase cascade. *Recent Prog Horm Res* **56**:127-55.

78. Campbell, S. L., R. Khosravi-Far, K. L. Rossman, G. J. Clark and C. J. Der. 1998. Increasing complexity of Ras signaling. *Oncogene* **17**:1395-413.

79. Superti-Furga, G. and S. A. Courtneidge. 1995. Structure-function relationships in Src family and related protein tyrosine kinases. *Bioessays* **17**:321-30.

80. Brown, M. T. and J. A. Cooper. 1996. Regulation, substrates and functions of src. *Biochim Biophys Acta* **1287**:121-49.

81. Brumell, J. H., A. L. Burkhardt, J. B. Bolen and S. Grinstein. 1996. Endogenous reactive oxygen intermediates activate tyrosine kinases in human neutrophils. *J Biol Chem* **271**:1455-61.

82. Hayashi, T., Y. Ueno and T. Okamoto. 1993. Oxidoreductive regulation of nuclear factor κ B. Involvement of a cellular reducing catalyst thioredoxin. *J Biol Chem* **268**:11380-8.

83. Nakamura, K. et al. 1993. Redox regulation of a src family protein tyrosine kinase p56lck in T cells. *Oncogene* **8**:3133-9.

84. Schoonbroodt, S. et al. 2000. Crucial role of the amino-terminal tyrosine residue 42 and the carboxyl-terminal PEST domain of I κ B alpha in NF-κ B activation by an oxidative stress. *J Immunol* **164**:4292-300.

85. Schreck, R., P. Rieber and P. A. Baeuerle. 1991. Reactive oxygen intermediates as apparently widely used messengers in the activation of the NF-κ B transcription factor and HIV-1. *Embo J* **10**:2247-58.

86. Akhand, A. A. et al. 1999. Nitric oxide controls src kinase activity through a sulfhydryl group modification-mediated Tyr-527-independent and Tyr-416-linked mechanism. *J Biol Chem* **274**:25821-6.

87. Lander, H. M. et al. 1997. A molecular redox switch on p21(ras). Structural basis for the nitric oxide-p21(ras) interaction. *J Biol Chem* **272**:4323-6.

88. Lander, H. M., A. T. Jacovina, R. J. Davis and J. M. Tauras. 1996. Differential activation of mitogen-activated protein kinases by nitric oxide-related species. *J Biol Chem* **271**:19705-9.

89. Lander, H. M. et al. 1996. Redox regulation of cell signalling. *Nature* **381**:380-1.

90. Lander, H. M., J. S. Ogiste, S. F. Pearce, R. Levi and A. Novogrodsky. 1995. Nitric oxide-stimulated guanine nucleotide exchange on p21ras. *J Biol Chem* **270**:7017-20.

91. Abe, J. and B. C. Berk. 1999. Fyn and JAK2 mediate Ras activation by reactive oxygen species. *J Biol Chem* **274**:21003-10.

92. Yordy, J. S. and R. C. Muise-Helmericks. 2000. Signal transduction and the Ets family of transcription factors. *Oncogene* **19**:6503-13.

93. Segal, R. A. and M. E. Greenberg. 1996. Intracellular signaling pathways activated by neurotrophic factors. *Annu Rev Neurosci* **19**:463-89.

94. Adler, V., Z. Yin, K. D. Tew and Z. Ronai. 1999. Role of redox potential and reactive oxygen species in stress signaling. *Oncogene* **18**:6104-11.

95. Abe, J., M. Kusuhara, R. J. Ulevitch, B. C. Berk and J. D. Lee. 1996. Big mitogen-activated protein kinase 1 (BMK1) is a redox-sensitive kinase. *J Biol Chem* **271**:16586-90.

96. Callsen, D., J. Pfeilschifter and B. Brune 1998. Rapid and delayed p42/p44 mitogen-activated protein kinase activation by nitric oxide: the role of cyclic GMP and tyrosine phosphatase inhibition. *J Immunol* **161**:4852-8.

97. Elbirt, K. K., A. J. Whitmarsh, R. J. Davis and H. L. Bonkovsky. 1998. Mechanism of sodium arsenite-mediated induction of heme oxygenase-1 in hepatoma cells. Role of mitogen-activated protein kinases. *J Biol Chem* **273**:8922-31.

98. Baas, A. S. and B. C. Berk. 1995. Differential activation of mitogen-activated protein kinases by H2O2 and O2- in vascular smooth muscle cells. *Circ Res* **77**:29-36.

99. Ushio-Fukai, M., R. W. Alexander, M. Akers and K. K. Griendling. 1998. Mitogen-activated protein kinase is a critical component of the redox-sensitive signaling pathways

activated by angiotensin II. Role in vascular smooth muscle cell hypertrophy. p38. *J Biol Chem* **273**:15022-9.

100. Pfeilschifter, J. and A. Huwiler. 1996. Nitric oxide stimulates stress-activated protein kinases in glomerular endothelial and mesangial cells. *FEBS Lett* **396**:67-70.

101. Yoshizumi, M., J. Abe, J. Haendeler, Q. Huang and B. C. Berk.C. 2000. Src and Cas mediate JNK activation but not ERK1/2 and p38 kinases by reactive oxygen species. *J Biol Chem* **275**:11706-12.

102. Pu, M. et al. 1996. Evidence of a novel redox-linked activation mechanism for the Src kinase which is independent of tyrosine 527-mediated regulation. *Oncogene* **13**:2615-22.

103. Adler, V. et al. 1999. Regulation of JNK signaling by GSTp. *Embo J* **18**:1321-34.

104. Hehner, S. P. et al. 2000. Enhancement of T cell receptor signaling by a mild oxidative shift in the intracellular thiol pool. *J Immunol* **165**:4319-28.

105. Doan, T. N., D. L. Gentry, A. A. Taylor and S. J. Elliott. 1994. Hydrogen peroxide activates agonist-sensitive Ca(2+)-flux pathways in canine venous endothelial cells. *Biochem J* **297 (Pt 1)**:209-15.

106. Dreher, D. and A. F. Junod. 1996. Role of oxygen free radicals in cancer development. *Eur J Cancer* **32A**:30-8.

107. Hallbrucker, C., M. Ritter, F. Lang, W. Gerok and D. Haussinger. 1993. Hydroperoxide metabolism in rat liver. K+ channel activation, cell volume changes and eicosanoid formation. *Eur J Biochem* **211**:449-58.

108. Kumasaka, S., H. Shoji and E. Okabe. 1999. Novel mechanisms involved in superoxide anion radical-triggered Ca2+ release from cardiac sarcoplasmic reticulum linked to cyclic ADP-ribose stimulation. *Antioxid Redox Signal* **1**:55-69.

109. Okabe, E. et al. 1987. Calmodulin participation in oxygen radical-induced cardiac sarcoplasmic reticulum calcium uptake reduction. *Arch Biochem Biophys* **255**:464-8.

110. Okabe, E. et al. 1991. The effect of ryanodine on oxygen free radical-induced dysfunction of cardiac sarcoplasmic reticulum. *J Pharmacol Exp Ther* **256**:868-75.

111. Okabe, E., M. Sugihara, K. Tanaka, H. Sasaki and H. Ito. 1989. Calmodulin and free oxygen radicals interaction with steady-state calcium accumulation and passive calcium permeability of cardiac sarcoplasmic reticulum. *J Pharmacol Exp Ther* **250**:286-92.

112. Roveri, A. et al. 1992. Effect of hydrogen peroxide on calcium homeostasis in smooth muscle cells. *Arch Biochem Biophys* **297**:265-70.

113. Saitoh, M. et al. 1998. Mammalian thioredoxin is a direct inhibitor of apoptosis signal-regulating kinase (ASK) 1. *Embo J* **17**:2596-606.

114. Matsui, M. et al. 1996. Early embryonic lethality caused by targeted disruption of the mouse thioredoxin gene. *Dev Biol* **178**:179-85.

115. Sachi, Y. et al. 1995. Induction of ADF/TRX by oxidative stress in keratinocytes and lymphoid cells. *Immunol Lett* **44**:189-93.

116. Taniguchi, Y., Y. Taniguchi-Ueda, K. Mori and J. Yodoi. 1996. A novel promoter sequence is involved in the oxidative stress-induced expression of the adult T-cell leukemia-derived factor (ADF)/human thioredoxin (Trx) gene. *Nucleic Acids Res* **24**:2746-52.

117. Gopalakrishna, R. and W. B. Anderson. 1989. Ca2+- and phospholipid-independent activation of protein kinase C by selective oxidative modification of the regulatory domain. *Proc Natl Acad Sci U S A* **86**:6758-62.

118. Konishi, H. et al. 1997. Activation of protein kinase C by tyrosine phosphorylation in response to H2O2. *Proc Natl Acad Sci U S A* **94**:11233-7.

119. Angel, P. and M. Karin. 1991. The role of Jun, Fos and the AP-1 complex in cell-proliferation and transformation. *Biochim Biophys Acta* **1072**:129-57.

120. Shaulian, E. and M. Karin. 2002. AP-1 as a regulator of cell life and death. *Nat Cell Biol* **4**:E131-6.

121. Karin, M. and E. Shaulian. 2001. AP-1: linking hydrogen peroxide and oxidative stress to the control of cell proliferation and death. *IUBMB Life* **52**:17-24.

122. Shaulian, E. and M. Karin. 2001. AP-1 in cell proliferation and survival. *Oncogene* **20**:2390-400.

123. Shaulian, E. et al. 2000. The mammalian UV response: c-Jun induction is required for exit from p53-imposed growth arrest. *Cell* **103**:897-907.

124. Shaulian, E. and M. Karin. 1999. Stress-induced JNK activation is independent of Gadd45 induction. *J Biol Chem* **274**:29595-8.

125. Kuge, S. and N. Jones. 1994. YAP1 dependent activation of TRX2 is essential for the response of Saccharomyces cerevisiae to oxidative stress by hydroperoxides. *Embo J* **13**:655-64.

126. Flohe, L., R. Brigelius, C. Saliou and M. Traber. 1997. Redox regulation of NF-kB activation. *Free Radic Biol Med* **22**:1115-1126.

127. Piette, J. et al. 1997. Multiple redox regulation in NF-κB transcription factor activation. *Biol Chem* **378**:1237-45.

128. Sen, C. and L. Packer. 1996. Antioxidant and redox regulation of gene transcription. *Faseb J* **10**:709-720.

129. Crawford, D., I. Zbinden, P. Amstad and P. Cerutti. 1987. *Expression of oxidant stress-related genes in tumor promotion of mouse epidermal cells JB6.*, pp. 183-190 Plenum, New York

130. Janssen, Y. M., S. Matalon and B. T. Mossman. 1997. Differential induction of c-fos, c-jun, and apoptosis in lung epithelial cells exposed to ROS or RNS. *Am J Physiol* **273**:L789-96.

131. Meyer, M., R. Schreck and P. A. Baeuerle. 1993. H₂O₂ and antioxidants have opposite effects on activation of NF-κ B and AP-1 in intact cells: AP-1 as secondary antioxidant-responsive factor. *Embo J* **12**:2005-15.

132. Maki, A., I. K. Berezesky, J. Fargnoli, N. J. Holbrook and B. F. Trump. 1992. Role of [Ca2+]i in induction of c-fos, c-jun, and c-myc mRNA in rat PTE after oxidative stress. *Faseb J* **6**:919-24.

133. Morris, B. J. 1995. Stimulation of immediate early gene expression in striatal neurons by nitric oxide. *J Biol Chem* **270**:24740-4.

134. Muehlematter, D., T. Ochi and P. Cerutti. 1989. Effects of tert-butyl hydroperoxide on promotable and non-promotable JB6 mouse epidermal cells. *Chem Biol Interact* **71**:339-52.

135. Shibanuma, M., T. Kuroki and K. Nose. 1988. Induction of DNA replication and expression of proto-oncogene c-myc and c-fos in quiescent Balb/3T3 cells by xanthine/xanthine oxidase. *Oncogene* **3**:17-21.

136. Weston, C. R. and R. J. Davis. 2002. The JNK signal transduction pathway. *Curr Opin Genet Dev* **12**:14-21.

137. Galter, D., S. Mihm and W. Droge. 1994. Distinct effects of glutathione disulphide on the nuclear transcription factor κ B and the activator protein-1. *Eur J Biochem* **221**:639-48.

138. Rebollo, A. et al. 2000. Bcl-3 expression promotes cell survival following interleukin-4 deprivation and is controlled by AP1 and AP1-like transcription factors. *Mol Cell Biol* **20**:3407-16.

139. Lei, K. et al. 2002. The Bax subfamily of Bcl2-related proteins is essential for apoptotic signal transduction by c-Jun NH(2)-terminal kinase. *Mol Cell Biol* **22**:4929-42.

140. Jenab, S. and V. Quinones-Jenab. 2002. The effects of interleukin-6, leukemia inhibitory factor and interferon-gamma on STAT DNA binding and c-fos mRNA levels in cortical astrocytes and C6 glioma cells. *Neuroendocrinol Lett* **23**:325-8.

141. Kolonics, A. et al. 2001. Unregulated activation of STAT-5, ERK1/2 and c-Fos may contribute to the phenotypic transformation from myelodysplastic syndrome to acute leukaemia. *Haematologia (Budap)* **31**:125-38.

142. Servidei, T. et al. 1998. Coordinate regulation of STAT signaling and c-fos expression by the tyrosine phosphatase SHP-2. *J Biol Chem* **273**:6233-41.

143. Simon, A. R., U. Rai, B. L. Fanburg and B. H. Cochran. 1998. Activation of the JAK-STAT pathway by reactive oxygen species. *Am J Physiol* **275**:C1640-52.

144. Monteiro, H. P. and A. Stern. 1996. Redox modulation of tyrosine phosphorylation-dependent signal transduction pathways. *Free Radic Biol Med* **21**:323-33.

145. Cramer, P. and C. W. Muller. 1999. A firm hand on NFκB: structures of the IκBalpha-NFκB complex. *Structure Fold Des* **7**:R1-6.

146. Beg, A. A., T. S. Finco, P. V. Nantermet. and A. S. Baldwin, Jr. 1993. Tumor necrosis factor and interleukin-1 lead to phosphorylation and loss of I κ B alpha: a mechanism for NF-κ B activation. *Mol Cell Biol* **13**:3301-10.

147. Kretz-Remy, C., E. E. Bates and A. P. Arrigo. 1998. Amino acid analogs activate NF-κB through redox-dependent IκB-alpha degradation by the proteasome without apparent IκB-alpha phosphorylation. Consequence on HIV-1 long terminal repeat activation. *J Biol Chem* **273**:3180-91.

148. Kretz-Remy, C., P. Mehlen, M. E. Mirault and A. P. Arrigo. 1996. Inhibition of I κ B-alpha phosphorylation and degradation and subsequent NF-κ B activation by glutathione peroxidase overexpression. *J Cell Biol* **133**:1083-93.

149. Manna, S. K., H. J. Zhang, T. Yan, L. W. Oberley and B. B. Aggarwal. 1998. Overexpression of manganese superoxide dismutase suppresses tumor necrosis factor-induced apoptosis and activation of nuclear transcription factor-κB and activated protein-1. *J Biol Chem* **273**:13245-54.

150. Traenckner, E. B., S. Wilk and P. A. Baeuerle. 1994. A proteasome inhibitor prevents activation of NF-κ B and stabilizes a newly phosphorylated form of I κ B-alpha that is still bound to NF-κ B. *Embo J* **13**:5433-41.

151. Mihm, S., J. Ennen, U. Pessara, R. Kurth and W. Droge. 1991. Inhibition of HIV-1 replication and NF-κ B activity by cysteine and cysteine derivatives. *Aids* **5**:497-503.

152. Romas, E., M. T. Gillespie and T. J. Martin. 2002. Involvement of receptor activator of NFκB ligand and tumor necrosis factor-alpha in bone destruction in rheumatoid arthritis. *Bone* **30**:340-6.

153. Said, S. I. and K. G. Dickman. 2000. Pathways of inflammation and cell death in the lung: modulation by vasoactive intestinal peptide. *Regul Pept* **93**:21-9.

154. Leong, K. G. and A. Karsan. 2000. Signaling pathways mediated by tumor necrosis factor alpha. *Histol Histopathol* **15**:1303-25.

155. Hatada, E. N., D. Krappmann and C. Scheidereit. 2000. NF-κB and the innate immune response. *Curr Opin Immunol* **12**:52-8.

156. Li, X. and G. R. Stark. 2002. NFκB-dependent signaling pathways. *Exp Hematol* **30**:285-96.

157. Glasgow, J. N. et al. 2001. Transcriptional regulation of the BCL-X gene by NF-κB is an element of hypoxic responses in the rat brain. *Neurochem Res* **26**:647-59.

158. Kuhnel, F. et al. 2000. NFκB mediates apoptosis through transcriptional activation of Fas (CD95) in adenoviral hepatitis. *J Biol Chem* **275**:6421-7.

Chapter 18

REDOX REGULATION OF GENE EXPRESSION: TRANSCRIPTIONAL INDUCTION OF HEME OXYGENASE-1

Timothy P. Dalton, Lei He, Howard G. Shertzer and Alvaro Puga

18.1 Introduction

Incomplete reduction of oxygen to water during normal aerobic metabolism generates reactive oxygen species (ROS) that pose a serious threat to all aerobic organisms. Oxygen, on the other hand, is an essential element of life. ROS such as singlet oxygen, superoxide anion, hydrogen peroxide and nitric oxide are crucial for many physiologic processes. The need for the reduction potential to maintain cells in a state of redox balance is supplied by biochemical antioxidants (i.e. glutathione, pyridine nucleotides, ascorbate, retinoic acid, tocopherols) and a host of enzymatic reactions. As a consequence of exposure to a changing environment, cells shift their redox status to a more oxidized state, characterizing a cellular condition known as *oxidative stress*, often accompanied by decreases in the concentrations of biochemical antioxidants. To return to a state of redox balance, cells mount an *oxidative stress response* that increases the activities of antioxidant enzymes and restores antioxidant defenses.

Excess ROS are harmful because they are very reactive and modify cellular macromolecules that are critical targets responsible for behavioral abnormalities, cytotoxicity and mutagenic damage.[1-5] Aerobic organisms wage a constant battle to maintain redox homeostasis in the face of a constant exposure to environmental oxidants that increase ROS production, such as ultraviolet and ionizing radiation, heavy metals, redox active chemicals, anoxia and hyperoxia.[4,6-8] It has become

H. J. Forman, J. Fukuto, and M. Torres (eds.), Signal Transduction by Reactive Oxygen and Nitrogen Species: Pathways and Chemical Principles, ©2003, Kluwer Academic Publishers. Printed in the Netherlands

apparent that antioxidant defenses exist in a balance with endogenous oxidants and that it is the disruption of this balance that characterizes the pathogenesis of many human diseases and aging.[9-11]

In this chapter, we will briefly discuss the activation of signal transduction pathways by ROS, focusing with greater detail on the transcriptional induction of the heme oxygenase-1 gene. We will use the term *oxidative stress* for any condition that increases the cellular oxidation state to produce an oxidative stress response. Although an oxidative stress response does not necessarily result in toxicity, it is a component in the molecular mechanism of action of many toxicants.

18.2 Reactive Oxygen and Signal Transduction

Reactive oxygen species are generated in mitochondria during respiration and cytosol. They constitute a common regulatory mechanism of protein conformation and function because they are important determinants of the redox state of protein cysteinyl residues.[12] Redox cycling of cysteinyl thiols is also critical for the establishment of the protein-protein and protein-DNA interactions that determine many aspects of the signal transduction pathways. In addition, ROS are also responsible for the induction of many genes[13-17] and for the perturbation of signal transduction pathways responsible for the concerted maintenance of gene expression patterns.[18,19] In particular, ROS are critical in the regulation of transcription factors in the AP1[20-23] and NF-κB[24-26] families, two transcription factors families with crucial functions in proliferation, differentiation, apoptosis and morphogenesis. ROS signaling pathways for AP1 and NF-κB are activated in enucleated cells and in the absence of protein synthesis, indicating that DNA damage or nuclear factors are not required for their activation.[27,28]

18.3 Activation of AP1 by Reactive Oxygen Species

Activator protein-1 (AP1) is a general term for a family of basic domain/leucine zipper (bZIP) transcription factors that specifically bind to and activate gene transcription through a *cis*-acting regulatory DNA element known as the TPA response element (TRE). AP1 is a heterodimer of the protein products of individual members of the *FOS* and *JUN* immediate early response gene families, or a homodimer of JUN proteins.[29,30] AP1 controls the expression of many genes, including those encoding collagenase, stromelysin, cyclin D, TGF-1β and many cytokines, by binding to TREs in the promoters of these genes.[31] Expression of c-*JUN* and c-*FOS* is quickly induced by mitogens and by phorbol esters, such as 12-O-tetradecanoyl phorbol-13-acetate (TPA).[32-34] AP1 is also induced by H_2O_2 either exogenously administered,[35-37] or endogenously generated H_2O_2,[38] UV-C,[20,21,27,39] UV-A,[40] and by ionizing radiation,[41] asbestos,[42] and dioxin.[43,44]

The activity of AP1 is controlled by both transcriptional and posttranslational mechanisms.[45,46] Exposure of HeLa cells to UV-C or to H_2O_2 causes a rapid increase in AP1 DNA-binding activity, independently of new protein synthesis, indicating that activation is the result of posttranslational modifications in the FOS and JUN proteins.[20,27] These modifications are the consequence of changes in the phosphorylation patterns of the AP1 subunits.[46] In addition, increased rates of synthesis and decreased rates of degradation of *FOS* and *JUN* mRNA species[47-49] contribute to the overall level of functional AP1 in the cell by increasing FOS and JUN levels from *de novo* sybthesis. Reversible oxidation and reduction of FOS/JUN proteins are key signaling mechanisms responsible for AP1 activation.

18.4 Redox Regulation of AP1

Evidence for the redox regulation of AP1 activity was derived from experiments in which AP1 DNA binding activity showed significant increases after thioredoxin treatment *in vitro*[50,51] and after transient overexpression of thioredoxin *in vivo*.[52] Redox regulation of AP1 DNA-binding activity depends on the presence of critical cysteine residues in the AP1 protein. Substitution of Cys-154 in FOS or Cys-272 in JUN for serine leads to loss of redox regulation and enhanced DNA binding.[51] These critical cysteine residues are in contact with DNA and, under oxidant conditions, interfere with binding, whereas treatment with reducing agents restores their binding activity. These observations led to the discovery of REF-1 (also known as HAP1 and APEX), a redox factor that cooperates with thioredoxin to regulate the redox status of these critical cysteines and hence, the DNA-binding activity of AP1. REF-1 promotes the cycling of these cysteines between the reduced form and an oxidation product tentatively identified as the sulfenic (R-SOH) or the sulfinic ($R-SO_2H$) acid derivatives.[22,23,53-56] There is not evidence for a biologically significant mechanism that would reduce sulfinic acid, whereas, the non-enzymatic reduction of GSH with sulfenic acid and the given ability of thioredoxin to restore DNA binding activity, would argue in favor of the formation of mixed disulfides. Interestingly, REF-1 is also an apurinic/apyrimidinic endonuclease, which may make REF-1 a link between regulation of transcription, oxidative stress damage and DNA repair processes. The biological role of REF-1 appears to be even more extensive. HeLa cells in which REF-1 expression was blocked by antisense Ref-1 RNA showed hypersensitivity to killing by a wide range of oxidants and DNA damaging agents, including methyl methanesulfonate, H_2O_2, menadione, paraquat, hypoxia (1% oxygen), hyperoxia (100% oxygen), and L-buthionine *S,R*-sulfoximine (BSO), an inhibitor of glutathione biosynthesis.[57] These results suggest that REF-1 may be instrumental in protecting cells against a wide range of cellular stressors, including ROS and changes in oxygen tension.

18.5 Activation of AP1 by UV and by Phenolic Antioxidants

UV-C irradiation of murine and human cells stimulates c-*FOS* transcription.[39,58] ROS generated by UV-C, and ionizing radiation induce c-*FOS* and c-*JUN* expression by a mechanism that could be inhibited by N-acetyl cysteine (NAC)[59-61] and that involved activation of MAP kinases [60]. Earlier work on the role of SRC tyrosine kinases,[21] p21^{RAS} (see above) and RAF-1 kinase[62] in the transactivation of *FOS* and *JUN* by UV-C was followed up by the demonstration that the UV-C signal starts at the cellular membrane with the generation of reactive oxygen and initiates a cascade of phosphorylation events by activation of growth factor receptors, in particular epidermal growth factor receptor (EGFR) and platelet-derived growth factor receptor (PDGFR), and follows through an obligatory cytoplasmic signal transduction pathway that involves p21^{RAS}, RAF-1 and the MAPK kinases ERK-1 and -2, which phosphorylate ELK-1/TCF.[63]   Interestingly, exposure of human neutrophils to the thiol-oxidizing agent diamide or to H_2O_2 caused a significant increase in the activities of MAPK/ERKs kinases (MEKs), the MAPKs that phosphorylate ERK-1 and -2, and inhibited CD45, a protein tyrosine phosphatase (PTP) that dephosphorylates and inactivates MAP kinase.[64]   ROS potentiate the MAP kinase cascade not only by activation of kinases, but also by inhibition of phosphatases; sublethal levels of H_2O_2 were also found to stimulate MAP kinases and to inhibit the activity of PTPs and of protein phosphatase PP2A in Jurkat cells.[65] MEKs have also been found to regulate the reactive oxygen-dependent, p53-independent activation of the cyclin-dependent kinase inhibitor p21,[WAF1/CIP1] pointing at a causal connection between immediate-early responses to mitogenic stimuli and cell cycle progression.[66]

Exposure to sublethal concentrations of phenolic antioxidants, such as butylated hydroxytoluene, butylated hydroxyanisole (BHA) and its metabolite, *tert*-butylhydroquinone (BHQ), protects cells from oxidative damage by up-regulating the levels of γ-glutamyltranspeptidase and γ-glutamylcysteine ligase and hence increasing cellular GSH concentrations.[67]   Phenolic antioxidants also increase the levels of c-*FOS* and c-*JUN* mRNA and induce AP1 DNA-binding activity.[68] Paradoxically, BHQ also inhibits the induction of AP1 transcriptional activity by TPA.   BHQ induces FRA-1, a FOS-related member of the FOS family, that heterodimerizes with JUN and forms inhibitory AP1 complexes that antagonize the active FOS/JUN AP1 complexes induced by TPA.[69]   BHA and BHQ have been shown to activate MEK, MAPKs, ERK2 and JNK1 by an oxidative stress pathway that involves the formation of phenoxyl radicals.[70]

18.6   Role of Reactive Oxygen Species in the Activation of NF-κB

NF-κB plays a central role in the regulation of many genes involved in cellular defense mechanisms, pathogen defenses, immunological responses and expression of cytokines and cell adhesion molecules.   Functional NF-κB binding sites are present in the promoters of all NF-κB-responsive genes.[71]   Retroviruses, such as HIV-1, have NF-κB binding sites in their long terminal repeat (LTR) promoter region and use this factor as one of their major regulatory transactivators for proviral expression and viral replication.

A transcriptionally active NF-κB factor is a dimer of two proteins.   One is a member of the NFKB-1/-2 family, also known by their molecular mass as p50/p52, and the other is a member of the REL/RELA/RELB family, or p65.[72]   An inactive NF-κB complex is formed by two p50 homodimers or a p50/p65 heterodimer bound to a member of the IκB family.[71,73]   The NF-κB/IκB complex resides in the cytoplasm of unstimulated cells and can be rapidly induced to enter the cell nucleus without a requirement for *de novo* protein synthesis.   Stimulatory signals induce the phosphorylation of IκB at Ser-32 and Ser-36 by the IκB protein kinase, followed by ubiquitination at nearby lysine residues and proteolytic degradation, probably while still bound to NF-κB, which rapidly translocates to the nucleus.[73-75]   Many pathogenic and proinflammatory stimuli activate NF-κB, including viral infections, bacterial lipopolysaccharide, UV-C, ionizing radiation, and the cytokines interleukin-1and tumor necrosis factor. Reactive oxygen species generated in the mitochondrial respiratory chain have been proposed as the intermediate second messengers to the activation of NF-κB by TNF and IL-1.[28,76,77] Furthermore, cells lacking functional mitochondrial electron transport as a result of drug treatment or organelle depletion also show significant suppression of NK-κB activation.[78]

A large number of studies have shown that virtually all stimuli known to activate NF-κB can be blocked by antioxidants, including L-cysteine, NAC, thiols, dithiocarbamates, and vitamin E and its derivatives.[18,79] Phosphorylation of IκB at Ser-32 and Ser-36 can also be inhibited by dithiocarbamates,[80] pointing at the possibility that the IκB kinase could be activated by ROS. A multisubunit high molecular weight (>700 kDa) kinase composed by two subunits, IκKα and IκKβ, has been identified as the IκB kinase that phosphorylates these serine residues, but its activation by ROS has not been demonstrated,[81] although formation of a mixed disulfide by S-thiolation of critical cysteine residues in this kinase has been proposed on theoretical grounds as a possible mechanism to explain the activation of this enzyme by ROS.[18] On the other hand, oxidation at Cys-179 of the activation loop of the IκKα/IκKβ complex by arsenite was shown to block the phosphorylation of IκB and the activation of NF-κB.[82,83]

NF-κB activation may be selectively induced by peroxides. Incubation of Jurkat T cells with H_2O_2 or butylperoxide, but not with various superoxide, hydroxyl radical and NO-generating compounds, resulted in the rapid activation of NF-κB and the induction of an HIV-LTR-driven chloramphenicol acetyltransferase.[84] In agreement with these results, activation of NF-κB was significantly reduced in cells engineered to overexpress catalase, whereas it was significantly increased by overexpression of Cu/Zn SOD.[85]

The effects of oxidative stress on NF-κB activation have come under scrutiny because of conflicting interpretations of data from studies that measure DNA binding rather than NF-κB transcriptional responses and from the contradictory results supporting both NF-κB stimulation and suppression by ROS.[24,86,87] DNA binding can be inhibited by the oxidation of a sensitive thiol at Cys-62 in the p50 subunit, which must be maintained in a reduced state for binding to take place.[88-90] Observations that antioxidants block activation cannot be construed as direct evidence that ROS are involved in the signaling mechanism, since many low molecular weight antioxidants may inhibit NF-κB activation and subsequent gene expression by non-antioxidant actions,[19,91] or by secondary effects of ROS on membrane stimuli responsible for NF-κB induction.[92] Optimal induction of NF-κB by H_2O_2 requires GSH, since GSH depletion suppresses activation,[93] but it may be restricted to some subclones of Jurkat T cells, since H_2O_2 only activates NF-κB in subclone JR, but not in other Jurkat clones or T-cell lines.[94,95] In addition, unlike in Jurkat (JR) T cells, transient overexpression of catalase failed to block induction of NF-κB by TNFα or TPA in COS-1 cells.[96]

18.7 Heme Oxygenase

There are three known heme oxygenase (HO) [EC 1.14.99.3] proteins. Heme oxygenase-2 (*HO2*) and the more recently-identified heme oxygenase-3 (*HO3*) are constitutive enzymes, whereas heme oxygenase-1 (*HO1*) is inducible. These membrane-bound enzymes, intrinsic to the endoplasmic reticulum, catalyze the oxygen-dependent mixed function oxidation of hemoproteins, dependent on reducing equivalents supplied by NADPH via NADPH-cytochrome P450 oxidoreductase. The initial hydroxylation on the carbon *alpha* to the pyrrole ring produces an unstable intermediate that releases carbon monoxide (CO). A second NADPH-requiring oxidation step releases the globin and the heme iron, and generates biliverdin. Biliverdin may be reduced to bilirubin and subsequently glucuronidated to produce the bile pigments. Recent reviews concerning heme oxygenases have focused on biological effects in pulmonary and cardiovascular diseases,[97] vascular smooth muscle cells,[98] endothelial cells,[99] neurodegenerative disorders,[100,101] kidney pathology,[102] systemic and various tissue disorders.[103,104] Other reviews have focused on methods that may be used to assess the protective roles of HO in various diseases.[105,106] Here, will examine the relationship between oxidative stress and the transcriptional regulation of *HO1*.

It is generally thought that the most important housekeeping activity of *HO2*, and perhaps *HO3*, is to mitigate the pro-oxidant properties of heme were it to be released from the large number of hemoproteins as free heme in the course of protein turnover and hemoprotein homeostasis. These considerations apply to systemic hemoglobin, as well as to a number of hemoproteins that vary in concentration in different cell types. These include myoglobin, mitochondrial cytochromes, cytochromes P450, guanylate cyclase and nitric oxide synthase.[107] *HO2* may have a distinct function in nitric oxide metabolism.[108] *HO2* was believed to have antioxidant properties, although the presence in its moiety of two heme-binding sites not involved in heme catalysis suggested the possibility that the antioxidant role of *HO2* resulted from heme sequestration, rather than degradation.[109] Targeted disruption of the *HO2* gene in mice uncovered a related role for this gene. Despite increased HO1 expression, *HO2* knockout mice exposed to >95% O_2 had a two-fold increase in lung glutathione relative to wild-type controls and showed a significant increase in lung hemoproteins and iron, in the absence of an increase in ferritin,[110] suggesting that the function of *HO2* in the lung is to increase iron turnover during oxidative stress and that this function cannot be compensated by *HO1*. The conditions that lead to *HO1* induction are typically pro-inflammatory, or more generally, conditions that produce an oxidative stress response. Although free heme is not released in the course of *HO* activity, potentially inflammatory non-heme iron is released; such iron is efficiently metabolized by iron regulatory proteins to prevent pro-oxidant pathology.[111,112]

In general, induction of *HO1* is correlated with protection against cardiovascular disease,[113] pulmonary dysfunction,[114] ischemia and ischemia-reperfusion injury,[115] neurological diseases and injury[116] and burns.[117] There are, however, exceptions to this rule and situations where "too much of a good thing" turns out to be deleterious (see below). *HO1* is also important in supporting fetal growth and maintaining pregnancy, possibly through the induction of placental growth factors and cytokines.[118,119] The potent anti-inflammatory effects of interleukin-10,[120] and protection of cardiac myocytes from hypoxia-induced stress,[121] appear to be mediated by induction of *HO1* and a p38 mitogen-activated protein kinase pathway. Many of the protective functions of *HO1* are believed to be associated with the products of *HO1*-mediated heme degradation, biliverdin and CO.[107]

Biliverdin is reduced to bilirubin by NADPH-biliverdin oxidoreductase [EC 1.3.1.24]. Biliverdin is an inhibitor of *HO* activity *in vivo*, and thus may be an important regulator of heme degradation. *HO1*-derived biliverdin and bilirubin are potent cellular antioxidants that can protect liver from acetaminophen toxicity,[122] cobalt chloride-induced oxidative injury[123] and ischemia-reperfusion injury.[124] These *HO* metabolites also protect smooth muscle cells[125] against oxidative damage including post-ischemic myocardial dysfunction,[126] and decrease the risk of coronary artery disease.[127] Although nanomolar concentrations of bilirubin protect neurons from hydrogen peroxide toxicity,[128] high levels of bilirubin are neurotoxic to the newborn and cause kernicterus.[129]

The known physiological roles for *HO1*-generated CO continue to grow in number. In many cases the protective function of CO suggests possible medicinal or pharmaceutical potential for therapy or intervention. For example, *HO1*-derived CO prevents hepatic ischemia-reperfusion injury, which is especially relevant to tissue preservation for liver transplants.[130] In this respect, *HO1*-generated CO has also been shown to protect pancreatic beta cells from apoptosis following transplantation.[131] This effect appears to be mediated by cGMP-dependent protein kinases. *HO1*, via CO, has also been shown to mitigate tissue graft rejection,[132-134] possibly by means of its action as a vasodilator, to maintain blood flow to regions that might otherwise become ischemic[135] and to prevent vasospasms.[136] Vasodilation may not always be beneficial, as it may contribute toward hypotension resulting from septic shock,[137] and in cerebral vasodilation during glutamate-induced cerebral seizures.[138] Similarly, CO enhancement of smooth muscle cell proliferation[139] may be either beneficial with respect to normal tissue remodeling and vascularization, or detrimental in the case of tumor growth.

18.8 Induction of the Heme Oxygenase-1 gene

Regulation of inducible *HO1* is critical in maintaining cellular homeostasis. Historically, heme oxygenase was one of the first enzymes shown to be upregulated by oxidative stress. A large number of agents, including singlet oxygen resulting from UVA radiation, hydrogen peroxide, cadmium chloride, menadione and sodium arsenite where shown to induce the appearance of a 32-kDa stress protein and the transcription of the *HO1* gene in human skin fibroblasts. Cloning of the cDNA coding for this protein identified it unequivocally as heme oxygenase.[13] Induction of *HO1* mRNA could be prevented by prior treatment of the cells with iron chelators, such as desferrioxamine, which also protected against cell killing by H_2O_2, suggesting that both killing and induction where connected by the generation of highly reactive hydroxyl radicals by an iron-catalyzed Fenton reaction.[140] Endogenous glutathione levels were also found to play a critical role in the modulation of constitutive as well as inducible levels of heme oxygenase, with glutathione depletion leading to an enhanced level of *HO1* gene expression.[141] Increased expression of *HO1* in hypoxic cardiomyocytes could be greatly reduced by replenishing glutathione with NAC in the culture medium while expression of *HO2* mRNA was unaffected by either hypoxia or glutathione levels.[142]

HO1 is also induced in a dose-dependent manner by many different kinds of mineral and asbestos fibers. Among these, chrysotile and crocidolite are the most efficient inducers, whereas tremolite and erionite are poorer inducers. Addition of superoxide dismutase or catalase to the culture medium inhibited *HO1* induction but not the clonogenic toxicity produced by the fibers, suggesting that HO1 induction and toxicity of oxidant stressors are not synonymous.[143] Peroxidation of membrane lipids is also an important endogenous source of *HO1* inducers. Lipid peroxidation products, such as 4-hydroxynonenal, induce *HO1* mRNA by 40-fold, whereas phospholipase metabolites, such as diacylglycerol and arachidonic acid induce HO1 by 3- to 6-fold above basal levels.[144]

Several signal transduction pathways have been implicated in the induction of *HO1* by oxidants. Generation of CO seems to be a critical second messenger in many of these pathways. CO production by degradation of heme in rat aortic vascular smooth muscle cells (VSMCs) is stimulated by treatment of the VSMCs with platelet-derived growth factor (PDGF), which also stimulates in a time- and concentration-dependent fashion an increase in levels of *HO1* mRNA and production of ROS.[145] Incubation of platelets with PDGF-treated VSMCs resulted in a significant increase in platelet cGMP concentration that could be reverse by addition of HO1 inhibitors, such as tin protoporphyrin-IX or by addition of a CO scavenger, such as hemoglobin. In contrast, effects on platelet cGMP could not be blocked by the NO inhibitor methyl-L-arginine.[145] Serum also stimulated HO1 expression and CO production in VSMCs and CO scavengers potentiated cell proliferation. Exogenous CO blocked the phosphorylation of the retinoblastoma protein and the expression of S-phase markers, arresting cells at the G1/S boundary.[146] CO also inhibited apoptosis in these cells, an effect that was partially dependent on the activation of soluble guanylate cyclase and was associated with the inhibition of mitochondrial cytochrome c release and with the suppression of p53 expression.[147] Identification of the actual target of the CO endogenously generated by *HO1* has proven elusive. CO appears to inhibit apoptosis in endothelial cells by activating the MAPK p38, but p38 is unaffected by CO treatment in VSMCs.[147] Regratably, many of the experiments with exogenous CO have been conducted at CO levels near or within the toxic range, which makes their results questionable from a physiologic stand point.

In macrophages, fibroblasts and liver cells, HO1 is also induced by antiinflamatory drugs, NO, TNFα and lipopolysaccharide (LPS).[148-150] Sodium nitroprusside, a NO donor, caused dose-dependent increases of *HO1* in HeLa cells by mechanisms that were dependent on activation of ERK and p38 MAP kinases, but independent of JNK and unrelated to cGMP signaling.[151] Induction by LPS appeared to be dependent on the activation of c-*fos* and c-*jun* and on the binding of LPS-upregulated AP1 to 5' distal enhancer elements in the regulatory domains of the HO1 gene.[148] On the other hand, induction by LPS in mouse liver was reduced to very low levels in TNFα knock out mice, but was unaffected by knocking out the IL-1β gene.[149] Liver nitrite or nitrate levels had no effect on the response, but phosphorylation levels of c-JUN N-terminal kinase (JNK) and p38 were very low in TNFα knock out mice administered LPS, suggesting that this MAPK kinases might be critical to trigger *HO1* induction in response to LPS.[149]

18.9 *HO1* Expression and Protection against Oxidative Stress Damage

Up-regulation of *HO1* is an adaptative mechanism that protects cells from oxidative damage in the presence of stressors. Formal proof for this hypothesis was obtained by analyzing the responses to oxidative challenges of cells and mice lacking functional *HO1*. Cultured mouse embryonic fibroblast from *HO1* knock out mice produced high levels of oxygen free radicals when they were exposed to hemin, H_2O_2, paraquat or $CdCl_2$, and they were hypersensitive to cytotoxicity caused by hemin and H_2O_2.[152] Furthermore, young adult *HO1*-null mice were prone to mortality and liver necrosis when challenged with LPS.[152] These data provide strong genetic evidence that *HO1* protects cells from oxidative damage during stress. In addition, HO1 knock out mice developed anemia associated with abnormally low serum iron levels and at the same time, showed large accumulations of iron in liver and kidney.[153] This iron overload contributed to oxidative damage, tissue injury and chronic inflamation, and suggested that *HO1* had also an important role in iron hemostasis, possibly by facilitating the release of iron from hepatic and renal cells.

Overexpression of *HO1* in cultured cells or in laboratory mice is protective against oxidative stress damage in many different tissues. Induction of HO-1 is an indispensable response to protect against acute heme protein toxicity in vivo. Kidney damage by a heme overload is rapidly induced in the kidney of HO-1+/+ mice as they sustain mild, reversible renal insufficiency without mortality. In contrast, HO-1-/- mice accumulate eight times more heme in their kidneys than the HO1+/+ mice and exhibit fulminant, irreversible renal failure and 100% mortality after heme exposure. Furthermore, doses of hemoglobin that exert no nephrotoxicity or mortality in HO-1+/+ mice, precipitate rapidly developing, acute renal failure and marked mortality in HO-1-/- mice.[154] Similarly, *HO1* protects against ozone-induced cell injury and pulmonary inflammation and contributes to the development of cellular adaptation to chronic ozone exposure.[155] *HO1* is also protective against oxidative stress induced injury of neuronal cells in culture[156] and of cerebellar granular neurons of transgenic mice overexpressing *HO1*.[157] *HO1* also protects against cardiac ischemia/reperfusion in the rat heart[158] and against hypoxia-reoxygenation in cardiac myocytes.[159] In both case, hemin or other *HO1* inducers were required for the full protective effect of HO1 in decreasing cardiac infarct area or attenuating the generation of reactive oxygen species.[158,159] *HO1* also protected against induction of DNA damage in lymphocytes of human subjects exposed to hyperbaric oxygen[160] and its overexpression prevented TNFα-induced apoptosis in mouse fibroblasts.[161] Inhibition was reversed by guanylate cyclase inhibition, suggesting that the antiapoptotic effect of *HO1* may be mediated by carbon monoxide.

Several reports have suggest that genetic polymorphisms in the human *HO1* gene contribute to disease in gene-gene and gene-environment interactions. A $(GT)_n$ dinucleotide repeat in the 5' flanking region of human *HO1* shows length polymorphism between 16 and 38 repeats and is associated with differential expression of reporter assays. Fusion of $(GT)_{16}$ or $(GT)_{20}$ to a luciferase reporter results in normal level of reporter expression, whereas fusion of $(GT)_{29}$ or $(GT)_{38}$ fails to express the reporter gene.[162] The frequency of the long microsatellite allele was significantly higher among smokers with chronic pulmonary emphysema than among smokers without the disease.[162] Similar association was found between the presence of long $(GT)_n$ alleles (n>25) and the incidence of restenosis after percutaneous transluminal angioplasty[163] and the susceptibility to coronary artery disease in type 2 diabetic patients,[164] suggesting that the ability to up regulate *HO1* is an important protective factor in cardiovascular disease.

The protective effect of *HO1* expression is not absolute. In fact, there is evidence that "too much of a good thing" might be deleterious. Transient over expression of *HO1* in rat fetal lung cells exposed to hyperoxia was protective against cell death, protein oxidation and lipid peroxidation, but it increased lactic dehydrogenase release and glutathione depletion, suggesting that there may be a beneficial threshold of over expression that, if exceeded, leads to damage.[165] Several experiments have demonstrated that the reversal of protection is due to the generation of reactive iron species. Cells overexpressing *HO1* became hypersensitive to UVA radiation after loading with hemin. Hypersensitivity was due to the increased accumulation of chelatable iron, because the iron chelator desferrioxamine greatly reduced it, suggesting that it is the release of iron from heme that causes the increased oxidative stress sensitivity of cells overexpressing *HO1*.[166] Elegant experiments using transient expression of a tetracycline receptor-regulated *HO1* demonstrated this to be the case. In cells exposed to hyperoxia, varying concentrations of the tetracycline analog doxycycline, upregulated *HO1* activity levels between 3- and 17-fold. Levels less than 5-fold were cytoprotective and showed low levels of heme as well, while levels of *HO1* greater than 15-fold were associated with significant oxidative damage, cellular injury and with high levels of non-heme reactive iron. Cellular cGMP was not changed by *HO1* activity, indicating that guanylate cyclase activation by CO had no role in the physiologic consequences of *HO1* overexpression. On the other hand, chelation of iron or inhibition of HO1 activity prior to hyperoxic exposure, significantly reduced oxygen toxicity.[167] Thus, overexpression of *HO1* beyond the ideal homeostatic level is also detrimental.

18.10 Transcription Factors and Control Elements that Mediate *HO1* Induction

HO1 is considered the prototypic oxidant stress-inducible gene. As a result, considerable research effort has focused on the transcriptional machinery responsible for *HO1* induction. Because *HO1* is transcriptionally upregulated by diverse classes of chemicals that as a common feature produce oxidative stress, oxidative stress has been proposed as the unifying mechanism underlying *HO1* induction by these compounds. In some instances this concept is supported by the observation that these diverse activators function through the same transcriptional elements and factors. However, it should be noted that response element mapping studies conducted on the *HO1* gene from several species often fail to identify consensus control elements.[168] In some instances the control elements identified in a particular species are inducer-specific. The following sections review the current understanding of the mechanisms of transcriptional activation of *HO1* from mouse, human, rat and chicken.

18.11 Mouse HO1

The transcriptional regulation of mouse *HO1* has been the one most extensively studied. Two powerful enhancer regions located approximately 4- and 10-kb upstream of the transcription start site have been identified.[169,170] Using many convergent approaches, including comparison of endogenous Ho1 expression with that of stably integrated reporter constructs, the E1 and E2 sequence elements were demonstrated to be powerful enhancers.[168,169,171] In addition, one or both of these enhancers have been shown to drive *HO1* expression in response to diverse inducers including H_2O_2,[169] LPS,[170,172] arsenite,[169] Cd^{2+},[169,171,173] Co^{2+},[174] TPA,[169,171] and hyperoxia.[175] These enhancers, referred to as E2 (formerly EB1[169]) and E1 (formerly SX2[171]) are shown in Figure 1 and contain 3 and 2 copies, respectively, of the stress response element (StRE).[168,176] Stress response elements are composite elements consisting of both the TPA response element (TRE) recognized by AP1 (see above) and the antioxidant response element (ARE[177]), also known as electrophile response element (EpRE[178]). The structure of the StRE is shown below:

```
123456789
TGACTCAGC(A/G)       StRE
TGACTCA TRE
TGACNNNGC(A/G)       ARE
```

The ARE is similar in consensus sequence to the NF-E2-binding site[179] and the MAF recognition element (MARE).[180] Heterodimeric transcription factor complexes containing nuclear factor-E2 p45-related factor-1 (NRF1) and NRF2 have been shown to be competent for transactivating through the ARE.[181] NRF1 and NRF2 are members of the CNC-bZIP family of transcription factors.[182] and form obligate heterodimers most avidly with the small MAF family of transcription factors, MAFF, MAFG, MAFK. In addition, they can also heterodimerize with Fos B, c-Jun, JunD, ATF2 and ATF4. However, before NRF2 can enter the nucleus and choose a dimerization partner, it must first escape interaction with its cytosolic tether KEAP1. Importantly, the interaction between NRF2 and KEAP1 is redox sensitive and NRF2 is released when cells are treated with electrophiles.[183] NRF1 and NRF2 are ubiquitously expressed, which fits well with the observation that *HO1* is expressed and regulated by oxidant stress in a vast number cell types.

In the early 1990's, when the mouse *HO1* enhancers were discovered, a role for AP1 in oxidant stress signaling was already established. Because the StRE encompasses a TRE, it was not surprising that it was found to support AP1-driven transcriptional activation.[169,171] In fact, similar findings were reported for several phase II drug metabolizing genes whose transcription is driven via AREs.[182] Concurrent with an appreciation for a role of AP1-independent regulation of the ARE responses to electrophiles, careful mutational analysis of the StRE revealed that it functions in response to heme and Cd^{2+} following mutations at positions 5 and 6 (see above) that change the TRE consensus while leaving intact the ARE.[176] Following this analysis, NRF2 was shown to be a transcriptional regulator of *HO1*.[184] In transient transfections, NRF2 was found to be a more powerful regulator of transcription through the *HO1* E1 enhancer than AP1. Furthermore, stable inducible expression of a dominant negative NRF2 mutant, incapable of transactivation but competent to dimerize, was found to block heme, Cd^{2+}, arsentite and tert-butlyhydroquinone mediated transcriptional activation of endogenous *HO1* mRNA.[184]   The possibility could not be excluded that NRF2 competed for interactions between non-NRF2 containing transcription factor complexes that were important for *HO1* induction.   The role of NRF2 in *HO1* induction has been established using NRF2 knockout mice. These mice are deficient in the induction of a battery of EpRE inducible genes[185] and in them, *HO1* induction is attenuated in response to several inducers.[186-189]   Caution must still be observed in interpreting these results, because induction of *HO1* may be downstream, but not a direct result of NRF2. Notwithstanding, since the NRF2 binding consensus is contained in the StRE motif, direct regulation of *HO1* by NRF2 seems likely.

NRF2 interacts most avidly with the small MAF proteins. While these proteins are the likely endogenous partners of NRF2, some evidence suggest that other bZIP proteins may also be NRF2 partners. In this regard, cJUN has been suggest to play a role in transactivation through the *NQO1* (quinone reductase) and *GCLC* (glutamate-cysteine ligase catalytic subunit) AREs,[190,191] while ATF4 has been shown to function in the context of the *HO1* gene,[192] although, in agreement with previous studies,[179,180] small MAF proteins were more avid binding partners for NRF2 than for ATF4. It must be stressed that caution should be exercised in interpreting results of transient transfections using small MAF proteins, because they can readily form homodimers.   When ectopically overexpressed, small MAF homodimers may compete for DNA binding sites and as a result, transcriptional responses could be inhibited, since small MAFs lack a transactivation domain. Furthermore, in the case of binding-incompetent small MAF dominant negative mutants, they may compete for their own inclusion in binding reactions.   These attributes of the small MAF proteins should be considered when evaluating their function in transcriptional complexes.

NRF2 interacts with KEAP1 and, provided that sufficient KEAP1 is present, NRF2 may be retained in the cytoplasm in a transcriptionally inactive form.[183] The mechanisms regulating the NRF2/KEAP1 interaction are poorly understood. Pro-oxidant conditions disrupt this interaction, but these results cannot be interpreted as evidence of a direct oxidant attack of the NRF2/KEAP1 interaction, although they are certainly consistent with this mode of action.   Several signal transduction pathways that are activated by pro-oxidant conditions intensify NFR2 dependent transcriptional signaling. In particular, protein kinase C (PKC) and its activators have been demonstrated to trigger NRF2 in vivo phosphorylation by PKC and nuclear translocation.[193]   In other studies, JNK1 (c-Jun N-terminal kinase-1) was demonstrated to potently increase NRF2 transcriptional activation.[194] In the context of the *HO1*, a p38 dominant negative mutant but not dominant negative mutants of ERK1/2 or JNK1/2, decreased E1-driven reporter gene expression in response to Cd^{2+}.[195] Ironically, as attention in the transcriptional regulation of *HO1* has shifted from the role of AP1 to the study of heterodimers containing NRF2, the signal transduction pathways so elegantly demonstrated to regulate AP1 signaling may be shown to be involved in a similar capacity in NRF2 signaling.

Not all the transcriptional responses of the mouse *HO1* gene are mediated by the E1 and E2 enhancer elements.   Hyperoxia-induced *HO1* reporter response is cooperatively regulated by E1 in the presence of two STAT binding sites within 600 bp upstream of the transcription start site.[175] The JAK-STAT pathway of transcriptional activation is best characterized in signals transduced by cytokines.[175,196] The relative importance of the STAT binding sites in the context of both enhancers was not addressed in those studies.

18.12 Human *HO1*

Several reports have identified control elements and transcription factors that may regulate human *HO1*. These analyses, with one notable exception (see below), examine human *HO1* 5'-flanking sequences that do not include those which are orthologous with E1 and E2 in the mouse gene, neglecting to consider the fact that all StRE elements in both E1 and E2 are conserved between the mouse and the human genes (Figure 1). Human *HO1* contains a consensus binding element for USF (upstream stimulatory factor) at position –156 to –147, near the transcription start site. USF is a factor originally identified for its ability to activate transcription from the adenovirus-2 major late promoter.[198] An orthologous sequence was also found to be functional in the regulation of the rat *HO1* gene (see below). USF was not shown to induce human HO1, even though it has been suggested that it is involved in oxidant stress-mediated regulation of the mouse metallothionein 1 gene.[199] Upstream of the USF motif there is a motif that becomes occupied following TPA-mediated differentiation of the monocytic cell line THP-1 to macrophage-like cells.[200] This motif was shown to interact with a novel zinc-finger protein termed MTB-Zf.[201] The involvement of this motif and of MTB-Zf in the acute transcriptional induction of *HO1* by TPA or in *HO1* expression changes as a result of cellular differentiation have not been fully determined.

An apparent discrepancy exists in the identification of control elements that regulate mouse and human *HO1* in response to Cd^{2+}. Cd^{2+} responsive regions within the human *HO1* promoter map to positions –4.5 to –4.0 kbp upstream of the transcription start site. Although this domain shares considerable relatedness to the mouse *HO1* E1 enhancer (Figure 1), including two StRE motifs. The StREs are necessary and sufficient to drive Cd^{2+}-inducible expression of mouse *HO1*, but they are dispensable in the human *HO1* response, while a novel 10 bp response element, poorly conserved between mouse and human *HO1* (Figure 1), appears to be essential.[197] This novel element specifically bound to factors present in a HeLa cell nuclear extract, but binding did not changed when cells were treated with Cd^{2+}. Thus far, the identity of the protein(s) that interact with this element is unknown.

18.13 Rat *HO1*

The promoter of rat *HO1* is regulated by USF.[202] In addition, rat *HO1* is regulated via a cAMP response element/TRE present at positions –665 to –654.[203,204] This element drives reporter gene expression in response to cGMP or cAMP [203,204]. Importantly, both of these cyclic nucleotides also induce the expression of endogenous HO1, which is attenuated by inhibitors of protein kinase G or protein kinase A. Consistent with these results, the phosphatase inhibitor okadaic acid also induced expression of the rat *HO1* gene.[204]

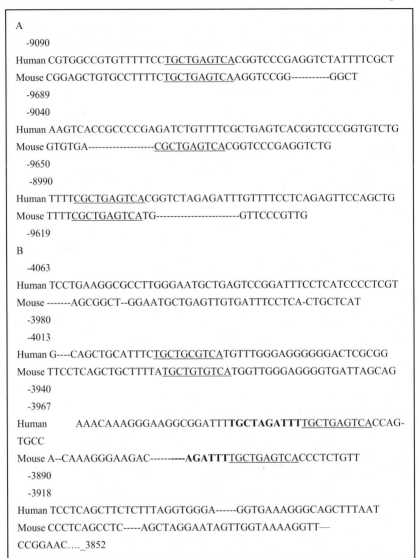

A

-9090

Human CGTGGCCGTGTTTTTCC<u>TGCTGAGTCA</u>CGGTCCCGAGGTCTATTTTCGCT

Mouse CGGAGCTGTGCCTTTTC<u>TGCTGAGTCA</u>AGGTCCGG-----------GGCT

-9689

-9040

Human AAGTCACCGCCCCGAGATCTGTTTTCGCTGAGTCACGGTCCCGGTGTCTG

Mouse GTGTGA------------------<u>CGCTGAGTCA</u>CGGTCCCGAGGTCTG

-9650

-8990

Human TTTT<u>CGCTGAGTCA</u>CGGTCTAGAGATTTGTTTTCCTCAGAGTTCCAGCTG

Mouse TTTT<u>CGCTGAGTCA</u>TG-----------------------GTTCCCGTTG

-9619

B

-4063

Human TCCTGAAGGCGCCTTGGGAATGCTGAGTCCGGATTTCCTCATCCCCTCGT

Mouse -------AGCGGCT--GGAATGCTGAGTTGTGATTTCCTCA-CTGCTCAT

-3980

-4013

Human G----CAGCTGCATTTC<u>TGCTGCGTCA</u>TGTTTGGGAGGGGGGACTCGCGG

Mouse TTCCTCAGCTGCTTTTA<u>TGCTGTGTCA</u>TGGTTGGGAGGGGTGATTAGCAG

-3940

-3967

Human   AAACAAAGGGAAGGCGGATTT**TGCTAGATTT**<u>TGCTGAGTCA</u>CCAG-
TGCC

Mouse A--CAAAGGGAAGAC----------**AGATTT**<u>TGCTGAGTCA</u>CCCTCTGTT

-3890

-3918

Human TCCTCAGCTTCTCTTTAGGTGGGA------GGTGAAAGGGCAGCTTTAAT

Mouse CCCTCAGCCTC-----AGCTAGGAATAGTTGGTAAAAGGTT—

CCGGAAC...._3852

Figure 1. Alignment of E1 and E2 enhancer regions of human and mouse HO1. The human and mouse HO1 genes were used in TRAFAC analysis to identify common regulatory elements. Shown is the alignment of the E2 (A) and E1 (B) enhancer regions previously identified in mouse [169,171]. StRE elements are underlined and the CdRE previously identified in human HO1 [197] is shown in bold.

18.14 Chicken *HO1*

The transcriptional response of the chicken *HO1* gene has been mapped within a 7.1 kbp region of the 5'flanking region. Within this region, the response to arsenite mapped to a TRE located at positions -1580 to -1573 bp and to a MRE/cMyc site located at positions -52 to -41bp, both relative to the site of transcription initiation.[205-207]   The MRE/cMyc site, but not the TRE, was also involved in regulation by Co^{2+}. Interestingly, the MRE/cMyc binding site is similar in sequence to the USF binding site previously described in the rat and human promoters (see above). Further efforts were made to identify the signal transduction pathways involved in activating the chicken *HO1* TRE. Arsenite but not heme, rapidly induced the activity of ERK, JNK and p38, and, using constitutively active and dominant negative mutants of these kinases, the ERK and p38 pathways were implicated in activation through the TRE. Interestingly, p38 was also found to activate through the mouse StRE in response to Cd^{2+} (see above).

The sequence of the chicken TRE is not only a TRE in the sense orientation but also an ARE in the antisense orientation. Although activation through this element was assumed to be due to AP1, the TRE mutations designed to map the site functionally also mutated the ARE core. Thus, ARE function in the context of chicken *HO1* regulation remains to be specifically tested.

A novel heme-regulated DNA response element has recently been mapped within the chicken *HO1* 5'-flanking region.[208] This element is localized to a 200 bp region between -3.8 and -3.6 kbp relative to the transcription start site. DNase-1 footprint analysis identified an 18 bp core sequence that was necessary for heme response. This element allowed transcriptional upregulation in response to heme and to other selected metalloporphyrins, but did not respond to arsenite.[208]

18.15 Summary

Over the past decade, considerable progress has been made toward understanding the transcriptional regulatory mechanisms of *HO1*. Regulation of *HO1* is clearly complex, but comparative genomics data from several species points at several common regulatory motifs. First, a USF site in the proximal promoter domain is conserved from chicken to human. Second, in both chicken and mouse, homodimers and heterodimers of the members of the bZIP superfamily of transcription factors are implicated in *HO1* regulation. It is noteworthy that both chicken and mouse response elements represent composite TRE/EpRE binding sites. In mouse, the two powerful enhancers E1 and E2, which together contain 5 StRE motifs, mediate *HO1* induction by a variety of inducers, including heme. The identification of NRF2 as the transcription factor that regulates *HO1* expression through these elements is bolstered by the demonstration that NRF2 knockout animals cannot regulate *HO1* properly. A unifying mechanism by which *HO1* is regulated as a result of cellular oxidative stress by numerous different stimuli is attractive, and, as noted above, NRF2 is a good candidate for this role. Certainly, the conservation of E1 and E2 enhancer motifs between mouse and human should be a good reason to warrant further evaluation of the function of these elements across species. On the other hand, the possibility that NRF2 signaling may act in synergy with other inducer-specific transcription factors should not be discounted, for in that manner, *HO1* gene expression may become more dynamic and perhaps better tailored to particular cellular insults.   HO1 is regulated by several different transcription factors whose activities are in turn regulated by interdependent signal transduction cascades. Unraveling the mechanisms of HO1 regulation promises to pose many challenges for future investigations for many years to come.

18.16 Concluding Remarks

In eukaryotic organisms, transcription factors exclusively activated by ROS have not been found and probably do not exist. This is probably because, unlike facultative anaerobic organisms, aerobic organisms have evolved with an absolute dependence on oxygen. Thus, facultative anaerobes may have oxygen sensors that redirect gene expression towards aerobic metabolism when oxygen is available, whereas aerobic organisms, restricted to live always in an oxygen-rich environment, have no need for oxygen sensors. Eukaryotic cells, on the other hand, require oxygen to produce ATP, to survive and to reproduce. Increases in mitochondrial respiration, cellular oxidative state and energy generation may be interpreted as a state of cellular well-being that may signal critical biological events, such as cell division or differentiation. This may in part explain the similarity in cellular signals (e.g. increased Ca^{2+}, arachidonic acid, activation of kinase cascades and transcription factors, etc.) elicited by both oxidants and mitogens, and why exposure to oxidant stressors shares so many components with the responses elicited by physiologic signals such as cytokines and growth factors.[18,41]

The same signals that trigger cellular events that increase oxygen consumption—and therefore ROS levels—may also signal a concomitant protective response to activate the expression of antioxidant enzymes. The extent of this response may depend on the duration of the oxidative episode, such that transient oxidant stress would result in modest and largely unnoticed changes in expression of protective genes. Mitogenic signals are generally rapid and transient and cause minimal if any damage to cells. Oxidative stress signals might be expected to be equally innocuous, but they are not; in fact, they are often pathogenic. Many environmental factors, such as redox-cycling xenobiotics or metals, are harmful because they increase the severity and length of the oxidative stress response. For this reason, a good working definition of oxidative stress must include an indication of the severity of the cellular redox change generated and of its duration. ROS contribute to disease or aging in controlled settings in which exposure to harmful environmental agents is unlikely. This leads us to the inescapable conclusion that there are conditions that signify "harmful oxidative stress" that we truly ignore and whose identification would be extremely beneficial. We believe that understanding ROS regulation of transcription factors and antioxidant proteins is a step in the right direction.

Acknowledgments

Preparation of this chapter was supported in part by grants NIEHS ES06273, NIEHS ES10807 and NIEHS ES10133 and the Center for Environmental Genetics NIEHS grant P30 ES06096. We thank Dr. B. Aronow and the Center for Bioinformatics of the Children's Hospital Research Foundation/University of Cinicnnati Medical Center for the use of the TRAFAC algorithm.

References

1. Fridovich, I. 1978. The biology of oxygen radicals. *Science* **201**:875-880.

2. Sohal, R. S. and R. G. Allen. 1990. Oxidative stress as a causal factor in differentiation and aging: a unifying hypothesis. *Exp. Gerontol.* **25**: 499-522.

3. Sahu, S. C. 1990. Oncogenes, oncogenesis, and oxygen radicals. *Biomed. Environ. Sci.* **3**:183-201.

4. Floyd, R. A. 1991. Oxidative damage to behavior during aging. *Science* **254**:1597.

5. Floyd, R. A. 1990. Role of oxygen free radicals in carcinogenesis and brain ischemia. *FASEB J.* **4**:2587-2597.

6. Carney, J. M., P. E. Starke-Reed, C. N. Oliver, R. W. Landum, M. S. Cheng, J. F. Wu and R. A. Floyd. 1991. Reversal of age-related increase in brain protein oxidation, decrease in enzyme activity, and loss in temporal and spatial memmory by chronic administration of the spin-trapping compound *N-tert*-butyl-α-phenylnitrone. *Proc. Natl. Acad. Sci. USA* **88**:3633-3636.

7. Becker, J., V. Mezger, A. M. Courgeon and M. Best-Belpomme. 1991. On the mechanism of action of H_2O_2 in the cellular stress. *Free Rad. Res. Comm.* **12-13 Pt 1**:455-460.

8. Troll, W. 1991. Prevention of cancer by agents that suppress oxygen radical formation. *Free Rad. Res. Comm.* **12**:751-757.

9. Ames, B. N., M. K. Shigenaga and T. M. Hagen. 1993. Oxidants, antioxidants, and the degenerative diseases of aging. *Proc. Natl. Acad. Sci. USA* **90**:7915-7922.

10. Guyton, K. Z. and T. W. Kensler. 1993. Oxidative mechanisms in carcinogenesis. *Br. Med. Bull.* **49**:523-544.

11. Ryrfeldt, A., G. Bannenberg and P. Moldéus. 1993. Free radicals and lung disease. *Brit. Med. Bull.* **49**:588-603.

12. Ziegler, D. M. 1985. Role of reversible oxidation-reduction of enzyme thiols-disulfides in metabolic regulation. *Annu. Rev. Biochem.* **54**:305-329.

13. Keyse, S. M. and R. M. Tyrrell. 1989. Heme oxygenase is the major 32-kDa stress protein induced in human skin fibroblasts by UVA radiation, hydrogen peroxide, and sodium arsenite. *Proc. Natl. Acad. Sci. USA* **86**:99-103.

14. Saran, M. and W. Bors. 1989. Oxygen radicals acting as chemical messengers: A hypothesis. *Free Rad. Res. Comm.* **7**:213-220.

15. Cerutti,P., Larsson,R. and Krupitza,G. Genetic Mechanisms in Carcinogenesis and Tumor Promotion. Haws,C.C. & Liotta,L.A. (eds.), pp. 69-82 (Wiley-Liss, New York,1990).

16. Pahl, H. L. and P. A. Baeuerle. 1994. Oxygen and the control of gene expression. *Bioessays* **16**:497-502.

17. Cowan, D. B., R. D. Weisel, W. G. Williams and D. G. Mickle. 1993. Identification of oxygen responsive elements in the 5'-flanking region of the human glutathione peroxidase gene. *J. Biol. Chem.* **268**:26904-26910.

18. Schulze-Osthoff, K., M. K. Bauer, M. Vogt and S. Wesselborg. 1997. Oxidative stress and signal transduction. *Int. J. Vit. Nut. Res.* **67**:336-342.

19. Suzuki, Y. J., H. J. Forman and A. Sevanian. 1997. Oxidants as stimulators of signal transduction. *Free Rad. Biol. Med.* **22**:269-285.

20. Devary, Y., R. A. Gottlieb, L. F. Lau and M. Karin. 1991. Rapid and preferential activation of the c-*jun* gene during the mammalian UV response. *Mol. Cell. Biol.* **11**:2804-2811.

21. Devary, Y., R. A. Gottlieb, T. Smeal and M. Karin. 1992. The mammalian ultraviolet response is triggered by activation of Src tyrosine kinases. *Cell* **71**:1081-1091.

22. Xanthoudakis, S., G. Miao, F. Wang, Y. C. Pan and T. Curran. 1992. Redox activation of Fos-Jun DNA binding activity is mediated by a DNA repair enzyme. *EMBO J.* **11**:3323-3335.

23. Okuno, H., A. Akahori, H. Sato, S. Xanthoudakis, T. Curran and H. Iba. 1993. Escape from redox regulation enhances the transforming activity of Fos. *Oncogene* **8**:695-701.

24. Toledano, M. B. and W. J. Leonard. 1991. Modulation of transcription factor NF-kB binding activity by oxidation-reduction in vitro. *Proc. Natl. Acad. Sci. USA* **88**:4328-4332.

25. Schreck, R., K. Albermann and P. A. Baeuerle. 1992. Nuclear factor kB: An oxidative stress-responsive transcription factor of eukaryotic cells (a review). *Free Rad. Res. Commun.* **17**:221-37.

26. Müller, J. M., H. W. Ziegler-Heitbrock and P. A. Baeuerle. 1993. Nuclear factor κ B, a mediator of lipopolysaccharide effects. *Immunobiology* **187**:233-256.

27. Buscher, M., H. J. Rahmsdorf, M. Liftin, M. Karin and P. Herrlich. 1988. Activation of the c-fos gene by UV and phorbol ester: different signal transduction pathways converge to the same enhancer element. *Oncogene* **3**:301-311.

28. Devary, Y., C. Rosette, J. A. DiDonato and M. Karin. 1993. NF-κ B activation by ultraviolet light not dependent on a nuclear signal. *Science* **261**:1442-1445.

29. Curran, T. 1992. Fos and Jun: oncogenic transcription factors. *Tohoku J. Exp. Med.* **168**:169-174.

30. Forrest, D. and T. Curran. 1992. Crossed signals: oncogenic transcription factors. *Curr. Opin. Genet. Dev.* **2**:19-27.

31. Angel, P. and M. Karin. 1991. The role of Jun, Fos and the AP-1 complex in cell proliferation and transformation. *Biochim. Biophys. Acta* **1072**:129-157.

32. Angel, P., M. Imagawa, R. Chiu, B. Stein, R. J. Imbra, H. J. Rahmsdorf, C. Jonat, P. Herrlich and M. Karin. 1987. Phorbol ester-inducible genes contain a common *cis* element recogmized by a TPA-modulated *trans*-acting factor. *Cell* **49**:729-739.

33. Lee, W., P. Mitchell and R. Tjian. 1987. Purified transcription factor AP-1 interacts with TPA inducible elements. *Cell* **49**:741-752.

34. Angel, P., E. A. Allegretto, S. T. Okino, K. Hattori, W. J. Boyle, T. Hunter and M. Karin. 1988. Oncogene *jun* encodes a sequence-specific *trans*-activator similar to AP-1. *Nature* **332**:166-171.

35. Hollander, M. C. and A. J. J. Fornace. 1989. Induction of fos RNA by DNA-damaging agents. *Cancer Res.* **49**:1687-1692.

36. Amstad, P. A., G. Krupitza and P. A. Cerutti. 1992. Mechanism of c-fos induction by active oxygen. *Cancer Res.* **52**:3952-3960.

37. Cerutti, P., G. Shah, A. Peskin and P. Amstad. 1992. Oxidant carcinogenesis and antioxidant defense. *Ann. N. Y. Acad. Sci.* **663**:158-166.

38. Iles, K. E., D. A. Dickinson, N. Watanabe, T. Iwamoto and H. J. Forman. 2002. AP-1 activation through endogenous H(2)O(2) generation by alveolar macrophages. *Free Radic. Biol. Med.* **32**:1304-1313.

39. Stein, B., H. J. Rahmsdorf, A. Steffen, M. Litfin and P. Herrlich. 1989. UV-induced DNA damage is an intermediate step in UV-induced expression of human immunodeficiency virus type 1, collagenase, c- fos, and metallothionein. *Mol. Cell. Biol.* **9**:5169-5181.

40. Djavaheri-Mergny, M., J. L. Mergny, F. Bertrand, R. Santus, C. Mazière, L. Dubertret and J. C. Mazière. 1996. Ultraviolet-A incudes activation of AP-1 in cultured human keratimocytes. *FEBS Lett.* **384**: 92-96.

41. Janssen, Y. M. W., B. Van Houten, P. J. A. Borm and B. T. Mossman. 1993. Cell and tissue responses to oxidative damage. *Laboratory Invest.* **69**:261-274.

42. Heintz, N. H., Y. M. W. Janssen and B. T. Mossman. 1993. Persistent induction of c-*fos* and c-*jun* protooncogene expression by asbestos. *Proc. Natl. Acad. Sci. USA* **90**:3299-3303.

43. Puga, A., D. W. Nebert and F. Carrier. 1992. Dioxin induces expression of c-*fos* and c-*jun* proto-oncogenes and a large increase in transcription factor AP-1. *DNA Cell Biol.* **11**:269-281.

44. Hoffer, A., C.-Y. Chang and A. Puga. 1996. Dioxin induces *fos* and *jun* gene expression by Ah receptor dependent- and independent- pathways. *Toxicol. Appl. Pharmacol.* **141**:238-247.

45. Karin, M. 1995. The regulation of AP-1 activity by mitogen-activated protein kinases. *J. Biol. Chem.* **270**:16483-16486.

46. Hunter, T. and M. Karin. 1992. The regulation of transcription by phosphorylation. *Cell* **70**:375-387.

47. Yang-Yen, H.-F., R. Chiu and M. Karin. 1990. Elevation of AP-1 activity during F9 cell differentiation is due to increased c-*jun* transcription. *New Biol.* **2**:351-361.

48. Abate, C., D. Luk and T. Curran. 1991. Transcriptional regulation by Fos and Jun in vitro: interaction among multiple activator and regulatory domains. *Mol. Cell. Biol.* **11**:3624-3632.

49. Edwards, D. R. and L. C. Mahadevan. 1992. Protein synthesis inhibitors differentially superinduce c-*fos* and c-*jun* by three distinct mechanisms: lack of evidence for labile repressors. *EMBO J.* **11**:2415-2424.

50. Abate, C., D. Luk, E. Gagne, R. G. Roeder and T. Curran. 1990. Fos and jun cooperate in transcriptional regulation via heterologous activation domains. *Mol. Cell. Biol.* **10**:5532-5535.

51. Abate, C., L. Patel, F. J. 3. Rauscher and T. Curran. 1990. Redox regulation of fos and jun DNA-binding activity in vitro. *Science* **249**:1157-1161.

52. Schenk, H., M. Klein, W. Erdbrügger, W. Dröge and K. Schulze-Osthoff. 1994. Distinct effect of thioredoxin and other antioxidants on the activation of NF-κB and AP-1. *Proc. Natl. Acad. Sci. USA* **91**:1672-1676.

53. Xanthoudakis, S. and T. Curran. 1992. Identification and characterization of Ref-1, a nuclear protein that facilitates AP-1 DNA-binding activity. *EMBO J.* **11**:653-665.

54. Ng, L., D. Forrest and T. Curran. 1993. Differential roles for Fos and Jun in DNA-binding: redox- dependent and independent functions. *Nucleic Acids Res.* **21**:5831-5837.

55. Xanthoudakis, S., G. G. Miao and T. Curran. 1994. The redox and DNA-repair activities of Ref-1 are encoded by nonoverlapping domains. *Proc. Natl. Acad. Sci. USA* **91**:23-27.

56. Nakamura, H., K. Nakamura and J. Yodoi. 1997. Redox regulation of cellular activation. *Annu. Rev. Immunol.* **15**:351-369.

57. Walker, L. J., R. B. Craig, A. L. Harris and I. D. Hickson. 1994. A role for the human DNA repair enzyme HAP1 in cellular protection against DNA damaging agents and hypoxic stress. *Nucleic Acids Res.* **22**:4884-4889.

58. Sachsenmaier, C., A. Radler-Pohl, A. Muller, P. Herrlich and H. J. Rahmsdorf. 1994. Damage to DNA by UV light and activation of transcription factors. *Biochem. Pharmacol.* **47**:129-136.

59. Datta, R., D. E. Hallahan, S. M. Kharbanda, E. Rubin, M. L. Sherman, E. Huberman, R. R. Weichselbaum and D. W. Kufe. 1992. Involvement of reactive oxygen intermediates in the induction of c-jun gene transcription by ionizing radiation. *Biochemistry* **31**:8300-8306.

60. Stevenson, M. A., S. S. Pollock, C. N. Coleman and S. K. Calderwood. 1994. X-irradiation, phorbol esters, and H_2O_2 stimulate mitogen-activated protein kinase activity in NIH-3T3 cells through the formation of reactive oxygen intermediates. *Cancer Res.* **54**:12-15.

61. Schreiber, M., B. Baumann, M. Cotten, P. Angel and E. F. Wagner. 1995. Fos is an essential component of the mammalian UV response. *EMBO J.* **14**:5338-5349.

62. Radler-Pohl, A., C. Sachsenmaier, S. Gebel, H. P. Auer, J. T. Bruder, U. Rapp, P. Angel, H. J. Rahmsdorf and P. Herrlich. 1993. UV-induced activation of AP-1 involves obligatory extranuclear steps including Raf-1 kinase. *EMBO J.* **12**:1005-1012.

63. Sachsenmaier, C., A. Radler-Pohl, R. Zinck, A. Nordheim, P. Herrlich and H. J. Rahmsdorf. 1994. Involvement of growth factor receptors in the mammalian UVC response. *Cell* **78**:963-972.

64. Fialkow, L., C. K. Chan, D. Rotin, S. instein and G. P. wney. 1994. Activation of the mitogen-activated protein kinase signaling pathway in neutrophils. Role of oxidants. *J. Biol. Chem.* **269**:31234-31242.

65. Whisler, R. L., M. A. Goyette, I. S. Grants and Y. G. Newhouse. 1995. Sublethal levels of oxidant stress stimulate multiple serine/threonine kinases and suppress protein phosphatases in Jurkat T cells. *Arch. Biochem. Biophys.* **319**:23-35.

66. Esposito, F., F. Cuccovillo, M. Vanoni, F. Cimino, C. W. Anderson, E. Appella and T. Russo. 1997. Redox-mediated regulation of p21(waf1/cip1) expression involves a post-transcriptional mechanism and activation of the mitogen-activated protein kinase pathway. *Eur. J. Biochem.* **245**:730-737.

67. Choi, J., R. M. Liu and H. J. Forman. 1997. Adaptation to oxidative stress: quinone-mediated protection of signaling in rat lung epithelial L2 cells. *Biochem. Pharmacol.* **53**:987-993.

68. Choi, H.-S. and D. D. Moore. 1993. Induction of *c-fos* and *c-jun* gene expression by phenolic antioxidants. *Mol. Endocrinol.* **7**:1596-1602.

69. Yoshioka, K., T. Deng, M. Cavigelli and M. Karin. 1995. Antitumor promotion by phenolic antioxidants: inhibition of AP-1 activity through induction of Fra expression. *Proc. Natl. Acad. Sci. USA* **92**:4972-4976.

70. Yu, R., T. H. Tan and A. T. Kong. 1997. Butylated hydroxyanisole and its metabolite tert-butylhydroquinone differentially regulate mitogen-activated protein kinases. The role of oxidative stress in the activation of mitogen-activated protein kinases by phenolic antioxidants. *J. Biol. Chem.* **272**:28962-28970.

71. Baeuerle, P. A. and T. Henkel. 1994. Function and activation of NF-κB in the immune system. *Annu. Rev. Immunol.* **12**:141-179.

72. Nabel, G. J. and I. M. Verma. 1993. Proposed NF-κB/IκB family nomenclature. *Genes Develop.* **7**:2063.

73. Piette, J., B. Piret, G. Bonizzi, S. Schoonbroodt, M.-P. Merville, S. Legrand-Poels and V. Bours. 1997. Multiple redox regulation in NF-κB transcription factor activation. *Biol. Chem.* **378**:1237-1245.

74. Rice, N. R. and M. K. Ernst. 1993. *In vivo* control of NF-κB activation by IκBα. *EMBO J.* **12**:4685-4695.

75. Beg, A. A. and A. S. Baldwin, Jr. 1993. The IκB proteins: multifunctional regulators of Rel/NF-κB transcription factors. *Genes Develop.* **7**:2064-2070.

76. Mohan, N. and M. M. Meltz. 1994. Induction of nuclear factor κB after low-dose ionizing radiation involves a reactive oxygen intermediate signaling pathway. *Rad. Res.* **140**:97-104.

77. Schulze-Osthoff, K., M. Los and P. A. Baeuerle. 1995. Redox signalling by transcription factors NF-κB an AP-1 in lymphocytes. *Biochem. Pharmacol.* **50**:735-741.

78. Schulze-Osthoff, K., R. Beyaert, V. Vandervoorde, G. Haegeman and W. Fiers. 1993. Depletion of the mitochondrial electron transport abrogrates the cytotoxic and gene-inductive effects of TNF. *EMBO J.* **12**:3095-3104.

79. Staal, F. J. T., M. Roederer and L. A. Herzenberg. 1990. Intracellular thiols regulate activation of nuclear factor kB and transcription of human immunodeficiency virus. *Proc. Natl. Acad. Sci. USA* **87**:9943-9947.

80. Traenckner, E. B. M., H. L. Pahl, K. N. Schmidt, S. Wilk and P. A. Baeuerle. 1995. Phosphorylation of human IκB on serines 32 and 36 controls IκB-α proteolysis and NF-κB activation in response to diverse stimuli. *EMBO J.* **14**:2876-2883.

81. Chen, Z. J., L. Parent and T. Maniatis. 1996. Site-specific phosphorylation of IκBα by a novel ubiquitination-dependent protein kinase activity. *Cell* **84**:853-862.

82. Roussel, R. R. and A. Barchowsky. 2000. Arsenic inhibits NF-κB-mediated gene transcription by blocking IκB kinase activity and IκBalpha phosphorylation and degradation. *Arch. Biochem. Biophys.* **377**:204-212.

83. Kapahi, P., T. Takahashi, G. Natoli, S. R. Adams, Y. Chen, R. Y. Tsien and M. Karin. 2000. Inhibition of NF-κ B activation by arsenite through reaction with a critical cysteine in the activation loop of Iκ B kinase. *J. Biol. Chem.* **275**:36062-36066.

84. Schreck, R., P. Rieber and P. A. Baeuerle. 1991. Reactive oxygen intermediates as apparently widely used messengers in the activation of the NF-κ B transcription factor and HIV-1. *EMBO J.* **10**:2247-2258.

85. Schmidt, K. N., P. Armstad, P. Cerutti and P. A. Baeuerle. 1995. The roles of hydrogen peroxide and superoxide as messengers in the activation of transcription factor NF-κB. *Biol. Chem.* **2**:13-22.

86. Droge, W., K. Schulze-Osthoff, S. Mihm, D. Galter, H. Schenk, H. P. Eck, S. Roth and H. Gmunder. 1994. Functions of glutathione and glutathione disulfide in immunology and immunopathology. *FASEB J.* **8**:1131-1138.

87. Galter, D., S. Mihm and W. Droge. 1994. Distinct effects of glutathione disulphide on the nuclear transcription factor κ B and the activator protein-1. *Eur. J. Biochem.* **221**:639-648.

88. Mahon, T. M. and L. A. O'Neill. 1995. Studies into the effect of the tyrosine kinase inhibitor herbimycin A on NF-κ B activation in T lymphocytes. Evidence for covalent modification of the p50 subunit. *J. Biol. Chem.* **270**:28557-28564.

89. Matthews, J. R., W. Kaszubska, G. Turcatti, T. N. Wells, R. T. Hay, N. Wakasugi, J. L. Virelizier and J. Yodoi. 1993. Role of cysteine 62 in DNA recognition by the P50 subunit of NF-κ B. *Nucleic Acids Res.* **21**:1727-1734.

90. Matthews, J. R., N. Wakasugi, J. L. Virelizier, J. Yodoi and R. T. Hay. 1992. Thioredoxin regulates the DNA binding activity of NF-κ B by reduction of a disulphide bond involving cysteine 62. *Nucleic Acids Res.* **20**:3821-3830.

91. Suzuki, Y. J. and L. Packer. 1993. Inhibition of NF-κ B activation by vitamin E derivatives. *Biochem. Biophys. Res. Commun.* **193**:277-283.

92. Israel, N., M. A. Gougerot-Pocidalo, F. Aillet and J. L. Virelizier. 1992. Redox status of cells influences constitutive or induced NF-κ B translocation and HIV long terminal repeat activity in human T and monocytic cell lines. *J. Immunol.* **149**:3386-3393.

93. Ginn-Pease, M. E. and R. L. Whisler. 1996. Optimal NFκB mediated transcriptional responses in Jurkat T cells exposed to oxidative stress are dependent on intracellular glutathione and costimulatory signals. *Biochem. Biophys. Res. Comm.* **226**:695-702.

94. Anderson, M. T., F. J. T. Staal, C. Gitler and L. A. Herzenberg. 1994. Separation of oxidant-initiated and redox-regulated steps in the NF-κB signal transduction pathway. *Proc. Natl. Acad. Sci. USA* **91**:11527-11531.

95. Brennan, P. and L. A. O'Neill. 1995. Effects of oxidants and antioxidants on nuclear factor κ B activation in three different cell lines: evidence against a universal hypothesis involving oxygen radicals. *Biochim. Biophys. Acta* **1260**:167-175.

96. Suzuki, Y. J., M. Mizuno and L. Packer. 1995. Transient overexpression of catalase does not inhibit TNF- or PMA-induced NF-κB activation. *Biochem. Biophys. Res. Comm.* **210**:537-541.

97. Morse, D. and A. M. Choi. 2002. Heme Oxygenase-1. The "emerging molecule" has arrived. *Am. J. Respir. Cell Mol. Biol.* **27**:8-16.

98. Dulak, J., A. Jozkowicz, R. Foresti, A. Kasza, M. Frick, I. Huk, C. J. Green, O. Pachinger, F. Weidinger and R. Motterlini. 2002. Heme oxygenase activity modulates vascular endothelial growth factor synthesis in vascular smooth muscle cells. *Antioxid. Redox. Signal.* **4**:229-240.

99. Soares, M. P., A. Usheva, S. Brouard, P. O. Berberat, L. Gunther, E. Tobiasch and F. H. Bach. 2002. Modulation of endothelial cell apoptosis by heme oxygenase-1-derived carbon monoxide. *Antioxid. Redox. Signal.* **4**:321-329.

100. Dore, S. 2002. Decreased activity of the antioxidant heme oxygenase enzyme: implications in ischemia and in Alzheimer's disease(1,2). *Free Radic. Biol. Med.* **32**:1276-1282.

101. Koehler, R. C. and R. J. Traystman. 2002. Cerebrovascular effects of carbon monoxide. *Antioxid. Redox. Signal.* **4**:279-290.

102. Hill-Kapturczak, N., S. H. Chang and A. Agarwal. 2002. Heme oxygenase and the kidney. *DNA Cell Biol.* **21**:307-321.

103. Tosaki, A. and D. K. Das. 2002. The role of heme oxygenase signaling in various disorders. *Mol. Cell Biochem.* **232**:149-157.

104. Kourembanas, S. 2002. Hypoxia and carbon monoxide in the vasculature. *Antioxid. Redox. Signal.* **4**:291-299.

105. Maines, M. D. 2002. Heme oxygenase 1 transgenic mice as a model to study neuroprotection. *Methods Enzymol.* **353**:374-388.

106. Yet, S. F., L. G. Melo, M. D. Layne and M. A. Perrella. 2002. Heme oxygenase 1 in regulation of inflammation and oxidative damage. *Methods Enzymol.* **353**:163-176.

107. Ryter, S. W. and R. M. Tyrrell. 2000. The heme synthesis and degradation pathways: role in oxidant sensitivity. Heme oxygenase has both pro- and antioxidant properties. *Free Radic. Biol. Med.* **28**:289-309.

108. Maines, M. D. and N. Panahian. 2001. The heme oxygenase system and cellular defense mechanisms. Do HO-1 and HO-2 have different functions? *Adv. Exp. Med. Biol.* **502**:249-272.

109. McCoubrey, W. K., Jr., T. J. Huang and M. D. Maines. 1997. Heme oxygenase-2 is a hemoprotein and binds heme through heme regulatory motifs that are not involved in heme catalysis. *J. Biol. Chem.* **272**:12568-12574.

110. Dennery, P. A., D. R. Spitz, G. Yang, A. Tatarov, C. S. Lee and M. L. Shehog. 1998. Oxygen toxicity and iron accumulation in the lungs of mice lacking heme oxygenase-2. *J. Clin. Invest.* **101**:1001-1011.

111. Cairo, G., S. Recalcati, A. Pietrangelo and G. Minotti. 2002. The iron regulatory proteins: targets and modulators of free radical reactions and oxidative damage(1,2). *Free Radic. Biol. Med.* **32**:1237-1243.

112. Roy, C. N., K. P. Blemings, K. M. Deck, P. S. Davies, E. L. Anderson, R. S. Eisenstein and C. A. Enns. 2002. Increased IRP1 and IRP2 RNA binding activity accompanies a reduction of the labile iron pool in HFE-expressing cells. *J. Cell Physiol* **190**:218-226.

113. Grabellus, F., C. Schmid, B. Levkau, D. Breukelmann, P. F. Halloran, C. August, N. Takeda, A. Takeda, M. Wilhelm, M. C. Deng and H. A. Baba. 2002. Reduction of hypoxia-inducible heme oxygenase-1 in the myocardium after left ventricular mechanical support. *J. Pathol.* **197**:230-237.

114. Minamino, T., H. Christou, C. M. Hsieh, Y. Liu, V. Dhawan, N. G. Abraham, M. A. Perrella, S. A. Mitsialis and S. Kourembanas. 2001. Targeted expression of heme oxygenase-1 prevents the pulmonary inflammatory and vascular responses to hypoxia. *Proc. Natl. Acad. Sci. U. S. A* **98**:8798-8803.

115. Akagi, R., T. Takahashi and S. Sassa. 2002. Fundamental role of heme oxygenase in the protection against ischemic acute renal failure. *Jpn. J. Pharmacol.* **88**:127-132.

116. Liu, Y., T. Tachibana, Y. Dai, E. Kondo, T. Fukuoka, H. Yamanaka and K. Noguchi. 2002. Heme oxygenase-1 expression after spinal cord injury: the induction in activated neutrophils. *J. Neurotrauma* **19**:479-490.

117. Nakae, H. and H. Inaba. 2002. Expression of heme oxygenase-1 in the lung and liver tissues in a rat model of burns. *Burns* **28**:305-309.

118. Kreiser, D., X. Nguyen, R. Wong, D. Seidman, D. Stevenson, S. Quan, N. Abraham and P. A. Dennery. 2002. Heme oxygenase-1 modulates fetal growth in the rat. *Lab Invest* **82**:687-692.

119. Zenclussen, A. C., R. Joachim, E. Hagen, C. Peiser, B. F. Klapp and P. C. Arck. 2002. Heme oxygenase is downregulated in stress-triggered and interleukin-12- mediated murine abortion. *Scand. J. Immunol.* **55**:560-569.

120. Lee, T. S. and L. Y. Chau. 2002. Heme oxygenase-1 mediates the anti-inflammatory effect of interleukin- 10 in mice. *Nat. Med.* **8**:240-246.

121. Kacimi, R., J. Chentoufi, N. Honbo, C. S. Long and J. S. Karliner. 2000. Hypoxia differentially regulates stress proteins in cultured cardiomyocytes: role of the p38 stress-activated kinase signaling cascade, and relation to cytoprotection. *Cardiovasc. Res.* **46**:139-150.

122. Chiu, H., J. A. Brittingham and D. L. Laskin. 2002. Differential induction of heme oxygenase-1 in macrophages and hepatocytes during acetaminophen-induced hepatotoxicity in the rat: effects of hemin and biliverdin. *Toxicol. Appl. Pharmacol.* **181**:106-115.

123. Llesuy, S. F. and M. L. Tomaro. 1994. Heme oxygenase and oxidative stress. Evidence of involvement of bilirubin as physiological protector against oxidative damage. *Biochim. Biophys. Acta* **1223**:9-14.

124. Yamaguchi, T., M. Terakado, F. Horio, K. Aoki, M. Tanaka and H. Nakajima. 1996. Role of bilirubin as an antioxidant in an ischemia-reperfusion of rat liver and induction of heme oxygenase. *Biochem. Biophys. Res. Commun.* **223**:129-135.

125. Clark, J. E., R. Foresti, C. J. Green and R. Motterlini. 2000. Dynamics of haem oxygenase-1 expression and bilirubin production in cellular protection against oxidative stress. *Biochem. J.* **348 Pt 3**:615-619.

126. Clark, J. E., R. Foresti, P. Sarathchandra, H. Kaur, C. J. Green and R. Motterlini. 2000. Heme oxygenase-1-derived bilirubin ameliorates postischemic myocardial dysfunction. *Am. J. Physiol Heart Circ. Physiol* **278**:H643-H651.

127. Mayer, M. 2000. Association of serum bilirubin concentration with risk of coronary artery disease. *Clin. Chem.* **46**:1723-1727.

128. Dore, S., M. Takahashi, C. D. Ferris, R. Zakhary, L. D. Hester, D. Guastella and S. H. Snyder. 1999. Bilirubin, formed by activation of heme oxygenase-2, protects neurons against oxidative stress injury. *Proc. Natl. Acad. Sci. U. S. A* **96**:2445-2450.

129. Gourley, G. R. 1997. Bilirubin metabolism and kernicterus. *Adv. Pediatr.* **44**:173-229.

130. Amersi, F., X. D. Shen, D. Anselmo, J. Melinek, S. Iyer, D. J. Southard, M. Katori, H. D. Volk, R. W. Busuttil, R. Buelow and J. W. Kupiec-Weglinski. 2002. *Ex vivo* exposure to carbon monoxide prevents hepatic ischemia/reperfusion injury through p38 MAP kinase pathway. *Hepatology* **35**:815-823.

131. Gunther, L., P. O. Berberat, M. Haga, S. Brouard, R. N. Smith, M. P. Soares, F. H. Bach and E. Tobiasch. 2002. Carbon monoxide protects pancreatic beta-cells from apoptosis and improves islet function/survival after transplantation. *Diabetes* **51**:994-999.

132. Soares, M. P., S. Brouard, R. N. Smith and F. H. Bach. 2001. Heme oxygenase-1, a protective gene that prevents the rejection of transplanted organs. *Immunol. Rev.* **184**:275-285.

133. Redaelli, C. A., Y. H. Tian, T. Schaffner, M. Ledermann, H. U. Baer and J. F. Dufour. 2002. Extended preservation of rat liver graft by induction of heme oxygenase- 1. *Hepatology* **35**:1082-1092.

134. Avihingsanon, Y., N. Ma, E. Csizmadia, C. Wang, M. Pavlakis, T. Giraldo, T. B. Strom, M. P. Soares and C. Ferran. 2002. Expression of protective genes in human renal allografts: a regulatory response to injury associated with graft rejection. *Transplantation* **73**:1079-1085.

135. Takeda, R., A. Tanaka, T. Maeda, Y. Yamaoka, K. Nakamura, K. Sano, M. Kataoka, Y. Nakamura, T. Morimoto and S. Mukaihara. 2002. Perioperative changes in carbonylhemoglobin and methemoglobin during abdominal surgery: Alteration in endogenous generation of carbon monoxide. *J. Gastroenterol. Hepatol.* **17**:535-541.

136. Ono, S., T. Komuro and R. L. Macdonald. 2002. Heme oxygenase-1 gene therapy for prevention of vasospasm in rats. *J. Neurosurg.* **96**:1094-1102.

137. Ou, H. S., J. Yang, L. W. Dong, Y. Z. Pang, J. Y. Su, C. S. Tang and N. K. Liu. 1999. Role of endogenous carbon monoxide in the pathogenesis of hypotension during septic shock. *Sheng Li Xue. Bao.* **51**:1-6.

138. Pourcyrous, M., H. S. Bada, H. Parfenova, M. L. Daley, S. B. Korones and C. W. Leffler. 2002. Cerebrovasodilatory contribution of endogenous carbon monoxide during seizures in newborn pigs. *Pediatr. Res.* **51**:579-585.

139. Peyton, K. J., S. V. Reyna, G. B. Chapman, D. Ensenat, X. M. Liu, H. Wang, A. I. Schafer and W. Durante. 2002. Heme oxygenase-1-derived carbon monoxide is an autocrine inhibitor of vascular smooth muscle cell growth. *Blood* **99**:4443-4448.

140. Keyse, S. M. and R. M. Tyrrell. 1990. Induction of the heme oxygenase gene in human skin fibroblasts by hydrogen peroxide and UVA (365 nm) radiation: evidence for the involvement of the hydroxyl radical. *Carcinogenesis* **11**:787-791.

141. Lautier, D., P. Luscher and R. M. Tyrrell. 1992. Endogenous glutathione levels modulate both constitutive and UVA radiation/hydrogen peroxide inducible expression of the human heme oxygenase gene. *Carcinogenesis* **13**:227-232.

142. Borger, D. R. and D. A. Essig. 1998. Induction of HSP 32 gene in hypoxic cardiomyocytes is attenuated by treatment with N-acetyl-L-cysteine. *Am. J. Physiol* **274**:H965-H973.

143. Suzuki, K. and T. K. Hei. 1996. Induction of heme oxygenase in mammalian cells by mineral fibers: distinctive effect of reactive oxygen species. *Carcinogenesis* **17**:661-667.

144. Basu-Modak, S., P. Luscher and R. M. Tyrrell. 1996. Lipid metabolite involvement in the activation of the human heme oxygenase-1 gene. *Free Radic. Biol. Med.* **20**:887-897.

145. Durante, W., K. J. Peyton and A. I. Schafer. 1999. Platelet-derived growth factor stimulates heme oxygenase-1 gene expression and carbon monoxide production in vascular smooth muscle cells. *Arterioscler. Thromb. Vasc. Biol.* **19**:2666-2672.

146. Peyton, K. J., S. V. Reyna, G. B. Chapman, D. Ensenat, X. M. Liu, H. Wang, A. I. Schafer and W. Durante. 2002. Heme oxygenase-1-derived carbon monoxide is an autocrine inhibitor of vascular smooth muscle cell growth. *Blood* **99**:4443-4448.

147. Liu, X., G. B. Chapman, K. J. Peyton, A. I. Schafer and W. Durante. 2002. Carbon monoxide inhibits apoptosis in vascular smooth muscle cells. *Cardiovasc. Res.* **55**:396-405.

148. Camhi, S. L., J. Alam, G. W. Wiegand, B. Y. Chin and A. M. Choi. 1998. Transcriptional activation of the HO-1 gene by lipopolysaccharide is mediated by 5' distal enhancers: role of reactive oxygen intermediates and AP-1. *Am. J. Respir. Cell Mol. Biol.* **18**:226-234.

149. Oguro, T., Y. Takahashi, T. Ashino, A. Takaki, S. Shioda, R. Horai, M. Asano, K. Sekikawa, Y. Iwakura and T. Yoshida. 2002. Involvement of tumor necrosis factor alpha, rather than interleukin-1alpha/beta or nitric oxides in the heme oxygenase-1 gene expression by lipopolysaccharide in the mouse liver. *FEBS Lett.* **516**:63-66.

150. Alcaraz, M. J., A. Habib, M. Lebret, C. Creminon, S. Levy-Toledano and J. Maclouf. 2000. Enhanced expression of haem oxygenase-1 by nitric oxide and antiinflammatory drugs in NIH 3T3 fibroblasts. *Br. J. Pharmacol.* **130**:57-64.

151. Chen, K. and M. D. Maines. 2000. Nitric oxide induces heme oxygenase-1 via mitogen-activated protein kinases ERK and p38. *Cell Mol. Biol. (Noisy. -le-grand)* **46**:609-617.

152. Poss, K. D. and S. Tonegawa. 1997. Reduced stress defense in heme oxygenase 1-deficient cells. *Proc. Natl. Acad. Sci. U. S. A* **94**:10925-10930.

153. Poss, K. D. and S. Tonegawa. 1997. Heme oxygenase 1 is required for mammalian iron reutilization. *Proc. Natl. Acad. Sci. U. S. A* **94**:10919-10924.

154. Nath, K. A., J. J. Haggard, A. J. Croatt, J. P. Grande, K. D. Poss and J. Alam. 2000. The indispensability of heme oxygenase-1 in protecting against acute heme protein-induced toxicity in vivo. *Am. J. Pathol.* **156**:1527-1535.

155. Li, L., R. F. Hamilton, Jr. and A. Holian. 2000. Protection against ozone-induced pulmonary inflammation and cell death by endotoxin pretreatment in mice: role of HO-1. *Inhal. Toxicol.* **12**:1225-1238.

156. Le, W. D., W. J. Xie and S. H. Appel. 1999. Protective role of heme oxygenase-1 in oxidative stress-induced neuronal injury. *J. Neurosci. Res.* **56**:652-658.

157. Chen, K., K. Gunter and M. D. Maines. 2000. Neurons overexpressing heme oxygenase-1 resist oxidative stress-mediated cell death. *J. Neurochem.* **75**:304-313.

158. Hangaishi, M., N. Ishizaka, T. Aizawa, Y. Kurihara, J. Taguchi, R. Nagai, S. Kimura and M. Ohno. 2000. Induction of heme oxygenase-1 can act protectively against cardiac ischemia/reperfusion in vivo. *Biochem. Biophys. Res. Commun.* **279**:582-588.

159. Foresti, R., H. Goatly, C. J. Green and R. Motterlini. 2001. Role of heme oxygenase-1 in hypoxia-reoxygenation: requirement of substrate heme to promote cardioprotection. *Am. J. Physiol Heart Circ. Physiol* **281**:H1976-H1984.

160. Speit, G., C. Dennog, U. Eichhorn, A. Rothfuss and B. Kaina. 2000. Induction of heme oxygenase-1 and adaptive protection against the induction of DNA damage after hyperbaric oxygen treatment. *Carcinogenesis* **21**:1795-1799.

161. Petrache, I., L. E. Otterbein, J. Alam, G. W. Wiegand and A. M. Choi. 2000. Heme oxygenase-1 inhibits TNF-alpha-induced apoptosis in cultured fibroblasts. *Am. J. Physiol Lung Cell Mol. Physiol* **278**:L312-L319.

162. Yamada, N., M. Yamaya, S. Okinaga, K. Nakayama, K. Sekizawa, S. Shibahara and H. Sasaki. 2000. Microsatellite polymorphism in the heme oxygenase-1 gene promoter is associated with susceptibility to emphysema. *Am. J. Hum. Genet.* **66**:187-195.

163. Exner, M., M. Schillinger, E. Minar, W. Mlekusch, G. Schlerka, M. Haumer, C. Mannhalter and O. Wagner. 2001. Heme oxygenase-1 gene promoter microsatellite polymorphism is associated with restenosis after percutaneous transluminal angioplasty. *J. Endovasc. Ther.* **8**:433-440.

164. Chen, Y. H., S. J. Lin, M. W. Lin, H. L. Tsai, S. S. Kuo, J. W. Chen, M. J. Charng, T. C. Wu, L. C. Chen, Y. A. Ding, W. H. Pan, Y. S. Jou and L. Y. Chau. 2002. Microsatellite polymorphism in promoter of heme oxygenase-1 gene is associated with susceptibility to coronary artery disease in type 2 diabetic patients. *Hum. Genet.* **111**:1-8.

165. Suttner, D. M., K. Sridhar, C. S. Lee, T. Tomura, T. N. Hansen and P. A. Dennery. 1999. Protective effects of transient HO-1 overexpression on susceptibility to oxygen toxicity in lung cells. *Am. J. Physiol* **276**:L443-L451.

166. Kvam, E., V. Hejmadi, S. Ryter, C. Pourzand and R. M. Tyrrell. 2000. Heme oxygenase activity causes transient hypersensitivity to oxidative ultraviolet A radiation that depends on release of iron from heme. *Free Radic. Biol. Med.* **28**:1191-1196.

167. Suttner, D. M. and P. A. Dennery. 1999. Reversal of HO-1 related cytoprotection with increased expression is due to reactive iron. *FASEB J.* **13**:1800-1809.

168. Choi, A. M. and J. Alam. 1996. Heme oxygenase-1: function, regulation, and implication of a novel stress-inducible protein in oxidant-induced lung injury. *Am. J. Respir. Cell Mol. Biol.* **15**:9-19.

169. Alam, J., S. Camhi and A. M. Choi. 1995. Identification of a second region upstream of the mouse heme oxygenase-1 gene that functions as a basal level and inducer-dependent transcription enhancer. *J. Biol. Chem.* **270**:11977-11984.

170. Camhi, S. L., J. Alam, L. Otterbein, S. L. Sylvester and A. M. Choi. 1995. Induction of heme oxygenase-1 gene expression by lipopolysaccharide is mediated by AP-1 activation. *Am. J. Respir. Cell Mol. Biol.* **13**:387-398.

171. Alam, J. and Z. Den. 1992. Distal AP-1 binding sites mediate basal level enhancement and TPA induction of the mouse heme oxygenase-1 gene. *J. Biol. Chem.* **267**:21894-21900.

172. Camhi, S. L., J. Alam, G. W. Wiegand, B. Y. Chin and A. M. Choi. 1998. Transcriptional activation of the HO-1 gene by lipopolysaccharide is mediated by 5' distal enhancers: role of reactive oxygen intermediates and AP-1. *Am. J. Respir. Cell Mol. Biol.* **18**:226-234.

173. Alam, J., J. Cai and A. Smith. 1994. Isolation and characterization of the mouse heme oxygenase-1 gene. Distal 5' sequences are required for induction by heme or heavy metals. *J. Biol. Chem.* **269**:1001-1009.

174. Gong, P., B. Hu, D. Stewart, M. Ellerbe, Y. G. Figueroa, V. Blank, B. S. Beckman and J. Alam. 2001. Cobalt induces heme oxygenase-1 expression by a hypoxia-inducible factor-independent mechanism in Chinese hamster ovary cells: regulation by Nrf2 and MafG transcription factors. *J. Biol. Chem.* **276**:27018-27025.

175. Lee, P. J., S. L. Camhi, B. Y. Chin, J. Alam and A. M. Choi. 2000. AP-1 and STAT mediate hyperoxia-induced gene transcription of heme oxygenase-1. *Am. J. Physiol Lung Cell Mol. Physiol* **279**:L175-L182.

176. Inamdar, N. M., Y. I. Ahn and J. Alam. 1996. The heme-responsive element of the mouse heme oxygenase-1 gene is an extended AP-1 binding site that resembles the recognition sequences for MAF and NF-E2 transcription factors. *Biochem. Biophys. Res. Commun.* **221**:570-576.

177. Rushmore, T. H., R. G. King, K. E. Paulson and C. B. Pickett. 1990. Regulation of glutathione S-transferase Ya subunit gene expression: identification of a unique xenobiotic-responsive element controlling inducible expression by planar aromatic compounds. *Proc. Natl. Acad. Sci. U. S. A* **87**:3826-3830.

178. Friling, R. S., A. Bensimon, Y. Tichauer and V. Daniel. 1990. Xenobiotic-inducible expression of murine glutathione S-transferase Ya subunit gene is controlled by an electrophile-responsive element. *Proc. Natl. Acad. Sci. U. S. A* **87**:6258-6262.

179. Andrews, N. C. 1998. The NF-E2 transcription factor. *Int. J. Biochem. Cell Biol.* **30**:429-432.

180. Motohashi, H., J. A. Shavit, K. Igarashi, M. Yamamoto and J. D. Engel. 1997. The world according to Maf. *Nucleic Acids Res.* **25**:2953-2959.

181. Hayes, J. D. and M. McMahon. 2001. Molecular basis for the contribution of the antioxidant responsive element to cancer chemoprevention. *Cancer Lett.* **174**:103-113.

182. Dalton, T. P., H. G. Shertzer and A. Puga. 1999. Regulation of gene expression by reactive oxygen. *Annu. Rev. Pharmacol. Toxicol.* **39**:67-101.

183. Itoh, K., N. Wakabayashi, Y. Katoh, T. Ishii, K. Igarashi, J. D. Engel and M. Yamamoto. 1999. Keap1 represses nuclear activation of antioxidant responsive elements by Nrf2 through binding to the amino-terminal Neh2 domain. *Genes Dev.* **13**:76-86.

184. Alam, J., D. Stewart, C. Touchard, S. Boinapally, A. M. Choi and J. L. Cook. 1999. Nrf2, a Cap'n'Collar transcription factor, regulates induction of the heme oxygenase-1 gene. *J. Biol. Chem.* **274**:26071-26078.

185. Itoh, K., T. Chiba, S. Takahashi, T. Ishii, K. Igarashi, Y. Katoh, T. Oyake, N. Hayashi, K. Satoh, I. Hatayama, M. Yamamoto and Y. Nabeshima. 1997. An Nrf2/small Maf heterodimer mediates the induction of phase II detoxifying enzyme genes through antioxidant response elements. *Biochem. Biophys. Res. Commun.* **236**:313-322.

186. Kwak, M. K., K. Itoh, M. Yamamoto, T. R. Sutter and T. W. Kensler. 2001. Role of transcription factor Nrf2 in the induction of hepatic phase 2 and antioxidative enzymes in vivo by the cancer chemoprotective agent, 3H-1, 2-dimethiole-3-thione. *Mol. Med.* **7**:135-145.

187. Kataoka, K., H. Handa and M. Nishizawa. 2001. Induction of cellular antioxidative stress genes through heterodimeric transcription factor Nrf2/small Maf by antirheumatic gold(I) compounds. *J. Biol. Chem.* **276**:34074-34081.

188. Ishii, T., K. Itoh, S. Takahashi, H. Sato, T. Yanagawa, Y. Katoh, S. Bannai and M. Yamamoto. 2000. Transcription factor Nrf2 coordinately regulates a group of oxidative stress-inducible genes in macrophages. *J. Biol. Chem.* **275**:16023-16029.

189. Cho, H. Y., A. E. Jedlicka, S. P. Reddy, T. W. Kensler, M. Yamamoto, L. Y. Zhang and S. R. Kleeberger. 2002. Role of NRF2 in protection against hyperoxic lung injury in mice. *Am. J. Respir. Cell Mol. Biol.* **26**:175-182.

190. Venugopal, R. and A. K. Jaiswal. 1998. Nrf2 and Nrf1 in association with Jun proteins regulate antioxidant response element-mediated expression and coordinated induction of genes encoding detoxifying enzymes. *Oncogene* **17**:3145-3156.

191. Jeyapaul, J. and A. K. Jaiswal. 2000. Nrf2 and c-Jun regulation of antioxidant response element (ARE)- mediated expression and induction of gamma-glutamylcysteine synthetase heavy subunit gene. *Biochem. Pharmacol.* **59**:1433-1439.

192. He, C. H., P. Gong, B. Hu, D. Stewart, M. E. Choi, A. M. Choi and J. Alam. 2001. Identification of activating transcription factor 4 (ATF4) as an Nrf2- interacting protein. Implication for heme oxygenase-1 gene regulation. *J. Biol. Chem.* **276**:20858-20865.

193. Huang, H. C., T. Nguyen and C. B. Pickett. 2000. Regulation of the antioxidant response element by protein kinase C- mediated phosphorylation of NF-E2-related factor 2. *Proc. Natl. Acad. Sci. U. S. A* **97**:12475-12480.

194. Kong, A. N., E. Owuor, R. Yu, V. Hebbar, C. Chen, R. Hu and S. Mandlekar. 2001. Induction of xenobiotic enzymes by the MAP kinase pathway and the antioxidant or electrophile response element (ARE/EpRE). *Drug Metab Rev.* **33**:255-271.

195. Alam, J., C. Wicks, D. Stewart, P. Gong, C. Touchard, S. Otterbein, A. M. Choi, M. E. Burow and J. Tou. 2000. Mechanism of heme oxygenase-1 gene activation by cadmium in MCF-7 mammary epithelial cells. Role of p38 kinase and Nrf2 transcription factor. *J. Biol. Chem.* **275**:27694-27702.

196. Darnell, J. E., Jr., I. M. Kerr and G. R. Stark. 1994. Jak-STAT pathways and transcriptional activation in response to IFNs and other extracellular signaling proteins. *Science* **264**:1415-1421.

197. Takeda, K., S. Ishizawa, M. Sato, T. Yoshida and S. Shibahara. 1994. Identification of a cis-acting element that is responsible for cadmium- mediated induction of the human heme oxygenase gene. *J. Biol. Chem.* **269**:22858-22867.

198. Sato, M., S. Ishizawa, T. Yoshida and S. Shibahara. 1990. Interaction of upstream stimulatory factor with the human heme oxygenase gene promoter. *Eur. J. Biochem.* **188**:231-237.

199. Dalton, T., R. D. Palmiter and G. K. Andrews. 1994. Transcriptional induction of the mouse metallothionein-I gene in hydrogen peroxide-treated Hepa cells involves a composite major late transcription factor/antioxidant response element and metal response promoter elements. *Nucleic Acids Res.* **22**:5016-5023.

200. Muraosa, Y. and S. Shibahara. 1993. Identification of a cis-regulatory element and putative trans-acting factors responsible for 12-O-tetradecanoylphorbol-13-acetate (TPA)- mediated induction of heme oxygenase expression in myelomonocytic cell lines. *Mol. Cell Biol.* **13**:7881-7891.

201. Muraosa, Y., K. Takahashi, M. Yoshizawa and S. Shibahara. 1996. cDNA cloning of a novel protein containing two zinc-finger domains that may function as a transcription factor for the human heme-oxygenase-1 gene. *Eur. J. Biochem.* **235**:471-479.

202. Sato, M., Y. Fukushi, S. Ishizawa, S. Okinaga, R. M. Muller and S. Shibahara. 1989. Transcriptional control of the rat heme oxygenase gene by a nuclear protein that interacts with adenovirus 2 major late promoter. *J. Biol. Chem.* **264**:10251-10260.

203. Immenschuh, S., V. Hinke, A. Ohlmann, S. Gifhorn-Katz, N. Katz, K. Jungermann and T. Kietzmann. 1998. Transcriptional activation of the haem oxygenase-1 gene by cGMP via a cAMP response element/activator protein-1 element in primary cultures of rat hepatocytes. *Biochem. J.* **334 (Pt 1)**:141-146.

204. Immenschuh, S., V. Hinke, N. Katz and T. Kietzmann. 2000. Transcriptional induction of heme oxygenase-1 gene expression by okadaic acid in primary rat hepatocyte cultures. *Mol. Pharmacol.* **57**:610-618.

205. Elbirt, K. K., A. J. Whitmarsh, R. J. Davis and H. L. Bonkovsky. 1998. Mechanism of sodium arsenite-mediated induction of heme oxygenase-1 in hepatoma cells. Role of mitogen-activated protein kinases. *J. Biol. Chem.* **273**:8922-8931.

206. Lu, T. H., R. W. Lambrecht, J. Pepe, Y. Shan, T. Kim and H. L. Bonkovsky. 1998. Molecular cloning, characterization, and expression of the chicken heme oxygenase-1 gene in transfected primary cultures of chick embryo liver cells. *Gene* **207**:177-186.

207. Lu, T. H., Y. Shan, J. Pepe, R. W. Lambrecht and H. L. Bonkovsky. 2000. Upstream regulatory elements in chick heme oxygenase-1 promoter: a study in primary cultures of chick embryo liver cells. *Mol. Cell Biochem.* **209**:17-27.

208. Shan, Y., J. Pepe, R. W. Lambrecht and H. L. Bonkovsky. 2002. Mapping of the chick heme oxygenase-1 proximal promoter for responsiveness to metalloporphyrins. *Arch. Biochem. Biophys.* **399**:159-166.

Chapter 19

APOPTOSIS MECHANISMS INITIATED BY OXIDATIVE STRESS

Irene E Kochevar

19.1 Introduction

Apoptosis or programmed cell death is an orderly process whereby cells destroy themselves. An orchestrated series of enzymatic and protein association/dissociation steps lead to morphologic signs of apoptosis, i.e., blebbing and vacuole formation, chromatin condensation and DNA fragmentation. By this program, an organism disposes of damaged cells or of cells that are no longer needed. Because of the critical nature of the "to live or not to live" question for cells, apoptosis is regulated at many of the steps from initiation to execution. Apoptosis of post-mitotic cells, possibly initiated by redox reactions, is believed to contribute to neurodegenerative diseases such as Alzheimer's and Parkinson's and to loss of function following ischemia in several organs. On the other hand, deletion of cells by apoptosis is an essential step in development and during maturation of the immune system.

Multiple sites of initiation and diverse pathways for inducing apoptosis exist (Figure 1). Oxidative stress has been shown to utilize most of these apoptosis pathways. Apoptosis is initiated at "death receptors" that are members of the TNF-receptor (TNFR)[1] family and are found in the plasma membrane. The receptors share similar cysteine-rich extracellular domains and homologous cytoplasmic sequences containing "death domains" that bind to proteins of the apoptotic machinery. Binding of Fas ligand (FasL) to Fas or of TNFα to TNFR1 induces activation of the receptors, which exist at least partially in preformed, inactive

365

H. J. Forman, J. Fukuto, and M. Torres (eds.), Signal Transduction by Reactive Oxygen and Nitrogen Species: Pathways and Chemical Principles, ©2003, Kluwer Academic Publishers. Printed in the Netherlands

multimers in the plasma membrane. Activation of Fas is associated with multimerization and capping of the receptors, possibly by a mechanism involving ceramide production[1] and binding of an adaptor protein, FADD (Fas-associating protein with death domain). Activation of TNFR1 recruits TRADD (TNFR-associated death domain) which subsequently binds FADD. Pro-caspase-8 binds to FADD and is then cleaved to form active caspase-8, which either cleaves Bid leading to release of cytochrome c and AIF (apoptosis inducing factor) from mitochondria or directly activates caspase-3 by proteolysis. Cytochrome c, in conjunction with Apaf1 (apoptotic protease-activating factor-1) and ATP, activates caspase-9, which subsequently activates caspase-3. Caspase-3 then cleaves multiple apoptotic substrates leading to the morphological signs of apoptosis.[2] These late, execution-phase steps are shared by most apoptotic pathways irrespective of the initiating signal.

Activation of mitogen activated protein kinases (MAPK) by stresses including redox reactions also leads to apoptosis involving critical mitochondrial steps. p38 kinase and JNK (c-Jun N-terminal kinase) are members of the MAPK family that are associated with apoptosis. JNK is also known as SAPK, stress-activated protein kinase, but the former term will be used for simplicity. JNK and p38 are directly activated by different MAP kinase kinases (MAPKKs), although both of these

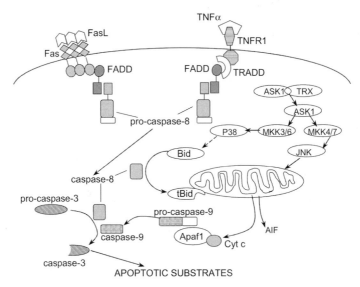

Figure 1. Simplified schematic diagram of signaling pathways leading to apoptosis. This figure is intended only to serve as a guide for the reports discussed in the text. The pathways shown are not complete and pathways leading from nuclear DNA damage, lysosomes, endoplasmic reticulum and Golgi apparatus are not included. Abbreviations are: Apaf1, apoptotic protease inducing factor-1; ASK1, apoptosis signal-regulating kinase-1; AIF, apoptosis-inducing factor; cyt c, cytochrome c; FasL, Fas ligand; FADD, Fas associated death domain protein; JNK, c-jun N-terminal c kinase; MKK, MAP kinase kinase; tBid, truncated Bid; TRADD, TNF receptor associated death domain protein; TRX, thioredoxin; TNFR1, TNF receptor-1.

MAPKKs are activated by apoptosis signal regulated kinase-1 (ASK1). JNK was recently reported to phosphorylate Bcl-2 family proteins in the mitochondrial membrane thereby activating the apoptotic pathway.[3] TNF☐ also induces apoptosis by a JNK-mediated pathway in which ASK1 is activated by TRAF2 (TNFR associated factor-2), a protein that associates with active TNFR2.[4]

Primary initiating events for apoptosis occur in several organelles including the lysosomes, nucleus, endoplasmic reticulum and Golgi apparatus in addition to mitochondria (reviewed in[5]). Lysosomal proteases have been shown to be available prior to release of cytochrome c from mitochondria and to cause cleavage of Bid. Several processes are activated in the endoplasmic reticulum that lead to activation of the caspase cascade and induction of mitochondrial membrane permeability. The induction of apoptosis by damage to nuclear DNA has been extensively studied.[6] Many agents including ionizing radiation, ultraviolet radiation and chemotherapeutic drugs cause DNA damage leading to apoptosis by mechanisms that involve new protein synthesis. This very large and important topic is outside of the range of this review but must be kept in mind since many agents are capable of inducing apoptosis by both DNA damage and by other mechanisms.

Oxidative stress and redox chemistry initiate most, if not all, of the pathways leading to apoptosis. This review focuses on the signaling pathways leading from activation of plasma membrane receptors, MAP kinases and mitochondrial and lysosomal initiation sites. The high level of activity in this area precludes an inclusive treatment. The chapter should be viewed as an update of previous reviews and covers studies published in 2001 and up to about July, 2002 with additional studies cited as necessary[2]. Recent reviews include those by Chandra *et al.*,[7] Curtin *et al.*,[8] and Chung *et al.*[9] My apologies are extended to the authors of many excellent publications that are not discussed.

A prevailing theme throughout the literature in this area is that oxidative stress induces both pro-apoptotic as well as pro-survival signals and that a cell's fate depends on the balance between these signals. Although this review does not focus on induction of pro-survival pathways by redox chemistry, when both pro and anti-apoptotic pathways are activated under the same conditions the contrast may be discussed. The term "reactive oxidizing species" (ROS) is intended to denote both oxygen- and nitrogen-containing oxidizing species and includes hydrogen peroxide (H_2O_2), hydroxyl radical ($\cdot OH$), superoxide anion (O_2^-), singlet oxygen ($^1O_2$), nitric oxide (NO), peroxynitrite ($ONOO^-$) and others.

19.2 Apoptotic Signaling Initiated At The Plasma Membrane

Reactive oxidizing species initiate apoptosis by stimulating plasma membrane death receptors, but ROS, particularly H_2O_2 and O_2^- are also produced as a consequence of activation of these receptors by their ligands, and these latter ROS are also believed to participate in the apoptotic process.

19.2.1   Activation of plasma membrane death receptors by oxidative stress

19.2.1.1 Fas apoptotic pathway

Activation of the Fas pathway to apoptosis was initiated in human leukemia HL-60 cells by $^1O_2$.[10]   Singlet oxygen is a short-lived ROS that is produced by photosensitization. In this process, the photosensitizer dye absorbs light energy and then transfers this energy to dissolved oxygen.   Initiation was very rapid with recruitment of FADD and pro-caspase-8 to a complex with Fas (the so-called DISC, death-inducing signaling complex) occurring within 5 min. Anti-Fas antibody did not influence recruitment of FADD and caspase-8 to the DISC or inhibit initiation of apoptosis. Downstream signaling events from activation of caspase-8 included those typically observed by activation of the Fas pathway, i.e., cleavage of Bid, release of mitochondrial cytochrome c and activation of caspase-3. The mechanism for $^1O_2$-induced activation of Fas is unclear.   Treatment with a PKC activator inhibited apoptosis although this treatment did not affect the binding of caspase-8 to activated Fas. These results suggest that PKC inhibits $^1O_2$ induced apoptosis by preventing activation of caspase-8.

Nitric oxide and H_2O_2 induce apoptosis by mechanisms involving the Fas apoptotic pathway, but generally direct activation of Fas is not involved. The NO donor SNAP initiated apoptosis in human and rat pulmonary artery smooth muscle cells (PASMCs).[11]   FasL induced apoptosis in rat PASMCs, and neutralizing anti-FasL antibody inhibited SNAP-induced apoptosis measured at 24 h. NO treatment did not appear to directly cause Fas activation, as found for $^1O_2$, but rather up-regulated FasL and Fas proteins and mRNA levels by increasing cGMP. In another study, treatment of colon cancer cells with NO by use of glyceryl trinitrate induced apoptotic cell death that correlated with decreased expression of inhibitors of apoptosis (IAP) and increased expression of Fas.[12]

Hydrogen peroxide-induced apoptosis appears to involve membrane death receptors only by induction of new protein expression. In PC12 cells and rat myocytes apoptosis was not reduced by inhibition of caspase-8 suggesting neither Fas nor TNFR1 apoptotic pathways were directly activated.[13,14]   Increased expression of Fas and/or FasL was associated with H_2O_2-induced apoptosis in astrocytoma cell lines and differentiated PC12 cells.[15,16]

19.2.2   Formation of ROS in response to ligand activation of death receptors

Binding between plasma membrane receptors and their respective ligands increases the intracellular ROS level, usually within a few minutes, not only for Fas and TNFR but also for growth factor receptors, T-cell receptors and G-protein coupled receptors.   The major sources of these intracellular ROS are the mitochondria and the plasma membrane NADPH oxidases that have now been identified in many non-phagocytic cells. Irrespective of their source, the question of whether these ROS mediate downstream signaling to apoptosis is still controversial. Results of studies both support and deny a role of ROS in death receptor-induced apoptosis.

The specific type of ROS produced by activation of a membrane receptor that induces apoptosis is not identified in many studies. Most often, the increase in ROS is detected by an increase in fluorescence intensity of a reporter dye for ROS that is not specific for a single type of ROS. However, the relative contribution of H_2O_2, O_2^- and $ONOO^-$ to apoptosis was addressed in mesangial cells treated with $TNF\alpha$.[17] The results indicated that O_2^- was the dominant ROS inducing apoptosis because transfection with MnSOD inhibited $TNF\alpha$-induced apoptosis as did treatment with the O_2^- scavenger, Tiron. Although both H_2O_2 and SIN-1, a $ONOO^-$ precursor, induced apoptosis when added separately, neither transfection with catalase nor use of the NO synthase inhibitor, L-NAME, inhibited apoptosis induced by $TNF\alpha$. These results were supported by use of pharmacological inhibitors of O_2^- and $ONOO^-$ as well as treatment with a O_2^- releasing agent, pyrogallol, that induced apoptosis in the presence of scavengers for H_2O_2 and $ONOO^-$. The results of this study raise an interesting question about the molecular target that was activated by O_2^- but not by H_2O_2, and $ONOO^-$. Perhaps the localized formation of O_2^- near the TNF receptor allowed it to react with a target that was not accessible to H_2O_2 and $ONOO^-$.

A study in T-cell derived lines extended the concept that different types of ROS initiate different signaling pathways to apoptosis.[18] Cross-linking of T-cell receptors using anti-CD3 antibody rapidly increased levels of both H_2O_2 and O_2^- although with somewhat different time courses. Interestingly, these two ROS appeared to be produced separately, i.e., dismutation of O_2^- was not the source of H_2O_2, consistent with the results of a previous study.[19] Superoxide anion mediated the T-cell receptor-stimulated activation of the FasL promoter and subsequent apoptosis. H_2O_2 was not involved since transfection with neither catalase nor thioredoxin peroxidase inhibited the response. On the other hand, H_2O_2 that was produced in response to receptor cross-linking elicited extracellular signal-regulated kinase 1/2 (ERK1/2) phosphorylation but O_2^- did not. Thus two distinct TNF signaling pathways involving ROS exist in T cells and are followed concurrently; however, only one signaling pathway, that involving O_2^-, led to apoptosis by up-regulating FasL gene expression. Interestingly, the H_2O_2-stimulated pathway activates a MAP kinase pathway that is generally associated with cell survival. The independent intracellular sources of H_2O_2 and O_2^- remain to be identified.

Other studies have also shown that $TNF\alpha$ induces activation of pro-apoptotic and anti-apoptotic signaling pathways via ROS-dependent mechanisms. Chandel *et al.* studied how the recruitment of TRAFs to the TNFR after ligand binding contribute to regulation of intracellular ROS levels in human embryonic kidney 293 cells.[20] TRAF2 was shown to signal to the mitochondria and regulate ROS formation by the mitochondrial electron transport chain. A divergence between subsequent pro-and anti-apoptotic pathways was observed. Activation of NF-κB, generally an anti-apoptotic signal, was stimulated by ROS but this response was not blocked by expression of Bcl-x$_L$, an anti-apoptotic mitochondrial membrane protein. In contrast, apoptosis was inhibited by Bcl-x$_L$ under the same conditions. The involvement of TRAF2 in $TNF\alpha$ induced ROS production was independently confirmed and the mechanism examined by Liu.[21] Expression of TRAF2 enhanced production of ROS. The mechanism for activation of ASK1 by TRAF2 was shown to involve ROS-induced oxidation of thioredoxin (TRX), an endogenous inhibitor of ASK1 that interferes with the TRAF2-ASK1 interaction. Dissociation of the oxidized TRX from ASK1 was proposed to allow homo-oligomerization and activation of ASK1.

The ROS produced upon activation of Fas also play a role in one of the characteristic features of apoptotic cells, the externalization of phosphatidyl serine. Both oxidation and externalization of phosphatidyl serine was observed when Fas was activated on Jurkat cells[22] and normal human keratinocytes.[23] In keratinocytes, these effects occurred much earlier than detectable changes in other apoptotic markers such as caspase-3 activation or DNA fragmentation. Phagocytosis of FasL-activated Jurkat cells was inhibited by superoxide dismutase and catalase, which also prevented oxidation of phosphatidylserine while allowing it to remain externalized on these cells. Cells possessing oxidized phosphatidylserine in their external leaflet were phagocytosed more readily that those containing phosphatidylserine indicating that this oxidized lipid on the cell surface is an important signal for macrophage clearance of apoptotic cells. A model had been proposed in which cytochrome c released from mitochondria by oxidative stress binds to phosphatidylserine located on the cytoplasmic surface of the plasma membrane and induces oxidation of this membrane lipid.[24]

19.3 Ros Activation of MAPK Signaling Pathways to Apoptosis

Studies in several cell types have established that activation of the stress-induced MAP kinases, JNK and p38, mediate pro-apoptotic signaling by H_2O_2 and other redox agents. Addition of external ROS to cells as well as formation of ROS by internal sources, e.g., mitochondrial oxidative phosphorylation, NADPH oxidase or monoamine oxidase, activate these MAPKs.

Two upstream kinases, MMK4 and MMK7, activate JNK in response to oxidative stress. MMK4, also called SEK1 (stress-activated protein kinase/extracellular signal-regulated kinase-1), which is known to upregulate JNK activity without affecting other MAPKs, was shown to be activated in H_2O_2-treated cardiac myocytes[14] and in human U937 leukemia cells.[25] Transfection with dominant negative SEK1/MMK4 inhibited H_2O_2-induced JNK activation[25] and apoptosis.[14] MMK7 is also an upstream activator of JNK in response to H_2O_2 treatment.[25] In addition, ASK1, a MAPKKK family member that acts directly upstream of MMK4 and MMK7, is well known to activate JNK as well as p38. A recent study showed definite use of this pathway in H_2O_2-induced apoptosis. Embryonic fibroblasts from mice deficient in ASK1 did not produce sustained activation of JNK and p38 in response to H_2O_2 and were resistant to H_2O_2-induced apoptosis.[26] Interestingly, TNFα- but not Fas-induced, apoptosis was dependent on activation of ASK1-JNK/p38 pathways by mechanisms involving ROS.

Apoptosis induced by nitric oxide (NO) was also shown to be mediated by ASK1 in PC12 cells using stable transfection of a dominant negative mutant of ASK1, which both decreased apoptosis and p38 activation.[27] Activation of p38 was shown to be essential for apoptosis initiated by NO as determined by use of dominant negative transfection and inhibitors of p38. Interestingly, the inhibitors did not substantially inhibit apoptosis induced by added H_2O_2 or menadione indicating that these ROS acted by other pathways to induce apoptosis.

The redox-initiated signaling mechanism mediated by ASK1 has received detailed study. ASK1 exists as an inactive complex with the reduced form of thioredoxin in unstressed cells. H_2O_2 and other oxidative stresses oxidize thioredoxin and disrupt this complex, thereby producing an activated ASK1. The mechanism for activation of ASK1 after its release was addressed by Tobiume *et al.*[28] Phosphorylation of Thr845 at the activation loop was shown to be essential for activation of ASK1 by H_2O_2. An unidentified Thr845 kinase by H_2O_2 was proposed to be activated and then to activate ASK1 in its oligomerized and activation-competent forms. Inactivation of a specific phosphatase is also likely since H_2O_2 is known to oxidize the ionized cysteine in the active site of phosphatases thereby inactivating the enzyme. Thus, H_2O_2 mediates at least two steps in the signaling pathway, namely, by oxidizing thioredoxin and by activating a kinase (or inactivating a phosphatase) specifically acting on Thr845. Down-regulation of ASK1 activity was shown to involve serine/threonine protein phosphatase 5 (PP5).[29] PP5 was identified as a binding partner of ASK1 from a yeast two-hybrid screening and deactivated ASK1 by dephosphorylating an essential threonine. Interaction between PP5 and ASK1 was induced by H_2O_2 treatment and followed as a decrease in ASK1 activity. PP5 inhibited not only H_2O_2-induced sustained activation of ASK1 but also ASK1-dependent apoptosis. CDC25A, a phosphatase that promotes cell cycle progression by activating G1 cyclin-dependent kinases, was also shown to inhibit ASK1 activity.[30] Binding was detected by yeast two-hybrid screening and co-immunoprecipitation. Interestingly, inhibition by CDC25A of activation of the kinases, ASK1, p38 and JNK1 and of H_2O_2-induced apoptosis did not require the phosphatase activity of this protein. The inhibition appeared to be due to a decrease in homo-oligomerization of ASK1.

Direct activation of the mitochondrial apoptotic machinery by JNK in response to H_2O_2 was reported in cardiac myocytes.[14] JNK was shown to reside on mitochondria and to be activated there by SED/MMK4 after treatment with H_2O_2. New protein synthesis was not required for apoptosis induced by H_2O_2 ruling against a mechanism involving JNK-induced expression of Fas, a mechanism previously proposed for JNK-mediated apoptosis induced by other stimuli. In a cell free system, activated JNK was shown to cause release of cytochrome c, AIF and other intermembrane proteins. Although phosphorylation of some membrane proteins was detected, inactivation of Bcl-2 or Bcl-x_L by phosphorylation was not observed. However, in another study, phosphorylation of Bcl-2-related proteins was shown to be essential for apoptotic signal transduction by JNK, although redox agents were not the initiators.[3]

Hydrogen peroxide initiated apoptosis by a Bid-mediated process in HL-60 cells.[31] In response to H_2O_2 and other ROS, p38 was activated, which led to subsequent cleavage of Bid. The truncated form of Bid binds to the outer mitochondrial membrane and mediates the membrane permeability transition. The mechanism for Bid-activation of this mitochondrial pathway has been reviewed.[32] Singlet oxygen also initiated rapid apoptosis by a mechanism involving p38 activation.[31] Under the same conditions, JNK was not activated suggesting that an ASK1-independent pathway was used by these ROS. Bid cleavage and subsequent activation of the mitochondrial pathway for apoptosis was observed. In this study, $^1O_2$ and H_2O_2 activated p38, but only $^1O_2$ activated caspase-8 again indicating that different ROS initiate divergent apoptotic pathways.

The ROS-induced activation of JNK and/or p38 may underlie the mechanism for apoptosis induced by stress agents such as UV, cytokines and chemotherapy agents. For example, the mechanism for cis-platin-induced apoptosis involving activation of JNK and p38 appears to involve ROS generated from mitochondrial oxidative phosphorylation as well as NAD(P)H oxidase activity in oncogene-transformed NIH/3T3 cells.[33] The greater sensitivity of some transformed cell types to stress was attributed to higher levels of production of ROS.

19.4   Initiation of Apoptosis by Ros in Organelles

19.4.1   *Mitochondria*

Mitochondria play several roles in oxidative stress-induced apoptosis. The various initiation pathways converge on the mitochondria and result in release of proteins that further the apoptotic process. In addition, ROS generated in response to activation of plasma membrane receptors are believed to derived from mitochondria in some cases (see Section 2.2). Also, various chemotherapeutic agents and toxic agents perturb the electron transfer chain of mitochondrial oxidative phosphorylation resulting in release of ROS from mitochondria and subsequent initiation of apoptosis.

Photosensitization with a dye that is located in mitochondria is another means of generating ROS in the mitochondria. Most common photosensitizers produce $^1O_2$, although this has not been assessed directly in cells. Recent work suggests that a major pathway may involve selective cleavage of anti-apoptotic Bcl-2 that subsequently leads to loss of mitochondrial membrane potential, and release of cytochrome c.[34,35] Loss of Bcl-2 was found in several cell types, including cells lacking caspase-3, and occurred in chilled cells and in the presence of protease inhibitors suggesting that a chemical reaction between $^1O_2$ and Bcl-2 was occurring. This proposition was supported by the detection of high molecular weight proteins by electrophoresis suggesting that cross-linked proteins are produced, a result expected from reactions of $^1O_2$ with membrane proteins. Other mitochondrial membrane proteins, e.g., Bcl-x_L, Bax, Bad, the voltage-dependent anion channel, or the adenine nucleotide translocator, were not affected. In contrast, Bcl-2 degradation in melanoma cells treated with H_2O_2 was shown to require caspase-3/-7 and to be involved in apoptosis in these cells.[36]

Changes in ROS levels and intracellular redox environment can occur prior to mitochondrial ROS production leading to apoptosis. For example, depletion of glutathione (GSH) using L-buthionine sulphoximine (BSO), which inhibits gamma glutamyl cysteine synthetase, lead to a increase in ROS production from mitochondrial Complex III.[37,38] Although the lowered GSH level in the absence of external apoptotic stimuli initiated the apoptotic process, the results indicated that apoptosis correlated with the mitochondrial production of ROS rather than with the initial GSH depletion. However, in another study redox imbalance appeared able to initiate apoptosis in the absence of ROS formation.[39] Diamide was used to treat of PC-12 cells with a concentration that oxidized thiols, but did not produce ROS. An increase in glutathione disulfide (GSSG) and loss of cytochrome c from mitochondria occurred rapidly followed by caspase-3 activation and apoptosis.

Monoamine oxidase (MOA), located on the cytoplasmic face of the outer mitochondrial membrane, represents another source of H_2O_2 that can lead to apoptosis. Inhibition of MAO with pargyline inhibited H_2O_2 production and apoptosis in a rat model of renal ischemia-reperfusion injury.[40] JNK phosphorylation that was induced by the treatment was also inhibited by pargyline suggesting that the H_2O_2 may induce apoptosis by a JNK-mediated mechanism.

19.4.2 Lysosomes

Lysosomes are frequently discussed as a possible target of redox or H_2O_2-induced apoptosis since released lysosomal proteases are capable of cleaving Bid to the truncated form that interacts with the outer mitochondrial membrane to initiate apoptosis. Treatment of various cell types with H_2O_2, either as a bolus or slow steady state amount, causes lysosomal disruption.[41-44] An decrease in mitochondrial membrane potential was found to follow the lysosome disruption[41] and H_2O_2-induced apoptosis has been reported to be inhibited by desferrioxamine which is taken into lysosomes after endocytosis.[42] An investigation of the mechanism for H_2O_2-induced lysosome disruption implicated activation of cytosolic phospholipase A_2 (PLA$_2$) since inhibition of this enzyme inhibited H_2O_2-induced lysosome disruption and apoptosis and microinjection of PLA$_2$ induced these processes.[44] A recent study of apoptosis initiated by a photosensitizer localizing in lysosomes also implicated lysosomal enzymes in the apoptotic pathway.[45] Photosensitization, which produces ROS, initiated lysosome disruption and cleavage of Bid. Caspase-8 was not activated suggesting that lysosomal enzymes were responsible for Bid cleavage.

19.5 Involvement of Ceramide in Ros-Induced Apoptosis

Ceramide has been implicated as a signaling molecule in apoptosis mechanisms induced by several agents, including ROS. The relationship between ROS and ceramide was recently reviewed.[46] Ceramide belongs to the sphingolipid class of membrane lipids and is composed of a sphingosine backbone, an amide-linked long chain fatty acid and a hydroxyl head group. Sphingomyelinase, which generates ceramide from sphingomyelin, is rapidly activated in response to H_2O_2 and other oxidative stress, and inhibition of sphingomyelinase blocks apoptosis in many cases. Glutathione was shown to inhibit ceramide production that was stimulated by H_2O_2 human alveolar epithelial cells and increasing the intracellular H_2O_2 by inhibiting catalase also increased ceramide and induced apoptosis.[47] In another study, overexpression of glutathione peroxidase was shown to inhibit ceramide production and apoptosis induced by daunomycin, a chemotherapeutic agent known to induce ROS and ceramide-mediated apoptosis.[48]

A mechanism for rapid activation of sphingomyelinase and involvement of ceramide in Fas-induced capping was recently proposed.[49] Ceramides have been shown to mediate steps in other apoptotic processes such as induction of mitochondrial permeability transition and inhibition of the mitochondrial respiratory chain leading to ROS overproduction.[46] Major questions in this area are how ROS can rapidly activate sphingomyelinase and how ceramide is involved in ROS-induced apoptotic signaling.

19.6 Summary

Many aspects of the pro-apoptotic signal transduction processes induced by oxidative stress are becoming clear. The information now available indicates that ROS initiate apoptosis by most of the known signaling pathways and that different ROS utilize different initiation mechanisms. Interestingly, ROS may act at more than one step of an apoptotis signaling cascade. Some evidence suggests that the same ROS may initiate a different signal transduction process when it is generated intracellularly by an endogenous biochemical pathway than when it is added externally. These observations lead to questions about the relative importance for the pro-apoptotic signaling processes of the chemical reactivity of particular ROS, the site of formation/addition of the ROS, and the concentration and time course for exposure to an ROS. These questions deserve further study so that we can understand the processes leading to apoptosis and use this information for design of therapies.

19.7 Endnotes

1. Abbreviations used are: AIF, apoptosis-inducing factor; Apaf1, apoptotic protease inducing factor-1; ASK1, apoptosis signal-regulating kinase-1; FasL, Fas ligand; FADD, Fas associated death domain protein; JNK, c-jun N-terminal kinase;   MAPK, mitogen activated protein kinase; ROS, reactive oxidizing species; TRADD, TNF receptor associated death domain protein;   TRAF, TNFR-associated factors; TNFR, TNF receptor;

2. Preparation of this manuscript was supported by National Institutes of Health Grant GM 30755.

References

1. Grassme, H., A. Jekle, A. Riehle, H. Schwarz, J. Berger, K. Sandhoff, R. Kolesnick and E. Gulbins. 2001. CD95 signaling via ceramide-rich membrane rafts. *J Biol Chem* **276**:20589-20596.

2. Adrain, C. and S. J. Martin. 2001. The mitochondrial apoptosome: a killer unleashed by the cytochrome seas. *Trends Biochem Sci* **26**:390-397.

3. Lei, K., A. Nimnual, W. X. Zong, N. J. Kennedy, R. A. Flavell, C. B. Thompson, D. Bar-Sagi and R. J. Davis. 2002. The Bax subfamily of Bcl2-related proteins is essential for apoptotic signal transduction by c-Jun NH(2)-terminal kinase. *Mol Cell Biol* **22**:4929-4942.

4. Hoeflich, K. P., W. C. Yeh, Z. Yao, T. W. Mak and J. R. Woodgett. 1999. Mediation of TNF receptor-associated factor effector functions by apoptosis signal-regulating kinase-1 (ASK1). *Oncogene* **18**:5814-5820.

5. Ferri, K. F. and G. Kroemer. 2001. Organelle-specific initiation of cell death pathways. *Nat Cell Biol* **3**:E255-263.

6. Rich, T., R. L. Allen and A. H. Wyllie. 2000. Defying death after DNA damage. *Nature* **407**:777-783.

7. Chandra, J., A. Samali and S. Orrenius. 2000. Triggering and modulation of apoptosis by oxidative stress. *Free Radic Biol Med* **29**:323-333.

8. Curtin, J. F., M. Donovan and T. G. Cotter. 2002. Regulation and measurement of oxidative stress in apoptosis. *J Immunol Methods* **265**:49-72.

9. Chung, H. T., H. O. Pae, B. M. Choi, T. R. Billiar and Y. M. Kim. 2001. Nitric oxide as a bioregulator of apoptosis. *Biochem Biophys Res Commun* **282**:1075-1079.

10. Zhuang, S., J. T. Demirs and I. E. Kochevar. 2001. Protein kinase C inhibits singlet oxygen-induced apoptosis by decreasing caspase-8 activation. *Oncogene* **20**:6764-6776.

11. Hayden, M. A., P. A. Lange and D. K. Nakayama. 2001. Nitric oxide and cyclic guanosine monophosphate stimulate apoptosis via activation of the Fas-FasL pathway. *J Surg Res* **101**:183-189.

12. Millet, A., A. Bettaieb, F. Renaud, L. Prevotat, A. Hammann, E. Solary, B. Mignotte and J. F. Jeannin. 2002. Influence of the nitric oxide donor glyceryl trinitrate on apoptotic pathways in human colon cancer cells. *Gastroenterology* **123**:235-246.

13. Yamakawa, H., Y. Ito, T. Naganawa, Y. Banno, S. Nakashima, S. Yoshimura, M. Sawada, Y. Nishimura, Y. Nozawa and N. Sakai. 2000. Activation of caspase-9 and -3 during H_2O_2-induced apoptosis of PC12 cells independent of ceramide formation. *Neurol Res* **22**:556-564.

14. Aoki, H., P. M. Kang, J. Hampe, K. Yoshimura, T. Noma, M. Matsuzaki and S. Izumo. 2002. Direct activation of mitochondrial apoptosis machinery by c-Jun N-terminal kinase in adult cardiac myocytes. *J Biol Chem* **277**:10244-10250.

15. Facchinetti, F., S. Furegato, S. Terrazzino and A. Leon. 2002. H_2O_2 induces upregulation of Fas and Fas ligand expression in NGF-differentiated PC12 cells: modulation by cAMP. *J Neurosci Res* **69**:178-188.

16. Kwon, D., C. Choi, J. Lee, K. O. Kim, J. D. Kim, S. J. Kim and I. H. Choi. 2001. Hydrogen peroxide triggers the expression of Fas/FasL in astrocytoma cell lines and augments apoptosis. *J Neuroimmunol* **113**:1-9.

17. Moreno-Manzano, V., Y. Ishikawa, J. Lucio-Cazana and M. Kitamura. 2000. Selective involvement of superoxide anion, but not downstream compounds hydrogen peroxide and peroxynitrite, in tumor necrosis factor-alpha-induced apoptosis of rat mesangial cells. *J Biol Chem* **275**:12684-12691.

18. Devadas, S., L. Zaritskaya, S. G. Rhee, L. Oberley and M. S. Williams. 2002. Discrete generation of superoxide and hydrogen peroxide by T cell receptor stimulation: selective regulation of mitogen-activated protein kinase activation and fas ligand expression. *J Exp Med* **195**:59-70.

19. Deshpande, S. S., P. Angkeow, J. Huang, M. Ozaki and K. Irani. 2000. Rac1 inhibits TNF-alpha-induced endothelial cell apoptosis: dual regulation by reactive oxygen species. *FASEB J* **14**:1705-1714.

20. Chandel, N. S., P. T. Schumacker and R. H. Arch. 2001. Reactive oxygen species are downstream products of TRAF-mediated signal transduction. *J Biol Chem* **276**:42728-42736.

21. Liu, H., H. Nishitoh, H. Ichijo and J. M. Kyriakis. 2000. Activation of apoptosis signal-regulating kinase 1 (ASK1) by tumor necrosis factor receptor-associated factor 2 requires prior dissociation of the ASK1 inhibitor thioredoxin. *Mol Cell Biol* **20**:2198-2208.

22. Kagan, V. E., B. Gleiss, Y. Y. Tyurina, V. A. Tyurin, C. Elenstrom-Magnusson, S. X. Liu, F. B. Serinkan, A. Arroyo, J. Chandra, S. Orrenius and B. Fadeel. 2002. A role for oxidative stress in apoptosis: oxidation and externalization of phosphatidylserine is required for macrophage clearance of cells undergoing fas-mediated apoptosis. *J Immunol* **169**:487-499.

23. Shvedova, A. A., J. Y. Tyurina, K. Kawai, V. A. Tyurin, C. Kommineni, V. Castranova, J. P. Fabisiak and V. E. Kagan. 2002. Selective peroxidation and externalization of phosphatidylserine in normal human epidermal keratinocytes during oxidative stress induced by cumene hydroperoxide. *J Invest Dermatol* **118**:1008-1018.

24. Tyurina, Y. Y., A. A. Shvedova, K. Kawai, V. A. Tyurin, C. Kommineni, P. J. Quinn, N. F. Schor, J. P. Fabisiak and V. E. Kagan. 2000. Phospholipid signaling in apoptosis: peroxidation and externalization of phosphatidylserine. *Toxicology* **148**:93-101.

25. Kim, D. K., E. S. Cho, J. K. Seong and H. D. Um. 2001. Adaptive concentrations of hydrogen peroxide suppress cell death by blocking the activation of SAPK/JNK pathway. *J Cell Sci* **114**:4329-4334.

26. Tobiume, K., A. Matsuzawa, T. Takahashi, H. Nishitoh, K. Morita, K. Takeda, O. Minowa, K. Miyazono, T. Noda and H. Ichijo. 2001. ASK1 is required for sustained activations of JNK/p38 MAP kinases and apoptosis. *EMBO Rep* **2**:222-228.

27. Han, O. J., K. H. Joe, S. W. Kim, H. S. Lee, N. S. Kwon, K. J. Baek and H. Y. Yun. 2001. Involvement of p38 mitogen-activated protein kinase and apoptosis signal-regulating kinase-1 in nitric oxide-induced cell death in PC12 cells. *Neurochem Res* **26**:525-532.

28. Tobiume, K., M. Saitoh and H. Ichijo. 2002. Activation of apoptosis signal-regulating kinase 1 by the stress-induced activating phosphorylation of pre-formed oligomer. *J Cell Physiol* **191**:95-104.

29. Morita, K., M. Saitoh, K. Tobiume, H. Matsuura, S. Enomoto, H. Nishitoh and H. Ichijo. 2001. Negative feedback regulation of ASK1 by protein phosphatase 5 (PP5) in response to oxidative stress. *Embo J* **20**:6028-6036.

30. Zou, X., T. Tsutsui, D. Ray, J. F. Blomquist, H. Ichijo, D. S. Ucker and H. Kiyokawa. 2001. The cell cycle-regulatory CDC25A phosphatase inhibits apoptosis signal-regulating kinase 1. *Mol Cell Biol* **21**:4818-4828.

31. Zhuang, S., J. T. Demirs and I. E. Kochevar. 2000. p38 mitogen-activated protein kinase mediates bid cleavage, mitochondrial dysfunction, and caspase-3 activation during

apoptosis induced by singlet oxygen but not by hydrogen peroxide. *J Biol Chem* **275**:25939-25948.

32. Korsmeyer, S. J., M. C. Wei, M. Saito, S. Weiler, K. J. Oh and P. H. Schlesinger. 2000. Pro-apoptotic cascade activates BID, which oligomerizes BAK or BAX into pores that result in the release of cytochrome c. *Cell Death Differ* **7**:1166-1173.

33. Benhar, M., I. Dalyot, D. Engelberg and A. Levitzki. 2001. Enhanced ROS production in oncogenically transformed cells potentiates c-Jun N-terminal kinase and p38 mitogen-activated protein kinase activation and sensitization to genotoxic stress. *Mol Cell Biol* **21**:6913-6926.

34. Xue, L. Y., S. M. Chiu and N. L. Oleinick. 2001. Photochemical destruction of the Bcl-2 oncoprotein during photodynamic therapy with the phthalocyanine photosensitizer Pc 4. *Oncogene* **20**:3420-3427.

35. Kessel, D. and M. Castelli. 2001. Evidence that bcl-2 is the target of three photosensitizers that induce a rapid apoptotic response. *Photochem Photobiol* **74**:318-322.

36. Del Bello, B., M. A. Valentini, F. Zunino, M. Comporti and E. Maellaro. 2001. Cleavage of Bcl-2 in oxidant- and cisplatin-induced apoptosis of human melanoma cells. *Oncogene* **20**:4591-4595.

37. Armstrong, J. S. and D. P. Jones. 2002. Glutathione depletion enforces the mitochondrial permeability transition and causes cell death in HL60 cells that overexpress Bcl-2. *FASEB J* **7**:7.

38. Armstrong, J. S., K. K. Steinauer, B. Hornung, J. M. Irish, P. Lecane, G. W. Birrell, D. M. Peehl and S. J. Knox. 2002. Role of glutathione depletion and reactive oxygen species generation in apoptotic signaling in a human B lymphoma cell line. *Cell Death Differ* **9**:252-263.

39. Pias, E. K. and T. Y. Aw. 2002. Apoptosis in mitotic competent undifferentiated cells is induced by cellular redox imbalance independent of reactive oxygen species production. *FASEB J* **16**:781-790.

40. Kunduzova, O. R., P. Bianchi, N. Pizzinat, G. Escourrou, M. H. Seguelas, A. Parini and C. Cambon. 2002. Regulation of JNK/ERK activation, cell apoptosis, and tissue regeneration by monoamine oxidases after renal ischemia-reperfusion. *FASEB J* **16**:1129-1131.

41. Dare, E., W. Li, B. Zhivotovsky, X. Yuan and S. Ceccatelli. 2001. Methylmercury and H_2O_2 provoke lysosomal damage in human astrocytoma D384 cells followed by apoptosis. *Free Radic Biol Med* **30**:1347-1356.

42. Antunes, F., E. Cadenas and U. T. Brunk. 2001. Apoptosis induced by exposure to a low steady-state concentration of H_2O_2 is a consequence of lysosomal rupture. *Biochem J* **356**:549-555.

43. Zhao, M., J. W. Eaton and U. T. Brunk. 2001. Bcl-2 phosphorylation is required for inhibition of oxidative stress-induced lysosomal leak and ensuing apoptosis. *FEBS Lett* **509**:405-412.

44. Zhao, M., U. T. Brunk and J. W. Eaton. 2001. Delayed oxidant-induced cell death involves activation of phospholipase A2. *FEBS Lett* **509**:399-404.

45. Reiners Jr, J. J., J. A. Caruso, P. Mathieu, B. Chelladurai, X. M. Yin and D. Kessel. 2002. Release of cytochrome c and activation of pro-caspase-9 following lysosomal photodamage involves bid cleavage. *Cell Death Differ* **9**:934-944.

46. Andrieu-Abadie, N., V. Gouaze, R. Salvayre and T. Levade. 2001. Ceramide in apoptosis signaling: relationship with oxidative stress. *Free Radic Biol Med* **31**:717-728.

47. Lavrentiadou, S. N., C. Chan, T. Kawcak, T. Ravid, A. Tsaba, A. van der Vliet, R. Rasooly and T. Goldkorn. 2001. Ceramide-mediated apoptosis in lung epithelial cells is regulated by glutathione. *Am J Respir Cell Mol Biol* **25**:676-684.

48. Gouaze, V., M. E. Mirault, S. Carpentier, R. Salvayre, T. Levade and N. Andrieu-Abadie. 2001. Glutathione peroxidase-1 overexpression prevents ceramide production and partially inhibits apoptosis in doxorubicin-treated human breast carcinoma cells. *Mol Pharmacol* **60**:488-496.

49. Kolesnick, R. 2002. The therapeutic potential of modulating the ceramide/sphingomyelin pathway. *J Clin Invest* **110**:3-8.

Chapter 20

REACTIVE OXYGEN AND NITROGEN SPECIES IN THE PRODUCTION OF CONGENITAL MALFORMATIONS BY KNOWN TERATOGENIC AGENTS AND MATERNAL CONDITIONS

Antonio F Machado, William J. Scott and Michael D. Collins

20.1 Introduction

The development of complex mammalian systems is a delicate interplay between the activation and inactivation of gene expression, between protein synthesis and degradation, between cellular proliferation and programmed cell death (apoptosis), and between the differentiation of tissues and the resorption of newly formed tissues. Not surprisingly, many of these intricate developmental processes have proven sensitive to alterations in oxygen or other free radical concentrations. The basic concept of reactive oxygen species (ROS) having the potential to perturb development has been developed and explored for at least two decades.[1-3] Numerous chemical teratogens have been proposed to work through the production of excessive ROS, and evidence supporting this idea has been collected for many of them. More recently, nitric oxide and other reactive nitrogen species (RNS) have been recognized as signaling molecules, modulating the DNA binding of redox-sensitive transcription factors, affecting the translation or stability of mRNA, or regulating posttranslational modifications, particularly those of immune-system components and adhesion molecules.[4] The following pages present some of the evidence that RONS are involved in the embryotoxicity of specific teratogenic

379

H. J. Forman, J. Fukuto, and M. Torres (eds.), Signal Transduction by Reactive Oxygen and Nitrogen Species: Pathways and Chemical Principles, ©2003, *Kluwer Academic Publishers. Printed in the Netherlands*

chemicals, as well as maternal diabetes. It should be recognized, however, that the precise teratogenic mechanisms remain unknown, despite years of scientific inquiry. At present, the data supporting a role for RONS in the production of congenital malformations is not sufficient to convince the authors that this is the specific mechanism by which these agents induce teratogenesis. The primary purpose of this chapter is to provide an overview of the field of RONS-mediated teratogenicity in order to stimulate thinking about congenital malformations as a toxicological end point resulting from the excessive generation of RONS.

20.2 Oxygen-Induced Teratogenesis

Oxygen, like many other nutrient substances, has been proposed to induce teratogenesis when it is in excess or in deficiency. However, what constitutes a normal oxygen level in the embryo is debatable, with recent work indicating that the entire period of organogenesis in the human embryo is a period of relative hypoxia.[5] Excess oxygen, and by implication ROS, perturb neurulation in cultured rodent embryos exposed to atmospheric oxygen levels or higher. In cultured rat embryos, the neural tube is the most sensitive tissue to this insult and, by maintaining sub-atmospheric oxygen levels until after closure of the neural tube, oxygen partial pressures can subsequently be increased to atmospheric concentrations (20%) or higher with no deleterious effects.[6-9] As demonstrated by electron microscopy, the mitochondria of embryos, grown *in vivo* or in 5% oxygen from gestational day 9.5 to 10.5, were of the anaerobic type (round in shape with few cristae), whereas embryos grown in either 20 or 40% oxygen during the culture period had mitochondria of the aerobic type (ovoid with many parallel cristae).[7] Using the same procedure with rat embryos of the same stage, Miki *et al.*[8] reported that the morphological characteristic found in embryos exposed to excessive oxygen was the formation of phagolysosomes in the cytoplasm. This difference in findings has not been resolved, but our own findings with mouse embryos have produced results comparable to those of Miki *et al.*

Rat embryos cultured in atmospheric oxygen levels on gestational day 9.5 showed increased intracellular hydrogen peroxide levels, as well as perturbations of the glutathione system; inhibitors of the glutathione pathway increased, and glutathione esters reduced the open neural tube phenotype.[9] Cultured rat embryos exposed to the combination of xanthine and xanthine oxidase to induce production of superoxide radical demonstrated irregular neural suture lines.[10] This is presumably a milder phenotype than an open neural tube. The use of various antioxidant species suggested that the damage was due to the effects of ROS. Thus, both excessive oxygen, which may be as little as atmospheric concentrations of O_2, and ROS may perturb the closure of the neural tube in cultured rat embryos (and our studies demonstrate that mouse embryos are likewise susceptible to atmospheric oxygen tensions).

The period of neurulation represents the initiation of a period of profound change in the rodent embryo which has been hypothesized to be related to the availability and utilization of oxygen.[3] This period is marked by increases in mitochondrial cristae and NADH oxidase activity and decreases in glucose utilization and lactate production, all of which are significantly affected by mbryonic oxygenation. Since the developing placental function is a dominant determinant of embryonic oxygenation, placentation may play a role in normal and abnormal development. This period of changing embryonic oxygenation is when most organogenesis occurs and is thus the period of highest sensitivity to chemically induced dysmorphogenesis. However, exposure to a number of human teratogens including ethanol, phenytoin, hydroxyurea and cocaine can induce hypoperfusion which is thought to cause limb reduction and possibly brain malformations when exposure occurs during the late fetal period.[11] Superoxide anion and hydrogen peroxide formation decreased during fetal development, but the hydroxyl radical increased. Tissue formation of ROS did not correspond to sensitivity to teratogenesis. There was an increase in the ratio of 8-hydroxy-2'-deoxyguanosine to 2-deoxyguanosine in tissues of fetuses after ischemia/reperfusion in utero. It is not clear if the reactive oxygen species resulted from ischemia or the reperfusion.

Relatively few studies have been able to demonstrate the *in vivo* teratogenicity of excess oxygen. Hamsters exposed to pressures between 3.0 and 4.0 atmospheres for 2-3 hours on gestational day 6, 7 or 8 had approximately 20% maternal central nervous system toxicity, and a small number of fetuses had the congenital malformations omphalocele, exencephaly, cleft lip, spina bifida, and hypoplasia of the hindlimb.[12] Resorption rates were not elevated after hyperbaric exposures. In the rabbit, exposures on a single day of gestation to 1.5-2.0 atmospheres for 5 hours (lower dose for a longer period) increased rates of both malformations and resorptions.[13] Oxygen concentrations of 97-100%, administered at 10-minute intervals over 15 hours, caused low incidences of eye defects, mortality and prematurity in rabbit fetuses.[14] Furthermore, the levels of ROS in various anatomical regions of midgestation mouse embryos correlated with those regions that were undergoing programmed cell death, such as interdigital regions of the limbs.[15] Interdigital apoptosis was inhibited by antioxidants.

Like increased oxygen, reduced oxygen concentrations or hypoxia has been shown to induce an increase in ROS (although some study conditions result in decreased ROS).[16] Reduced oxygen levels also have been found to cause congenital malformations. Reduced atmospheric pressure (260 to 280 mm Hg) or reduced oxygen concentration (6%) given on various days during murine gestation have been shown to induce hemivertebrae, fused ribs, cleft palate or cranioschisis.[17,18] The vertebral effects produced by hypoxia in this species were confirmed.[19] The rabbit has been reported to be susceptible to oxygen deficiency, with rib and vertebral malformations being produced.[20] Hypoxia during the fetal period produced by a number of techniques has been shown to initiate with an edema followed by hemorrhaging that leads to digital defects.[21] In addition, there are a number of studies demonstrating that increased CO_2 (hypercapnia), which has a reciprocal relationship with oxygen concentration, produces malformations in rats and rabbits, including cardiac, vertebral, limb and dentition defects.[22] The epidemiological literature has shown an association between births at high altitude and patent ductus arteriosus,[23] which is consistent with the idea that the increase in local oxygen tension supposedly induces the ductus arteriosus to constrict after birth.[24] Additional epidemiological evidence indicates an association between ventricular septal defects and reduced atmospheric pressure.[23]

20.3   Teratogenesis from Alterations in RNS

Only a few investigations have sought to examine the direct effects of RNS on the developing embryo. Lee and Juchau[25] microinjected the NO donor, sodium nitroprusside (SNP), into the amniotic fluid of cultured rat embryos and reported dose-dependent decreased survival, decreased somite counts, abnormalities in cephalic and caudal development, abnormal axial rotation, open otic vesicles, and slight increases in open rostral and caudal neuropores. The most striking observation was the presence of distinct patches of dead cells in the mesencephalon, closely associated with the neural tube. These patches were not seen in vehicle controls or controls injected with SNP breakdown products and may have been cells that had undergone apoptosis secondary to NO exposure.

The NOS inhibitor, L-NAME, induces prolonged hypertension and has been used to model preeclampsia.[26] The fetal effects of L-NAME, maternally administered in rats either orally or via subcutaneously implanted osmotic minipump, include weight reduction, hemorrhagic necrosis of the hindlimbs and limb reduction defects.[27,28] Similar results were obtained with another NOS inhibitor, N^{ω}-nitro-L-arginine (NO-Arg).[29] Protection against the embryotoxic effects of L-NAME were observed with oral L-arginine or with SNP or S-nitroso-acetylpenicillamine (SNAP), administered by subcutaneously implanted osmotic minipump.[27,28] Intraperitoneal (ip) administration of L-NAME to rat dams resulted in hemorrhaging and limb-reduction malformations in both fore- and hindlimbs of fetuses, effects that were ameliorated with the free radical spin trapping agent α-phenyl-N-*t*-butylnitrone (PBN).[30] More recent work shows that hemorrhaging and limb reduction defects produced by NOS inhibitors also can be ameliorated with allopurinol, a xanthine oxidase inhibitor.[31] Taken together, these studies suggest a role for excess ROS in the production of limb malformations by L-NAME.

HYDROXYUREA

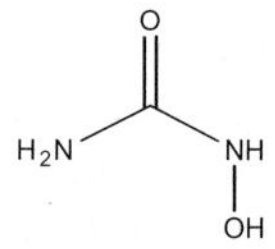

Figure 1. Chemical structure of hydroxyurea

The concept that chemical agents can cause congenital malformations via the initiation of free-radical chain reactions was first proposed in 1979 for the chemotherapeutic agent, hydroxyurea.[32] At that time, hydroxyurea was already a well-established teratogen and had proven capable of producing gross malformations in a variety of species.[33-36] Studies of the tumoricidal properties of hydroxyurea had revealed that this compound strongly inhibited DNA synthesis[37] and selectively killed cells in S-phase of the cell cycle.[38] These toxic properties of hydroxyurea were proposed as the mechanistic basis for the teratogenic effects.[39] Later studies, however, revealed that the time between maternal administration of hydroxyurea and cell death in embryonic limbs was as little as 30 minutes,[40,41] a time frame inconsistent with the proposed mechanism.[32] Thus, it was proposed[32] that since, at physiologic pH, hydroxyurea can participate in free radical chemistry,[42,43] rapid oxidative damage to macromolecules was a mechanism more consistent with the observed time course.

The identification and precise timing of the responses of the embryonic cardiovascular system to teratogenic doses of hydroxyurea were studied in rabbit embryos in situ by direct, *in vivo*, microscopic observation[44]. In nearly 90% of the embryos of exposed dams, the microvasculature of the craniofacial region exhibited vasodilation and hemorrhaging within 4 minutes. Pericardial hemorrhaging was observed shortly thereafter, followed by cardiac tamponade. Vasodilation and hemorrhaging were also observed in the forelimb buds. These types of cardiovascular changes were not observed in control animals or in animals treated with a sub-teratogenic dose of hydroxyurea.[44] Teratogenic doses of hydroxyurea also affect the maternal vasculature, resulting in a transient 77% reduction in uterine blood flow, which peaks at 3-4 minutes and returns to normal within 10 minutes.[45] Thus, embryotoxicity of this compound, as well as the structurally related compound, hydroxylamine, may be maternally mediated by uterine ischemia/reperfusion.[45,46] That teratogenicity may be directly mediated in the embryo was demonstrated by direct injection of hydroxyurea into the exocoelomic cavities of embryos resulting in similar cardiovascular effects.[47] The free-radical hypothesis has since been supported by experiments demonstrating that the radical scavenger propyl gallate reduces the embryotoxicity of hydroxyurea in a dose-dependent manner.[47,48] The antioxidants D-mannitol,[49] ethoxyquin, and nordihydroguaiaretic acid also ameliorate the developmental toxicity of hydroxyurea.[50]

Most recently, a structure-activity analysis of several developmental toxicants containing a hydroxylamine functional group, including hydroxyurea, was conducted in order to evaluate the hypothesis that the hydroxylamine moiety contributes to teratogenesis.[51] Specifically, the embryotoxicity of five hydroxylamine-containing compounds was compared to that of five structural analogues in which the hydroxylamine moiety had been replaced with an amine. These compounds were injected subcutaneously into rabbit dams or directly into the exocoelomic cavities of embryos. In all cases, the hydroxylamine-bearing compounds caused early abnormal cell death in embryonic forelimb buds, while the structurally similar, amine-bearing compounds had no observable effect. Propyl gallate ameliorated the cytotoxicity of the hydroxylamine-bearing compounds.[51] The authors proposed that the generation of excessive free radicals in embryonic tissues by the hydroxylamine moiety contributes to the observed developmental toxicity, but it must be recognized that data available thus far do not permit differentiation between a secondary effect of free radicals and a direct effect of hydroxylamine.

THALIDOMIDE

Figure 2. Chemical structure of thalidomide.

Thalidomide, (R)/(S)-α-phthalimidoglutarimide, is a potent human teratogen that was originally marketed as a sedative-hypnotic and antiemetic agent and that was epidemiologically associated with causation of the limb reduction defect, phocomelia.[52] Over 12,000 affected children were born with skeletal defects[53] after being exposed, between gestational days 23 and 38, to as little as a single dose.[54,55] Subsequent studies determined that thalidomide produced defects of the limbs, spine, eyes, ears, cranial nerves, lungs, larynx, trachea, heart and vasculature, gastrointestinal tract, and genitourinary tract, as well as other systems.[56] However, the fact that thalidomide embryopathy did not include cleft palate indicates that this potent teratogen does not perturb all developing systems.

Recently, thalidomide has been reintroduced as a therapeutic agent. Due to the broad range of biological effects of thalidomide on cytokine secretion, immune function, angiogenesis, cell adhesion, and cell proliferation, it is efficacious in treating a wide variety of dermatologic, infectious, autoimmune, and neoplastic diseases.[53] Thalidomide has been approved for the treatment of erythema nodosum leprosum, an inflammatory condition in patients with Hansen's disease.[57]

Although the primary biotransformation of thalidomide is nonenzymatic hydrolysis,[58] many studies have demonstrated that various biological activities are dependent on microsomal CYP-catalyzed metabolism.[59] The metabolism of thalidomide is complex. In humans, twelve hydrolysis products have been identified, hydroxylation is theoretically possible on five different carbon atoms, and there is a single asymmetric carbon yielding an R and an S enantiomer, thus producing more than 100 metabolites and degradation products.[60] Inexplicably, the individual enantiomers of thalidomide have significantly different biological activities than the racemic mixture.[61] The monooxygenase isoform CYP2C19, which is polymorphic in the human population, has been shown to hydroxylate thalidomide at the 5- and 5'- positions.[59] In rats, hydroxylation of thalidomide at the same positions was accomplished by CYP2C6, as well as the sex-specific enzyme CYP2C11. Despite the cumulative data indicating that hydroxylation is required for various biological activities of thalidomide, including teratogenesis, it is difficult to reconcile the fact that the rodent species that are not sensitive to teratogenesis produce more CYP metabolites than the humans and rabbits that are sensitive to thalidomide-induced embryopathy.[59]

Despite many years of investigation, the mechanism of action of thalidomide-induced birth defects is unknown. There are a number of hypothesized mechanisms for the teratogenic activity,[62] which include the same mechanisms as the anti-tumor activity of this agent. The leading hypotheses regarding anti-tumor activity are anti-angiogenesis and immune modulation, which may be related through the effects of thalidomide on cytokine secretion. One of the hypotheses regarding the mechanism of thalidomide activity in teratogenesis is that thalidomide perturbs (either increases or decreases) the formation of reactive oxygen species. The ideas that thalidomide produces biological damage by the mechanisms of anti-angiogenesis or reactive oxygen species are not mutually exclusive. It has been reported that thalidomide inhibits angiogenesis via the formation of hydroxyl radicals.[63] Early work in the reactive oxygen species theory of thalidomide activity did not yield uniform results. Thalidomide increased the production of the superoxide radical in normal mononuclear phagocytes,[64] but decreased superoxide radical, hydrogen peroxide and hydroxyl radical production in zymosan-stimulated polymorphonuclear leukocytes.[65] At the time of this research, it was not possible to suggest how a reduction of ROS could induce toxic sequelae, because the ROS were considered the proximate toxin. But it is now well documented that ROS function as physiologically relevant signaling molecules.

One of the major unresolved issues concerning thalidomide teratogenesis is the species-specificity of the response to this agent. Primates are generally sensitive, rodents are generally not sensitive, and rabbits have intermediate sensitivity. Namely, some strains of rabbits get thalidomide-like limb defects if exposed to doses several orders of magnitude higher that the doses required to produce the embryopathy in humans. In a species comparison, it was shown that thalidomide exposure results in the oxidation of embryonic DNA from rabbits but not from mice.[66] This implicates reactive oxygen processes which were substantiated by inhibiting both DNA oxidation and teratogenicity caused by thalidomide with PBN. However, the mechanism of action of thalidomide is not proposed to be genotoxicity.[67,68] As opposed to thalidomide metabolism by spontaneous hydrolysis or CYP activity, ROS were produced by prostaglandin H synthase in the rabbit model.[66,69] In a second study on this issue, rabbit limb bud cells produced quantitatively more ROS than rat limb bud cells, and, moreover, ROS were concentrated in the nuclei of rabbit cells but uniformly distributed in the rat.[70] Previous work had shown that thalidomide caused a depletion of glutathione in the visceral yolk sac of rabbit embryos but not in rat embryos grown in culture.[71] Glutathione depletion in the rat limb cells occurred in the cytosol but not the nuclei, whereas for the rabbit it occurred in both compartments.[70] Thus, thalidomide may induce dysmorphology via an increase in reactive oxygen species.

20.4 Ethanol

Fetal alcohol syndrome (FAS) is a well-recognized (if not easily diagnosed) syndrome that includes pre- and postnatal growth retardation, psychomotor abnormalities, including mental retardation,[72] and a characteristic set of craniofacial defects including an indistinct philtrum and narrow upper lip, a small chin, a low nasal bridge, and a short palpebral fissure.[73,74] It is not uncommon for children with FAS to exhibit malformations of the heart, joints and external genitalia, as well as urinary tract abnormalities.[73,75,76]

While the teratogenicity of ethanol in humans usually derives from chronic alcohol abuse throughout gestation,[77] and numerous animal models have been developed to mimic this pathology,[78-80] reports of certain specific structural malformations suggest that acute doses of ethanol are also toxic to the embryo during gastrulation and organogenesis. Animal studies in mice support this observation in that maternal treatment with one or two doses of ethanol intraperitoneally during gastrulation produced increased resorptions and the malformations, holoprosencephaly and exencephaly.[81] Exposure to ethanol in the same manner later in organogenesis resulted in increased resorptions, as well as craniofacial malformations analogous to those associated with FAS,[81] eye anomalies,[82] specific brain malformations, heart dysmorphology,[83] urinary tract abnormalities,[84] limb defects[85] cleft lip and palate, exencephaly, and skeletal malformations.[81,86] Fetal growth retardation and a spectrum of malformations also have been observed to result from maternal ethanol treatment, either orally or intraperitoneally, in pigs,[79] guinea pigs,[87] beagles,[88] rats,[89] and nonhuman primates.[90] Models for acute ethanol-induced embryopathy have also been developed in chick[91] and rodent embryo culture.[92] Dose dependencies for various ethanol-induced teratogenic effects have been reported both *in vivo*[86] and *in vitro*.[92]

Although it has been suggested that different mechanisms may be responsible for different developmental defects observed in FAS,[93] the generation of excessive reactive intermediates seems to be a unifying hypothesis that may underlie many, if not all, of the deleterious effects of ethanol on the developing embryo.[78,94] Ethanol treatment of pregnant rats by gastric intubation results in increased membrane lipid peroxidation, as indicated by increases in malondialdehyde levels and conjugated dienes, as well as decreased GSH levels in fetal brain and liver.[95] Reduced GSH levels in fetal brain and liver also results from administration of alcohol to pregnant dams in a liquid diet.[96] In cultured mouse embryos, ethanol induces superoxide generation, lipid peroxidation, excessive cell death, and dysmorphogenesis.[97] These effects can be partly ameliorated by the addition of CuZn SOD to the culture medium.

The generation of ROS secondary to ethanol treatment may be related to the effect of this teratogen on iron metabolism. Alcohol-dehydrogenase oxidation of ethanol requires the cofactor NAD^+, which is reduced to NADH. Additionally, the product of this reaction, acetaldehyde, is further oxidized by aldehyde oxidase or xanthine oxidase, both of which reactions generate superoxide.[98,99] Increased cellular levels of NADH or superoxide resulting from these reactions have been shown to mobilize cellular iron stores from ferritin.[100] Consistent with this observation, fetal[101] and neonatal liver[99] iron levels in the rat were found to be increased secondary to maternal ethanol treatment. Increases in free iron levels have been implicated in ethanol toxicity to cultured neural crest cells in that the iron chelators deferoxamine and phenanthroline are both protective.[99] This same study found that toxicity secondary to iron loading in neural crest cells could be ameliorated with the antioxidant N-acetylcysteine.

Many of the malformations identified in FAS are of structures derived from neural crest cells.[87,103] Neuronal cells, including premigratory neural crest cells, are particularly sensitive to ethanol toxicity both *in vivo*[104,105] and *in vitro*, and ethanol has been shown to induce the generation of superoxide, hydrogen peroxide and hydroxyl anions in primary cultures of chick neural crest cells.[106] The addition of superoxide dismutase to the cell culture reduced both ethanol toxicity and the generation of ROS. Interestingly, these investigators assayed neural crest cells for SOD activity and found none. Primary cultures of murine neural crest cells treated with teratogenic levels of ethanol demonstrate the production of the superoxide anion radical as evidenced by nitroblue tetrazolium staining.[107] In this same study, neural crest-cell death secondary to ethanol treatment was reduced by co-treatment with CuZn SOD, catalase, or α-tocopherol. In later developmental stages, microglia (the CNS macrophage) could be a significant source of ethanol-induced ROS/RNS. In cultured neonatal hamster microglia, ethanol causes a dose-dependent increase in superoxide release, as well as a depression of NO levels in microglia stimulated to produce NO by polyinosinic:polycytidylic acid[108]. Ethanol has been shown *in vivo* to inhibit iNOS protein levels post-transcriptionally in rat alveolar macrophages.[109]

PHENYTOIN

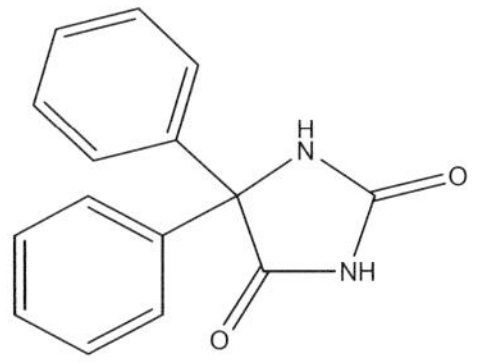

Figure 3. Chemical structure of phenytoin.

Phenytoin (5,5-diphenylhydantoin; Dilantin®) is a commonly prescribed anticonvulsant drug that exerts well-established teratogenic effects in humans (termed the fetal hydantoin syndrome),[110] as well as in a wide variety of experimental animal systems.[111] The fetal hydantoin syndrome includes a broad array of craniofacial defects, including microcephaly and cleft lip and/or palate, cardiovascular malformations, limb defects, including hypoplasia of the nails and distal phalanges, and mental retardation.[110] Most of these effects have been mimicked in animal models, and the behavioral effects of exposure to phenytoin *in utero* have also been studied extensively in animals.[112]

The metabolism of phenytoin *in vivo* is extensive, and very little of the parent compound is excreted unaltered.[113] Phenytoin metabolism produces an epoxide intermediate, formation of which is catalyzed by various CYP450s,[114] prostaglandin H synthase[115,116] and lipoxygenases.[117] This phenytoin reactive intermediate has the potential for electrophilic attack on embryonic macromolecules,[113,114] and maternally administered, radiolabeled phenytoin has been shown to bind covalently to both placental and embryonic tissues.[114] *In vitro*, protein binding of radiolabeled phenytoin correlates linearly with time, as well as protein and phenytoin concentrations.[118] Pretreatment of mice with inhibitors of cytochromes P450 or epoxide hydrolase also decreased covalent binding of phenytoin to proteins *in vitro*; pretreatment with P450 inducers had the opposite effect.[118] In human fetal liver slices, incubation with phenytoin results in decreased GSH concentrations, increased GSSG, and increases in the mRNA transcripts of *hGSTA1* and *p53*.[119] The significance of the transcriptional changes is that both genes are involved in cellular protection mechanisms against oxidative stress.

In addition to metabolism via an epoxide intermediate, maternally administered phenytoin may contribute to embryonic oxidative stress through one-electron oxidation to a free radical intermediate and/or the subsequent production of ROS.[115,116] This bioactivation of phenytoin can be catalyzed by prostaglandin H synthase,[115,120] lipoxygenases[117] and various other peroxidases.[115,121] The free radical intermediate, again, can bind to embryonic proteins, lipids or DNA.[116,120] This mechanism of phenytoin bioactivation may be favored in organogenesis-stage embryos, in which most cytochromes P450 are expressed at low levels relative to adult expression,[122] and the peroxidase enzymes noted above are expressed at levels commensurate with adult expression.[122,123]

Consistent with the hypothesis of a pro-oxidant mechanism for phenytoin teratogenicity is the observation that pretreatment with compounds that reduce cellular glutathione levels, including diethyl maleate,[124] acetaminophen[125] and buthionine sulfoximine,[124] all enhance the embryotoxic effects of phenytoin both *in vivo* and *in vitro*.[126] The peroxidase (prostaglandin synthetase and lipoxygenase) inhibitor, 5, 8, 11, 14-eicosatetraynoic acid, has been shown to ameliorate phenytoin embryotoxicity.[126]

RETINOIDS

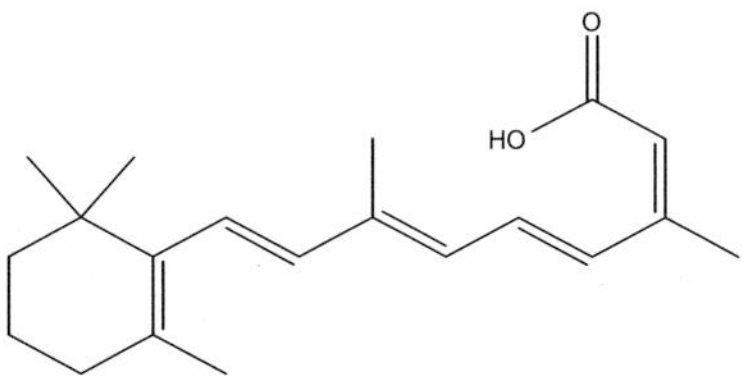

Figure 4. Chemical structure of retinoic acid.

Retinoids are vitamin A derivatives that regulate a range of biological processes including cell proliferation, differentiation, development and apoptosis.[127] Because of the important functions of retinoids during development, they are teratogenic in either excess or deficiency; thus, either hypovitaminosis A or hypervitaminosis A can lead to abnormal development.[128] Many of the activities of retinoids are believed to be mediated by the various forms of retinoic acid that serve as ligands for the two families of nuclear receptors which are members of the steroid hormone receptor superfamily.[129] Each of the two families of nuclear receptors, retinoic acid receptors (RARs , subclass NR1B) and retinoid X receptors (RXRs, subclass NR2B), has at least three genes coding for the nuclear proteins designated subtypes RAR-α, -β, -γ and RXR-α, -β, -γ. Each subtype has isoforms produced by alternative splicing and differential promoter usage. These receptors have different ligand specificity as indicated by the fact that RARs are activated by all-*trans*- and 9-*cis*-retinoic acid whereas the RXRs are only activated by 9-*cis*-retinoic acid.[130] Following ligand binding, the receptors form dimers (both homo- and hetero-) and function as transcriptional regulators. The ligand-activated dimers bind to response elements in the promoter region of target genes, with specific response elements for RARs called retinoic acid response elements (RAREs) and for RXRs called retinoid X response elements (RXREs).

Some retinoids seem to function without binding to the nuclear receptors. For example, specific retinoids can have unique cellular functions that are not shared with retinoic acid.[131] Candidates for cytoplasmic receptors include proteins of the serine/threonine kinase family, including cRaf and protein kinase C. These molecules can bind retinoids with high affinity and initiate signal transduction.[132] Retinoids that bind these cytoplasmic proteins can either enhance or inhibit the oxidative activation of these serine/threonine kinases.[133] Signal transduction has been shown to be mediated by ROS, with oxidative conditions leading to the release of zinc ions and initiation of a phosphorylation cascade.

The most potent natural retinoid teratogens are the acidic forms. RAR agonists are the most powerful at inducing dysmorphogensis, RXR agonists are relatively ineffective, and the mixed agonists have intermediate potency.[134] The agents that do not activate the nuclear receptors also have lower capacity to cause malformations.[135] Although it has been established that teratogenic retinoids bind to the nuclear RARs, the series of events following this binding have not been delineated although a number of hypotheses have been suggested including perturbed apoptosis, repatterning, differentiation, neural crest cell migration, proliferation, cellular induction, and inflammation.[128]

It is well documented that the pharmaceutical agents 13-*cis*-retinoid acid (isotretinoin, Accutane®), used for the treatment of cystic acne, and etretinate (Tigason®), used for the treatment of psoriasis, are human teratogens.[136,137] It has also been proposed that excess vitamin A intake can cause congenital malformations in humans,[138] because it has been teratogenic in every species tested. Laboratory animal studies have established that many and perhaps all developing organ systems can be malformed by excess retinoids, with a critical factor in determining the type of malformation being the time of administration as related to the developmental stage of a given organ.[139] Despite this, there is a human syndrome of retinoid embryopathy that includes craniofacial, cardiac, thymic and central nervous system structures.[136]

One of the inherent problems in trying to make an association between retinoid induction of ROS and retinoid induction of teratogenesis is that some of the retinoids that have been shown definitively to induce ROS are relatively weak teratogens. Thus, N-(4-hydroxyphenyl)retinamide (4HPR), a synthetic retinoid that has been found to effective against a variety of cancer cells,[140] has been shown to induce apoptosis via the formation of ROS in neoplastic cells,[141] as well as other mechanisms.[142] Furthermore, 4HPR induces apoptosis in cells with a point mutation in a retinoic acid receptor so that the cells are unresponsive to retinoic acid, indicating that this retinamide operates by a different mechanism than retinoic acid.[141] However, 4HPR has been shown to be a relatively weak teratogen in rats and rabbits.[144] Likewise, the *retro*-retinoid anhydroretinol, a physiological metabolite of vitamin A, has been shown to kill cells by generating ROS.[145] Once again, this agent is a relatively weak teratogen in mice.[135] If these agents are inducing teratogenesis via the formation of ROS then it may be expected that more potent retinoid teratogens would cause a more potent ROS response. In contrast to this expectation, it was shown that retinol, like alpha-tocopherol, inhibited anhydroretinol-induced oxidative stress.[145]

Even if the effects of retinol and anhydroretinol are secondary to ROS production, the finding that these agents have opposing actions is not unfathomable, given that many redox-active compounds may function as both pro- and anti-oxidants depending on such factors as dose, cell type, and redox status or reduction potential of the cell.[146] Retinoids have been shown to have both prooxidant[135,141,147] and antioxidant activity,[148-150] and it has been demonstrated that a single compound can have either function depending on the environment.[151] All-trans-retinol (like βcarotene) is a more effective antioxidant at low oxygen partial pressures.[150] This may be relevant because embryos develop in a relatively hypoxic environment.[5]

A number of transcription factors are thought to be controlled by redox status because of experiments which indicate that the level of DNA binding is altered by oxidative processes. The list of transcription factors includes NF-κB, AP-1, Pax5 and 8, p53, and others.[152] Although the *in vivo* relevance of this phenomenon is debatable, recent experiments have shown that oxidative conditions reduce the DNA binding of the RAR/RXR heterodimer and that reducing conditions enhance binding.[153] This indicates that ROS may be an upstream, in addition to the previously cited downstream impacts, regulator of retinoid activity. Furthermore, there may be cross-talk between ligand activity and receptor function for the retinoid pathway, as well as pathways for other transcription factors.

A primary target of retinoid teratogenesis is the population of embryonic neural crest cells. It has been demonstrated that isotretinoin and its 4-oxo- metabolite generated ROS in chick neural crest cells, and that the viability of the cells was reduced.[154] It was further shown that the neural crest cells were deficient in superoxide dismutase and catalase, and that supplementation with these antioxidant enzymes could reduce or ameliorate the effects of these retinoids. This is consistent with the more general principle that undifferentiated cells are less effective at detoxifying reactive oxygen species than are more differentiated cells.[155] It has been suggested that the differentiation-inducing activities of retinoids may be associated with oxidative processes.[156] Thus, although undifferentiated cells are unable to defend against ROS, the creation of these species pushes them to differentiate.

20.5 Maternal Diabetes

Human birth defects associated with maternal type 1 or type 2 diabetes is a well-established phenomenon. The risk of a congenital malformation resulting from a diabetic pregnancy is at least two and as much as ten times higher than normal.[157-161] Discrepancies in risk may be at least partly attributable to differences in racial backgrounds of the various populations being studied, a factor which has been reported to affect the rates of specific congenital malformations, particularly neural tube defects.[162,163]

Numerous investigations have examined the mechanism underlying diabetic embryopathy. Most investigators seem to agree that the etiology is probably multifactorial, with contributions from genetic background, the excessive generation of ROS, and numerous metabolic disturbances, including hyper- and hypoglycemia and alterations in the concentrations of various intermediary metabolites.[164-167] In recent years, however, many lines of inquiry have focused on oxidative stress-related hypotheses, which have been proposed as a unifying concept for the mechanism of diabetic embryopathy.[159] This concept essentially proposes that increased nutrients and altered intermediary metabolites supplied to the embryo during organogenesis can lead to the excessive generation of ROS[168] and/or impaired ROS scavenging,[169] further causing enhanced protein damage, lipid peroxidation, alterations in prostaglandin metabolism, and DNA damage.[170,171]

Several lines of evidence suggest that diabetic embryopathy may be mediated by ROS. First, both the increased resorptions and the developmental anomalies caused by maternal diabetes can be ameliorated by numerous antioxidants in a variety of experimental systems. For rats *in vivo*, the embryopathic effects of maternal diabetes can be reduced by dietary supplementation with vitamin C,[172] vitamin E,[174] butylated hydroxytoluene[174], GSH[175], or a combination of vitamins C and E.[176] Diabetic embryopathy in rats can also be ameliorated with daily, intraperitoneal injections throughout organogenesis of the antioxidant, lipoic acid.[177] Many of the foregoing studies also demonstrate decreases in lipid peroxidation and other biomarkers of oxidative stress, as well as cellular and subcellular indicators of embryopathy, secondary to antioxidant supplementation of diabetic dams. Treatment of diabetic dams with buthionine sulfoxamine, a specific inhibitor of γ-glutamylcysteine synthetase, the rate-limiting enzyme in GSH synthesis, depletes GSH and exacerbates diabetic embryopathy in rats.[175]

In both murine and rat embryo culture, D-glucose,[178] sera from diabetic animals[179] or pregnant women with diabetes,[180] and several intermediary metabolites that are elevated in diabetic serum are all teratogenic in a dose-dependent fashion. The latter metabolites include pyruvate, α-ketoisocaproate[164] and β-hydroxybutyrate,[181] as well as somatomedin inhibitors[181] and the α-oxoaldehyde 3-deoxyglucosone.[182] Increased resorption and malformation rates have also been correlated with elevated triglycerides, creatinine and various branched chain amino acids,[183] the levels of which are elevated in diabetic serum. A high level of glucose (50 mM) in rat embryo culture results in decreased oxygen uptake by embryos and increased generation of superoxide in neural tissue.[168] In rat embryos cultured in high glucose, the antioxidant enzymes superoxide dismutase (SOD), glutathione peroxidase and catalase,[165,184] and the antioxidants N-acteylcysteine,[185] citiolone (an SOD inducer)[166] and GSH,[169] reduce glucose-induced malformations. Additionally, expression of a Cu-Zn SOD transgene in mouse embryos results in increased SOD activity and decreased embryonic dysmorphology induced by high glucose *in vitro* or maternal diabetes *in vivo*.[186] SOD also ameliorates embryopathy caused by the above-noted intermediary metabolites, as does the inhibitor of mitochondrial pyruvate uptake, α-cyano-4-hydroxycinnamic acid.[164]

One of the expected effects of increased production of ROS in embryos is lipid peroxidation, which can lead to alterations in fatty acid metabolism[187] and prostaglandin synthesis.[188] Normal pregnancy in rats is associated with increased lipid peroxidation and protein carbonylation, and these effects are exacerbated by maternal diabetes.[188] Additionally, excess glucose has been shown in several embryo-culture studies to decrease concentrations of the membrane lipid, *myo*-inositol, in a dose-related fashion.[189] Supplementation of embryo culture medium with *myo*-inositol reduces the teratogenicity of hyperglycemia in cultured mouse[190] and rat[189,191] embryos. Likewise, dietary supplementation with *myo*-inositol reduces the teratogenic effect of maternal diabetes in rats.[192] Arachidonic acid has a similar effect in murine[193] and rat[187] embryo culture, as does dietary supplementation in rats with the polyunsaturated fatty acid, safflower oil (which has been shown to increase arachidonic acid levels).[194] Oral supplementation of diabetic rat dams with a cocktail containing *myo*-inositol, safflower oil, and α-tocopherol significantly reduced malformed embryos.[195]

A further line of evidence that diabetic embryopathy is mediated by ROS derives from differences in susceptibility of two different substrains of Sprague-

Dawley rat. The Uppsala (U) strain is highly susceptible to diabetic embryopathy, demonstrating increased resorptions and malformations, particularly micrognathia and sacral dysgenesis, whereas the Hanover (H) strain is completely resistant to the teratogenic effects of maternal diabetes.[196] This difference is correlated with the observation that the H strain has higher levels of catalase activity in the embryo proper, as well as the extraembryonic membranes, and that catalase activity is unaffected by maternal diabetes.[197] In contrast, the U strain not only has lower catalase activity normally, but these levels are significantly depressed by maternal diabetes in both the embryo and extraembryonic membranes. A further study with these strains[198] has demonstrated increased mRNA levels of Mn-SOD and catalase in the malformation-resistant H strain in response to maternal diabetes but no such increase in the susceptible U strain.

20.6 Concluding Remarks

The various sections of this chapter have outlined substantive evidence of the teratogenicity of oxygen and RNS, as well as that RONS may play a role in the induction of congenital malformations by hydroxyurea, thalidomide, ethanol, phenytoin, retinoids, and maternal diabetes. The evidence for RONS-mediated teratogenicity is not equally strong for all of the agents in this chapter, and other mechanisms have been proposed for all of them with varying degrees of supporting evidence. Nonetheless, the production of reactive intermediates is being increasingly recognized as a potential mechanism for the production of congenital malformations by a wide variety of previously identified teratogenic agents. Even if the initial teratogenic mechanism does not involve RONS, many teratogens could induce excessive RONS at some point in the pathogenic pathway. Since the majority of human birth defects are idiopathic[199] it seems likely that multiple insults, each impacting the embryo at a sub-threshold level for the production of a structural malformation, are combining to cause birth defects in genetically sensitive individuals. The production of reactive intermediates thus constitutes a reasonable common pathway by which structurally unrelated chemicals might additively or synergistically combine to produce a teratogenic effect.

References

1. DeSesso, J. M. 1979. Cell death and free radicals: a mechanism for hydroxyurea teratogenesis. *Med Hypotheses* **5(9)**:937-951.

2. Johnson, M. H. and M. H. Nasr-Esfahani. 1994. Radical solutions and cultural problems: could free oxygen radicals be responsible for the impaired developments of preimplantation mammalian embryos *in vitro*? *BioEssays* **16(1)**:31-38.

3. Fantel, A.G. and R. E. Person. 2002. Involvement of mitochondria and other free radical sources in normal and abnormal fetal development. *Ann NY Acad Sci* **959**:424-433.

4. Bogdan, C. 2001. Nitric oxide and the regulation of gene expression. *TRENDS in Cell Biology* **11(2)**:66-75.

5. Burton, G. J. and E. Jauniaux. 2001. Maternal vascularisation of the human placenta: does the embryo develop in a hypoxic environment. *Gynecol Obstet Fertil* **29**:1-6.

6. New, D. A. T., P. T. Coppola and D. L. Cockroft. 1976. Improved development of head-fold rat embryos in culture resulting from low oxygen and modifications of the culture serum. *J Reprod Fert* **48**:219-222.

7. Morriss, G. M. and D. A. T. New. 1979. Effect of oxygen concentration on morphogenesis of cranial neural folds and neural crest in cultured rat embryos. *J Embryol Exp Morph* **54**:17-35.

8. Miki, A., E. Fujimoto, T. Ohsaki and H. Mizoguti. 1988. Effects of oxygen concentration on embryonic development in rats: a light and electron microscopic study using whole-embryo culture techniques. *Anat Embryol* **178**:337-343.

9. Ishibashi, M., S. Akazawa, H. Sakamaki, K. Matsumoto, H. Yamasaki, Y. Yamaguchi, S. Goto, Y. Urata, T. Kondo and S. Nagataki. 1997. Oxygen-induced embryopathy and the significance of glutathione-dependent antioxidant system in the rat embryo during early organogenesis. *Free Rad. Biol Med* **22**:447-454.

10. Jenkinson, P. C., D. Anderson and S. D. Gangolli. 1986. Malformations induced in cultured rat embryos by enzymically generated active oxygen species. *Teratog Carcinog Mutagen* **6**:547-554.

11. Fantel, A. G., B. Mackler, L. D. Stamps, T. T. Tran and R. E. Person. 1998. Reactive oxygen species and DNA oxidation in fetal rat tissues. *Free Rad Biol Med* **25**:95-103.

12. Ferm, V. H. 1964. Teratogenic effects of hyperbaric oxygen. *Proc Soc Exp Biol Med* **116**:975-976.

13. Grote, W. and W. D. Wagner. 1973. Malformations in rabbits following hyperbaric oxygenation. *Klin Wochenschr* **51**:248-250.

14. Fujikura, R. 1964. Retrolental fibroplasia and prematurity in newborn rabbits induced by maternal hyperoxia. *Am J Obstet Gynecol* **90**:854-858.

15. Salas-Vidal, E., H. Lomelí, S. Castro-Obregón, R. Cuervo, D. Escalante-Alcalde and L. Covarrubias. 1998. Reactive oxygen species participate in the control of mouse embryonic cell death. *Exp Cell Res* 238:136-147.

16. Chandel, N. S. and P. T. Schumacker. 2000. Cellular oxygen sensing by mitochondria: old questions, new insights. *J Appl Physiol* **88**:1880-1889.

17. Ingalls, T. H., F. J. Curley and R. A. Prindle. 1952. Experimental production of congenital abnormalities: Timing and degree of anoxia as factors causing fetal deaths and congenital abnormalities in the mouse. *N Engl J Med* **247**:758-768.

18. Ingalls, T. H. and F. J. Curley. 1957. Principles governing the genesis of congenital malformations induced in mice by hypoxia. *N Engl J.Med* **257**:1121-1127.

19. Murakami, U. and Y. Kameyama. 1963. Vertebral malformations in the mouse foetus caused by maternal hypoxia during early stages of pregnancy. *J Embryol Exp Morphol* **11**:107-118.

20. Degenhardt, K. H. and E. Knoche. 1959. Analysis of intrauterine malformations of vertebral column induced by oxygen deficiency. *Can Med Assoc J* **80**:441-445.

21. Petter, , C., J. Bourbon, J. Maltier and A. Jost. 1971. Production d'hemorragies des extremites chez le foetus de rat soumis a une hypoxie in utero. *C. R. Acad. Sci. (D) (Paris)* **272**:2488-2490.

22. Schardein, J. L. 1993. *Chemically induced birth defects.* 2nd edition. Marcel Dekker, Inc., New York.

23. Warkany, J. 1971. *Congenital Malformations: Notes and Comments.* Yearbook Medical Publishers, Inc., Chicago.

24. Larsen, W. J. 2001. Human embryology. 3rd edition. Churchill Livingstone, New York.

25. Lee, Q. P. and M. R. Juchau. 1994. Dysmorphogenic effects of nitric oxide (NO) and NO-synthase inhibition: studies with intra-amniotic injections of sodium nitroprusside and NG-monomethyl-L-arginine. *Teratology* **49**:452-464.

26. Yallampalli, C. and R. E. Garfield. 1993. Inhibition of nitric oxide synthesis in rats during pregnancy produces signs similar to those of preeclampsia. *Am J Obstet Gynecol* **169**:1316-1320.

27. Diket, A. L., M. R. Pierce, U. K. Munshi, C. A. Voelker, S. Eloby-Childress, S. S. Greenberg, X. J. Zhang, D. A. Clark and M. J. Miller. 1994. Nitric oxide inhibition causes intrauterine growth retardation and hind-limb disruptions in rats. *Am J Obstet Gynecol* **171**:1243-1250.

28. Pierce, R. L., M. R. Pierce, H. Liu, P. J. Kadowitz and M. J. S. Miller. 1995. Limb reduction defects after prenatal inhibition of nitric oxide synthase in rats. *Pedatr Res* **38**:905-911.

29. Salas, S. P., F. Altermatt, M. Campos, A. Giacaman and P. Rosso. 1995. Effects of long-term nitric oxide synthesis inhibition on plasma volume expansion and fetal growth in the pregnant rat. *Hypertension* **26**:1019-1023.

30. Fantel, A. G., L. D. Stamps, T. T. Tran, B. Mackler, R. E. Person and N. Nekahi. 1999. Role of free radicals in the limb teratogenicity of L-NAME (N^{ω}-nitro-L-arginine methyl ester): a new mechanistic model of vascular disruption. *Teratology* **60**:151-160.

31. Fantel, A. G. and R. E. Person. 2002. Further evidence for the role of free radicals in the limb teratogenicity of L-NAME. *Teratology* **66**:24-32.

32. DeSesso, J. M. 1979. Cell death and free radicals: a mechanism for hydroxyurea teratogenesis. *Med Hypotheses* **5**:937-951.

33. Murphy, M. L. and S. Chaube. 1964. Preliminary survey of hydroxyurea (NSC-32065) as a teratogen. *Cancer Chemother Rep* **40**:1-7.

34. Chaube, S. and M. L. Murphy. 1966. The effects of hydroxyurea and related compounds on the rat fetus. *Cancer Res* **26**:1448-1457.

35. Wilson, J. G. 1971. Use of rhesus monkeys in teratological studies. *Fed Proc* **30**:104-109.

36. DeSesso, J. M. and R. L. Jordan. 1977. Drug-induced limb dysplasias in fetal rabbits. *Teratology* **15(2)**:199-212.

37. Philips, F. S., S. S. Sternberg, H. S. Schwartz, A. P. Cronin, J. E. Sodergren and P. M. Vidal. 1967. Hydroxyurea: I. Acute cell death in proliferating tissues in rats. *Cancer Res* **27**:61-74.

38. Sinclair, W. K. 1965. Hydroxyurea: differential lethal effects on cultured mammalian cells during the cell cycle. *Science* **150**:1729-1731.

39. Scott, W. J., E. J. Ritter and J. G. Wilson. 1971. DNA synthesis inhibition and cell death associated with hydroxyurea teratogenesis in rat embryos. *Dev Biol* **26(2)**:306-315.

40. Sadler, T. W. and R. R. Cardell. 1977. Ultrastructural alterations in neuroepithelial cells of mouse embryos exposed to cytotoxic doses of hydroxyurea. *Anat Rec* **188**:103-123.

41. DeSesso, J. M. 1978. Histologic and ultrastructural analysis of the actions of hydroxyurea and methotrexate on the rabbit limb bud. *Anat Rec* **190**:381.

42. Freese, E. and E. B. Freese. 1965. The oxygen effects on deoxyribonucleic acid inactivation by hydroxylamines. *Biochem* **4**:2419-2433.

43. Freese, E. B., J. Gerson, H. Taber, H. J. Rhaese and E. Freese. 1967. Inactivating DNA alterations induced by peroxides and peroxide-inducing agents. *Mutat Res* **4**:517-531.

44. Millicovsky, G. and J. M. DeSesso. 1980. Cardiovascular alterations in rabbit embryos in situ after a teratogenic dose of hydroxyurea: an in vivo microscopic study. *Teratology* **22(1)**:115-124.

45. Millicovsky, G., J. M. DeSesso, L. I. Kleinman and K. E. Clark. 1981. Effects of hydroxyurea on hemodynamics of pregnant rabbits: a maternally mediated mechanism of embryotoxicity. *Am J Obstet Gynecol* **140**:747-752.

46. DeSesso, J. M. and G. C. Goeringer. 1990. Developmental toxicity of hydroxylamine: an example of a maternally mediated effect. *Toxicol Ind Health* **6**:109-121.

47. DeSesso, J. M. and G. C. Goeringer. 1990. The nature of the embryo-protective interaction of propyl gallate with hydroxyurea. *Reprod Toxicol* **4(2)**:145-152.

48. DeSesso, J. M. 1981. Amelioration of teratogenesis. I. Modification of hydroxyurea-induced teratogenesis by the antioxidant propyl gallate. *Teratology* **24**:19-35.

49. DeSesso, J. M., A. R. Scialli and G. C. Goeringer. 1994. D-mannitol, a specific hydroxyl free radical scavenger, reduces the developmental toxicity of hydroxyurea in rabbits. *Teratology* **49(4)**:248-259.

50. DeSesso, J. M. and G. C. Goeringer. 1990c. Ethoxyquin and nordihydroguaiaretic acid reduce hydroxyurea developmental toxicity. *Reprod Toxicol* **4(4)**:267-275.

51. DeSesso, J. M., C. F. Jacobson, A. R. Scialli and G. C. Goeringer. 2000. Hydroxylamine

moiety of developmental toxicants is associated with early cell death: a structure-activity analysis. *Teratology* **62**:346-355.

52. Eriksson, T., S. Björkman and P. Höglund. 2001. Clinical pharmacology of thalidomide. *Eur J Clin Pharmacol* **57**:365-376.

53. Mujagić, H., B. A. Chabner and Z. Mujagić. 2002. Mechanisms of action and potential therapeutic uses of thalidomide. *Croat Med J* **43**:274-285.

54. Lenz, W. and K. Knapp. 1962. Foetal malformations due to thalidomide. *Ger Med Monthly* **7**:253-258.

55. Neubert, R. and D. Neubert. 1997. Peculiarities and possible mode of action of thalidomide. In *Handbook of Experimental Pharmacology* vol. **124** (Kavlock, R. J. and G. P. Daston, eds.), pp. 41-119, Springer, Berlin.

56. Newman, C. G. H. 1985. Teratogen update: Clinical aspects of thalidomide embryopathy – A continuing preoccupation. *Teratology* **32**:133-144.

57. Friedman, J. M. and C. A. Kimmel. 1999. Teratology Society 1998 Public Affairs Committee Symposium: The new thalidomide era: Dealing with the risks. *Teratology* **59**:120-123.

58. Teo, S. K., P. J. Sabourin, K. O'Brien, K. A. Kook and S. D. Thomas. 2000. Metabolism of thalidomide in human microsomes, cloned human cytochrome P-450 enzymes, and Hansen's disease patients. *J Biochem Mol Toxicol* **14(3)**:140-147.

59. Ando, Y., E. Fuse and W. D. Figg. 2002. Thalidomide metabolism by the CYP2C subfamily. *Clin Cancer Res* **8**:1964-1973.

60. Schumacher, H., R. L. Smith and R. T. Williams. 1965. The metabolism of thalidomide: the spontaneous hydrolysis of thalidomide in solution. *Br J Pharmacol* **25**:324-337.

61. Eriksson, T., S. Björkman, B. Roth and P. Höglund. 2000. Intravenous formulations of the enantiomers of thalidomide: pharmacokinetic and initial pharmacodynamic characterization in humans. *J Pharm Pharmacol* **52**:807-817.

62. Stephans, T. 1988. Proposed mechanisms of action of thalidomide. *Teratology* **38**:229-239.

63. Sauer, H., J. Günther, J. Hescheler and M. Wartenberg. 2000. Thalidomide inhibits angiogenesis in embryoid bodies by the generation of hydroxyl radicals. *Am J Pathol* **156(1)**:151-158.

64. Nielsen, H. and T. Bennike. 1984. Thalidomide enhances defective monocyte function in lepromatous leprosy. *Lancet* **2**:98-99.

65. Miyachi, Y., M. Ozaki, K. Uchida and Y. Niwa. 1982. Effects of thalidomide on the generation of oxygen intermediates by zymosan-stimulated normal polymorphonuclear leukocytes. *Arch Dermatol Res* **274**:363-367.

66. Parman, T., M. J. Wiley and P. G. Wells. 1999. Free radical-mediated oxidative DNA

damage in the mechanism of thalidomide teratogenicity. *Nat Med* **5(5)**:582-585.

67. Ashby, J., H. Tinwell, R. D. Callander, I. Kimber, P. Clay, S. M. Galloway, R. B. Hill, S.. K. Greenwood, M. E. Gaulden, M. J. Ferguson, E. Vogel, M. Nivard, J. M. Parry and J. Williamson. 1997. Thalidomide: lack of mutagenic activity across phyla and genetic endpoints. *Mutat Res* **396(1-2)**:45-64.

68. Teo, S., M. Morgan, D. Stirling and S. Thomas. 2000. Assessment of the in vitro and in vivo genotoxicity of Thalomid (thalidomide). *Teratog Carcinog Mutagen* **20(5)**:301-311.

69. Arlen, R. R. and P. G. Wells. 1996. Inhibition of thalidomide teratogenicity by acetylsalicylic acid: evidence for prostaglandin H synthase-catalyzed bioactivation of thalidomide to a teratogenic reactive intermediate. *J Pharmacol Exp Ther* **277(3)**:1649-1658.

70. Hansen, J. M., K. K. Harris, M. A. Philbert and C. Harris. 2002. Thalidomide modulates nuclear redox status and preferentially depletes glutathione in rabbit limb versus rat limb. *J Pharmacol Exp Ther* **300(3)**:768-776.

71. Hansen, J. M., E. W. Carney, and C. Harris. 1999. Differential alteration by thalidomide of the glutathione content of rat vs. rabbit conceptuses in vitro. *Reprod Toxicol* **13**:547-554.

72. Mattson, S. N., A. M. Schoenfeld and E. P. Riley. 2001. Teratogenic effects of alcohol on brain and behavior. *Alcohol Res Health* **25(3)**:185-191.

73. Jones, K. L. and D. W. Smith. 1973. Recognition of the fetal alcohol syndrome in early infancy. *Lancet* **2(7836)**:999-1001.

74. Jones, K. L., D. W. Smith, C. N. Ulleland and A. P. Streissguth. 1973. Pattern of malformation in offspring of chronic alcoholic mothers. *Lancet* **1(7815)**:1267-1271.

75. Havers, W., F. Majewski, H. Olbing and H. U. Eickenberg. 1980. Anomalies of the kidneys and genitourinary tract in alcohol embryopathy. *J Urol* **124**:108-110.

76. Sandor, G. G., D. F. Smith and P. M. MacLeod. 1981. Cardiac malformations in the fetal alcohol syndrome. *J Pediatr* **98**:771-773.

77. Abel, E. L. 1998. *Fetal alcohol abuse syndrome*. Plenum Press, New York.

78. Chernoff, G. F. 1977. The fetal alcohol syndrome in mice: an animal model. *Teratology* **15(3)**:223-230.

79. Dexter, J. D., M. E. Tumbleson, J. D. Decker and C. C. Middleton. 1979. Morphologic comparisons of piglets from first and second litters in chronic ethanol-consuming Sinclair (S-1) miniature dams. *Alcohol Clin Exp Res* **3**:171-

80. Henderson, G. I., A. M. Hoyumpa, C. McClain and S. Schenker. 1979. The effects of chronic and acute alcohol administration on fetal development in the rat. *Alcohol Clin Exp Res* **3**:99-106.

81. Sulik, K. K., M. C. Johnston and M. A. Webb. 1981. Fetal alcohol syndrome: embryogenesis in a mouse model. *Science* **214**:936-938.

82. Cook, C. S., A. Z. Nowotny and K. K. Sulik. 1987. Fetal alcohol syndrome. Eye malformations in a mouse model. *Arch Ophthalmol* **105(11)**:1576-1581.

83. Daft, P. A., M. C. Johnston and K. K. Sulik. 1986. Abnormal heart and great vessel development following acute ethanol exposure in mice. *Teratology* **33(1)**:93-104.

84. Gage, J. C. and K. K. Sulik. 1991. Pathogenesis of ethanol-induced hydronephrosis and hydroureter as demonstrated following in vivo exposure of mouse embryos. *Teratology* **44(3)**:299-312.

85. Kotch, L. E., D. B. Dehart, A. J. Alles, N. Chernoff and K. K. Sulik. 1992. Pathogenesis of ethanol-induced limb reduction defects in mice. *Teratology* **46(4)**:323-332.

86. Kotch, L. E. and K. K. Sulik. 1992. Experimental fetal alcohol syndrome: proposed pathogenic basis for a variety of associated facial and brain anomalies. *Am J Med Genet* **44**:168-176.

87. Papara-Nicholson, D. and I. R. Telford. 1975. Effects of alcohol on reproduction and fetal development in the guinea pig. *Anat Rec* **127**:438-439.

88. Ellis, F. W. and J. R. Pick. 1980. An animal model of the fetal alcohol syndrome in beagles. *Alcohol Clin Exp Res* **4**:123-133.

89. Ismail, M. and M. Z. Janjua. 2001. Craniofacial alterations in adult rats after acute prenatal alcohol exposure. *J Ayub Med Coll Abbottabad* **13(3)**:7-10.

90. Astley, S. J., S. I. Magnuson, L. M. Omnell and S. K. Clarren. 1999. Fetal alcohol syndrome: changes in craniofacial form with age, cognition, and timing of ethanol exposure in the macaque. *Teratology* **59(3)**:163-172.

91. Fang, T.-T., H. J. Bruyere Jr., S. A. Kargas, T. Nishikawa, Y. Takagi and E. F. Gilbert. 1987. Ethyl alcohol-induced cardiovascular malformations in the chick embryo. *Teratology* **35**:95-103

92. Wynter, J. M., D. A. Walsh, W. S. Webster, S. E. McEwen and A. H. Lipson. 1983. Teratogenesis after acute alcohol exposure in cultured rat embryos. *Teratogen Carcinogen Mutagen* **3**:421-428

93. Randall, C. L. 2001. Alcohol and pregnancy: highlights from three decades of research. *J Stud Alcohol* **62(5)**:554-561

94. Henderson, G. I., B. G. Devi, A. Perez and S. Schenker. 1995. In utero ethanol exposure elicits oxidative stress in the rat fetus. *Alcohol Clin Exp Res* **19(3)**:714-720

95. Reyes, E., S. Ott and B. Robinson. 1993. Effects of in utero administration of alcohol on glutathione levels in brain and liver. *Alcohol Clin Exp Res* **17(4)**:877-881

96. Kotch, L. E., S.-Y. Chen and K. K. Sulik. 1995. Ethanol-induced teratogenesis: free radical damage as a possible mechanism. *Teratology* **52(3)**:128-136

97. Shaw, S., E. Jayatilleke, V. Herbert and N. Colman. 1989. Cleavage of folates during ethanol metabolism: role of acetaldehyde-xanthine oxidase generated superoxide.

Biochem J **257**:277-280

98. Shaw, S. and E. Jayatilleke. 1990. The role of aldehyde oxidase in ethanol-induced hepatic lipid peroxidation in the rat. *Biochem J* **268**:579-583

99. Shaw, S., E. Jayatilleke and C. S. Lieber. 1988. Lipid peroxidation as a mechanism of alcoholic liver injury: role of iron mobilization and microsomal induction. *Alcohol* **5**:135-140

100. Mendelson, R. A. and A. M. Huber. 1994. The effect of duration of alcohol administration on the deposition of trace elements in the fetal rat. *Adv Exp Med Biol* **132**:295-304

101. Sanchez, J., M. Casas and R. Rama. 1988. Effect of chronic ethanol administration on iron metabolism in the rat. *Eur J Haematol* **41**:321-325

102. Chen, S.-Y. and K. K. Sulik. 2000. Iron-mediated free radical injury in ethanol-exposed mouse neural crest cells. *J Pharmacol Exp Ther* **294(1)**:134-140

103. Cartwright, M. M. and Smith S. M. 1995. Increased cell death and reduced neural crest cell numbers in ethanol-exposed embryos: partial basis for the fetal alcohol syndrome phenotype. *Alcohol Clin Exp Res* **19(2)**:378-386

104. Kotch, L. E. and K. K. Sulik. 1992. Patterns of ethanol-induced cell death in the developing nervous sytem of mice; neural fold states through the time of anterior neural tube closure. *Int J Devl Neuroscience* **10(4)**:273-279

105. Dunty Jr., W. C., S. Y. Chen, R. M. Zucker, D. B. Dehart and K. K. Sulik. 2001. Selective vulnerability of embryonic cell populations to ethanol-induced apoptosis: implications for alcohol-related birth defects and neurodevelopmental disorder. *Alcohol Clin Exp Res* **25(10)**:1523-1535

106. Davis, W. L., L. A. Crawford, O. J. Cooper, G. R. Farmer, D. L. Thomas and B. L. Freeman. 1990. Ethanol induces the generation of reactive free radicals by neural crest cells in vitro. *J Craniofac Genet Dev Biol* **10**:277-293

107. Chen, S. Y. and K. K. Sulik. 1996. Free radicals and ethanol-induced cytotoxicity in neural crest cells. *Alcohol Clin Exp Res* **20(6)**:1071-1076

108. Colton, C. A., J. Snell-Callanan and O. N. Chernyshev. 1998. Ethanol induced changes in superoxide anion and nitric oxide in cultured microglia. *Alcohol Clin Exp Res* **22(3)**:710-716

109. Zhao, X., O. Jie, H. Li, J. Xie, T. D. Giles, S. S. Greenberg. 1997. Ethanol inhibits inducible nitric oxide synthase transcription and post-transcriptional processes in vivo. *Alcohol Clin Exp Res* **21(7)**:1246-1256

110. Hanson, J. W. 1986. Teratogen update: fetal hydantoin effects. *Teratology* **33**:349-353.

111. Finnell, R. H. and L. V. Dansky. 1991. Parental epilepsy, anticonvulsant drugs, and reproductive outcome: epidemiologic and experimental findings spanning three decades.

1. Animal studies. *Reprod Toxicol* **5**:281-299

112. Tachibana, T., Y. Terada, K. Fukunishi and T. Tanimura. 1996. Estimated magnitude of behavioral effects of phenytoin in rats and its reproducibility: a collaborative behavioral teratology study in Japan. *Physiol Behav* **60(3)**:941-952

113. Wells, P. G. and R. D. Harbison. 1980. Significance of the phenytoin reactive arene oxide intermediate, its oxepin tautomer, and clinical factors modifying their roles in phenytoin-induced teratology. In *Phenytoin-Induced Teratology and Gingival Pathology* (Hassell, T. M., M. C. Johnston, and K. H. Dudley, eds.), pp. 83-108, Raven Press, New York

114. Martz, F., C. Failinger III and D. A. Blake. 1977. Phenytoin teratogenesis: correlation between embryopathic effect and covalent binding of putative arene oxide metabolite in gestational tissue. *J Pharmacol Exp Ther* **203(1)**:231-239

115. Kubow, S. and P. G. Wells. 1989. *In vitro* bioactivation of phenytoin to a reactive free radical intermediate by prostaglandin synthetase, horseradish peroxidase, and thyroid peroxidase. *Mol Pharmacol* **35**:504-511

116. Parman, T., G. Chen and P. G. Wells. 1998. Free radical intermediates of phenytoin and related teratogens. Prostaglandin H synthase-catalyzed bioactivation, electron paramagnetic resonance spectrometry, and photochemical product analysis. *J Biol Chem* **273(39)**:25079-25088

117. Yu, W. K. and P. G. Wells. 1995. Evidence for lipoxygenase-catalyzed bioactivation of phenytoin to a teratogenic reactive intermediate: *in vitro* studies using linoleic acid-dependent soybean lipoxygenase, and *in vivo* studies using pregnant CD-1 mice. *Toxicol Appl Pharmacol* **131**:1-12

118. Roy, D. and W. R. Snodgrass. 1990. Covalent binding of phenytoin to protein and modulation of phenytoin metabolism by thiols in A/J mouse liver microsomes. *J Pharmacol Exp Ther* **252(3)**:895-900

119. Gallagher, E. P. and K. M. Sheehy. 2001. Effects of phenytoin on glutathione status and oxidative stress biomarker gene mRNA levels in cultured precision human liver slices. *Toxicol Sci* **59**:118-126

120. Liu, L. and P. G. Wells. 1995. DNA oxidation as a potential molecular mechanism mediating drug-induced birth defects: phenytoin and structurally related teratogens initiate the formation of 8-hydroxy-2'-deoxyguanosine in vitro and in vivo in murine maternal hepatic and embryonic tissues. *Free Rad Biol Med* **19(5)**:639-648

121. Uetrecht, J. and N. Zahid. 1988. *N*-chlorination of phenytoin by myeloperoxidase to a reactive metabolite. *Chem Res Toxicol* **1**:148-151

122. Wells, P. G. and L. M. Winn. 1996. Biochemical toxicology of chemical teratogenesis. *Crit Rev Biochem Mol Biol* **31(1)**:1-40

123. Parman, T. and P. G. Wells. 2002. Embryonic prostaglandin H synthase-2 (PHS-2) expression and benzo[*a*]pyrene teratogenicity in PHS-2 knockout mice. *FASEB J* **16**:1001-1009

124. Wong, M., L. M. J. Helston and P. G. Wells. 1989. Enhancement of murine phenytoin teratogenicity by the gamma-glutamylcysteine synthetase inhibitor buthionine sulfoximine and the glutathione depletor diethylmaleate. *Teratology* **40**:127-141

125. Lum, J. T. and P. G. Wells. 1986. Pharmacological studies on the potentiation of phenytoin teratogenicity by acetaminophen. *Teratology* **33**:53-72

126. Miranda, A. F., M. J. Wiley and P. G. Wells. 1994. Evidence for embryonic peroxidase-catalyzed bioactivation and glutathione-dependent cytoprotection in phenytoin teratogenicity: modulation by eicosatetraynoic acid and buthionine sulfoximine in murine embryo culture. *Toxicol Appl Pharmacol* **124**:230-241

127. Rogers, M. B. 1997. Life-and-death decisions influenced by retinoids. *Curr Top Dev Biol* **35**:1-46

128. Collins, M. D. and G. E. Mao. 1999. Teratology of retinoids. *Annu Rev Pharmacol Toxicol* **39**:399-430

129. Nuclear Receptors Committee. 1999. A unified nomenclature system for the nuclear receptor superfamily. *Cell* **97(2)**:161-163

130. Chambon, P. 1996. A decade of molecular biology of retinoic acid receptors. *FASEB J* **10**:940-954

131. Chen, Y., F. Derguini and J. Buck. 1997. Vitamin A in serum is a survival factor for fibroblasts. *Proc Natl Acad Sci USA* **94(19)**:10205-10208

132. Imam, A., B. Hoyos, C. Swenson, E. Levi, R. Chua, E. Viriya and U. Hammerling. 2001. Retinoids as ligands and coactivators of protein kinase C alpha. *FASEB J.* **15**:28-30

133. Hoyos, B., A. Imam, I. Korichneva, E. Levi, R, Chua and U. Hammerling. 2002. Activation of c-Raf kinase by ultraviolet light: regulation by retinoids. *J. Biol. Chem.* **277(26)**:23949-23957

134. Kochhar, D. M., H. Jiang, J. D. Penner, R. L. Beard and R. A. S. Chandraratna. 1996. Differential teratogenic response of mouse embryos to receptor selective analogs of retinoic acid. *Chem Biol Interact* **100**:1-12

135. Mao, G. E., M. D. Collins and F. Derguini. 2000. Teratogenicity, tissue distribution, and metabolism of the retro-retinoids: 14-hydroxy-4,14-retro-retinol and anhydroretinol, in the C57BL/6J mouse. *Toxicol Appl Pharmacol* **163**:38-49.

136. Lammer, E. J., D. T. Chen, R. M. Hoar, N. D. Agnish, P. J. Benke, J. T. Braun, C. J. Curry, P. M. Fernhoff, A. W. Grix Jr. and I. T. Lott. 1985. Retinoic acid embryopathy. *N Engl J Med* **313**:837-841

137. Rosa, F. W., A. L. Wilk and F. O. Kelsey. Teratogen update: vitamin A congeners. *Teratology* **33**:355-364

138. Rothman, K. J., L. L. Moore, M. R. Singer, U. S. Nguyen, S. Mannino and A. Milunsky. 1995. Teratogenicity of high vitamin A intake. *N. Engl. J. Med.* **333**:1369-1373

139. Shenefelt, R. E. 1972. Morphogenesis of malformations in hamsters caused by retinoic

acid: relation to dose and stage of treatment. *Teratology* **5**:103-118.

140. Fontana, J. A. and A. K. Rishi. 2002. Classical and novel retinoids: their targets in cancer therapy. *Leukemia* **16**:463-472

141. You, K. R., J. Wen, S. T. Lee and D. G. Kim. 2002. Cytochrome c oxidase subunit III: a molecular marker for N-(4-hydroxyphenyl)retinamide-induced oxidative stress in hepatoma cells. *J Biol Chem* **277(6)**:3870-3877.

142. Wu, J. M., A. M. DiPietrantonio and T. C. Hsieh. 2001. Mechanism of fenretinide (4-HPR)-induced cell death. *Apoptosis* **6(5)**:377-388

143. Delia, D., A. Aiello, L. Lombard, P. G. Pelicci, F. Grignani, F. Formelli, S. Menard, A. Costa, U. Veronesi and M Pierroti. 1993. N-(4-hydroxyphenyl)retinamide induces apoptosis of malignant hemopoietic cell lines including those unresponsive to retinoic acid. *Cancer Res* **53**:6036-6041

144. Kenel, M. F., J. H. Krayer, E. A. Merz and J. F. Pritchard. 1988. Teratogenicity of N-(4-hydroxyphenyl)-all-trans-retinamide in rats and rabbits. *Teratogen. Carcinogen. Mutagen.* **8**:1-11

145. Chen, Y., J. Buck and F. Derguini. 1999. Anhydroretinol induces oxidative stress and cell death. *Can. Res.* **59(16)**:3985-3990

146. Halliwell B. 2000. The antioxidant paradox. *Lancet* **355(9210)**:1179-1180

147. Murata, M. and S. Kawanishi. 2000. Oxidative DNA damage by vitamin A and its derivative via superoxide generation. *J Biol Chem* **275(3)**:2003-2008

148. Ahlenmeyer, B., E. Bauerbach, M. Plath, M. Steuber, C. Heers, F. Tegtmeier and J. Krieglstein. 2001. Retinoic acid reduces apoptosis and oxidative stress by preservation of SOD protein level. *Free Rad Biol Med* **30(10)**:1067-1077

149. Samokyszyn, V. M. and L. J. Marnett. 1990. Inhibition of liver microsomal lipid peroxidation by 13-cis-retinoic acid. *Free Rad Biol Med* **8(5)**:491-496

150. Tesoriere, L., D. D'Arpa, R. Re and M. A. Livrea. 1997. Antioxidant reactions of all-*trans*-retinol in phospholipid bilayers: effect of oxygen partial pressure, radical fluxes, and retinol concentration. *Arch Biochem Biophys* **343(1)**:13-18

151. Samokyszyn, V. M., M. A. Freyaldenhoven, H. C. Chang, J. P. Freeman and R. L. Compadre. 1997. Regiospecificity of peroxyl radical addition to (E)-retinoic acid. *Chem Res Toxicol* **10**:795-801

152. Marshall, H. E., K. Merchant and J. S. Stamler. 2000. Nitrosation and oxidation in the regulation of gene expression. *FASEB J* **14**:1889-1900.

153. Demary, K., L. Wong, J. S. Liou, D. V. Faller and R. A. Spanjaard. 2001. Redox control of retinoic acid receptor activity: a novel mechanism for retinoic acid resistance in melanoma cells. *Endocrinology* **142(6)**:2600-2605

154. Davis, W. L., B. H. Jacoby, G. R. Farmer and O. J. Cooper. 1991. Changes in cytosolic

calcium, bleb formation, and cell death in neural crest cells treated with isotretinoin and 4-oxo-isotretinoin. *J Craniofac Genet Dev Biol* **11(2)**:105-118

155. Covacci, V., A. Torsello, P. Palozza, A. Sgambato, G. Romano, A. Boninsegna, A. Cittadini and F. I. Wolf. 2001. DNA oxidative damage during diffentiation of HL-60 human promyelocytic leukemia cells. *Chem Res Toxicol* **14(11)**:1492-1497

156. Mäntymaa, P., T. Guttorm, T. Siitonen, M. Säily, E. R. Savolainen, A. L. Levonen, V. Kinnula and P. Koistinen. 2000. Cellular redox state and its relationship to the inhibition of clonal cell growth and the induction of apoptosis during all-trans retinoic acid exposure in acute myeloblastic leukemia cells. *Haematologica* **85**:238-245

157. Kucera, J. 1971. Rate and type of congenital anomalies among offspring of diabetic women. *J Reprod Med* **7(2)**:61-70

158. Freinkel, N. 1988. Diabetic embryopathy and fuel-mediated organ teratogenesis: lessons from animal models. *Horm Metab Res* **20**:463-475

159. Eriksson, U. J. 1995. The pathogenesis of congenital malformations in diabetic pregnancy. *Diabetes Metab Rev* **11**:63-82

160. Casson, I. F., C. A. Clarke, V. Howard, O. McKendrick, S. Pennycook, P. O. D. Pharoah, M. J. Platt, M. Stanisstreet, D. van Velszin and S. Walkinshaw. 1997. Outcomes of pregnancy in insulin dependent diabetic women: results of a five year population cohort study. *BMJ* **315**:275-278

161. Suhonen, L., V. Hiilesmaa and K. Teramo. 2000. Glycaemic control during early pregnancy and fetal malformations in women with type I diabetes mellitus. *Diabetologia* **43**:79-82

162. Shaw, G. M., N. G. Jensvold, C. R. Wasserman and E, J. Lammer. 1994. Epidemiologic characteristics of phenotypically distinct neural tube defects among 0.7 million California births, 1983-1987. *Teratology* **49(2)**:143-149

163. Canfield, M. A., J. F. Annegers, J. D. Brender, S. P. Cooper and F. Greenberg. 1996. Hispanic origin and neural tube defects in Houston/Harris County, Texas. I. Descriptive epidemiology. *Am J Epidemiol* **143(1)**:1-11

164. Eriksson, U. J. and L. A. H. Borg. 1993. Diabetes and embryonic malformations: role of substrate-induced free oxygen radical production for dysmorphogenesis in cultured rat embryos. *Diabetes* **42**:411-419

165. Sadler, T. W., E. S. Hunter III, R. E. Wynn and L. S. Phillips. 1989. Evidence for multifactorial origin of diabetes-induced embryopathies. *Diabetes* **38**:70-74

166. Eriksson, U. J. and L. A. H. Borg. 1991. Protection by free oxygen radical scavenging enzymes against glucose-induced embryonic malformations in vitro. *Diabetologia* **34**:325-331

167. Reece, E. A., C. J. Homko and Y. K. Wu. 1996a. Multifactorial basis of the syndrome of diabetic embryopathy. *Teratology* **54**:171-182

168. Yang, X., L. A. H. Borg and U. J. Eriksson. 1997. Altered metabolism and superoxide generation in neural tissue of rat embryos exposed to high glucose. *Am J Physiol* **272**:E173-E180

169. Trocino, R. A., S. Akazawa, M. Ishibashi, K. Matsumoto, H. Matsuo, H. Yamamoto, S. Goto, Y. Urata, T. Kondo and S. Nagataki. 1995. Significance of glutathione depletion and oxidative stress in early embryogenesis in glucose-induced rat embryo culture. *Diabetes* **44**:992-998

170. Lee, A. T., D. Reis and U. J. Eriksson. 1999. Hyperglycemia-induced embryonic dysmorphogenesis correlates with genomic DNA mutation frequency in vitro and in vivo. *Diabetes* **48**:371-376

171. Eriksson, U. J., L. A. H. Borg, J. Cederberg, H. Nordstrand, C. M. Siman, C. Wentzel and P. Wentzel. 2000. Pathogenesis of diabetes-induced congenital malformations. *Upsala J Med Sci* **105**:53-84

172. Siman, C. M. and U. J. Eriksson. 1997. Vitamin C supplementation of the maternal diet reduces the rate of malformation in the offspring of diabetic rats. *Diabetologia* **40(12)**:1416-1424

173. Siman, C. M., A. C. Gittenberger-de Groot, B. Wisse and U. J. Eriksson. 2000. Malformations in offspring of diabetic rats: morphometric analysis of neural crest-derived origins and effects of maternal vitamin E treatment. *Teratology* **61**:355-367

174. Eriksson, U. J. and C. M. Siman. 1996. Pregnant diabetic rats fed the antioxidant butylated hydroxytoluene show decreased occurrence of malformations in offspring. *Diabetes* **45**:1497-1502

175. Sakamaki, H., S. Akazawa, M. Ishibashi, K. Izumino, H. Takino, H. Yamasaki, Y. Yamaguchi, S. Goto, Y. Urata, T. Kondo and S. Nagataki. 1999. Significance of glutathione-dependent antioxidant system in diabetes-induced embryonic malformations. *Diabetes* **48**:1138-1144

176. Cederberg, J., C. M. Siman and U. J. Eriksson. 2001. Combined treatment with vitamin E and vitamin C decreases oxidative stress and improves fetal outcome in experimental diabetic pregnancy. *Pediatr Res* **49(6)**:755-762

177. Wiznitzer, A., N. Ayalon, R. Hershkovitz, M. Khamaisi, E. A. Reece, H. Trischler and N. Bashan. 1999. Lipoic acid prevention of neural tube defects in offspring of rats with streptozotocin-induced diabetes. *Am J Obstet Gynecol* **180**:188-193

178. Zusman, I., P. Yaffe and A. Ornoy. 1987. Effects of metabolic factors in the diabetic state on the in vitro development of preimplantation mouse embryos. *Teratology* **35**:77-85

179. Sadler, T. W. 1980. Effects of maternal diabetes on early embryogenesis. I. The teratogenic potential of diabetic serum. *Teratology* **21**:339-347

180. Ornoy, A., D. Kimyagarov, P. Yaffee, R. Abir, I. Raz and R. Kohen. 1996. Role of reactive oxygen species in diabetes-induced embryotoxicity: studies on pre-implantation mouse embryos cultured in serum from diabetic pregnant women. *Isr J Med Sci* **32**:1066-1073

181. Freinkel, N., D. L. Cockroft, N. J. Lewis, L. Gorman, S. Akazawa, L. S. Phillips and G. E. Shambaugh III. 1986. The 1986 McCollum award lecture. Fuel-mediated teratogenesis during early organogenesis: the effects of increased concentrations of glucose, ketones, or somatomedin inhibitor during rat embryo culture. *Am J Clin Nutr* **44**:986-995

182. Eriksson, U. J., P. Wentzel, H. S. Minhas and P. J. Thornalley. 1998. Teratogenicity of 3-deoxyglucosone and diabetic embryopathy. *Diabetes* **47**:1960-1966

183. Styrud, J., L. Thunberg, O. Nybacka and U. J. Eriksson. 1995. Correlations between maternal metabolism and deranged development in the offspring of normal and diabetic rats. *Pedatr Res* **37(3)**:343-353

184. Eriksson, U. J., L. A. H. Borg, H. Forsberg and J. Styrud. 1991. Diabetic embryopathy. Studies with animal and in vitro models. *Diabetes* **40(Suppl 2)**:94-98

185. Wentzel, P. and U. J. Eriksson. 1998. Antioxidants diminish developmental damage induced by high glucose and cyclooxygenase inhibitors in rat embryos in vitro. *Diabetes* **47**:677-684

186. Hagay, Z. J., Y. Weiss, I. Zusman, M. Peled-Kamar, E. A. Reece, U. J. Eriksson and Y. Groner. 1995. Prevention of diabetes-associated embryopathy by overexpression of the free radical scavenger copper zinc superoxide dismutase in transgenic mouse embryos. *Am J Obstet Gynecol* **173**:1036-1041

187. Pinter, E., E. A. Reece, P. L. Ogburn, S. Turner, J. C. Hobbins, M. J. Mahoney and F. Naftolin. 1988. Fatty acid content of yolk sac and embryo in hyperglycemia-induced embryopathy and effect of arachidonic acid supplementation. *Am J Obstet Gynecol* **159**:1484-1490

188. Cederberg, J., S. Basu and U. J. Eriksson. 2001. Increased rate of lipid peroxidation and protein carbonylation in experimental diabetic pregnancy. *Diabetologia* **44**:766-774

189. Hashimoto, M., S. Akazawa, M. Akazawa, M. Akashi, H. Yamamoto, Y. Maeda, Y. Yamaguchi, H. Yamasaki, D. Tahara, T. Nakanishi and S. Nagataki. 1990. Effects of hyperglycemia on sorbitol and *myo*-inositol contents of cultured embryos: treatment with aldose reductase inhibitor and *myo*-inositol supplementation. *Diabetologia* **33**:597-602

190. Baker, L., R. Piddington, A. Goldman, J. Egler and J. Moehring. 1997. *Myo*-inositol and prostaglandins reverse the glucose inhibition of neural tube fusion in cultured mouse embryos. *Diabetologia* **33**:593-596

191. Hod, M., S. Star, J. Passonneau, T. G. Unterman and N. Freinkel. 1990. Glucose-induced dysmorphogenesis in the cultured rat conceptus: prevention by supplementation with *myo*-inositol. *Isr J Med Sci* **26**:541-544

192. Reece, E. A., M. Khandelwal, Y. K. Wu and M. Borenstein. 1997. Dietary intake of *myo*-inositol and neural tube defects in offspring of diabetic rats. *Am J Obstet Gynecol* **176**:536-539

193. Goldman, A. S., L. Baker, R. L. Piddington, B. Marx, R. Herold and J. Egler. 1985. Hyperglycemia-induced teratogenesis is mediated by a functional deficiency of arachidonic acid. *Proc Natl Acad Sci USA* **82**:8227-8231

194. Reece, E. A., Y. K. Wu, A. Wiznitzer, C. J. Homko, J. Yao, M. Borenstein and G. Sloskey. 1996c. Dietary polyunsaturated fatty acids prevent malformations in offspring of diabetic rats. *Am J Obstet Gynecol* **175**:818-823

195. Reece, E. A. and Y. K. Wu. 1997. Prevention of diabetic embryopathy in offspring of diabetic rats with use of a cocktail of deficient substrates and an antioxidant. *Am J Obstet Gynecol* **176**:790-798

196. Eriksson, U. J., J. Styrud and R. S. M. Eriksson. 1989. Diabetes in preganancy: genetic and temporal relationships of maldevelopment in the offspring of diabetic rats. In *Carbohydrate Metabolism in Pregnancy and the Newborn IV* (Sutherland, H. W, J. M. Stowers, and D. W. M. Pearson, eds.), pp. 51-63, Springer-Verlag, London

197. Cederberg, J. and U. J. Eriksson. 1997. Decreased catalase activity in malformation-prone embryos of diabetic rats. *Teratology* **56**:350-357.

198. Cederberg, J., J. Galli, H. Luthman and U. J. Eriksson. 2000. Increased mRNA levels of Mn-SOD and catalase in embryos of diabetic rats from a malformation-resistant strain. *Diabetes* **49**:101-107

199. Hunter, A. G. W., R. H. Cleveland, J. G. Blickman and L. B. Holmes. 1996. A study of level of lesion, associated malformations and sib occurrence risks in spina bifida. *Teratology* **54**:213-218.

Index